Advances in
Quantum Phenomena

NATO ASI Series

Advanced Science Institutes Series

A series presenting the results of activities sponsored by the NATO Science Committee, which aims at the dissemination of advanced scientific and technological knowledge, with a view to strengthening links between scientific communities.

The series is published by an international board of publishers in conjunction with the NATO Scientific Affairs Division

A	**Life Sciences**	Plenum Publishing Corporation
B	**Physics**	New York and London
C	**Mathematical and Physical Sciences**	Kluwer Academic Publishers Dordrecht, Boston, and London
D	**Behavioral and Social Sciences**	
E	**Applied Sciences**	
F	**Computer and Systems Sciences**	Springer-Verlag
G	**Ecological Sciences**	Berlin, Heidelberg, New York, London,
H	**Cell Biology**	Paris, Tokyo, Hong Kong, and Barcelona
I	**Global Environmental Change**	

PARTNERSHIP SUB-SERIES

1. **Disarmament Technologies**	Kluwer Academic Publishers
2. **Environment**	Springer-Verlag
3. **High Technology**	Kluwer Academic Publishers
4. **Science and Technology Policy**	Kluwer Academic Publishers
5. **Computer Networking**	Kluwer Academic Publishers

The Partnership Sub-Series incorporates activities undertaken in collaboration with NATO's Cooperation Partners, the countries of the CIS and Central and Eastern Europe, in Priority Areas of concern to those countries.

Recent Volumes in this Series:

Volume 345 — Core Level Spectroscopies for Magnetic Phenomena: Theory and Experiment
edited by Paul S. Bagus, Gianfranco Pacchioni, and Fulvio Parmigiani

Volume 346 — Hot Hadronic Matter: Theory and Experiment
edited by Jean Letessier, Hans H. Gutbrod, and Johann Rafelski

Volume 347 — Advances in Quantum Phenomena
edited by Enrico G. Beltrametti and Jean-Marc Lévy-Leblond

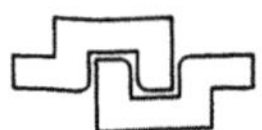

Series B: Physics

Advances in Quantum Phenomena

Edited by

Enrico G. Beltrametti

University of Genova
Genova, Italy

and

Jean-Marc Lévy-Leblond

University of Nice
Nice, France

Springer Science+Business Media, LLC

Proceedings of a NATO Advanced Study Institute on
Advances in Quantum Phenomena,
held February 16–18, 1994,
in Erice, Sicily, Italy

NATO-PCO-DATA BASE

The electronic index to the NATO ASI Series provides full bibliographical references (with keywords and/or abstracts) to about 50,000 contributions from international scientists published in all sections of the NATO ASI Series. Access to the NATO-PCO-DATA BASE is possible in two ways:

—via online FILE 128 (NATO-PCO-DATA BASE) hosted by ESRIN, Via Galileo Galilei, I-00044 Frascati, Italy

—via CD-ROM "NATO Science and Technology Disk" with user-friendly retrieval software in English, French, and German (©WTV GmbH and DATAWARE Technologies, Inc. 1989). The CD-ROM contains the AGARD Aerospace Database.

The CD-ROM can be ordered through any member of the Board of Publishers or through NATO-PCO, Overijse, Belgium.

```
           Library of Congress Cataloging-in-Publication Data


Advances in quantum phenomena / edited by Enrico G. Beltrametti and
  Jean-Marc Lévy-Leblond.
       p.   cm. -- (NATO ASI series. Series B, Physics ; v. 347)
     "Published in cooperation with NATO Scientific Affairs Division."
     "Proceedings of a NATO Advanced Study Institute Advances on
  Quantum Phenomena held February 16-18, 1994, in Erice, Sicily,
  Italy."--Verso t.p.
     Includes bibliographical references and index.
     ISBN 978-1-4613-5813-8    ISBN 978-1-4615-1975-1 (eBook)
     DOI 10.1007/978-1-4615-1975-1
     1. Quantum theory--Congresses.  2. Quantum chemistry--Congresses.
  3. Solid state physics--Congresses.  4. Quantum electronics-
  -Congresses.   I. Beltrametti, Enrico G.  II. Lévy-Leblond, Jean
  Marc.  III. North Atlantic Treaty Organization.  Scientific Affairs
  Division.  IV. NATO Advanced Study Institute on Advances in Quantum
  Phenomena (1994 : Erice, Italy) V. Series.
  QC173.96.A38  1995
  530.1'2--dc20                                          95-302
                                                         CIP
```

A precursory experiment in interferometry (Filippo Bonanni, Gabinetto Armonico, 1716; reprinted in F. Bonanni, *Antique Musical Instruments and Their Players*, Dover, N.Y., 1964.)

CONTENTS

SPECIFIC TOPICS

Advances in
Quantum Phenomena

INTRODUCTION

Enrico Beltrametti[1] and Jean-Marc Lévy-Leblond[2]

[1]Dipartimento di Fisica, Università di Genova
Via Dodecaneso 33, I-16146 Genova, Italy
[2]Physique Théorique, Université de Nice Sophia-Antipolis
Parc Valrose, F-06108 Nice Cedex, France

As customary, the idea of this meeting was born out during another one. The present authors were attending some colloquium among the numerous meetings devoted to the endless discussion of the foundational problems of quantum theory, when it struck their minds that physics, after all, was also an experimental science. Maybe, they thought to themselves, our very understanding of quantum concepts would gain from more thoroughly taking into account the impressive accumulation of modern experimental results on the fundamental aspects of quantum behaviour.

Experiments once deemed as 'gedankenexperiments' now become routinely feasible, and far-ranging progresses in the practical mastering of quantum phenomena have been accomplished in the recent years. The variety of domains where quantum phenomena arc currently investigated is extending beyond expectations, while their essential unity is ever more obvious. As no physicist can seriously believe that "quantum" is the last word about Nature, the drive to check the limits of validity of quantum ideas is a powerful motive behind the ever increasing precision of its experimental tests. But no experimental fact up to now has surfaced which seems to challenge the main concepts of quantum theory. The robustness and reliability of quantum theory is all the more surprising when one considers thc cxtension of its range of validity during the near century of its existence — from the atomic scale down to the subnuclear one ("particle" physics), and up to the astronomical one (white dwarfs, neutron stars, etc.). On the other hand, as in most fields of physics when they reach maturity, yesterday's problems become today's tools. Many examples are exhibited in this book of exquisite experimental devices based on sophisticated quantum effects, which are used today for new investigations. As a result, more and more technological applications are coming out of fundamental quantum physics, from magnetic resonance imaging and supraconductivity-based magnetometers, to electronic nanodevices and new cryptographic systems.

The purpose of this Colloquium thus was to offer a broad review of the main aspects of contemporary experimental work on quantum phenomena, beyond the specialized knowledge of the separate fields of investigation. Accordingly, the chosen speakers were mainly experimentalists coming from these widely different domains, allowing a few theoreticians to discuss possible extensions of the experimental investigations. We would like to think that the resulting book will be used both as a kind of introductory manual to quantum experimental physics for the young researchers, and as an appeal to the theoreticians for refreshing their analyses and grounding them in the present state-of-the art.

GENERAL REVIEWS

EXPERIMENTS WITH SINGLE ATOMS IN CAVITIES AND TRAPS

H. Walther

Sektion Physik der Universität München and
Max-Planck-Institut für Quantenoptik
85748 Garching, Fed. Rep. of Germany

INTRODUCTION

In this paper experiments performed with the one-atom maser and with a few ions stored in a Paul trap are reviewed. Special emphasis is placed on experiments for the test of quantum physics.

The modern techniques of laser spectroscopy allow to investigate single atoms. In this paper two groups of those experiments will be reviewed with special emphasis on applications to study quantum phenomena. The first one deals with the one-atom maser and the second one with trapped ions (for a review see e.g. Ref. 1). In recent years there has also been considerable progress in trapping neutral atoms.[2] These techniques are undergoing a rapid development in the moment. However, experiments with single isolated atoms using those techniques have not yet been described.

REVIEW OF THE ONE-ATOM MASER

The most fundamental system to study the generation process of radiation in lasers and masers is to drive a single mode cavity by single atoms. This system, at first glance, seems to be another example of a Gedankenexperiment but such a one-atom maser[3] really exists and can in addition be used to study the basic principles of radiation-atom interaction. The main features of the setup are:

(1) It is the first maser which sustains oscillations with less than one atom on the average in the cavity.

(2) The maser allows one to study the dynamics of the energy exchange between an atom and a single mode of the cavity field treated in the Jaynes-Cummings model.[4]

(3) The setup allows one to study in detail the conditions necessary to obtain nonclassical radiation, especially radiation with sub-Poissonian photon statistics in a maser system directly.

(4) It is possible to study a variety of phenomena of a quantum field including the quantum measurement process.

What are the tools that make this device work? It is the enormous progress in constructing superconducting cavities together with the laser preparation of highly excited

Advances in Quantum Phenomena, Edited by E.G. Beltrametti
and J.-M. Lévy-Leblond, Plenum Press, New York, 1995

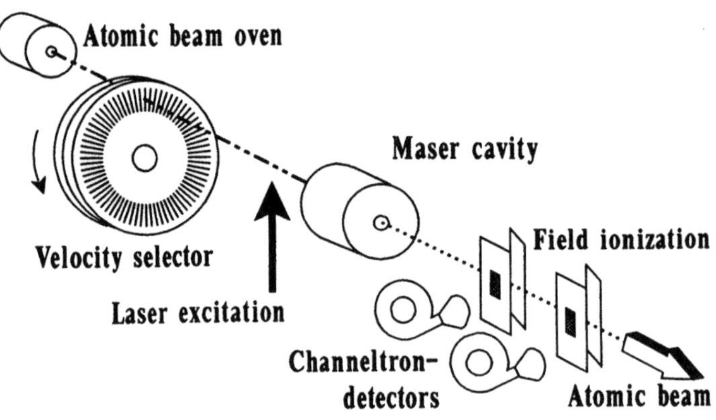

Figure 1. Scheme of the one-atom maser. To suppress blackbody-induced transitions to neighbouring states, the Rydberg atoms are excited inside the liquid-Helium-cooled environment.

atoms - Rydberg atoms - that have made the realisation of such a one-atom maser possible.[3] Rydberg atoms are obtained when one of the outermost electrons of an atom is excited into a level close to the ionization limit. The main quantum number of the electron is then typically of the order of 60 - 70. Those atoms have quite remarkable properties[5] which make them ideal for the maser experiments: The probability of induced transitions between neighbouring states of a Rydberg atom scales as n^4, where n denotes the principle quantum number. Consequently, a few photons are enough to saturate the transition between adjacent levels. Moreover, the spontaneous lifetime of a highly excited state is very large. We obtain a maser by injecting these Rydberg atoms into a superconducting cavity with a high quality factor. The injection rate is such that on the average there is much less than one atom present inside the resonator.

The experimental setup of the one-atom maser is shown in Fig.1. A highly collimated beam of rubidium atoms passes through a Fizeau velocity selector. Before entering the superconducting cavity, the atoms are excited into the upper maser level $63p_{3/2}$ by the frequency-doubled light of a cw ring dye laser. The laser frequency is stabilised onto the atomic transition $5s_{1/2} \rightarrow 63p_{3/2}$, which has a width determined by the laser linewidth and the transit time broadening corresponding to a total of a few MHz. In this way, it is possible to prepare a very stable beam of excited atoms. The superconducting niobium maser cavity is cooled down to a temperature of 0.5 K by means of a ^{3}He cryostat. At this temperature the number of thermal photons in the cavity is about 0.15 at a frequency of 21.5 GHz. The cryostat is carefully designed to prevent room temperature microwave photons from leaking into the cavity. This would considerably increase the temperature of the radiation field above the temperature of the cavity walls. The quality factor of the cavity can be as high as 3×10^{10} corresponding to a photon storage time of about 0.2 s. Two maser transitions from the $63p_{3/2}$ level to the $61d_{3/2}$ and to the $61d_{5/2}$ level are studied. In a new setup equipped with a dilution refrigerator temperatures in the range of 0.1 K are obtained. Some of the experiments described in this review have been performed with the latter setup.

The Rydberg atoms in the upper and lower maser levels are detected by two separate field ionisation detectors. The field strength is adjusted to ensure that in the first detector the atoms in the upper level are ionised, but not those in the lower level; the lower level atoms are then ionised in the second field.

To demonstrate maser operation, the cavity is tuned over the respective transition and the flux of atoms in the excited state is recorded simultaneously. Figure 2 shows the result for $63p_{3/2}$ - $61d_{3/2}$. Transitions from the initially prepared state to the $61d_{3/2}$ level (21.50658 GHz) are detected by a reduction of the electron count rate.

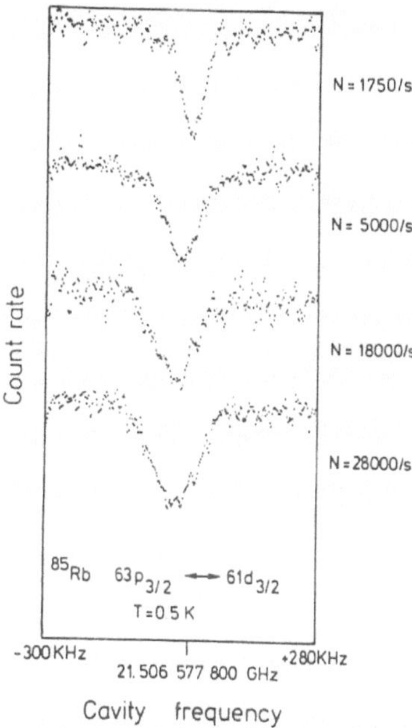

Figure 2. A maser transition of the one-atom maser manifests itself in a decrease of atoms in the excited state. The flux of excited atoms N governs the pump intensity. Power broadening of the resonance line demonstrates the multiple exchange of a photon between the cavity field and the atom passing through the resonator.

In the case of measurements at a cavity temperature of 0.5 K, shown in Fig.2, a reduction of the $63p_{3/2}$ signal can be clearly seen for atomic fluxes as small as 1750 atoms/s. An increase in flux causes power broadening and a small shift. This shift is attributed to the ac Stark effect, caused predominantly by virtual transitions to neighbouring Rydberg levels. Over the range from 1750 to 28000 atoms/s the field ionisation signal at resonance is independent of the particle flux which indicates that the transition is saturated. This, and the observed power broadening show that there is a multiple exchange of photons between Rydberg atoms and the cavity field.

For an average transit time of the Rydberg atoms through the cavity of 50 μs and a flux of 1750 atoms/s we estimate that approximately 0.09 Rydberg atoms are in the cavity on the average. According to Poisson statistics this implies that more than 90 % of the events are due to single atoms. This clearly demonstrates that single atoms are able to maintain a continuous oscillation of the cavity with a field corresponding to a mean number of photons between unity and several hundred.

Among the studies performed with the one-atom maser are the measurements of the dynamics of the photon exchange between a single atom and a cavity mode.[6,7] Before we discuss some experiments with the one-atom maser the theory will be briefly reviewed.

THEORY OF THE ONE-ATOM MASER

The simplest form of interaction between a two-level atom and a single quantized mode of the electromagnetic field is described by the Jaynes-Cummings Hamiltonian.[4,8]

$$H = \frac{1}{2}\,\hbar\,\omega_0\sigma_z + \hbar\,\omega\,a^{\dagger}a + \hbar\,(ga^{\dagger}\sigma_- + \text{adj.})$$
$$= H_0 + V,$$

where

$$H_o = \frac{1}{2}\hbar\,\omega_o\sigma_z + \hbar\,\omega\,a^\dagger a,$$

and

$$V = \hbar\,(ga^\dagger\sigma_- + \mathrm{adj.}).$$

Here, ω_o is the atomic transition frequency, ω is the field frequency, a and $a^\dagger$ are the photon annihilation and creation operators of the field mode, with $[a, a^\dagger] = 1$, σ_z, and σ_- and σ_+ are atomic pseudo-spin operators, with $[\sigma_+,\sigma_-] = \sigma_z$, and

$$g = \frac{p\,E_\omega}{2\hbar}\ \sin KZ$$

is the electric dipole matrix element at the location Z of the atom, where E_ω is the "electric field per photon", $E_\omega = (\hbar\,\omega/\varepsilon_o V)^{1/2}$.

The Jaynes-Cummings model plays a central role in quantum optics owing to several reasons: it gives the simplest description of Rabi-flopping in a quantised field and the simplest illustration of spontaneous emission; furthermore, it can be solved exactly and thus describes the true quantum dynamics observed with the one-atom maser such as collapse and revivals of the atomic inversion. The model describes the situation realised in the one-atom maser and allows to investigate in detail the complexities of the atom-field dynamics in the simplest of all situations.

In the following a few results following from the Jaynes-Cummings model are reviewed being relevant for the discussions in this paper

The eigenstates of the Jaynes-Cummings Hamiltonian are

$$E_{1n} = \hbar\,[-\frac{1}{2}\,\omega_o + (n{+}1)\,\omega + \frac{1}{2}\,(\Omega_n + \delta)]$$

$$E_{2n} = \hbar\,[\ \frac{1}{2}\,\omega_o + n\omega - \frac{1}{2}\,(\Omega_n + \delta)]$$

where $\delta = \omega_o - \omega$ is the atom-field frequency detuning. Ω_n is the generalized n-photon Rabi flopping frequency

$$\Omega_n = \sqrt{4g^2(n + 1) + \delta^2}$$

The corresponding eigenvalues are the one of the dressed atom:

$$|1,n\rangle = \sin\theta_n\,|a,n\rangle + \cos\theta_n\,|b,n + 1\rangle$$

and

$$|2,n\rangle = \cos\theta_n\,|a,n\rangle - \sin\theta_n\,|b,n + 1\rangle\,,$$

where the states $|a\rangle$ and $|b\rangle$ are the upper and lower atomic states, respectively, and $|n\rangle$ are the number states of the field mode with $a^\dagger a|n\rangle = n\,|n\rangle$. The angle θ is defined by means of the relations

$$\cos\theta_n = \frac{\Omega_n - \delta}{\sqrt{(\Omega_n - \delta)^2 + 4g^2(n + 1)}}\,,$$

$$\sin\theta_n = \frac{2g\sqrt{n + 1}}{\sqrt{(\Omega_n - \delta)^2 + 4g^2(n + 1)}}\,.$$

Note, in particular, that

$$\cos 2\theta_n = -\delta / \Omega_n$$

and

$$\sin 2\theta_n = 2g\sqrt{n+1}/\Omega_n$$

In the vacuum field $n = 0$ and on resonance $\omega_0 = \omega$ the dressed states are separated by the frequency $\Omega_0 = 2g$ generally called vacuum Rabi splitting.

One of the interesting phenomena described by the Jaynes-Cummings model is the dynamical behaviour. When we assume that the atom is initially in the upper state $|a\rangle$ and the field in the number state $|n\rangle$ it follows for the probability of an atom to be in the upper state

$$|C_{an}(t)|^2 = \cos^2(g\sqrt{n+1}\ t).$$

This shows that the upper state population oscillates periodically at the Rabi flopping frequency, similar to the case of classical fields. If the field is initially described by the photon statistics p_n, the above results have to be generalised to:

$$|C_a(t)|^2 = \sum_n p_n \cos^2(g\sqrt{n+1}\ t).$$

It has been shown that in the case where the field mode is initially in a coherent state $|C_a(t)|^2$ undergoes a collapse followed by a series of revivals.[6] The collapse is due to the destructive interference of quantum Rabi floppings at different frequencies, a similar phenomenon may also occur with a classical field, however, the revivals are a purely quantum mechanical effect that originates in the discreteness of the quantum field. Collapse and revivals have been observed in a micromaser experiment.[7] In this experiment the interaction time of the atoms with the cavity was varied and the probability was investigated that the atoms leave the cavity in the excited state. As will be shown below the photon statistics in the maser cavity changes when the interaction time is varied, therefore the photon statistics p_n is not a pure distribution. Nevertheless the revival shows up as was also confirmed in a computer simulation of the results on the basis of the Jaynes-Cummings model.[7]

In the following we would like to summarise the results of the quantum-theory of the one-atom maser. Since the atom-field interaction takes place in a closed single mode cavity, there is no spontaneous emission rate into free space modes. Owing to the extremely high quality factors achieved in the superconducting cavities the photon lifetime is extremely long compared to the transit time of the atoms through the resonator. This means that the cavity damping can be practically ignored while an atom interacts with the field. Since the atomic flux is kept small to have at most one atom present inside the cavity at a time, hence the cavity is empty most of the time, therefore cavity damping can be neglected during the rare instances when an atom interacts with the cavity mode.

The one-atom maser theory is therefore based on the following strategy:[9] while an atom is in the cavity, the coupled atom-field system is described by the Jaynes-Cummings Hamiltonian and for the interval between atoms the evolution of the field density matrix is described by a master equation considering damping and in addition the mean number of thermal photons in the cavity.

Besides this microscopic theory there is also a macroscopic theory based on the quantum theory of the laser.[10] The resulting probability distribution of the photons depends characteristically on the pump rate and on the interaction time t_{int} of the atoms with the cavity field. One obtains the following result for the probability of finding n photons in the maser cavity $P(n)$ in steady state:

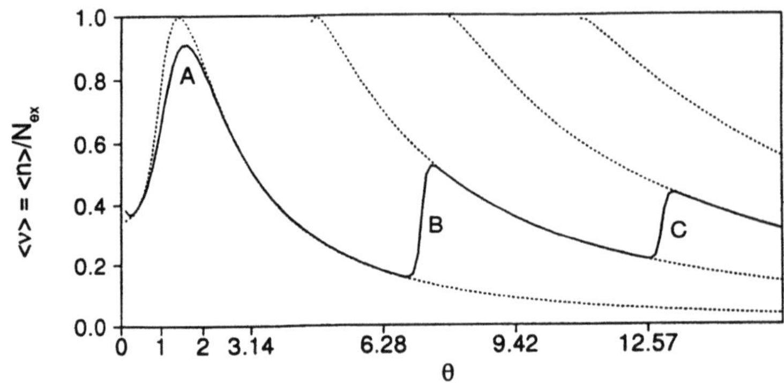

Figure 3. Mean value of the photon number $\nu = n/N_{ex}$ versus the pump parameter Θ, where the value of Θ is changed via N_{ex}. The solid line represents the micromaser solution for $\Omega = 36$ kHz, $t_{int} = 35$ μs, and temperature $T = 0.15$ K. The dotted lines are semiclassical steady-state solutions corresponding to fixed stable gain=loss equilibrium photon numbers. The crossing points between a line $\Theta = $ const. and the dotted lines correspond to the values where minima in the Fokker-Planck potential $V(\nu)$ occur. For details see text.

$$P(n) = P_0 \left(\frac{n_b}{n_b + 1} \right)^n \prod_{m=1}^{n} \left(1 + \frac{N}{n_b \gamma} \frac{\sin^2(g\sqrt{m}\, t_{int})}{m} \right),$$

with N being the atomic pump rate, n_b the thermal photon number, and γ the cavity decay rate. P_0 is determined by the normalization condition $\Sigma\, P(n) = 1$. One can now evaluate the mean photon number $\langle n \rangle$ and the field variance in the form of the Q_f-parameter[9]:

$$Q_f = \frac{\langle n^2 \rangle - \langle n \rangle^2 - \langle n \rangle}{\langle n \rangle}.$$

Figure 3 shows the mean photon number as a function of the interaction time of the atoms with the cavity. The photon number is scaled to N_{ex}; here N_{ex} is the average number of atoms that enter the cavity during the cavity decay time $N_{ex} = N/\gamma$. The maser thresholds occurs at $\Theta = 1$ and regions of sub-Poissonian photon statistics are between $\Theta = \pi$ and 2π as well as between 3π and 4π. The sub-Poissonian character leads to a negative Q_f. A large negative value is obtained close before 2π.[8,9]

To get a more intuitive insight into this effect we recall that the Fizeau velocity selector preselects the velocity of the atoms. Hence, the interaction time is well defined, which leads to conditions usually not achievable in standard masers. This has a very important consequence when the intensity of the maser field grows as more and more atoms transfer their excitation energy to the field: even in the absence of dissipation this increase in photon number is halted when the increasing Rabi frequency leads to a situation where the atoms reabsorb the photon and leave the cavity in the upper state. This situation is close to a quantum-nondemolition case. For any photon number, this situation can be achieved by appropriately adjusting the velocity of the atoms. Then the number distribution of the photons in the cavity is sub-Poissonian.

Unfortunately, the measurement of the nonclassical photon statistics in the cavity is not straightforward. The measurement process of the field involves the coupling to a measuring device, whereby losses lead inevitably to a destruction of the nonclassical properties. The ultimate technique to obtain information about the field employs the Rydberg atoms themselves: for this purpose the population and the statistics of the atoms in the upper and lower maser levels are probed when they leave the cavity. Accordingly, the atoms play a double role: (i) they pump the cavity and (ii) they are used for the diagnostics. These two roles interfere with one another because the detection of the atom in a known final state leads to a quantum mechanical reduction of the photon state inside the resonator. Frequent detection is accompanied by quasipermanent state reduction which can prevent the cavity field from relaxing to the steady state that would be reached if the atoms were left unobserved. Nevertheless, the steady-state properties determine the statistics of the clicks recorded by the atom detectors.

The theoretical treatment of the one-atom maser produces predictions about both the photon field and the emerging atoms. Only the latter can be tested experimentally, and the success of such tests feeds our confidence in the predictions about the quantized radiation field inside the cavity. The pump atoms are statistially independent in the standard one-atom-maser experiments, so that their arrival times are subject to a Poissonian statistics; we shall, therefore, restrict the discussion to this standard Poissonian situation.

Inasmuch as the atoms arrive at random, they are recorded at equally random times, and so the only reproducible data are statistical. Consequently, one is led to studying the statistics of the detector clicks. Numerical simulations investigating the effect of repeatd atomic measurements on the evolution of the cavity field have been performed by Meystre[11] as well as Meystre and Wright.[12] The relation between the counting statistics of the detected atoms emerging from the resonator and the photon-number statistics of the field inside the cavity has been studied analytically by Rempe and Walther[13] as well as by Paul and Richter.[14] In Ref. 13 the results are also compared with numerical simulations showing good agreement; experimental results are reported by Rempe et al.[15]

In a recent paper a general method for the computation of various statistical properties of the click distribution were presented.[16] This method does not resort to numerical simulations. Naturally, the efficiencies of the detectors - far from the ideal 100 %, unfortunately - are taken into account. A central tool used in those calulations is a nonlinear master equation that governs the dynamics of the photon field in periods between detector clicks. The nonlinearity arises from the necessity to distinguish between the notions of observation and detection. When the detectors are active, all emerging atoms are observed but only a fraction is actually detected. Most of the time the experimenter is observing but does not detect - he is listening but does not hear.

In another approach on the basis of the concept of the counting statistics of light beams, the atomic counting probability, the waiting-time distribution and the "two-atom correlation" function for a Poissonian atomic beam exciting the micro-maser cavity are calculated also. In an analytic treatment it is shown how the waiting-time distribution converges into the atom correlation function for vanishing detection efficiency.[17]

Under steady-state conditions, as mentioned above, the photon number and the photon statistics of the maser field is essentially determined by the dimensionless parameter Θ. The quantity $\langle \nu \rangle = \langle n \rangle / N_{ex}$ shows the following generic behavior (see Fig. 3): It suddenly increases at the threshold value $\Theta = 1$ and reaches a maximum for $\Theta \approx 2$. As Θ further increases, $\langle \nu \rangle$ decreases and reaches a minimum at $\Theta \approx 2\pi$, and then abruptly increases to a second maximum. This general type of behavior recurs roughly at integer multiples of 2π, but becomes less pronounced with increasing Θ. The reason for the periodic maxima of $\langle \nu \rangle$ is that for $\Theta \approx 2\pi$ and multiples thereof the pump atoms perform an almost integer number of full Rabi flopping cycles, and start to flip over at a slightly larger value of Θ, this leading to enhanced photon emission. The maser threshold at $\Theta = 1$ shows the characteristics of a continuous phase transition, whereas the subsequent maxima in $\langle \nu \rangle$ can be interpreted as first-order phase transitions. In the intervals between the phase-transition points, the photon statistics is mostly sub-Poissonian. The field is super-Poissonian for all phase transitions,[8,9] the large photon number fluctuations above $\Theta = 2\pi$ and multiples thereof being caused by the presence of two maxima in the photon number distribution P(n).

They result from the fact that atoms in the upper maser level may or may not tip over to the lower level.[8,9]

The phenomenon of the two coexisting neighbouring maxima in P(n) was also studied in a semi-heuristic Fokker-Planck (FP) approach.[9] There, the photon number distribution P(n) is replaced by a probability function $P(\nu,\tau)$ with continuous variables $\tau = t/\tau_{cav}$ and $\nu(n) = n/N_{ex}$, the latter replacing the photon number n. The steady-state solution obtained for $P(\nu,\tau),\tau \gg 1$, can be constructed by means of an effective potential $V(\nu)$ showing minima at positions where maxima of $P(\nu,\tau),\tau \gg 1$, are found. Close to $\Theta = 2\pi$ and multiples thereof, the effective potential $V(\nu)$ exhibits two equally attractive minima located at stable gain-loss equilibrium points of maser operation. The mechanism at the phase transitions mentioned is always the same: a minimum of $V(\nu)$ loses its global character when Θ is increased, and is replaced in this role by the next one. This reasoning is a variation of the Landau theory of first-order phase transitions, with $\sqrt{\nu}$ being the order parameter. This analogy actually leads to the notion that in the limit $N_{ex} \rightarrow \infty$ the change of the micromaser field around integer multiples of $\Theta = 2\pi$ can be interpreted as first-order phase transitions.

In the region of the first-order phase transitions long field evolution time constants τ_{field} are expected. This phenomenon was experimentally demonstrated, as well as related phenomena, such as spontaneous quantum jumps between equally attractive minima of $V(\nu)$, and bistability and hysteresis.[18] Some of those phenomena are also predicted in the two-photon micromaser.[19]

If there are no thermal photons in the cavity - a condition achievable by cooling the resonator to temperatures below 100 mK - very interesting features such as trapping states show up.[20] The investigation of the trapping states is discussed in detail in a recent review.[21]

In the following we would like to review two experiments, one on the measurement of the photon statistics of the one-atom-maser and another one onthe observation of quantum jumps and bistability in the maser field at the first order phase transitions points.

THE PHOTON STATISTICS OF THE ONE-ATOM MASER FIELD

As discussed above in the experiment on the photon statistics of the one-atom maser field[15] the number N of atoms in the lower maser level is counted for a fixed time interval T roughly equal to the storage time τ_{cav} of the photons in the cavity. By repeating this measurement many times the probability distribution of finding N atoms in the lower level is obtained. The normalized variance $Q_a = [\langle N^2 \rangle - \langle N \rangle^2 - \langle N \rangle]/\langle N \rangle$ is evaluated and is used to characterise the deviation from Poissonian statistics. A negative (positive) Q_a value indicates sub-Poissonian (super-Poissonian) statistics, while $Q_a = 0$ corresponds to a Poisson distribution with $\langle N^2 \rangle - \langle N \rangle^2 = \langle N \rangle$. The atomic Q_a can be related to the normalised variance Q_f of the photon number (for details see Refs. 13,15,16).

As an example we discuss one result for the $63p_{3/2}$ - $61d_{5/2}$ maser transition with g = 44 rd/s; it is shown in Fig. 4. Velocity selected atoms with an atom-cavity interaction time of $t_{int} = 35$ μs were used. A very low flux of atoms of $N_{ex} > 1$ is already sufficient to generate a nonclassical maser field. This is the case since the vacuum field initiates a transition of the atom to the lower maser level, thus driving the maser above threshold.

The sub-Poissonian statistics of the atoms near $N_{ex} = 30$, corresponding to $Q_a = -4$ % is generated by a photon field with a variance $\langle n^2 \rangle - \langle n \rangle^2 = 0.3 \langle n \rangle$, which is 70 % below the shot noise level.[15] Again, this result agrees with the prediction of the theory.[9,10] The mean number of photons in the cavity is about 12 and 13 in the regions $N_{ex} \approx 3$ and $N_{ex} \approx 30$, respectively. Near $N_{ex} \approx 15$, the photon number changes abruptly between these two values. The next maser phase transition with a super-Poissonian photon number distribution occurs above $N_{ex} \approx 50$.

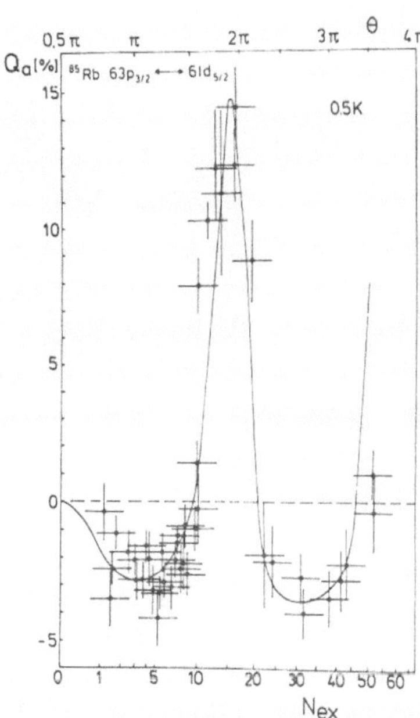

Figure 4 Variance Q_a of the atoms in the lower maser level as a function of the flux N_{ex} (see also Ref. 15).

We emphasise that the reason for the sub-Poissonian atomic statistics is the following: A changing flux of atoms changes the Rabi-frequency via the stored photon number in the cavity. By adjusting the interaction time, the phase of the Rabi-nutation cycle can be chosen so that the probability for the atoms leaving the cavity in the upper maser level increases when the flux and therefore the photon number in the cavity is raised. We observe sub-Poissonian atomic statistics in the case where the number of atoms in the lower state is decreasing with increasing flux and photon number in the cavity. This feedback mechanism is also demonstrated when the anticorrelation of atoms leaving the cavity in the lower state is investigated. Measurements of this 'antibunching' phenomena have also been performed (see Ref. 21 for a detailed review and also Ref. 14).

The fact that anticorrelation is observed shows that the atoms in the lower state are more equally spaced than expected for a Poissonian distribution. It means when two atoms enter the cavity close to each other the second one performs a transition to the lower state with a reduced probability.

The experimental results presented here clearly show the sub-Poissonian photon statistics of the one-atom maser field. In addition, the maser experiment leads to an atomic beam with atoms in the lower maser level showing number fluctuations which are up to 40 % below those of a Poissonian distribution found in usual atomic beams. This is interesting, because atoms in the lower level have emitted a photon to compensate for cavity losses inevitably present. Although this is a purely dissipative process giving rise to fluctuations; nevertheless the atoms still obey sub-Poissonian statistics.

QUANTUM JUMPS OF THE MICROMASER FIELD

The setup used for these measurements is similar to that described above. As before, ^{85}Rb atoms were used to pump the maser. They are excited from the $5S_{1/2}$, $F = 3$ ground state to 63 $P_{3/2}$, $m_J = \pm 1/2$ states by linearly polarized light of a frequency-doubled c.w. ring dye laser. The polarisation of the laser light is linear and parallel to the likewise linearly polarised maser field, and therefore only $\Delta m_J = 0$ transitions are excited. Superconducting

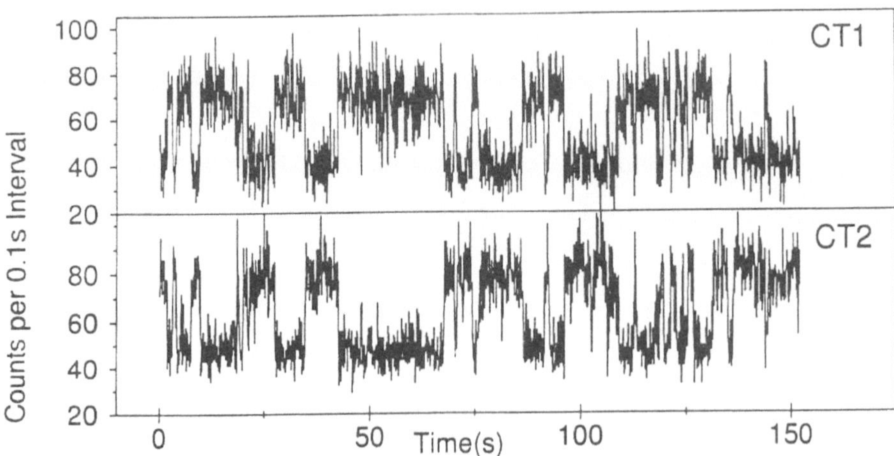

Figure 5. Quantum jumps between two equally stable operation points of the maser field. The channeltron counts are plotted versus time (CT1 = upper state, CT2 = lower state).

niobium cavities resonant with the transition to the 61 $D_{3/2}$, $m_J = \pm 1/2$ states were used. The experiments were performed in a ^{3}He/^{4}He-dilution refrigerator with cavity temperatures $T \approx 0.15$ K. The cavity Q-values ranged from $4 \cdot 10^9$ to $8 \cdot 10^9$. The velocity of the Rydberg atoms and thus their interaction time t_{int} with the cavity field were preselected by exciting a particular velocity subgroup with the laser. For this purpose, the laser beam irradiated the atomic beam at an angle of approximately 82°. As a consequence, the UV laser light (linewidth $\approx$ 2 MHz) is blue-shifted by 50-200 MHz by the Doppler effect, depending on the velocity of the atoms.[18]

As before, information on the maser field and interaction of the atoms in the cavity can be obtained solely by state-selective field ionization of the atoms in the upper or lower maser level after they have passed through the cavity. The field ionization detector was recently modified, so that there is now a detection efficiency of $\eta = (35 \pm 5)$ %. For different t_{int} the atomic inversion has been measured as a function of the pump rate; the coupling constant is about g = 40 rd/s.

Depending on the parameter range, essentially three regimes of the field evolution time constant τ_{field} can be distinguished.[18] We restrict the discussion here only to the results for intermediate time constants. The maser was operated under steady-state conditions close to the second first-order phase transition (C in Fig. 3). The interaction time was $t_{int} = 47$ μs and the cavity decay time $\tau_{cav} = 60$ ms. The value of N_{ex} necessary to reach the second first-order phase transition was $N_{ex} \approx 200$. For these parameters, the two maxima in P(n) are manifested in spontaneous jumps of the maser field between the two maxima with a time constant of $\approx$ 5s. This fact and the relatively large pump rate lead to the clearly observable field jumps shown in Fig. 5. Due to the large cavity field decay time, the average number of atoms in the cavity was still as low as 0.17. The two discrete values for the counting rates correspond to the metastable operating points of the maser, which correspond to ≈ 70 and ≈ 140 photons. In the Focker-Planck description, the two values correspond to two equally attractive minima in the Focker-Planck potential $V(v)$. If one considers, for instance, the counting rate of lower-state atoms (CT2 in Fig. 5), the lower (higher) plateaus correspond to time intervals in the low (high) field metastable operating point. If the actual photon number distribution is averaged over a time interval containing many spontaneous field jumps, the steady-state result P(n) of the micromaser theory is recovered.

In a parameter range where switching occurs much faster than in the case shown in Fig. 5, the individual jumps cannot be resolved any more owing to the reduced N_{ex} necessary in this case. Therefore, a different procedure has to be chosen for the investigation. Those experiments will not be discussed here, details are described in Ref. 18.

LINEWIDTH AND PHASE DIFFUSION OF THE ONE-ATOM MASER

In the following we would like to discuss another special feature of the one-atom maser: the spectrum. This is determined by the decay of the expectation value of the electric field[21]

$$\langle E(t) \rangle \sim \sum_{n=0}^{\infty} (n+1)^{1/2} \rho_{n,n+1}(t)$$

Hence the micromaser spectrum is different from the other effects discussed so far since it involves the off-diagonal elements $\rho_{n,n+1}$ of the radiation density matrix rather than e.g. the photon statistics, that is, the diagonal elements $\rho_{n,n}$; and it requires their time dependence rather than their steady state values. It could be shown that the linewidth D of the maser is given by

$$D = 4r \sin^2 \left[\frac{g\,t_{int}}{4\sqrt{\langle n \rangle}} \right] + \frac{\gamma(2n_b+1)}{4\langle n \rangle}$$

For small $g\tau/4\sqrt{\langle n \rangle}$ the sine function can be expanded; this leads to the familiar Schawlow-Townes linewidth

$$D = \frac{\alpha + \gamma(2n_b+1)}{4\langle n \rangle}$$

where

$$\alpha = \gamma(\sqrt{N_{ex}}\ g t_{int}).$$

The complicated pattern of the micromaser linewidth results, in part, from the complicated dependence of $\langle n \rangle$ on the pump parameter, which enters in the denominator.

We emphasize that the one-atom maser linewidth goes beyond the standard Schawlow-Townes linewidth. The sine function suggests in the limit of large pumping parameters an oscillatory behaviour of the linewidth confirmed in an exact numerical treatment. (For details see also Refs. 22 and 23).

It was pointed out that the maser linewidth can be measured when two phase coupled microwave fields are used; one before the atoms enter the microwave cavity and one after corresponding, in principle, to a modified Ramsey setup. The first micromaser field creates a superposition of the two maser states being then probed by the second one. The first field is only applied for an initial period in order to "seed" a phase in the micromaser cavity, in the second field the phase diffusion is probed. This technique will provide the basis for future experiments to measure the linewidth. It was tested by computer simulations and analytical calculations which showed that coherent pumping of the maser leads to new and interesting phenomena which cannot be discussed here. The first time coherent injection for the one atom maser was proposed in the paper by Krause et al.[24] and discussed later in connection with the measurement of the phase diffusion linewidth.[25] Also the phase dynamics of the maser field in steady-state operation was discussed by Wagner et al.[26] in the latter case only one microwave field is used for probing after the micromaser cavity. The latter scheme was further pursued in connection with the entanglement of states and symmetry breaking by Wagner et al.[27]

A NEW PROBE OF COMPLEMENTARITY IN QUANTUM MECHANICS - THE ONE-ATOM MASER AND ATOMIC INTERFEROMETRY

In one of the preceding sections we discussed how a nonclassical field inside the maser cavity can be generated. Such a field is extremely fragile because any attenuation causes a considerable broadening of the photon number distribution. Therefore it is difficult to couple the field out of the cavity while preserving its nonclassical character. But what is the use of such a field? In the present section we want to propose a new set of experiments performed inside the maser cavity to test the 'wave-particle' duality of nature, or better said 'complementarity' in quantum mechanics.

Complementarity lies at the heart of quantum mechanics: Matter sometimes displays wave-like properties manifesting themselves in interference phenomena, and at other times it displays particle-like behaviour thus providing 'which-path' information. No other experiment illustrates this wave-particle duality in a more striking way than the classic Young double-slit experiment. Here we find it impossible to tell which slit light went through while observing an interference pattern. In other words, any attempt to gain 'which-path' information disturbs the light so as to wash out the interference fringes. This point has been emphasised by Bohr in his rebuttal to Einstein's ingenious proposal of using recoiling slits to obtain 'which-path' information while still observing interference. The physical positions of the recoiling slits, Bohr argued, are only known to within the uncertainty principle. This error contributes a random phase shift to the light beams which destroys the interference pattern.

Such random-phase arguments, illustrating in a vivid way how the 'which-path' information destroys the coherent wave-like interference aspects of a given experimental setup, are appealing. Unfortunately, they are incomplete: In principle, and in practice, it is possible to design experiments which provide 'which-path' information via detectors which do not disturb the system in any noticeable way. Such 'Welcher Weg'- (German for 'which-path') detectors have been recently considered within the context of studies involving spin coherence.[28] In this section we describe a quantum optical experiment[29] which shows that the loss of coherence occasioned by 'Welcher Weg' information, that is, by the presence of a 'Welcher Weg'-detector, is due to the establishment of quantum correlations. It is in no way associated with large random-phase factors as in Einstein's recoiling slits.

The details of this application of the micromaser are discussed in Ref. 30. Here only the essential features are given. We consider an atomic interferometer where the two particle beams pass through two maser cavities before they reach the two slits of the Youngs interferometer. The interference pattern observed is then also determined by the state of the maser cavity. The corresponding interference term is given by:

$$\langle \Phi_1^{(f)}, \Phi_2^{(i)} | \Phi_1^{(i)}, \Phi_2^{(f)} \rangle,$$

where $|\Phi_j^{(i)}\rangle$ and $|\Phi_j^{(f)}\rangle$ denote the initial and final states of the maser cavity j.

Let us prepare, for example, both one-atom masers in coherent states $|\Phi_j^{(i)}\rangle = |\alpha_j\rangle$ of large average photon number $\langle n \rangle = |a_j|^2 \gg 1$. The Poissonian photon number distribution of such a coherent state is very broad, $\Delta m \approx \alpha \gg 1$. Hence the two fields are not changed much by the addition of a single photon associated with the two corresponding transitions. We may therefore write

$$|\Phi_j^{(f)}\rangle \approx |\alpha_j\rangle,$$

which to a very good approximation yields

$$\langle \Phi_1^{(f)}, \Phi_2^{(i)} | \Phi_1^{(i)}, \Phi_2^{(f)} \rangle \approx \langle \alpha_1, \alpha_2 | \alpha_1, \alpha_2 \rangle = 1.$$

Thus there is a 'radiation' interference cross term different from zero.

When, however, we prepare both maser fields in number states $|n_j\rangle$ the situation is quite different. Since the number states are orthogonal, the interference term disappears whenever a passing atom deposits a photon in one of the cavities therefore no interferences are observed in such a case. (For the case that thermal fields are present see Ref. 29).

At first sight this result might seem a bit surprising when we recall that in the case of a coherent state the transitions did not destroy the coherent cross term, i.e. did not affect the interference fringes. However, in the example of number states we can, by simply 'looking' at the one-atom maser state, tell which 'path' the atom took.

The atomic interference experiment in connection with one-atom maser cavities is a rather complicated scheme for a 'Welcher Weg'-detector. There is a much simpler possibility which we will discuss briefly in the following. This is based on the logic of the famous 'Ramsey fringe' experiment. In this experiment two microwave fields are applied to the atoms one after the other. The interference occurs since the transition from an upper state to a lower one may occur either in the first or in the second interaction region. In order to calculate the transition probability, we must sum the two amplitudes and then square, thus leading to an interference term in the same way as the two slit experiment. (For details see Ref. 31).

We conclude this section by emphasising again that this new and potentially experimental example of wave-particle duality and observation in quantum mechanics displays a feature which makes it distinctly different from the Bohr-Einstein recoiling-slit experiment. In the latter the coherence, that is the interference, is lost due to a phase disturbance of the light beams. In the present case, however, the loss of coherence is due to the correlation established between the system and the one-atom maser. Random-phase arguments never enter the discussion. We emphasise that the argument of the number state not having a well-defined phase is not relevant here; the important dynamics are due to the atomic transition. It is the fact that 'which-path' information is made available which washes out the interference cross terms.[30]

EXPERIMENTS WITH TRAPPED IONS - THE PAUL-TRAP

In contrast to neutral atoms, ions can easily be influenced by electromagnetic fields because of their charge. Therefore it is possible to isolate a single ion quite easily from its surroundings. In most of the experiments performed so far the Paul trap is used. It consists of a ring electrode and two end caps as shown in Fig. 6. Trapping can be achieved if time varying electric fields[32] are applied between the ring and caps (the two caps are electrically connected). A dc voltage in addition changes the relation of the potential depth along the symmetry axis (vertical direction in Fig. 6) to that in perpendicular direction. The equation of motion of an ion in such a situation is the Mathieu differential equation, well known in classical mechanics, which - depending on the voltages applied to the trap (dc and radio-frequency voltages) - allows stable and unstable solutions. Another way to achieve trapping is the use of a constant magnetic field aligned along the symmetry axis leading to the Penning trap.[1,33] In this case only a dc voltage has to be applied between ring and cap electrodes.

In order to produce the ions in the Paul trap a neutral atomic beam is directed through the trap centre and ionised by electrons. Unfortunately, the resulting trapped ions have a lot of kinetic energy rendering them useless for most applications, such as spectroscopy; therefore the ions have to be cooled. This is done by laser light. For this purpose the laser frequency ν is tuned below the resonance frequency, so that the energy of the photon is not sufficient to excite the atom. Crudely, the ion can extract the missing energy from its motion and thus reduce its kinetic energy. In other words, the atomic velocity Doppler-shifts the

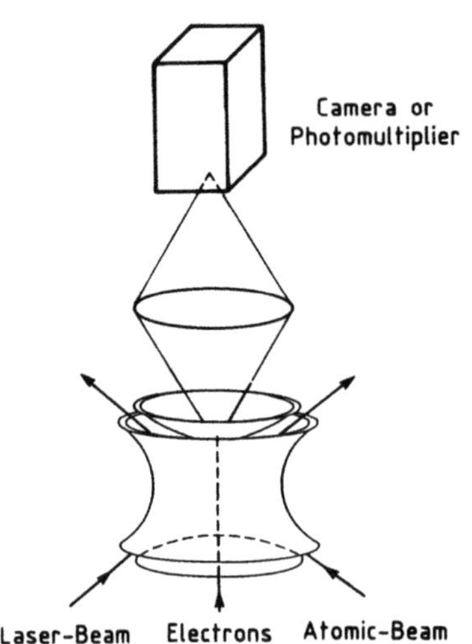

Figure 6. Sketch of the Paul trap. The fluorescence light is observed through a hole in the upper electrode.

atom into resonance to bridge the detuning gap Δ between laser and resonance frequency and the atom absorbs the photon of momentum $\hbar k = h\nu/c$. After the absorption process the momentum of the atom in direction opposite to the laser beam is reduced. This leads to a net cooling effect since the reemission of the energy due to atomic fluorescence is isotropic in space. The lowest temperature achievable is determined by the Doppler-limit[34] which is in the milli-Kelvin region. The low temperatures can be obtained within a fraction of a second.

The results discussed in this paper were obtained using a Paul trap with a ring diameter of 5 mm and an end-cap separation of 3.54 mm.[35] This trap is larger than most of the ion traps used in laser experiments.[1,36] The radio frequency of the field used for dynamic trapping is 11 MHz. The trap is mounted inside a stainless steel, ultrahigh vacuum chamber. We can obtain storage times of hours using a background gas at a pressure of 10^{-10} mbar. The ions are loaded into the trap by means of a thermal beam of neutral atoms (magnesium atoms in our case) which are then ionised close to the centre of the trap by an electron beam entering the trap through a small hole in the lower end-cap (see Fig. 6). In order not to distort the trap potential, the hole is covered by a fine molybdenum mesh. The neutral Mg beam and the laser beam pass through the gaps between the end-cap and the ring electrodes. The laser frequency is shifted by an amount Δ below the $3S_{1/2} \rightarrow 3P_{3/2}$ resonance-transition of $^{24}Mg^+$ at 280 nm to extract kinetic energy from the ions by radiation pressure as discussed above. In this way, a single ion can be cooled to a temperature below 10 mK. The fluorescence from the ions is observed through a hole in the upper end-cap, again covered by a molybdenum mesh. The large size of the trap allows a large solid angle for detecting the fluorescence radiation, either with a photomultiplier, or by means of a photon counting imaging system. To observe the ions, the cathode of the imaging system is placed in the image plane of the microscope objective attached to the trap; in this way images of the ions could be obtained.[37,38]

PHASE TRANSITIONS OF TRAPPED IONS

The existence of phase transitions in a Paul trap manifests itself by significant jumps in the fluorescence intensity of the ions as a function of the detuning Δ between the laser fre-

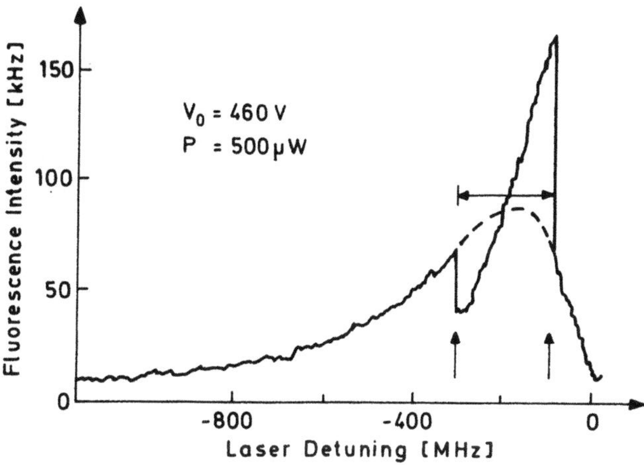

Figure 7. Fluorescence intensity, that is, photon counts per second, from five ions as a function of the laser detuning Δ. The vertical arrows indicate the detunings where phase transitions occur. The horizontal arrow shows the range of detunings in which a stable five-ion crystal is observed. The spectrum was scanned from left to right.

quency and the atomic transition frequency. These discontinuities are indicated in Fig. 7 by vertical arrows, and occur between two types of spectra: a broad and a narrow one, analogous to the fluorescence spectrum of a single, cooled ion. We interpret the broad spectrum as a fingerprint of an ion cloud and the narrow spectrum as being characteristic for an ordered many-ion situation with a 'single-ion signature'. Thus the jumps clearly indicate a transition from a state of erratic motion within a cloud to a situation where the ions arrange themselves in regular structures. In such a crystalline structure the mutual ion-Coulomb repulsion is compensated by the external, dynamic trap potential. The regime of detunings in which such crystals exist is depicted in Fig. 7 by the horizontal arrow. The existence of the two phases - crystal and cloud - can be verified experimentally by direct observation with the help of a highly sensitive imaging system, and theoretically by analysing ion trajectories via Monte Carlo computer simulations.[37,38]

The excitation spectrum of Fig. 7 and the jumps in it were recorded by altering the detuning Δ from large negative values to zero. When we scan the spectrum in the opposite direction the jumps occur at different values of Δ, which means there is hysteresis associated with these phase transitions.[37] Such a hysteresis behaviour can be expected with laser-cooled ions because the cooling power of the laser strongly depends upon the details of the velocity distribution of the ions. The behaviour of the ions in the trap is governed by the trap voltage, the laser detuning, and the laser power. Hysteresis loops appear whenever one of these parameters is changed up or down whilst the others are kept constant.

Fig. 8 shows ion structures as measured with the imaging system. For the measurements only a radio-frequency voltage was applied to the trap electrodes: in this case the potential is a factor of two deeper along the symmetry axis than perpendicular to it, and therefore plane ion structures are observed being perpendicular to the symmetry axis (for details see Refs. 37 and 38).

After we have accepted the existence of the ion crystals and the corresponding phase transitions, how do they actually occur? Would the cooling laser not force any cloud immediately to crystallise? A heating mechanism balancing the cooling effect of the laser must be the answer to this puzzle, but what heating mechanism? Since the early days of Paul traps this so-called radio-frequency heating has repeatedly been cited.[1] A deeper understanding however was missing and was provided only recently by a detailed study of the dependence of the cloud → crystal and crystal → cloud phase transitions on the relevant parameters.[37-39]

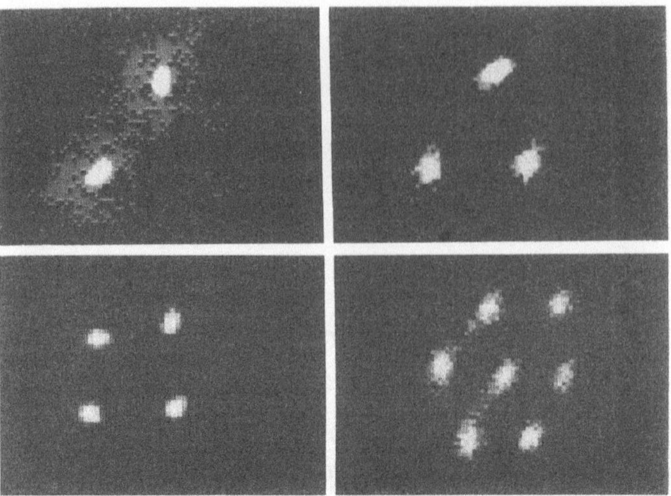

Figure 8. Two, three, four, five and seven trapped ions in the crystallized configuration. The distance between the ions is roughly 20 μm. The scale of the pictures is not completely identical. The trap potential is such that the ions arrange themselves practically in two-dimensional structures. There is a slight distortion of the ion structures by a contact potential caused by a coating on the ring electrode with Mg atoms.

The ions are subjected to essentially four different forces: the first one arising from the trapping field, then the Coulomb interaction between the ions, the laser cooling force, and finally a random force, arising from the spontaneously emitted photons. Using these forces computer simulations of the motion of the ions can be performed.[37] Depending on the external parameters such as the laser power, the laser detuning, and the radio-frequency voltage, the experimentally observed phenomena could be reproduced. Some of the results of the simulations are summarised in Fig. 9. Plotted is the radio-frequency heating parameter κ of five ions versus their mean separation.[39] For zero laser power and large r, we did not observe any net heating of the ions. This is confirmed by our experiments, in which, even in the absence of a cooling laser, large clouds of ions can be stored in a Paul trap over several hours without being heated out of the trap. When the ions are far apart, the Coulomb force is small, and on short time scales the ions behave essentially like independent single stored ions. For this reason, we call this part of the heating diagram the Mathieu regime.[38,39] Turning on a small laser, the rms radius r reduces drastically, but comes to a halt at about 14 μm. At this distance the nonlinear Coulomb force between the ions plays an important role and the motion of the ions gets chaotic. In this situation the power spectrum of the ions becomes a continuum leading to a radio-frequency heating process.

Increasing the laser power further results in an even smaller cloud. The smaller cloud produces more chaotic radio-frequency heating, as seen clearly by the negative slope of the heating curve in the range 8 < r < 14 μm. Finally, in the range 4 < r < 8μm there is still chaotic heating but the slope of the heating curve is positive. As a consequence of the resulting triangular shape of the heating curve at a laser power of about P = 150 μW, corresponding to r $\approx$ 8 μm, the chaotic heating power can no longer balance the cooling power of the laser light and the cloud collapses into the crystalline state located at r $\approx$ 3.8 μm. At this point the amplitude of the ion motion is so small that the nonlinear part of the repulsive Coulomb force is negligible again, therefore chaotic heating disappears and the phase transition occurs.

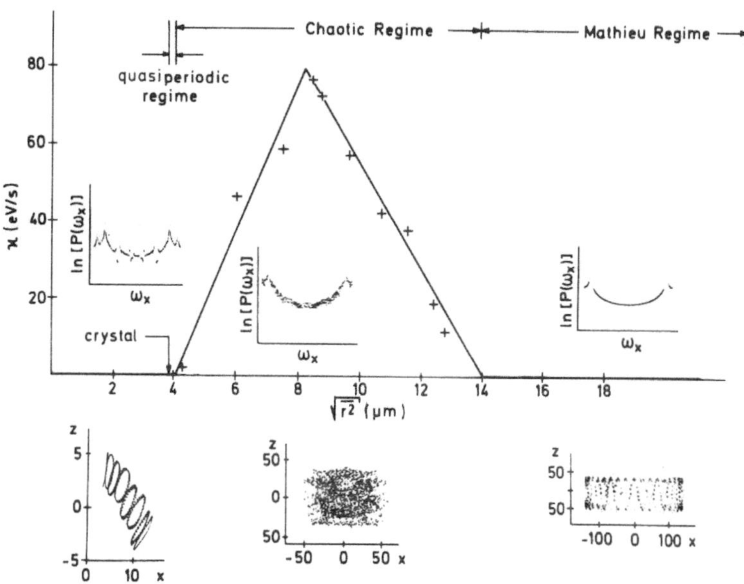

Figure 9. Average heating rate κ, of five ions in a Paul trap versus mean ion separation. The insets show the power spectrum and the corresponding stroboscopic Poincaré sections (plane perpendicular to the symmetry axis) of relative separation of two ions in three characteristic domains: The crystal state, the chaotic regime, and the Mathieu-regime. The units on the axes are in μm. In order to calculate the power spectrum of the 'crystal' shown on the left hand side, the distance of the two ions was displaced by 1 μm from the equilibrium position. The Mathieu-regime shown on the right is dominated by the secular motion.

THE ION STORAGE RING

A completely new era of accelerator physics could begin if it were possible to produce, store and accelerate Coulomb crystals in particle accelerators and storage rings. To work with crystals instead of the usual dilute, weakly coupled particle clouds has at least one advantage: the luminosity of accelerators (storage rings) could be greatly enhanced, and (nuclear) reactions whose cross sections are too small to be investigated in currently existing accelerators would become amenable to experimentation.

In the following, we would like to discuss very briefly our recent experiments using a miniature quadrupole storage ring. The storage ring is similar to the one described by Drees and Paul[40] or by Church.[41] We can observe phase transitions of the stored ions and the observed ordered ion structures are quite similar to the ones expected in relativistic storage rings, however, much easier to achieve. The motivation for building this small storage ring came from the fact that micromotion perturbs the ion structures in a Paul trap and only a single trapped ion is free of micromotion.[1] The ring trap used consists of a quadrupole field, leading to harmonic binding of the ions in a plane transversal to the electrodes of the quadrupole and no confinement along the axis (see Fig. 10). Confinement along the axis is achieved, however, by the Coulomb interaction between the ions when the ring is filled; then the total number of ions in the ring determines the average distance between them.

A scheme of the ring trap used for our experiment is shown in Fig. 10.[42,43] It consists of four electrode rings shown in the insert on the right. The hyperbolic cross section of the electrodes required for an ideal quadrupole field was approximated by a circular one. The experiments were also performed with Mg^+ ions. The ions were produced between the ring electrodes by ionising the atoms of a weak atomic beam produced in an oven which injected

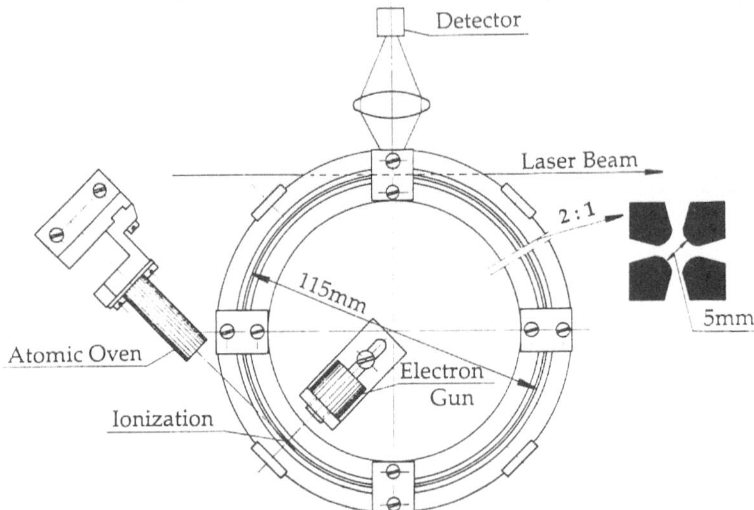

Figure 10. Quadrupole storage ring. The cross section of the electrode configuration is shown on the insert on the right hand side of the figure. The diameter of the ring is 113 mm and the distance between the electrodes is 5 mm.

the atoms tangentially into the trap region. The electrons used for the ionisation came from an electron gun the electron beam of which was perpendicular to the direction of the atomic beam. A shutter in front of the atomic beam oven allowed the interruption of the atomic flux. The ultrahigh vacuum chamber was pumped by an ion getter pump. After baking the chamber a vacuum of 10^{-10} mbar could be reached. Under these conditions the number of ions stored in the trap stayed practically constant for several hours.

When laser cooling of the ions is started a sudden change in the fluorescence intensity is observed, resembling very much that seen with stored ions in a Paul trap (Fig.7) which indicates a phase transition and the formation of an ordered structure of ions. The ion structure can also be observed using an ultrasensitive imaging system. Pictures of typical ion structures are shown in Fig. 11. The ions are excited by a frequency tunable laser beam which enters the storage ring tangentially. In the linear configuration the ions are all sitting in the center of the quadrupole field; therefore they do not show micromotion and it is possible to cool them further to temperatures in the micro-Kelvin region. The new cooling methods proposed by Dalibard et al.[34] can be applied to the Mg^+ ions so that the single-photon recoil limit can be achieved for the cooling process. At this limit the kinetic energy of the ions is smaller than the energy resulting from the 'zero-energy' motion of the harmonically bound ions; the ion structure then reaches its vibrational ground state, i.e. a Mössbauer situation is generated. The observed ion configurations in the quadrupole ring trap are described in a recent paper by Birkl et al.[44]. We will review the major results reported in this paper in the next section and compare the observed ordered structure to the results of molecular dynamic calculations.[45]

ORDERED STRUCTURES IN THE STORAGE RING

In the molecular dynamics calculations,[45] a cylindrically symmetric, static harmonic potential is assumed to describe the confining field. Each particle is subjected to the Coulomb forces of all other particles and to the confining field. The classical equations of motion are integrated for a system of several thousand particles, starting with random positions and velocities, to give the time evolution of the system. Cooling of the stored

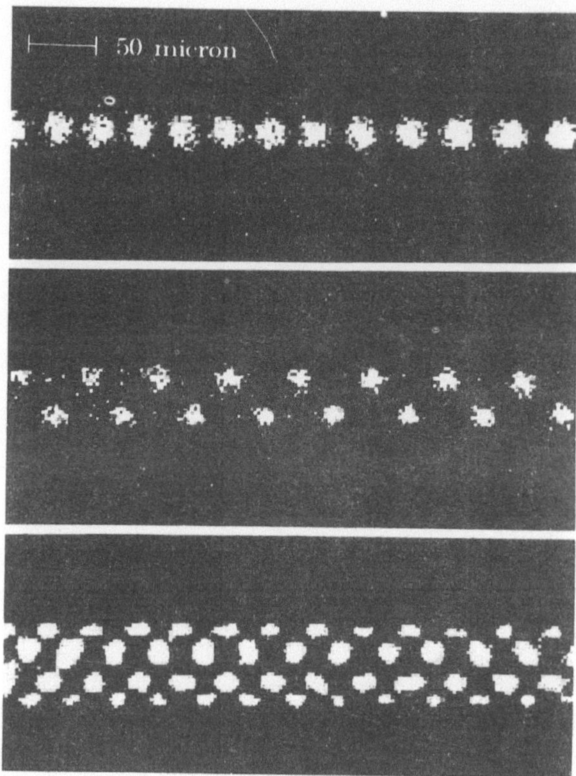

Figure 11. Crystalline structures of laser-cooled $^{24}Mg^+$ ions in the quadrupole storage ring. At a low ion density ($\lambda = 0.29$) the ions form a string along the field axis (a, upper). Increasing the ion density transforms the configuration to a zigzag (b, middle; $\lambda = 0.92$). At still higher ion densities, the ions form ordered helical structures on the surface of a cylinder, e.g., three interwoven helices at $\lambda = 2.6$ (c, lower). As the fluorescent light is projected onto the plane of observation in this case the inner spots are each created by two ions seated on opposite sides of the cylindrical surface, resulting in a single, bright spot.

particles is simulated by scaling down the resulting velocities of the stored particles at defined instants of the integration process. After sufficient cooling, ordered structures such as strings, zigzags, shells, and multiple shells should arise owing to the confining field's harmonic potential. Our experiments are well-suited to checking these predictions. To compare the experimental results with theory, we can use the normalised 'linear particle density' λ which is given by the ion density multiplied by the ratio of Coloumb repulsion and confining force of the trap.[45] Low λ values correspond to a deep potential well or a small number of ions, resulting in an equilibrium structure closely confined to the field axis comprising a string of ions (Fig. 11a). This is the micromotion-free configuration discussed above, and the analogue of the single stored ion in a Paul trap, as in both cases the ions sit in the potential minimum and show no micromotion. For higher values of λ, the structures extend more and more into the off-axis region, giving rise to (in the order of increasing λ) a plane zigzag structure (Fig. 11b) and cylindrical structures with the ions forming helices on cylindrical surfaces. The structure in Fig. 11c consists of three interwoven helices with six ions per pitch. The string and the zigzag have also been observed with laser-cooled Hg^+ ions in a linear trap[46].

Increasing further the number of ions leads to structures with smaller spacings between the ions where we cannot optically resolve individual ions any more. Images of these

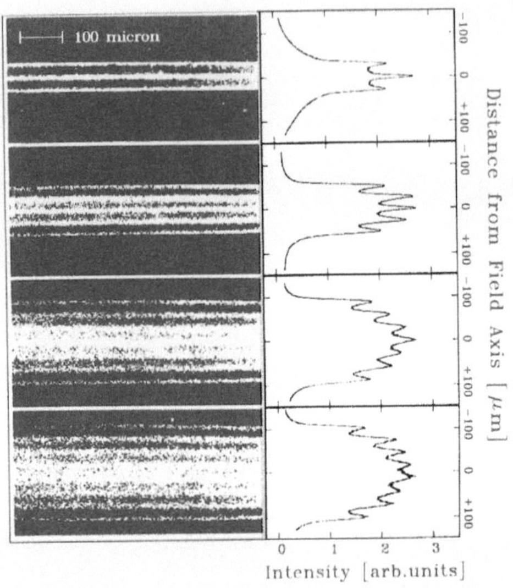

Figure 12. Images and intensity profiles of (from the top): one shell plus string (a; $\lambda \approx$ 4.3), two shells plus string (b: $\lambda \approx$ 12.2), three shells plus string (c; $\lambda \approx$ 26), and four shells (d; $\lambda \approx$ 34). There are up to $\approx 8 \times 10^5$ ions stored in the ring for the four-shell structure.

structures are presented in Fig. 12. The radial intensity profiles displayed on the right-hand side of the figure provide information about the structures as they give a measure of the radial distribution of the ions. For increasing λ it becomes energetically more favourable to create a string inside the first ion shell (Fig. 12a) to provide space for more particles. This results in a structure which is a three-dimensional analogue of the plane seven-ion crystal for a Paul trap (Fig. 8). The inner string turns into a second shell at higher densities: a string then develops inside this second shell (Fig. 12b), and so on. Figs. 12c and 12d show structures consisting of three shells plus string, and four shells, respectively. We have been able to observe all possible structures, from the string up to four shells plus string. The formation of multiple-shell crystalline structures in the quadrupole storage ring contrasts with the observation of shell structures in Penning traps, where the ions do not occupy fixed positions inside the shells.[47]

COMPARISON WITH THEORY

Fig. 13a gives a summary of experimental data for all recorded images in which the ions were individually resolved. The depth of ψ_0 of the potential well and the ion density per unit length are the experimental parameters. The theoretical boundaries between the different shell structures, predicted in Ref. 45 in terms of the functional dependence of λ on ψ_0 and the ion density, are given by the straight lines with constant λ. String structures are expected for $\lambda < 0.709$, zigzag structures in the range $0.709 < \lambda < 0.964$ and single shells in the range $0.964 < \lambda < 3.10$. Many different structures which are degenerate in energy are expected within the single-shell regime. We obtained stable configurations near $\lambda = 1.3$ and $\lambda = 2.0$ (two interwoven helices) and near $\lambda = 3.0$ (three interwoven helices - Fig. 13c). The observed structures agree with the predicted scheme for a large range of experimental parameters, thus confirming the theoretical results.

A summary of the experimental observations for ordered shell structures with up to four shells plus string and without resolution of individual ions is presented in Fig. 13b. The

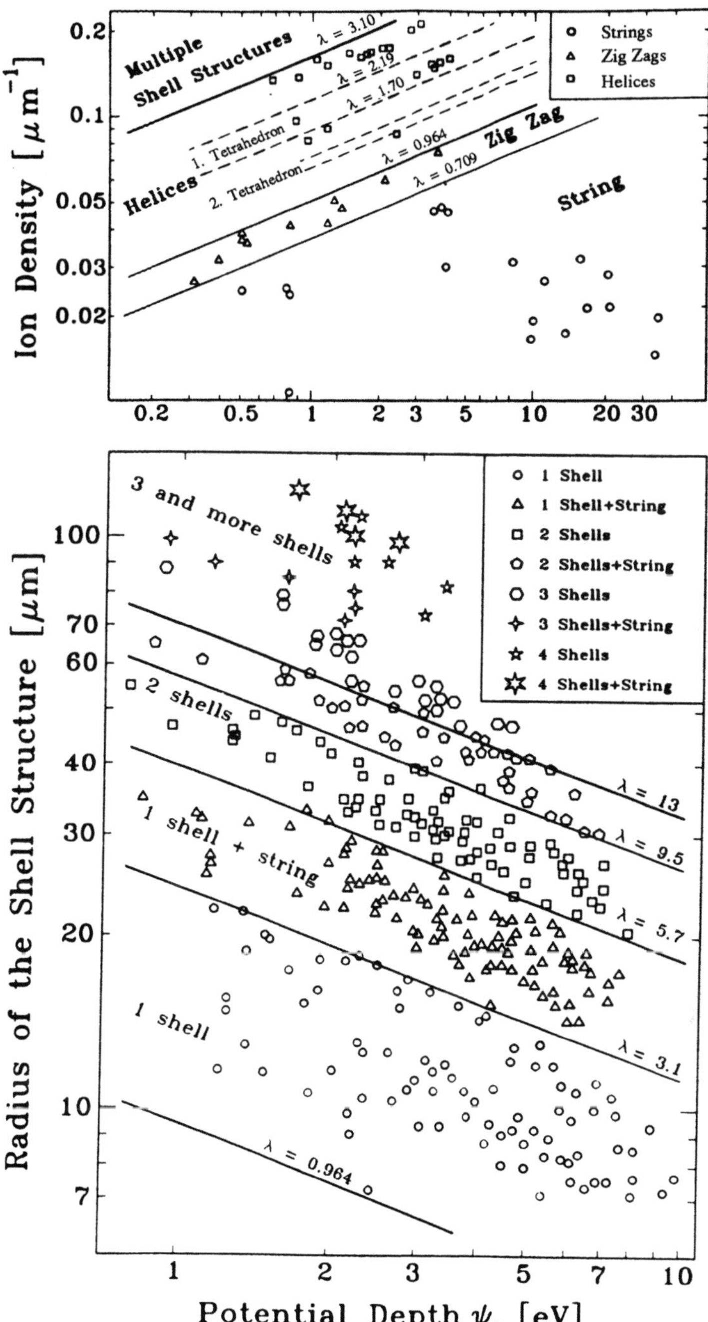

Figure 13. Summary of the experimental results. (a, top) Individual ions resolved, where the observed structures are chracterised by the ion density per unit length and the depth of the potential well ψ_0. These two parameters can be combined to give the normalised linear particle density λ which fully determines the ion configuration. The straight lines show critical λ values separating the regions corresponding to the various theoretically expected structures. The observed configurations are labelled with different symbols for each structure (b, down) Individual ions unresolved, where the observed shell structures with up to four shells plus string are characterised by their radius ρ and the potential depth ψ_0. The various observed structures are again separated by lines of theoretically determined critical λ.

depth ψ_0 of the confining potential well is again one of the experimental parameters. As the ion density cannot be determined directly from the images, the radius ρ of the structures is used instead as the second parameter. The theoretically predicted boundaries between the different shell structures are again given as straight lines of constant λ following[45] where the dependence of ρ on λ and ψ_0 was established. The observed ion configurations are seen once more to be determined by λ for a wide range of potential depths and ion densities.

Our results have important implications for two very different fields. Consider first the physics of low-energy particles: an ion string in a quadrupole ring, being free of micromotion, can be cooled to its vibrational ground state in the confining potential using recently proposed laser cooling techniques.[34] This would place the string in the Lamb-Dicke regime with a vanishing first-order Doppler effect because the spatial amplitude of the motion is smaller than the wavelength of the atomic transition. Furthermore, the second-order Doppler effect, which can only be reduced by further cooling, also disappears making the stored ions very interesting for frequency standards. The large number of ultra-cold ions available in the ring will lead to a high signal-to-noise ratio. Finally, cooled ions in the ring represent a quantum object of macroscopic dimensions (a Wigner crystal).

Second, with the experimental confirmation that the ordered structures expected in high-energy ion storage rings can indeed be formed, a completely new era of accelerator physics will emerge if it is possible to reproduce such Coulomb crystals in these rings. The enhanced luminosity of the corresponding beams would allow studies of ionic reactions whose cross-sections are too small for investigations in existing accelerators.

REFERENCES

1. D.J. Wineland, W.M. Itano, J.C. Bergquist, J.J.Bollinger and J.D. Prestage, Spectroscopy of stored atomic ions, in: "Atomic Physics 9," R.S. van Dyck, Jr., and E.N. Fortson, eds., World Scientific Publishing, Singapore (1984).

2. S. Chu, J.E. Bjorkholm, A. Ashkin and A. Cable, Demonstration and laser cooling and trapping of atoms, in: "Atomic Physics 10", H. Narumi and S. Shimamura, eds., North-Holland, Amsterdam (1986); and D.E. Pritchard, K. Helmerson, A.G. Martin, Atom traps, in: "Atomic Physics 11", S. Haroche, J.C. Gay and G. Grynberg, eds., World Scientific Publishing, Singapore (1988).

3. D. Meschede, H. Walther and G. Müller, The one-atom maser, *Phys. Rev.Lett.* 54:551 (1985).

4. E.T. Jaynes and F.W. Cummings, Comparison of quantum and semiclassical radiation theories with application to the beam maser, *Proc. IEEE* 51: 89 (1963).

5. For a review, see the articles by S. Haroche and J.M. Raimond, Radiative properties of Rydberg states in resonant cavities, in: "Advances in Atomic and Molecular Physics 20," D. Bates and B. Bederson, eds., Academic Press, New York (1985); J.A.Gallas, G. Leuchs, H. Walther and H. Figger, Rydberg atoms: high resolution spectroscopy and radiation interaction - rydberg molecules, in: "Advances in Atomic and Molecular Physics 20," D. Bates and B. Bederson, eds. Academic Press, New York (1985).

6. See, for example: J.H. Eberly, N.B. Narozhny and J.J. Sanchez-Mondragon, Periodic spontaneous collapse and revival in a simple quantum model, *Phys. Rev. Lett.* 44: 1323 (1980) and references therein.

7. G. Rempe, H. Walther and N. Klein, Observation of quantum collapse and revival in a one-atom maser, *Phys. Rev. Lett.* 58:353 (1987).

8. P. Meystre, Cavity quantum optics and the quantum measurement process, in: "Progress in Optics 30", E. Wolf, ed., North Holland, Amsterdam (1992).

9. P. Filipowicz, J. Javanainen and P. Meystre, The microscopic maser, *Opt. Comm.* 58: 327 (1986); Theory of a microscopic maser, *Phys. Rev.* A34:3077 (1986); Quantum

and semiclassical steady states of a kicked cavity mode, *J. Opt. Soc. Am.* B 3:906 (1986).

10. L. Lugiato, M.O. Scully and H. Walther, Connection between microscopic and macroscopic maser theory, *Phys. Rev.* A36:740 (1987).

11. P. Meystre, Repeated quantum measurements on a single-harmonic oscillator, *Opt. Lett.* 12:669 (1987); P. Meystre, Generation and detection of subpoissonian fields in micromasers, in: "Squeezed and Nonclassical Light," P. Tombesi and E.R. Pike, eds., Plenum, New York and London (1988).

12. P. Meystre, E.M. Wright, Measurement induced dynamics of a micromaser, *Phys. Rev.* A37:2524 (1988).

13. G. Rempe and H. Walther, Sub-Poissonian Atomic Statistics in a Micromaser, *Phys. Rev.* A42: 1650 (1990).

14. H. Paul, T. Richter, Bunching and antibunching of de-excited atoms leaving a micromaser, *Opt. Comm.* 85:508 (1991).

15. G. Rempe, F. Schmidt-Kaler and H. Walther, Observation of sub-Poissonian photon statistics in a micromaser, *Phys. Rev. Lett.* 64:2783 (1990).

16. H.J. Briegel, B.G. Englert, N. Sterpi, H. Walther, One-atom maser: statistics of detector clicks, *Phys. Rev.* A49:2962 (1994).

17. C. Wagner, A. Schenzle, H. Walther, Atomic waiting-times and correlation functions, *Opt. Comm.* 107:318 (1994).

18. O. Benson, G. Raithel, H. Walther, Quantum jumps of the micromaser field - dynamic behavior close to phase transition points, Phys. Rev. Lett., in print.

19. L. Davidovich, J.M. Raimond, M. Brune, S. Haroche, Quantum theory of the two-photon micromaser, *Phys. Rev.* A36:3771 (1987); M. Brune, J.M. Raimond, P. Goy, L. Davidovich, S. Haroche, Realisation of a two-photon maser oscillator, *Phys. Rev. Lett.* 59:1899 (1987).

20. P. Meystre, G. Rempe and H. Walther, Very-low temperature behaviour of a micromaser, *Opt. Lett.* 13:1078 (1988).

21. G. Raithel, Ch. Wagner, H. Walther, L.M. Narducci, M.O. Scully, The micromaser: a proving ground for quantum physics, in: "Advances in Atomic, Molecular,and Optical Physics, Supplement 2," P. Berman, ed., Academic Press, New York (1994)

22. M.O. Scully, H. Walther, G.S. Agarwal, T. Quang, W. Schleich, Micromaser spectrum, *Phys. Rev.* A44:5992 (1991).

23. T. Quang, G.S. Agarwal, J. Bergou, M.O. Scully, H. Walther, K. Vogel, W.P. Schleich, Calculation of the micromaser spectrum I. Green's-function approach and approximate analytical techniques, *Phys. Rev.* A48:803 (1993); K. Vogel, W.P. Schleich, M.O. Scully, H. Walther, Calculation of the micromaser spectrum II. Eigenvalue approach, *Phys. Rev.* A48:813 (1993).

24. J. Krause, M.O. Scully and H. Walther, Quantum theory of the micromaser: symmetry breaking via off-diagonal atomic injection, *Phys. Rev.* A34:2032 (1986).

25. R.J. Brecha, A. Peters, C. Wagner, H. Walther, Micromaser and separated-oscillatory-field measurements, *Phys. Rev.* A46:567 (1992).

26. C. Wagner, R.J. Brecha, A. Schenzle, H. Walther, Phase diffusion and continuous quantum measurements in the micromaser, Phys. Rev. A46:R5350 (1992).

27. C. Wagner, R.J. Brecha, A. Schenzle, H. Walther, Phase diffusion, entangled states, and quantum measurements in the micromaser, *Phys. Rev.* A47:5068 (1993).

28. B.-G. Englert, J. Schwinger and M.O. Scully, Is spin coherence like humpty-dumpty? I. simplified treatment, *Found. Phys.* 18:1045 (1988); J. Schwinger, M.O. Scully and B.-G. Englert, Is spin coherence like humpty-dumpty?, *Z. Phys.* D10:135 (1988); M.O. Scully, B.-G. Englert and J. Schwinger, Spin coherence and humpty-dumpty. II. The effects of observation, *Phys. Rev.* A40:1775 (1989).

29. M.O. Scully and H. Walther, Quantum optical test of observation and complementarity in quantum mechanics, *Phys. Rev.* A39:5229 (1989).

30. M.O. Scully, B.-G. Englert and H. Walther, Quantum optical tests of complementarity, *Nature* 351:III (1991).

31. B.-G. Englert, H. Walther and M.O. Scully, Quantum optical Ramsey fringes and complementarity, *Appl. Phys.* B54:366 (1992).

32. W. Paul, O. Osberghaus and E. Fischer, Ein Ionenkäfig, *Forschungsberichte des Wirtschafts- und Verkehrsministeriums Nordrhein-Westfalen* 415 (1958); E. Fischer, Die dreidimensionale Stabilisierung von Ladungsträgern in einem Vierpolfeld, *Z. Phys.* 156:1 (1959).

33. H.G. Dehmelt, Radiofrequency spectroscopy of stored ions, I: storage, in: "Adv. Atom. Molec. Phys. 3," D.R. Bates and I. Estermann, eds., Academic Press, New York (1967); F.M. Penning, Die Glimmentladung bei niedrigem Druck zwischen koaxialen Zylindern in einem axialen Magnetfeld, *Physica* 3:873 (1936).

34. J. Dalibard, C. Salomon, A. Aspect, E. Arimondo, R. Kaiser, N. Vansteenkiste and C. Cohen-Tannoudji, New schemes in laser cooling, in: "Atomic Physics 11", S. Haroche, J.C. Gay and G. Grynberg, eds., World Scientific Publishing, Singapore (1988).

35. F. Diedrich, E. Peik, J.M. Chen, W. Quint and H. Walther, Observation of a phase transition of stored laser-cooled ions, *Phys. Rev. Lett.* 59:2931 (1987).

36. W. Neuhauser, M. Hohenstatt, P. Toschek and H. Dehmelt, Optical-sideband cooling of visible aotm cloud confined in a parabolic well, *Phys. Rev. Lett.* 41:233 (1978); W. Neuhauser, M. Hohenstatt, P. Toschek and H. Dehmelt, Localized visible Ba^+ mono-ion oscillator, *Phys. Rev.* A22:1137 (1980).

37. R. Blümel, J.M. Chen, W. Quint, W. Schleich, Y.R. Shen and H. Walther, Phase transitions of stored laser-cooled ions, *Nature* 334:309 (1988).

38. R. Blümel, J.M. Chen, F. Diedrich, E. Peik, W. Quint, W. Schleich, Y.R. Shen and H. Walther, Phase transitions of stored laser-cooled ions, in: "Atomic Physics 11," S. Haroche, J.C. Gay and G. Grynberg, eds., World Scientific Publishing, Singapore (1988).

39. R. Blümel, C. Kappler, W. Quint and H. Walther, Chaos and order of laser-cooled ions in a Paul trap, *Phys. Rev.* A40:808 (1989).

40. J. Drees and W. Paul, Beschleunigung von Elektronen in einem Plasmabetatron, *Z. Phys.* 180:340 (1964).

41. D.A. Church, Storage-ring ion trap derived from the linear quadrupole radio-frequency mass filter, *Journal of Appl. Phys.* 40:3127 (1969).

42. H. Walther, chaos and order of laser-cooled ions in a Paul trap, in: "Proc. of the Workshop on Light Induced Kinetic Effects on Atoms, Ions and Molecules", L. Moi, S. Gozzini, C. Gabbanini, E. Arimondo and F. Strumia, eds., ETS Editrice, Pisa (1991).

43. I. Waki, S. Kassner, G. Birkl and H. Walther, Observation of ordered structures of laser-cooled ions in a quadrupole storage ring, *Phys. Rev. Lett.* 68:2007 (1992).

44. G. Birkl, S. Kassner and H. Walther, *Nature* 357:310 (1992).

45. R.W. Hasse and J.P. Schiffer, The structure of the cylindrically confined Coulomb lattice, *Ann. Physics* 203: 419 (1990); A. Rahman and J.P. Schiffer, Structure of a one-component plasma in an external field: a molecular-dynamics study of particle arrangement in a heavy-ion storage ring, *Phys. Rev. Lett.* 57: 1133 (1986).

46. M.G. Raizen et al., Linear trap for high-accuracy spectroscopy of stored ions, *J. Mod. Opt.* 39:233 (1992).

47. S.L. Gilbert, J.J. Bollinger and D.J. Wineland, Shell-structure phase of magnetically confined strongly coupled plasmas, *Phys. Rev. Lett.* 60:2022 (1988).

EXPERIMENTS WITH SINGLE ATOMS, MOLECULES

OR PHOTONS

Serge Haroche

Laboratoire Kastler Brossel de l'Ecole Normale Supérieure
24 rue Lhomond, 75231, Paris Cedex 05, France

INTRODUCTION. "IN VIVO" SINGLE PARTICLE EXPERIMENTS

Atoms, molecules or photons have been known for a long time, but only recently has it become possible to manipulate them as single entities and to observe their behaviour directly. Among many other feats, one can now "see" the individual atoms at the surface of a metal with a scanning tunneling microscope (STM)[1] or an atomic force microscope (ATM)[2], one can manipulate a single DNA molecule with "optical tweezers"[3], one can confine an isolated ion in an electromagnetic trap[4] or else single out one molecule embedded in a crystal lattice and observe it as it scatters photons from a laser beam[5]. Not only can we observe single atomic entities, but we can also employ them to perform experiments. Single atoms can for example be used to modify the properties of electromagnetic cavities, to produce non classical fields in "micromaser" devices, and to manipulate and measure the field stored in these cavities, at the single photon, or even at subphoton level[6].

Single particle detection is of course, in a sense, a very old topic. Particle detectors and photon counters have been around for a very long time, but the events they record are passed when they are observed. At variance, the single particle physics we are interested in here deals with the observation and continuous manipulation of isolated quantum systems, with the possibility of influencing events as they occur. We can say that these experiments are performed "in vivo", on continuously observed particles, by opposition with the particle detection experiments performed in accelerator physics which could be called"in vitro".

"In vivo" single particle experiments have been made possible by advances in microstructure technology as well as by progresses in laser manipulation of atoms. They have opened the way to a wide variety of applications, ranging from nanostructure fabrication, surface analysis at the atomic scale, to the building of new kind of detectors and sensors of various forces and fields. At the fundamental level, which interests us here, the manipulation of single atomic entities offers a unique opportunity to observe the quantum behaviour of nature. Quantum jumps[7] of atoms, ions, molecules and even of photon fields

Advances in Quantum Phenomena. Edited by E.G. Beltrametti
and J.-M. Lévy-Leblond. Plenum Press, New York, 1995

from one state to another can now be observed. These jumps are closely related to the concept of the "wave function collapse" occuring in a measurement process. Single particles can be followed along their quantum trajectories, exhibiting intrinsic quantum fluctuations, which can be observed and studied very directly. Various subtle aspects of quantum theory such as non local correlations and entanglement[8] can now be tested with single particle systems and the fuzzy boundary[9] between the macroscopic classical world and the microscopic quantum one can be explored with new tools.

These lectures were intended to present to a theoretically oriented audience an overview of the single atom physics experiments which have been performed in various laboratories, starting about fifteen years ago. This is now a wide field, expanding very quickly, which I could not cover extensively within the frame of two lectures. I have thus chosen to describe some experiments and to give extended references in which more details could be found. Most of these references are either tutorial (Scientific American, Science, Nature, Physics Today, La Recherche) or short Letters reporting important and illustrative experiments.

EXPERIMENTS SENSITIVE TO SINGLE PARTICLES IN CONDENSED MATTER PHYSICS

In the field of "in vivo" single particle physics, we can distinguish between condensed matter and atomic physics type experiments. Many recent breakthroughs in single particle condensed matter physics have been made during the last ten years. Novel microscopy techniques (STM, AFM and other variants) allow us to observe directly and manipulate matter with an atomic resolution of a few angstoems. In short, the STM method consists in scanning a very thin metallic tip above a conducting surface and recording the tunneling electron current which flows across the gap when a voltage is applied between the tip and the surface. This current is extremely sensitive to the distance between the atoms located at the end of the tip and the surface. The tip is moved by piezo-stack devices and a feed-back circuit is used to lock the current between the tip and the surface to a constant value by adjusting the driving voltage of the piezostack which controls the tip motion normal to the surface. A computer processes the current and voltage information to yield a "picture" of the surface at the atomic scale. The AFM method consists in measuring directly the force experienced by a tip scanned over a surface. It applies to non conducting surfaces for which the STM method cannot work. The tip is then mounted at the end of a flexible cantilever whose displacements are recorded by various means (one way being to use a STM tip to measure the position of the top metallic side of the cantilever). STM and AFM experiments make it possible to see single atoms at the surface of a crystal lattice where they interact with many other atoms. Using STM it is also possible to displace and manipulate single atoms on a surface[10] and to fabricate structures and possibly to store information at the atomic scale.

It has also become recently possible to count electrons one at a time and to control their flow in low temperature turnstile microscopic electronic gates[11]. These experiments constitute another very important domain of single particle physics. Here again, the condensed matter surrounding is important, since it determines the electrical characteristics of the microscopic circuit in which the electrons are flowing and the "Coulomb blockade effect" which is essential for the electron flow control.

These condensed matter experiments provide direct illustrations of the discrete atomic character of matter and open the way to many fascinating applications. I will not discuss them here in any more details. The reader interested in STM and AFM physics could find informations in many tutorial reviews[1,2]. As for the beautiful condensed matter-single electron experiments[11], they are described by Devoret's lectures in these proceedings.

ATOMIC PHYSICS AT THE SINGLE PARTICLE LEVEL

Atomic physics experiments are of a different kind. Here the single particle (atom, ion, molecule, or even electron) is trapped in a small volume, usually (but not always) far from any other perturbing matter. The particle under study is detected by some kind of electromagnetic interaction, either by photon scattering from a laser beam, or possibly by mere induction effect in the trap walls, if it is a charged particle. The detected signals exhibit fluctuations which reveal the quantum jumps of the system between its quantum states. These experiments have been developed mainly for very high resolution purposes, in order to measure fundamental quantities (such as the electron g-2 factor)12 or in view of developing new frequency standards for metrology (ion trap experiments of various kinds)13. In the process of carrying out these experiments, many interesting effects have been investigated and used for various applications (laser cooling mechanisms, interaction between individual ions leading to ordered structures14). It is important to notice that in these experiments, one does not achieve the spatial resolution of the STM or AFM methods. The resolution is usually limited by the optical detection methods to the micrometer scale, which is sufficient to distinguish individual atoms as long as they are isolated or at least well separated from each other.

Charged Particle Traps

Single charged particles have been trapped for a very long time using various electromagnetic field configurations. The Penning12 and the Paul4 traps have been the most popular ones. The Penning trap uses a combination of a quadrupolar static electric field and of a static magnetic field. The particle undergoes cyclotron motion around the magnetic field while it is repelled by two end caps along the magnetic field direction. Penning traps have been used to perform spectacular experiments on isolated electrons12 and positrons and are now used to trap antiprotons15. The detection of these particles is usually performed by picking up the tiny induction currents produced by the moving charge in the cap electrodes. The same method can also be applied to detect trapped ions. By comparing the cyclotron frequencies of different ion species, trapped one at a time in a Penning trap, different groups16 are now measuring with an unprecedented precision the ratios of the masses of various elements, in an effort to replace the mass standard by a new atomic definition.

The Paul trap makes use exclusively of electric fields to confine the particles. Since a purely static configuration cannot provide an absolute minimum in free space for the electric potential, the trap must combine a static field and an oscillating one, which alternates the directions in which the particle may escape the trap, in such a way so that it undergoes a bound motion around the trap center. Paul traps, which do not need the high magnetic fields of Penning traps, are very popular to confine ions for high resolution spectroscopy experiments.

Single Ion Fluorescence

Fluorescence detection is an extremely sensitive way of single particle observation. It applies of course only to systems which have an electronic structure, such as atoms, ions or molecules. The method consists in irradiating the particle with a laser beam resonant with a transition between its electronic ground state and an excited one and recording the photons scattered by the particle in a direction different from the laser beam. If this beam is intense enough, it saturates the transition which means that the particle is excited from its ground state in a characteristic time short compared to the excited state spontaneous emission time. In this case, the rate of scattered photons is determined by the spontaneous decay of the excited state. Typically, 10^7 to 10^8 photons can be scattered by second by a single atom or ion which is rapidely recycled by the laser beam between its two relevant energy eigenstates. Even if the photon detection efficiency is generally rather small (only one in a thousand or so of these photons are usually collected by a microscope lens in a limited solid angle), tens of thousands of photons per second can be observed, making a single atom or ion visible even by the naked eye. The spatial resolution is of course limited by diffraction laws to the optical wavelength, making the particle appear as a blob of light having a dimension of the order of one micrometer.

Figure 1, from4, shows the stepwise variation of the fluorescence from Ba+ ions in a Paul trap, as the number of ions in the trap is increased. The single ion increments are clearly observable and the single ion signal is visible. Ions can also be counted by observing the spatial distribution of the fluorescence blobs in the trap, as will be discussed below. The fluorescence intensity of an ion varies resonantly when the laser frequency is tuned around the ion transition frequency. Recording the amount of fluorescence versus the laser frequency is thus a very sensitive way of performing single particle spectroscopy.

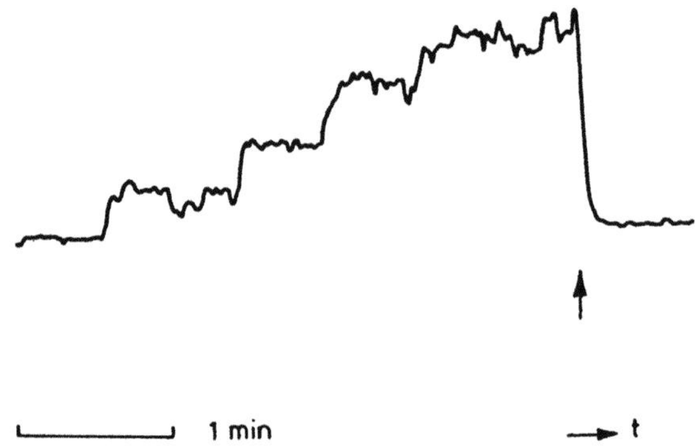

Figure 1. Stepwise variation of Ba+ laser-induced fluorescence when 1,2,3 and 4 ions enter the Paul trap. The fluorescence is suppressed at the end of the sequence (vertical arrow) by transferring the ions to a metastable level in which they do not inreact anymore with the laser (from ref4).

Laser Cooling of Single Particles in Traps

One of the main goals of single ion experiments is to perform high resolution single particle spetroscopy for metrology application13. The aim is to lock a laser to a narrow ion transition in order to realize a time standard more precise and more reproducible than the current one, based on the detection of the hyperfine transition of cesium atoms in an atomic beam machine. Single ions in a trap are indeed observable for a long time, without perturbation from neighbouring particles. This is a big asset for achieving high precision and high reproducibility in spectroscopy. The widths of the spectral lines are usually determined by the Doppler effect related to the motion of the particle in the trap. Achieving high resolution is thus conditionned by the ability of cooling the ion to low temperatures, corresponding to small particle velocities.

Lasers can be used very efficiently to achieve this cooling process17. Radiative cooling itself is based on the Doppler effect. Assume that a laser beam sent across the trap is slightly detuned in the red wing of the ion transition frequency. Due to the Doppler effect, the ion will then be more likely to absorb the light when it moves against the beam direction than when it moves along it. When it absorbs light, the ion takes the photon momentum and thus it is more likely to receive a momentum kick opposite to its velocity than along it. Since the ion is moving in the trap with equal probabilities along and against the laser beam, this results in a net cooling process. The cooling efficiency is proportional to the rate of fluorescence, since there is one photon absorbed from the laser beam for each photon scattered by the ion (the fluorescence photons, emitted randomly in all directions, do not participate to the average momentum exchange between the ion and radiation). Radiative heating can be achieved in a symmetric way, by irradiating the trap with light slightly detuned towards the blue of the ion transition. By heating the ions, one can make them "boil" out of the trap. Laser cooling and heating provides thus a very useful and flexible way of ion manipulation in a trap.

Under optimal conditions, laser cooling can bring the external degrees of freedom of a single ion to the lowest possible level. Idealizing the trap as a three dimension harmonic oscillator, the ion ends up in the oscillator ground state, where his motion is reduced to the zero point fluctuations. This condition can be fulfilled only if the transition natural linewidth (reciprocal of its spontaneous lifetime) is smaller than the trap oscillation frequency, which determines the separation between the harmonic oscillator levels. Such a condition has been achieved in small size Paul traps and the cooling to the trap ground state has been observed[18]. The ion is then at rest in free space, to within the quantum-mechanical limits imposed by the surrounding apparatus (zero point motion), an ideal situation to perform high resolution spectroscopy.

Ion Lattices in Traps

When several ions are trapped together, they tend to repel each other via Coulomb interaction. The ion dynamics is then determined by the balance between the ion repulsion and the trap confinement forces. The situation is not unlike the one prevailing in a molecule or in a solid lattice where the positive ions are bound together by the valence electrons, but the orders of magnitude are quite different. If the ions are cold enough, they organize in an ordered lattice structure with typical ion distances of the order of a few microns[14,19,20]. These structures give rise to a spatially resolved pattern in the ion fluorescence, so that the "ion crystal" can be observed directly. In this way, structures of two, three, four..., up to a few thousand ions have been observed. Figure 2 (from[20]) shows a linear structure of six mercury ions in a Paul trap. At a given transition temperature, the order disappears and the ions start to move randomly in the trap. The fluorescence is no longer spatially resolved and a structureless cloud of scattered light is observed. The phase transition like behaviour of this system has been extensively studied[19], as well as many other very interesting features. For example, the light scattered by two ions staying at a fixed distance from each other has been found to exhibit the fringe structure of the Young double hole eperiment, a striking illustration of interference effects at the single atom level[21].

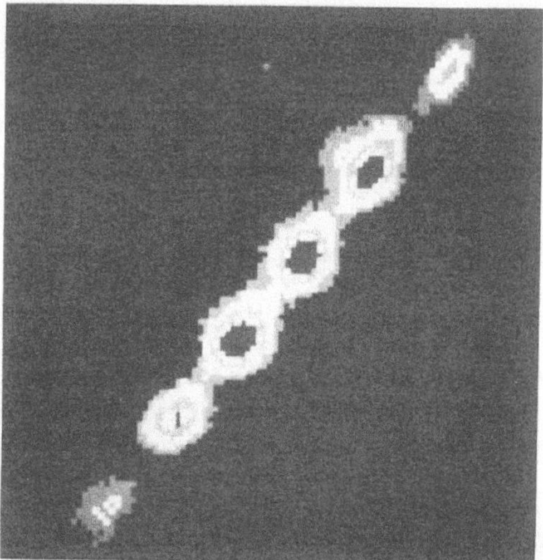

Figure 2. Six cold mercury ions forming an ordered linear structure in a Paul trap are observed by their spatially resoved fluorescence pattern (from[20]).

The cooling and heating processes of ion ensembles in traps is quite complex. Radiative mechanisms combine with Coulomb interaction in an intricate way. Ions can be cooled even if they do not interact directly with the light (sympathetic cooling)22. In this way, crystal made of different isotopes of a given element can be observed. The ions of the isotope quasi resonant with the laser beam are detected by their fluorescence and directly cooled in the process, while the other isotope ions are cooled by Coulomb interaction and remain invisible. They are revealed by "black spots" in the spatialy resolved fluorescence pattern of the ion crystal. When the goal of the experiment is to achieve very high spectroscopic resolution, the presence of many ions is a drawback since they tend to perturb each other, and single ion traps are preferable in this case.

Quantum Jumps

The observation of single ion fluorescence can reveal one of the most striking behaviour of a microscopic system, the occurence of quantum jumps[7,23,24,25]. The ion fluorescence intensity may exhibit sudden changes at random times, as shown in Figure 3a (lower trace) in the case of a single Hg ion[25]. These fluorescence intermittencies occur when the two levels between which the ion undergoes its fluorescence cycles are coupled by weaker transitions to metastable states. The relevant energy levels in Hg are shown in Figure 3b. The ion fluorescence is driven by a strong laser excitation, resonant with the $2S_{1/2} \to 2P_{1/2}$ transition and photons scattered at the same wavelength (194 nm) are observed. When the ion makes the transition to the metastable $2D_{3/2}$ state by emitting a 11 μm photon, the fluorescence cycles at 194 nm are suddenly interrupted and the fluorescence signal drops to zero. When the ion returns to the ground state by spontaneous emission (wavy line arrows in Figure 3b), the fluorescence goes back to the high level. The observation of such quantum jumps requires the existence of two different time scales in the ion evolution. The interval between two successive fluorescence photons (which defines the time resolution with which the jumps can be observed) must be much shorter than the lifetimes of the metastable levels involved in the process (which define the average duration of the steady fluorescence states).

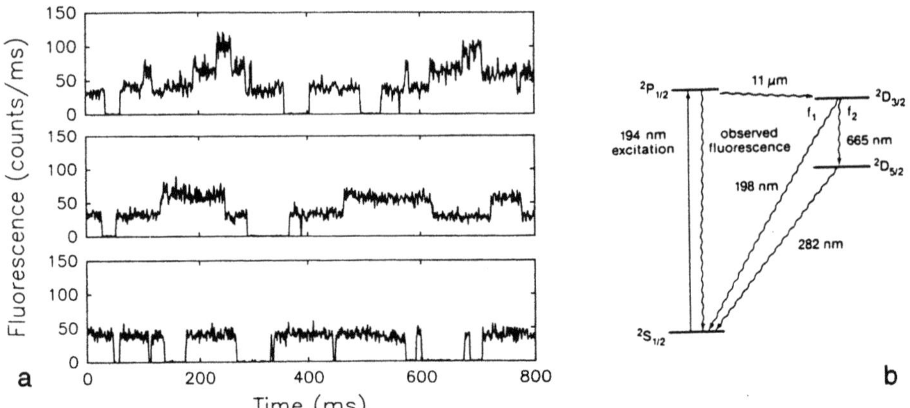

Figure 3. a) Intensity of 194-nm induced fluorescence as a function of time for a single mercury ion (lower trace), two ions (middle trace) and three ions (upper trace) in the trap. Quantum jumps are clrearly revaled by the discontinuous changes of the fluorescence level. b) Relevant energy levels of the Hg ion (from ref25b).

These sudden changes of the fluorescence intensity, which can be monitored with the naked eye, occur at random times, ruled by quantum mechanical chance. The average length of the fluorescence "on" periods is the inverse of the $2P_{1/2} \to 2D_{3/2}$ spontaneous transition rate whereas the average of the dark periods reflects the lifetime of the metastable states. The observation of a single ion fluorescence   signal corresponds to what is called a "quantum trajectory", which exhibits a noise of an inherent quantum mechanical nature. The probability distributions of the durations of the fluorescence "on" and "off" periods are exponential functions reflecting the well known exponential decay laws of quantum mechanics.

Quantum jumps are observed most clearly on single ion systems, with the signal dropping to zero during the "off" periods. When several ions are simultaneously present in the trap, they undergo quantum jumps at uncorrelated times and the fluorescence switches randomly between stable levels corresponding to integer numbers of single ion fluorescence levels (see Figure 3a, middle and upper traces corresponding to two and three ions respectively). As the number of ions is increased, the discrete nature of the phenomenon is progressively washed out and the fluorescence evolves into a steady state with a noise proportional to the square root of the number of particles. One can then predict the outcome of any measurement on the system by using standard quantum mechanical averaging procedures (density matrix approach). The quantum noise exhibited by the single quantum trajectories is then hidden by statistical averaging.

Aside from their fundamental interest, quantum jumps are practically very useful for high resolution spectroscopy of single particles in a trap. Ideally, one wants to perform spectroscopy on a highly forbidden transition which has an intrinsically very narrow linewidth. The $^2S_{1/2} \to {}^2D_{3/2}$ transition in Hg at 198 nm is a good example (see Figure 2b). Such a transition has however a very low fluorescence level (typically a few photon per second or even less), so that a direct detection of the fluorescence signal produced on such a transition would be unpractical. One can however switch on an auxiliary laser beam on the coupled $^2S_{1/2} \to {}^2P_{1/2}$ transition and observe on the strong fluorescence at 194 nm the quantum jumps of the ion when it is "shelved" by the 198 nm laser in the metastable state. The frequency of these jumps varies resonantly when the wavelength of the laser exciting the weak transition is tuned around 198 nm. By measuring the quantum jump frequency versus the wavelength of this laser, one can record the weak transition line on a strong fluorescence signal. The method is akin to an amplification process, the weak photon yield of the forbidden transition being multiplied by a huge factor. This method, now generally applied on a variety of ions isolated in Paul or Penning traps, had been first proposed by H.Dehmelt long ago under the name of "shelving spectroscopy". It will certainly be the choice method in the design of single ion clock standards.

Photon Antibunching

The occurence of quantum jumps in an ion's fluorescence is a non classical phenomenon whose time scale is characterized by the relatively long lifetimes of the ion metastable states. The phenomenon of photon antibunching[26,27,28] is another striking nonclassical effect, also observed on the radiation of a single particle, but at the much shorter scale corresponding to the time interval between two successive fluorescence photons. Let us consider the statistics of the detection times of the photons scattered by a single ion driven at resonance by a strong laser beam. If the photons were emitted in an uncorrelated way, the probability of detecting one of them would be constant in time, independent of the time origin (Poisson process). It turns out however that the probability of collecting a photon immediately after one has been detected is smaller than it is for longer time delays. This is expressed by saying that the photons are "antibunched" (the single atom photon emission process is also said to be subPoissonian). This effect is observed by recording the scattered field intensity correlation function, obtained by performing double photon counts. A first photon is registered at a time origin and one measures, by repeating the experiment many times, the conditional probability that a second photon will be registered a delay t later. The photon correlation function g2(t) obtained under these conditions after proper normalization and background suppression is shown in Figure 4 (from28). A correlation g2(t) equal to one corresponds to a Poisson process. The antibunching effect, corresponding to the minimum of g2(t) around t = 0, is quite clear.

The observation of photon antibunching can be considered as a direct evidence of the ion "wave function collapse" during the fluorescence cycle. Immediately after the detection of a photon, the ion must indeed be projected into the transition lower state, from which it cannot promptly emit another photon. A recovery time must ellapse to let the ion be excited again to the upper state by the laser beam before another photon can be detected. This recovery mechanism is a coherent process, the ion undergoing a transient precession between its lower and upper electronic states, known as a Rabi oscillation. The frequency of this precession is proportional to the laser field amplitude. This precession is the cause of the oscillatory behaviour observed in Figure 4: the conditional probability of detecting a photon after a first one has been observed oscillates indeed when the delay between the two photons is varied, reflecting the corresponding oscillation in the probability to find the atom excited. Figure 4 clearly shows that the frequency of the Rabi oscillation increases with the laser beam amplitude (from bottom to top).

The photon antibunching effect cannot be understood in a classical way. A source model which does not include the inherently quantum nature of the atom jumps between states cannot explain it. In fact, classical source theories can only predict positive correlations (photon bunching effects) such as the one observed in the famous Hanburry-Brown and Twiss experiment[29]. The quantum photon antibunching effect is also a typical single particle effect. If the fluorescence from many atoms is observed, the photons emitted by different radiators are uncorrelated in time and the antibunching effect is washed out. For samples with large absolute atom numbers, the radiation properties become classical.

The observation of quantum jumps as well as photon antibunching are not caracteristic features of ion traps. These phenomena can indeed be observed on untrapped atoms (the first photon antibunching experiment has been performed on an atomic beam[27]), but the trap configuration which allows one to observe a single particle for a long time is very convenient to study these effects.

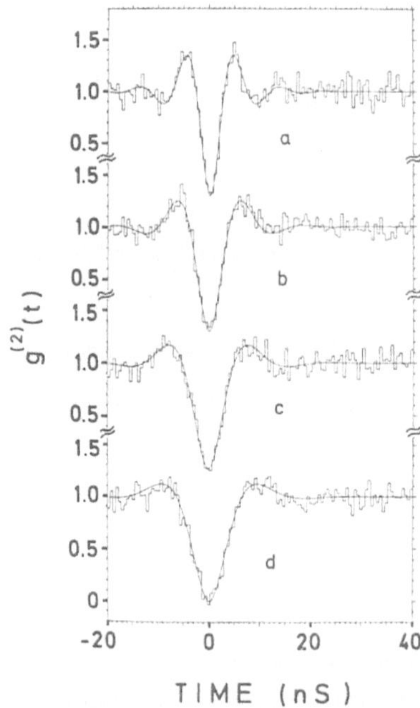

Figure 4. Photon correlation fluorescence signal g$^{(2)}$(t) from a single Mg ion driven by a resonant laser field in a Paul trap. The antibunching effect around t = 0 is observed. Traces a to d correspond to decreasing laser field intensities. The signal modulation reflects the transcient Rabi oscillation of the ion level populations following the wave function collapse after the emission of a photon. The frequency of this modulation varies linearly with the laser amplitude. The data have been corrected to eliminate irrelavant effects due to the ion micromotion and background due to accidental coincidences (from ref [28]).

Observing a Single Molecule in a Crystal

The ultrasensitive fluorescence spectroscopy methods can also be used to detect single molecules in a crystal lattice. A typical example is the detection of individual pentacene molecule located in a paraterphenyl lattice and cooled to liquid helium temperature[5,30]. Each molecule is then studied over a huge background of undected moelcules, by taking advantage of the frequency selectivity of the laser absorption process.

When spectroscopy is performed on a sample having a relatively large molecular concentration, the absorption spectrum is usually very broad, over hundreds of GHz in width, because different molecules in the sample have different internal electric field environments in the crystal and thus have their absorption spectra shifted over a wide frequency range. This is known as the "inhomogeneous broadening" of the absorption spectrum. When the concentration of the molecule under study in the host crystal is reduced below a certain value, the inhomogeneously broadened absorption line breaks into an ensemble of narrow absorption peaks, lorentzian in shape and randomly distributed in frequency. Each of these peaks has a width in the few MHz range only and corresponds to the absorption from a single molecule, with no appreciable Doppler effect since the molecule is trapped at its location in the crystal. The laser frequency being locked to one of these peaks, one can record the fluorescence of the molecule, basically in the same way as one does for ion fluorescence in a trap.

The energy level configuration of a pentacene molecule is much more complex than an atom or a ion one, due to the existence of molecular vibration states and phonon lattice bands. At room temperature, a large number of levels are rapidly populated while the molecule undergoes a very fast relaxation process. The coupling to the phonon modes also produces a large homogeneous broadening in the absorption spectrum. The absorption of the laser at any given frequency is relatively weak, due to the dispersion of the molecule oscillator strength over a large number of transitions. When the crystal is cooled to liquid Helium temperature, the phonons are "frozen" and the homogeneous width is greatly reduced. The laser absorption from the system ground electronic state at the bottom of the phonon band becomes then very strong[31], leading to a large fluorescence yield when each single molecule is resonantly excited[30].

The fluorescence occurs at a frequency different fom the laser one, on a transition from the upper electronic level of the laser transition to an intermediate metastable state relaxing relatively rapidly to the molecule ground state. The molecule undergoes a large number of cycles per second, from the ground state to the electronic upper state by laser absorption, then down to the ground state via the metastable level and the detection of its fluorescence yields precious information on its dynamical evolution, on the properties of the molecule local surroundings in the crystal and on molecular relaxation processes. Photon antibunching has been observed by performing correlation experiments on the fluorescence of single molecules, with results quite similar to the ones obtained on single trapped ions[32].

Resonance Ionization Spectroscopy

Fluorescence spectroscopy is not the only method with single atom sensitivity and high background discrimination ability. The general method known as Resonance Ionization Spectroscopy (RIS)[33] is also very efficient to detect and manipulate single atomic particles. Instead of relying on fluorescence, it combines selective laser absorption to an upper excited level with an ionization of the atom from this upper level by the action of a static electric field or a microwave field. The resulting ion or electron is detected by an electron multiplier or a channeltron device, yielding a large pulse of current for each particle, with a near unity efficiency. The process is highly selective, different isotopes of the same element being easily separated since they absorb the laser light at different frequency. As in fluorescence spectroscopy, very low atomic concentrations can be detected on a huge background of undetected atoms.

The detection of very excited atomic Rydberg states by selective dc field ionization[34] is a particular case of RIS which is of great interest for the experiments discussed in the last part of this chapter. In a typical experiment, an atomic beam is excited by stepwise laser excitation into a very excited atomic state, having a large principal quantum number n, known as a Rydberg state. These states, in which the valence electron occupies a very large orbital, are ionized by electric fields of moderate amplitude (typically a hundred V/cm for an atom with principal quantum number n = 50). Moreover, the threshold for field ionization depends on the level of excitation and varies as $1/n^4$. By conveniently adjusting the amplitude of the ionizing field, one can selectively detect the atom in different Rydberg states and thus monitor transitions between these states induced by a microwave field. In this way, atoms can be detected one at a time with a high efficiency and the absorption of a single microwave photon by an atom can be monitored, since it results in a change of the Rydberg level detected by the state selective field ionization process[35].

EXPERIMENTS WITH SINGLE ATOMS IN HIGH Q CAVITIES: "THE PHOTON TRAP"

In single atom experiments, photons are usually employed to manipulate the atomic system and gain information on its evolution. A new kind of experiments is now rapidly developing in which the roles of matter and radiation are switched. A photon field stored in a high Q cavity is the object of the investigation and atoms, sent one at a time across the cavity are used to manipulate and detect the field. These experiments can be viewed as "photon trap" experiments since the photons confined in the cavity for a relatively long time play now the role which was the one of the ions in the usual trap experiments. "Photon traps" constitute a subfield of "Cavity Quantum Electrodynamics" (Cavity QED)[36,37], which is itself a rapidly expanding domain of Quantum Optics dealing with the radiative behaviour of isolated atoms in cavities. Some experiments demonstrating the potentialities of "photon traps" have already been realized, but many experiments in this field are still in progress or projected. This section is thus more prospective than the previous one.

Tools and Orders of magnitude of Cavity QED experiments

The intrinsic coupling between the atoms and the field in a Cavity QED experiment is defined by the "vacuum Rabi frequency" Ω which represents the rate at which the atom and the empty cavity exchange a photo[36]. The coupling $\hbar\Omega$ is equal to the product of the atom electric dipole by the electric field per photon in the cavity. Atoms with as large electric dipoles as possible have thus to be employed and the cavity size must be as small as possible to achieve a maximal photon confinement and the largest possible field per photon. Furthermore, the quantum correlations produced by the atom-field interaction are destroyed in a time of the order of the atom and field relaxation times T_{at} and T_{cav}. The experiments must thus be carried out within a time shorter than T_{at} and T_{cav} and the conditions ΩT_{at}, $\Omega T_{cav} \gg 1$ must ideally be fulfilled. Very high Q cavities with extremely low losses have to be used, in conjonction with atoms whose relevant energy levels have very long radiative damping times.

Two strategies are employed to optimize all these parameters. One kind of Cavity QED experiment[38] is performed with very small optical cavities (less than a millimeter in size) coupled to atoms on a ground to electronic excited state transition. The field per photon is then large, but the atomic dipole is small and the photon confinement time does not exceed a microsecond. Furthermore, the excited state of the transition has a rather short decay time, in the 10^{-8} s range typically. Ω values of the order of 1 to 10 MHz can be achieved with these systems, corresponding to ΩT_{cav} of the order of a few tens. We will not discuss here these optical experiments.

In the other kind of experiment, superconducting microwave cavities are coupled to very large Rydberg atoms having a huge electric dipole resonant microwave frequencies[36]. The atomic transitions fall in the 10 to 100 GHz range (microwaves) and the cavities have a centimetric size which makes the field per photon relatively small (of the order of a mV/cm typically). The intrinsic atom field coupling Ω is of the order of 100Khz, large enough to observe single photon effects during the time (about 10µs) that the atom spends to cross the cavity at thermal velocity.

The microwave cavities storing the photons are made of superconducting Niobium, cooled at a temperature of the order of 1K or less. The losses are very low and Q factors in the 10^8 to 10^{10} range can be obtained, corresponding to T_{cav} in the 10^{-3} to 10^{-1} s range. The ΩT_{cav} factor can thus reach a value of the order of 10^2 to 10^4. The cavity is either closed, with a cylindrical shape or open, of the Fabry Perot type, with two carefully polished spherical Niobium mirrors facing each other.

The atoms used in these experiments are Rubidium atoms excited by lasers to a Rydberg state with a principal quantum number value ranging from 40 to 70. The radiative decay time of these atoms is of the order of a few tens of microseconds, i.e. of the order of the atom transit time across the cavity. This time fails to achieve the $\Omega T_{at} >> 1$ condition which is required in some experiments. In order to overcome this problem, one can prepare circular Rydberg atoms[39], which correspond not only to a large principal quantum number n, but also to the maximum possible value of the electron angular momentum projection along the quantization direction. In these atoms, the electron revolves on a circle around the nucleus, as described by the old Bohr theory of quanta. Circular Rydberg atoms have a very long dipole damping time (in the hundredth of a second range) and a very large electric dipole on a transition between adjacing n and n-1 states (about 1000 a.u.). These atoms are prepared by combining laser excitation with radiofrequency adiabatic transfer between Stark substates in an electric and a magnetic field[40], or by having the Rydberg atom interact with static crossed electric and magnetic fields[41]. The various circular states with different n values ionize in different electric fields, which provides, as discussed above, a selective and very sensitive way of detecting them, one atom at a time. Circular Rydberg atoms appear thus as ideal tools to probe the field in a "photon trap" experiment.

The Micromaser

The first example of Cavity QED experiments we discuss is the micromaser[42,43]. In this device, atoms prepared in a Rydberg state $| e >$ are sent one at a time across a very high Q superconducting microwave cavity, resonant with a transition connecting $| e >$ to a lower level $| g >$. Each atom has a probability of releasing a photon, which has a large probability of surviving until the next atom interacts with the cavity. A steady state results from the competing effects of photon emission and relaxation of the field in the cavity walls. Such a system is the limit of ordinary maser/laser oscillators when the number of active atoms in the medium is reduced to unity. Such a low threshold is achievable because Rydberg atoms have a very large electric dipole and a very strong coupling to microwave radiation. Rydberg atom masers have been operated for about fifteen years, first with relatively large numbers of atoms in a pulsed regime[44]. Then, with the progresses in the development of high Q superconducting cavities, single atom cavity induced effects could be observed[45] and steady state operation of a micromaser achieved[42]. The atom-field coupling is so strong in these devices that a maser operation on a two-photon transition has also been observed[43]. In this case, the Rydberg atoms crossing the cavity release pairs of photons while they undergo a transition between two electronic levels of the same parity.

The experimental study of the micromaser operation relies on the detection by field ionization of the atoms which leave the cavity. A very simple analyzis of the process can be given if we neglect for the time being the field relaxation in the cavity walls. Assume that at a given time the cavity field contains exactly N photons. If an atom then crosses the cavity and is detected after exiting it in the upper level of the maser transition, the number of photons in the cavity has not changed. If this atom is detected in the lower state instead, we know that the photon number in the cavity has increased to N+1. Before the measurement is performed, the atom and the field are in an "entangled" state of the form $\alpha | e,N> +$

$\beta \mid g, N+1 >$ with the first and second symbol in each ket representing the atom and the field states respectively. This entanglement survives when the atom leaves the cavity, provided the relaxation times of both systems are long enough. The measuring process appears thus as a reduction "at a distance" of the system wave packet which "collapses" the state of the field into a Fock state (N or N+1).

The type of nonlocal entanglement we have described here is reminiscent of the one discussed in the "Einstein-Podolski-Rosen" experiments[8] with two spins 1/2. The mere difference is that, in the micromaser case, a two level system (the atom) is correlated to a harmonic oscillator like system (the field). The correlation between the two subsystems can be expressed in different basis sets. The entangled wave function written above can as well be expressed as an entangled product of the $\mid e> \pm \mid g >$ atomic states with linear combinations of N and N+1 Fock states. If, instead of setting the atomic detector to measure the atom's energy, one performs a measurement whose eigenstates are $\mid e > \pm \mid g >$ (detection of "atomic coherences"), one prepares the field in the cavity in a coherent state, superposition of Fock states. The decision to detect the atomic energy or coherence can be made after the atom has left the cavity (delayed choice) and the state of the field is affected in a very profound way by a decision made after the two subsystems have ceased to interact. This is the fact of nature which bothered so much Einstein, Podolski and Rosen[8] and which has been confirmed in a different context by the beautiful photon correlation experiments demonstrating a violation of Bell's inequalities[46]. Micromasers, if they could be operated with negligible losses, would be good systems to perform new tests of this fascinating aspect of the quantum theory.

Real micromasers cannot be described as simply as above, because we have completely neglected the field relaxation in the cavity walls and the atomic decay during the time of flight between the cavity and the detector. Moreover, thermal photons in the cavity must also be taken into account, even at temperatures of the order of 1K. More complete theories[47,48,49], which incorporate these factors, have predicted some very interesting features of these systems. Among the predicted phenomena which have been experimentally observed, let us mention the generation of fields with subPoissonian photon statistic[50] and the occurence of a bistable behaviour, the micromaser switching randomly between two operating points due to quantum fluctuations[51].

The micromaser experiments performed so far have employed low angular momentum relatively short lived Rydberg atoms. The use of long lived circular Rydberg atoms in conjonction with very high Q cavities at extremely low temperatures (around 0.1K, in order to suppress totally thermal photons) will certainly open the way to new fundamental studies of this very interesting quantum system[52].

Dispersive Cavity QED Experiments

In the micromaser, Rydberg atoms are resonant with the cavity field and can exchange photons with it. Let us now consider a totally different situation, in which the atoms are non resonant with the field. Their interaction with the cavity, which cannot change the photon number, results then in a phase shift of the atom's wave function and of the field itself (dispersive single atom index effect)[35,53]. The dispersive interaction between the atoms and the "trapped" photon makes it possible to realize subtle "Quantum Nondemolition (QND)"[54] experiments, in which the atoms measure the field intensity without perturbing it, quite at variance with usual photodetection processes.

A sketch of the experimental set up used at ENS in Paris to study these dispersive effects is shown in Figure 5. The Rubidium atoms effusing from an oven O are prepared in the n=51 circular Rydberg state in the cylindrical box CB, before crossing the cavity C which is fed through a wave guide by a microwave source S. The cavity mode frequency is slightly detuned from the $n=51 \rightarrow n=50$ Rydberg state transition, so that no energy exchange can occur between the atoms and the cavity field mode. The cavity C is sandwiched between two auxiliary cavities R1 and R2 fed by an auxiliary tunable microwave source (not shown), in which the atoms undergo $\pi/2$ microwave pulses admixing the n=51 and n =50 circular

Rydberg states. After interacting with R1, C and R2, the atoms are detected by the field ionization detector D, which counts them in the n=51 and n= 50 levels, allowing us to measure the transition probability between these states which is induced by the separated oscillatory fields in R1 and R2. This probability is an oscillating function of the frequency ν of the field applied in R1 and R2 (Ramsey fringes[55]). The modulation of the transition probability reveals an interference effect between two probability amplitudes, one corresponding to a transition undergone by the atom in R1, the other to a transition undergone in R2. The quantum state of the atom when it crosses C is different for these two channels. The field stored in C shifts by different amounts the energy of the atom in the n=51 and n=50 circular states, resulting in a differential phase shift of the atomic wave function in these two states and leading to a phase shift of the fringe pattern proportional to the field intensity. The measurement of this fringe shifts should thus yields a direct determination of the photon number in C[56].

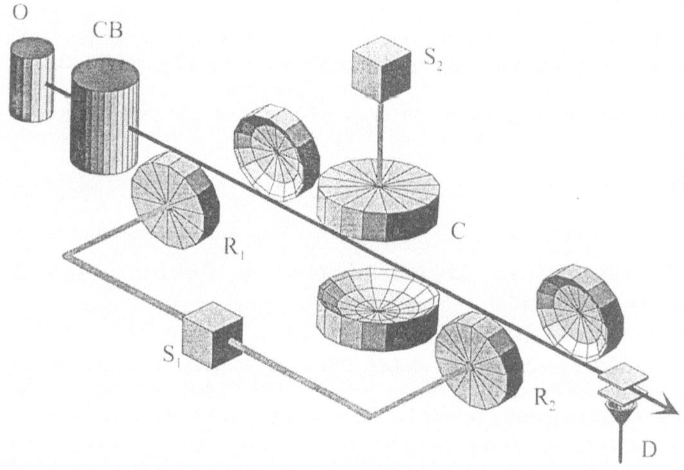

Figure 5. Sketch of the ENS dispersive Cavity QED experiment.

Sub-Photon Fields Measured by Atomic Interferometry

We now discuss the first very recent results obtained with this set up[57]. Figure 6 shows a typical fringe pattern, obtained with no field in C (bottom trace) compared to the fringe pattern recorded when a field containing on average one photon is fed into C (upper trace). The detuning between the cavity mode and the atomic transition is here 150 kHz. A shift of 315 Hz of the fringe pattern is clearly observable when a single photon is injected in C. The sensitivity of this interferometer to a shift as small as 25 Hz makes field containing only one tenth of a photon on average detectable. When C is empty, the atomic transition undergoes a residual shift, equal to the shift produced by half a photon, which is due to the effect of the vacuum field fluctuations in the cavity mode on the upper level of the atomic transition. We have also observed this effect, which can be considered as a Lamb-shift induced by a single cavity mode[57]. The calibration of this interferometric photon-meter is absolute, since the shift per photon of the atomic transition is entirely determined by the geometry of the cavity and by the quantum numbers of the atomic levels involved.

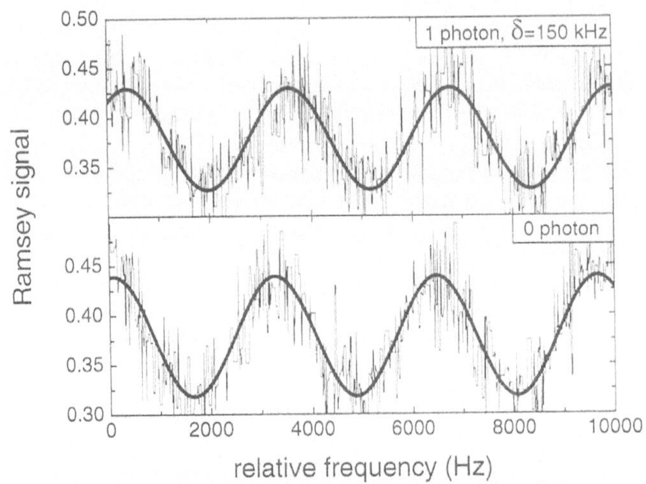

Figure 6. Ramsey fringe signal showing the population transfer between the circular Rydberg levels n = 51 and n =50 as a function of the frequency applied in the microwave zones R_1 and R_2. The cavity (detuned by 150 Khz from the atomic transition) contains zero photon (lower trace) or one photon (upper trace) on average. The translation of the fringe pattern can reveal the dispersive light shift produced by subphoton fields (from ref[57]).

Quantum NonDemolition Measurements of Photons and Field Jumps (experiments in progress)

We now turn to the discussion of experiments in progress with this set up. The Rydberg atom interferometer can be set to produce a π phase shift per photon, by adjusting the atom cavity detuning. The frequency of the R_1-R_2 auxiliary field can also be adjusted so that the probability of detecting the atom in the n=51 level is unity if there is zero (or an even number of photons) in C, and equal to zero if there is one (or an odd number of photons). In this way the information yielded by the atom detector fixes the parity of the photon number in the cavity. Assume that the field is weak enough so that it contains on average much less than one photon, with a vansihing probability of containing two or more. In this case, the single atom detection fixes unambiguously the photon number in C: if the atom is detected in n=51, there is no photon and if it is detected in n=50, the field is in in a one photon state[58]. This field measurement is a quantum nondemolition process (QND), since the photon number in C is not altered by energy absorption or emission. Information is acquired on the system by performing the atomic measurement and this process results in the "collapse" of the field state into an energy eigenstate, in what might be considered as a prototype of a pure quantum measurement. The procedure, described here for very small fields, can be generalized to fields containing more than one photon (several atoms must then be sent across C and detected[53,56]).

Once the photon number is pinned down to a given value by the atomic detection, this number will not change (provided the cavity is lossless). In a real cavity however the photon damping time is finite and the absorption of a photon in the cavity walls should result in a sudden change of the atomic detection signal. If, as discussed above, the n= 51 atomic state is correlated to a one photon state initially present in the cavity, a train of atoms crossing the apparatus must all be detected in this level, until the photon disappears and the atomic state suddenly switches from n = 51 to n=50, an event which should be readily detected by the atomic detection method. Similarly, if the cavity contains a larger number of photons, the detection of a train of atoms should allow us to pin this number down to a definite value and to follow how the photon number evolves randomly in time due to field relaxation[53]. Figure 7 shows the numerical simulation of a typical quantum trajectory corresponding to the decay of a field containing initially 8 photons. The steps associated to the decay of the field are random in time and exhibit all the characters of a quantum jump phenomenon. In this case, this is the field which undergoes the jumps and the atoms which are employed to detect them. The roles of photons and atoms are reversed in the ion quantum jump experiments. Of course, the usual exponential decay of the field in the cavity is recovered by averaging a large number of simulations.

The QND experiments require the cavity relaxation to be negligible during the atomic information acquisition process. This condition was not fulfilled in the experiment described in the previous section, which was performed with a preliminary cavity whose Q factor was poor. The field relaxation time (about 2 µs) was then shorter than the atom flight time across the cavity (20 µs), so that the atomic wave phase shift was resulting from the average effect of a fluctuating photon number in C. We are now resuming this experiment with an improved cavity having a 500µs photon damping time and we expect to be able soon to perform the quantum nondemolition and field quantum jumps experiments described in this section.

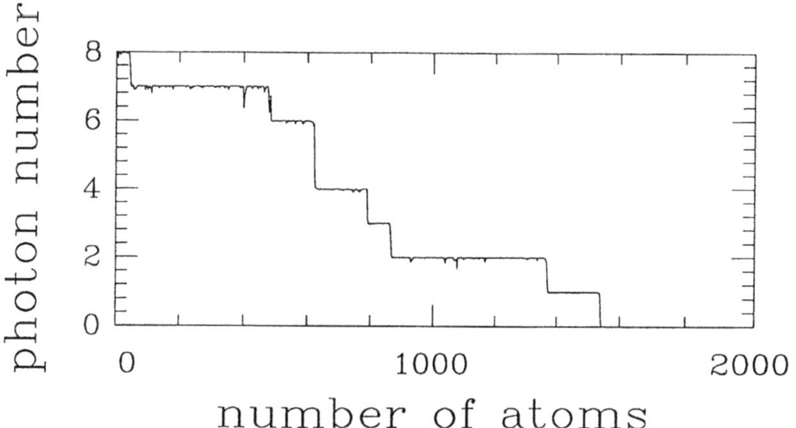

Figure 7. Photon number evolution in a simulation of a continuously monitored relaxing field in a cavity. The initial field is the N=8 Fock state. The horizontal scale is the number of detected atoms. The cavity relaxation time T_{at} corresponds to the passage of 1000 atoms across the cavity (from[53]).

Non Local Correlations, Schrödinger Cat States and Applications

The circular Rydberg atom-superconducting cavity interferometer could be used for many interesting applications, that we will only list here with references to the articles were more can be learned about them. A train of atoms exiting the cavity one by one could be prepared in a multiparticle entangled state. By exploiting the dispersive atom-field correlation effects, one could for example generate a linear superposition of a state in which all the atoms are in the n=51 level and of another state where all the atoms are in the n=50 level[59]. Similar superpositions have been considered recently in the context of gedanken experiments aiming at demonstrating the non-local character of quantum mechanics60. Such experiments performed in the Cavity QED context would constitute striking illustrations of the Einstein-Podolski-Rosen paradox.

One could also use the dispersive atom-field interaction to prepare nonclassical superpositions of field states. It is possible to realize a situation such that a single atom crossing C shifts the phase of the field by $\pm \pi/2$ depending upon whether it is in the n=51 or n=50 state[53]. Since the atom is prepared in R₁ in a superposition of these two states, the system thus evolves into a superposition of two states in which the phase of the field differs by π. After the atom has interacted with R₂ and has been detected, the field will collapse into a state which is the linear superposition of these two phase opposite field states, a highly non classical field which has been dubbed a "Schrödinger cat"[61] in reference to the discussion by Schrödinger[62] of one of the famous paradoxes of Quantum Mechanics. The coherence between the two parts of this state should vanish within a very short time, inversely proportional to the number of photons it contains[53]. Experiments in which this decoherence could be monitored are planned[63].

By sending Rydberg atoms across two identical cavities, it should also be possible to prepare non local entangled fields, for example a field which would be either in the first or in the second cavity, with a quantum coherence between the two[64,65] As it has been proposed recently[66], such non-local fields could be used for realizing particle "teleportation"[67] between the two cavities. The state of a two-level atom sent across the first cavity could be replicated on a second atom interacting with the field of the second cavity, even if this state is unknown. This experiment starts with a measurement to be performed on the first atom-first cavity system. This measurement reacts on the state of the second cavity since the fields in the two cavities are correlated. One could then transfer by appropriate manipulations the information contained in the second cavity field on the state of a second atom crossing the second cavity and reconstruct on this atom the initial unknown atomic state of the first atom. This transfer requires that the result of the measurement performed on the first cavity is known to the experimenter who manipulates the second atom-second cavity system. Thus this teleportation scheme is based on two channels[67], a quantum one (the non local field correlation between the two cavities) and a classical one (which carries the information between the two cavities).

Single atoms in linear superpositions of Rydberg states could also be used as dispersive "plungers" to control the flow of photons in high Q cavities, realizing quantum switches which would remain in a linear superposition of their open and closed states[65]. Such devices present the features required for the operation of quantum computers[68]. The experiments which are now planned to demonstrate all these effects, even if they do not lead to practical applications, will constitute stringent tests of the quantum measurement theory .

CONCLUDING REMARKS

We have discussed in these lectures a novel kind of atomic phyics experiments, performed on isolated quantum systems, which are at variance with traditional experiments in this field, usually performed on a large ensemble of particles. It is now possible to isolate single atoms, ions or molecules either litterally, by keeping them away from any other form of matter in vacuum, or by making them interact selectively with a detection apparatus which remains insensitive to a huge background of undetected particles. It is also possible to use single atoms as detectors of photon fields made of one or a few photons only and to keep this field virtually unperturbed in a high Q cavity. The cavity can then be considered as a photon trap which keeps the field available for continuous observation over long periods of time. Matter (at the single atom scale) and fields (at the single photon level) tend to play a dual role in these experiments, one system serving as a probe of the other.

When experiments are performed at this scale, the trapped matter or field exhibits a quantum behaviour, markedly different from the behaviour of large ensembles of atoms or photons. Quantum jumps reflecting the collapse of the system wave function are observed or predicted. These jumps (see figures 2 and 7) are random in time and can be predicted only in a probabilistic way.

The experimental study of these systems, now actively pursued in several laboratories, will provide new very interesting tests of quantum mechanics. By studying how the behaviour of the system evolves as the number of particles in it is increased, new interesting insight in the understanding of the boundary between classical and quantum physics will certainly be obtained.

REFERENCES

1. G.Binnig and H.Rohrer, "The Scanning tunneling microscope", Scientific American, (August 1985); C.F.Quate, "Vacuum tunneling: a new technique for microscopy", Phys.Today, 39 (8): 26 (August 1986).
2. H.K.Wickramasinghe, "Scanned-probe microscopes", Scientific American, (October 1989).

3. T.T.Perkins, D.E.Smith and S.Chu, "Direct observation of tube like motion of a single polymer chain", Science. 264:819 (1994). T.T.Perkins, S.R.Quate, D.E.Smith and S.Chu, "Relaxation of a single DNA molecule observed by optical microscopy", Science. 264:822 (1994).

4. P.E.Toschek and W.Neuhauser, "Spectroscopy on localized and cooled ions" in: "Atomic Physics 7", D.Kleppner and F.Pipikin editors, Plenum Press, New York (1981); W.Neuhauser, M.Hohenstatt, P.E.Toschek and H.Dehmelt, "Localized visible Ba^+ mono-ion oscillator", Phys.Rev. A22: 1137 (1980).

5. M.Orrit, J.Bernard and R.Brown, "Faire de la spectrocopie moléculaire par molécule", La Recherche, p.1395 (December 1993).

6. S.Haroche and J.M.Raimond, "Cavity Quantum Electrodynamics", Scientific American (April 1993).

7. R.J.Cooke and H.J.Kimble, "Possibility of direct observation of quantum jumps", Phys.Rev.Lett. 54:1023 (1985).

8. E.Einstein, B.Podolski and N.Rosen, "Can quantum-mechanical description of physical reality be considered complete?", Phys.Rev. 47:777 (1935).

9. W.Zurek, "Decoherence and the transition from quantum to classical", Physics Today, 44(10):36 (October 1991).

10. P.Zeppenfeld, D.M.Eigler and E.K.Schweizer, "On manipule même les atomes", La Recherche, p362 (March 1992).

11. L.J.Geerligs, V.F.Anderegg, P.A.M.Holweg, J.E.Mooij, H.Pothier, D.Esteve, C.Urbina and M.H.Devoret, "Frequency-locked turnstile device for single electrons", Phys.Rev.Lett. 64:2691 (1990).

12. P.Ekstrom and D.Wineland, "The isolated electron", Scientific American (August 1994).

13. W.M.Itano, J.C.Bergquist and D.J.Wineland, "Laser spectroscopy of trapped atomic ions", Science, 237:612 (1987); D.J.Wineland, "Trapped ions, laser cooling and better clocks", Science, 226:395 (1984).

14. B.Goss Levi "Clouds of trapped cooled ions condense into crystals", Physics Today (september 1988).

15. G. Gabrielse, X.Fei, L.A.Orozco,R.J.Tjoelker, J.Haas, H.Kalonowsky, T.A.Trainor and W.Kells, "Thousandfolf improvement in the measured antiproton mass", Phys.Rev.lett. 65:1317 (1990).

16. R.J.Van Duck, D.L.Farnham and P.B. Schwinberg, "Tritium-Helium 3 mass difference using the Penning trap mass spectroscopy", Phys.Rev.Lett., 70:2888 (1993); V.Natarajan, K.R.Boyce, F.DiFilippo and D.E.Pritchard, "Precision Penning trap comparison of Nondoublets: atomic masses of H,D, and the neutron", Phys.Rev.Lett. 71:1998 (1993).

17. D.J.Wineland and W.M.Itano, "Laser cooling", Physics Today (June 1987).

18. F.Diedrich, J.C.Bergquist, W.M.Itano and D.J.Wineland, "Laser cooling to the zero-point energy of motion", Phys.Rev.Lett. 62:403 (1989).

19. R.Blümel, J.M.Chen, E.Peik, W.Quint, W.Schleich, Y.R.Shen and H.Walther, "Phase transition of stored laser cooled ions", Nature, 334:309 (1988).

20. J.J.Bollinger and D.J.Wineland, "Microplasmas", Scientific American (January 1990).

21. U.Eichmann, J.C.Bergquist, J.J.Bollinger, J.M.Gilligan, W.M.Itano and D.J.Wineland, "Young's interference experiment with light scattered from two atoms", Phys.Rev.Lett. 70:2359 (1993).

22. D.J.Larson, J.C.Bergquist, J.J.Bollinger, W.M.Itano and D.J.Wineland, "Sympathetic cooling of trapped ions: a laser cooled two species non neutral plasma", Phys.Rev.lett. 57:70 (1986).

23. W.Nagourney, J.Sandberg and H.Dehmelt, "Shelved optical electron amplifier: observation of quantum jumps", Phys.Rev.Lett. 56:2797 (1986);

24. T.Sauter, W.Neuhauser, R.Blatt and P.E.Toschek, "Observation of quantum jumps", Phys.Rev.Lett. 57:1696 (1986).

25.a) J.C.Bergquist, Randall G.Hulet, WM.Itano and D.J.Wineland, "Observation of quantum jumps in a single atom", Phys.Rev.Lett. 57:1699 (1986);b) W.M.Itano, J.C.Bergquist, R.G.Hulet and D.Wineland, Phys.Rev.Lett. 59:2752 (1987).

26. H.J.Carmichael and D.F.Walls, "Proposal for the measurement of the resonant Stark effect by photon correlation techniques", J.Phys B9:L43 (1976); C.Cohen-Tannoudji, "Atoms in strong resonant fields" in "Frontiers in Laser Spectroscopy", Les Houches summer school XXVII, R.Balian, S.Haroche and S.Liberman editors, North Holland (1977).

27. H.J.KImble, M.Dagenais and L.Mandel, "Photon antibunching in resonance fluorescence", Phys.Rev.lett. 39:691 (1977).

28. F.Diedrich and H.Walther, "Nonclassical radiation of a single stored ion", Phys.Rev.Lett. 58:203 (1987).

29. R.Loudon, "The Quantum Theory of Light" (2nd edition), chapters 3 and 6, Clarendon Press (1985).

30. M.Orrit and J.Bernard, "Single pentacene molecules detected by fluorescence excitation in a p-terphenyl crystal", Phys.Rev.Lett. 65:2716 (1990).

31. W.E.Moerner and L.Kador, "Optical detection and spectroscopy of single molecules in a solid", Phys.Rev.Lett. 62:2535 (1989).

32. T.Basché, W.E.Moerner, M.Orrit and H.Talon, "Photon antibunching in the fluorescence of a single dye molecule trapped in a solid", Phys.Rev.Lett. 69:1516 (1992).

33. G.S.Hurst, M.G.Payne,S.D.Kramer and C.H.Chen, "Counting the atoms", Physics Today 33(9): 24 (September 1990).

34. D.Kleppner, M.G.Littman and M.L.Zimmerman, "Highly excited atoms, Scientific American p108 (May 1981).

35. S.Haroche, "Des atomes géants", La Recherche, p737 (September 1978).

36. S.Haroche, "Cavity Quantum Electrodynamics" in "Fundamental Systems in Quantum Optics", Les Houches Summer School LIII, J.Dalibard, J.M.Raimond and J.Zinn-Justin editors, North-Holland (1992).

37. P.R.Berman (editor), "Cavity Quantum Electrodynamics", a supplement issue to Adv.in Atom, Mol and Opt. Physics, Academic Press (1994).

38.H.J.Kimble, "Structure and dynamics in Cavity Quantum Electrodynamics"in ref[37].

39. R.J.Hulet and D.Kleppner, "Rydberg atoms in circular states", Phys.Rev.Lett. 51:1430 (1983).

40. P.Nussenzveig, F.Bernardot,M.Brune, J.Hare, J.M.Raimond, S.Haroche and W.Gawlik, "Preparation of high-principal-quantum number "circular" states of rubidium, Phys.RevA48:3991 (1993).

41R.J.Brecha, G.Raithel, C.Wagner and H.Walther, "Circular Rydberg states with very large n", Opt.Comm. 102: 257 (1993).

42. D.Meschede, H.Walther and G.Müller, "One atom-maser", Phys.Rev.Lett. 54:551 (1985).

43. M.Brune, J.M.Raimond, P.Goy, L.Davidovich and S.Haroche, "Realization of a two-photon maser oscillator", Phys.Rev.Lett. 59:1899 (1987).

44. M.Gross, P.Goy, C.Fabre, S.Haroche and J.M.Raimond, "Maser oscillation and microwave superradiance in small systems of Rydberg atoms", Phys.Rev.Lett. 43:343 (1979); S.Haroche, P.Goy, J.M.Raimond, C.Fabre and M.Gross, "Exploration of radiative properties of very excited atoms, Philo.Trans.Roy.Soc. London, 307:659 (1982).

45.P.Goy,J.M.Raimond,M.Gross and S.Haroche."Observation of cavity enhanced single atom spontaneous emission", Phys.Rev.Lett.50:1903 (1983).

46. A.Aspect, J.Dalibard and G.Roger, "Experimental test of Bell's inequalities using time varying analyzers", Phys.Rev.Lett. 49:1804 (1982).

47. P.Filipowicz, J.Javanainen and P.Meystre, "Theory of a microscopic maser", Phys.RevA34, 3077 (1986).

48. L.A.Lugiato, M.O.Scully and H.Walther, "Connection between microscopic and macroscopic maser theory", Phys.Rev.A 36:740 (1987).

49 L.Davidovich, J.M.Raimond, M.Brune and S.Haroche, "Quantum theory of a two-photon micromaser", Phys.Rev.A 36;3771 (1987).

50. G.Rempe, F.Schmidt-Kaler and H.Walther, "Observation of subPoissonian statistics in a micromaser", Phys.Rev.Lett. 64:2783 (1990).

51. O.Benson, G.Raithel and H.Walther, "Quantum jumps of the micromaser field: dynamic behaviour close to the transition points", Phys.rev.Lett. 72:3506 (1994).

52. P.Meystre, G.rempe and H.Walther, "Very low temperature behaviour of a micromaser", Opt.Lett. 13:1078 (1988).

53. M.Brune, S.Haroche, J.M.Raimond, L.Davidovich and N.Zagury, "Manipulation of photons in a cavity by dispersive atom-field coupling: Quantum-nondemolition measurement and generation of "Schrödinger cat states", Phys.Rev.A45: 5193 (1992).

54. C.M.Caves, K.S.Thorne, R.W.D.Drever, V.D.Sandberg and M.Zimmerman, "On the measurement theory of a weak classical force coupled to a quantum mechanical oscillator- Issues of principle." Rev.Mod.Phys. 52:341 (1980).

55.N.Ramsey, "Molecular Beams", Oxford University Press (1985).

56. M.Brune, S.Haroche, V.Lefèvre, J.M.Raimond and N.Zagury, "Quantum Nondemolition measurement of small photon numbers by Rydberg-atom phase-sensitive detection", Phys.Rev.Lett. 65:976 (1990).

57. M.Brune, P.Nussenzveig, F.Schmidt-Kaler, F.Bernardot, A.Maali, J.M.Raimond and S.Haroche, "From Lamb-shift to light shifts: vacuum and subphoton cavity fields measured by atomic phase sensitive detection", Phys.Rev.Lett. 72:3339 (1994).

58. S.Haroche, M.Brune and J.M.Raimond, "Measuring photon numbers in a cavity by atomic interferomtry: optimizing the convergence procedure", J.Phys.II France 2:659 (1992).

59. S.Haroche, "Atoms and photons in high Q cavities: new tests of quantum theory" in "Fundamental Problems in Quantum Theory", D.Greenberger (editor), New York Academy of Sciences (1994).

60. D.Mermin , "What is wrong with these elements of reality", Physics Today, 43(6):7 (June 1990).

61. B.Yurke and D.Stoler, "Generating quantum mechanical superpositions of macroscopically distinguishable states via amplitude dispersion", Phys.Rev.Lett. 57:13 (1986).

62. E.Schrödinger in Naturwissenschaften 23:807 and 23:844 (1935); english translation: J.D.Trimmer, Proc.Am.Phys.Soc. 124:3325 (1980).

63. S.Haroche, M.Brune, J.M.Raimond and L.Davidovich, "Mesoscopic quantum coherences in Cavity QED" in "Fundamentals of Quantum Optics III", F.Ehlotzky (editor), Springer Verlag (1993).

64. P.Meystre, "Cavity quantum optics", Progress in OpticsXXX edited by E.Wolf (Elsevier Science 1992).

65. L.Davidovich,A.Maali,M.Brune, J.M.Raimond and S.Haroche, "Quantum switches and nonlocal microwave fields", Phys.Rev.Lett. 71:2360 (1993).

66. L.Davidovich, N.Zagury, M.Brune, J.M.Raimond and S.Haroche, "Teleportation of an atomic state between two cavities using nonlocal microwave fileds", Phys.Rev A (August 1994).

67. C.H.Bennet, G.Brassard, C.Crepeau, R.Josza, A.Peres and W.Wooters, "Teleporting an unknown quantum state via dual classical and Einstein-Podolski-Rosen channels", Phys.Rev.Lett.70:1895 (1993).

68. D.Deutsch, "Quantum theory, the Church-Turing principle and the universal quantum computer", Proc.Roy.Soc. London, A400:97 (1985).

QUANTUM EFFECTS WITH ULTRACOLD ATOMS

Y. Castin, J. Dalibard and C. Cohen-Tannoudji

Laboratoire de Spectroscopie Hertzienne de l'E.N.S.[1]
et Collège de France
24, rue Lhomond, F-75231 Paris Cedex 5, France

INTRODUCTION

Efficient techniques using laser beams are available to control the motion of neutral atoms and to cool them down to ultralow temperatures, in the μKelvin range. Two different schemes are discussed in this lecture, the Sisyphus cooling in the so-called "optical lattices" and the velocity selective coherent population trapping (VSCPT). The basic processes involved in each scheme are presented on simple models. An outline of treatments of laser cooling with a quantization of the atomic motion in the laser field is given. These treatments allow the derivation of fundamental limits on the cooling efficiency, which we compare to the lowest temperatures obtained experimentally.

THE BASIC CONCEPTS

In this section, we discuss briefly the effect of the laser on the atomic state. Two types of atomic variables are involved. The first ones, called "internal" variables, are related to the electronic motion relative to the nucleus. The other ones are "external" variables describing the motion of the atomic center of mass.

Effect of light on the atomic internal state

Since the atoms in laser cooling are illuminated by quasi-resonant light, they can be considered as two-level atoms, with a metastable state g, called "ground state" in what follows, and an excited state e. The transition between g and e has a resonant frequency ω_A and an electric

---

[1] Unité de recherche de l'Ecole Normale Supérieure et de l'Université Paris 6, associée au CNRS

Advances in Quantum Phenomena. Edited by E.G. Beltrametti
and J.-M. Lévy-Leblond. Plenum Press. New York. 1995

dipole d. The excited state can decay by spontaneous emission with a rate Γ. In this section, we consider the simple case of atoms with no sublevels in the ground state g. The influence of the presence of Zeeman sublevels in g on laser cooling is the subject of the next section.

The laser electric field is a superposition of traveling waves of frequency ω_L close to ω_A and of complex amplitude $\mathcal{E}_0$. An important parameter is $\delta = \omega_L - \omega_A$, detuning of the laser frequency ω_L from the atomic frequency ω_A. The dipolar coupling between atom and light leads to a time dependent amplitude of transition from g to e, noted $(\Omega/2)e^{-i\omega_L t}$, and from e to g, $(\Omega^*/2)e^{i\omega_L t}$. The parameter $\Omega = -d\mathcal{E}_0/\hbar$ is called the Rabi frequency. The amplitude of transition can be made time independent by the following time dependent unitary transformation $S(t)$ changing the excited state:

$$S(t) = e^{i\omega_L t|e\rangle\langle e|} \tag{1}$$

which gives to the $|g\rangle \longrightarrow |e\rangle$ transition the effective Bohr frequency $\omega_A - \omega_L = -\delta$ instead of ω_A. This leads to the time independent effective hamiltonian H_{eff} for the evolution of the internal atomic state:

$$H_{\text{eff}} = \hbar \begin{pmatrix} -\delta - i\Gamma/2 & \Omega/2 \\ \Omega^*/2 & 0 \end{pmatrix} \tag{2}$$

This matrix is given in the $\{|e\rangle, |g\rangle\}$ basis and the $-i\Gamma/2$ term accounts for the instability of the excited state.

The effect of the coupling will be considered in the regime:

$$s = \frac{|\Omega|^2/2}{\delta^2 + \Gamma^2/4} \ll 1 \tag{3}$$

The dimensionless quantity s is called the saturation parameter and gives the relative amount of time spent by the atoms in the excited state. In this regime, one of the eigenvectors of H_{eff} remains close to g, with the eigenvalue $\hbar(\delta - i\Gamma/2)s/2$. The corresponding real part $\hbar\delta' = \hbar\delta s/2$ is the signature of a reactive effect: it describes the shift of the energy of the ground state by the incoming light (lightshift). The imaginary part is the signature of a dissipative effect: the ground state g becomes unstable by contamination by the excited state through the Rabi coupling; the corresponding rate $\Gamma' = \Gamma s/2$ is the excitation rate of the ground state by the laser. Note that the following discussion can be generalized to arbitrary laser intensities (see e.g. the so-called "dressed atom" picture in [1]).

Effect of light on the atomic external state

In laser cooling the important external variables are the position $\vec{r}$ and the momentum $\vec{p}$ of the atomic center of mass. These external variables are also changed by the interaction with the laser. For example, the momentum $\vec{p}$ shifts to $\vec{p} \pm \hbar\vec{k}$ after the absorption or the emission of one photon of momentum $\hbar\vec{k}$. The corresponding atomic recoil velocity $\hbar k/M$, where M is the mass of the atom, is an important parameter in laser cooling, as we shall see; it is on the order of 3 mm/s for cesium. It is convenient to split the effect of light on the atomic motion in a dissipative part and a reactive part, depending on the contribution or the non contribution of the momentum of spontaneously emitted photons.

For the simple case of an atom in a plane running wave, the dissipative part corresponds to the usual radiation pressure; it is due to the absorption of one laser photon of momentum $\hbar\vec{k}_L$ and the spontaneous emission of one fluorescence photon of momentum $\hbar\vec{k}_S$. The corresponding average change in atomic momentum is $\langle\delta\vec{p}\rangle = \langle\hbar\vec{k}_L - \hbar\vec{k}_S\rangle = \hbar\vec{k}_L$; the mean contribution of $\hbar\vec{k}_S$ is zero because spontaneous emission occurs with the same probability in two opposite directions. The mean radiative force is then simply $\hbar\vec{k}_L$ times the rate of absorption-spontaneous emission cycles. This rate, on the order of $\Gamma s/2$ at low saturation, cannot exceed $\Gamma/2$ at high saturation. The dissipative force thus saturates at high laser intensity to $\hbar k_L\Gamma/2$, a limit almost 10^4 higher than the gravitational force Mg for cesium.

The reactive effect on the atomic motion derives from a potential, which is the position dependent lightshift of the atomic level g; it is due to successive absorption-stimulated emission cycles. Consider for example the lightshift U of the atomic ground state in the vicinity of a gaussian laser beam. When the laser is detuned to the red ($\delta < 0$), $U(\vec{r})$ is negative and maximal in absolute value in the laser beam, and close to 0 out of the laser beam. The effect of $U(\vec{r})$ is thus a trapping of the atoms around the maxima of intensity of the laser. This has been used to make a trap, at a detuning large enough for the dissipative effects (heating due to spontaneous emission) to be negligible [2]. An improved version of the optical trap, using stabilization by radiative pressure forces, is presented in [3].

Doppler cooling and the Doppler limit

The Doppler cooling scheme has been proposed independently by Hänsch and Schawlow, Wineland and Dehmelt in 1975 [4, 5]. It strongly relies on the atomic velocity dependence of the radiation pressure due to the Doppler effect. In this scheme, the cooling of the atomic velocity along a given direction z is obtained by the superposition of two counterpropagating running waves along z. The two waves have the same weak amplitude $\mathcal{E}_0$ and the same frequency ω_L, detuned to the red ($\delta = \omega_L - \omega_A < 0$). In a simple classical picture, which is sufficient here, consider an atom of velocity $\vec{v}$, and call $\vec{k}$ the wave vector of the laser wave copropagating with the atom ($\vec{k} \cdot \vec{v} > 0$). In the atomic rest frame, this wave has an apparent frequency $\omega_L - \vec{k} \cdot \vec{v} < \omega_L < \omega_A$, so it is moved farther from resonance by the Doppler effect $-\vec{k} \cdot \vec{v}$. The counterpropagating wave, with wave vector $-\vec{k}$, is on the contrary put closer to resonance by the Doppler effect (see Figure 1). The atom will then absorb photons preferentially in the counterpropagating wave, so that it feels a mean radiative force opposed to its velocity along z. This force vanishes for an atom at rest, and can be shown to behave as a linear friction force:

$$F_z = -\alpha v_z \tag{4}$$

where α is a friction coefficient, for slow enough atoms. Cooling is provided along the other directions x and y by use of two additional standing waves. The corresponding light field acts as a viscous medium on the atomic motion, the so-called "optical molasses"[6].

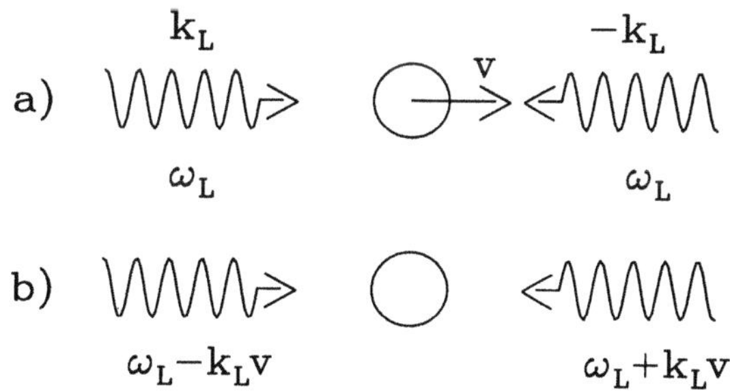

Figure 1: The laser waves of the Doppler cooling configuration, in the lab (a) and in the atomic rest frame (b).

Spontaneous emission plays an essential role in the cooling mechanism: it allows the dissipation of energy by emission of fluorescence photons with a frequency $\omega_S > \omega_L$. However, the random momentum recoil of the atoms after spontaneous emission is responsible for a heating, i.e. an increase of the mean atomic kinetic energy, described as in brownian motion theory by a momentum diffusion coefficient D and which counterbalances the effect of the damping force. When the equilibrium is reached, the atomic velocity distribution is gaussian, with a "temperature" T given by Einstein's law:

$$k_B T = \frac{D}{\alpha} \tag{5}$$

where k_B is the Boltzmann constant. After an explicit derivation of α and D from the two-level atom model, one finds that the minimal temperature T_D, called the Doppler limit, is obtained for a detuning $\delta = -\Gamma/2$ and is given by [7, 8]:

$$k_B T_D = \frac{\hbar \Gamma}{2} \tag{6}$$

It corresponds to $T_D = 120 \ \mu K$ for cesium atoms.

Note that the radiation pressure force can also be made position dependent, through a gradient of magnetic field, so that the atoms are also trapped, in the so-called magneto-optical trap (see [9] for the first experimental evidence, and [10, 11, 12, 13] for more recent experiments).

BEATING THE DOPPLER LIMIT

Since 1988, precise experimental measurements of the temperatures in optical molasses have been performed [14, 15, 16, 17, 18, 19, 20]. The lowest temperatures are found to be well below the Doppler limit (6). On a heavy atom like cesium, the lowest measured temperatures are on the order of 3 μK, a factor 40 smaller than the Doppler limit, and are observed at large atom-laser detuning: $|\delta| \gg \Gamma$, instead of $\delta = -\Gamma/2$ as expected from the Doppler cooling theory. All these results are experimental evidence of very efficient cooling mechanisms.

These mechanisms have been identified theoretically on simple 1D models [21, 22]. They strongly rely on the existence of several degenerate sublevels in the atomic state g, a feature left out in the two-level atom model used for Doppler cooling but present in experimental configurations, because of the hyperfine and Zeeman structures of the atomic ground state. A second important feature is the spatial variation of the laser field polarization at the optical wavelength scale, a condition which is automatically fulfilled in 3D laser configurations, and which is obtained in the simple 1D models by giving different polarizations to the two counterpropagating waves.

In this section, we will analyze in some details only one of the presently known polarization gradient cooling mechanisms, relying on the so-called "Sisyphus effect". The corresponding minimal kinetic energies no longer scale as $\hbar\Gamma$, as for Doppler cooling, but as the recoil energy $E_R = \hbar^2 k^2/2M$, i.e. the mean increase of the atomic kinetic energy after the spontaneous emission of a single photon. Note that E_R is well below the Doppler limit $\hbar\Gamma/2$ for the typical atomic transitions used in laser cooling. In the case of cesium atoms, for example, $\hbar\Gamma/2$ is one thousand times larger than the recoil energy. In this very cold regime, it has been predicted theoretically and checked experimentally that quantization of atomic motion plays an important role in Sisyphus cooling. This has led to a new picture for the laser cooled atomic samples, the so-called "optical lattices".

Optical pumping and lightshifts

As we have seen for the two-level atom case in the low saturation regime, the effect of light on the internal atomic state is to shift the energy levels (reactive effect) and to give a finite lifetime to the ground state (dissipative effect). When the atom has several Zeeman sublevels in the ground state, the reactive effect can lead to different lightshifts of the sublevels, and the dissipation effect can lead to real transitions between the various sublevels, through optical pumping cycles.

The last two features play an important role in laser cooling. They are present in the simple 1D scheme depicted in Figure 2. The light field is obtained by superposition of two coherent counterpropagating running waves along z, with the same amplitude $\mathcal{E}_0$ and frequency ω_L, but with orthogonal linear polarizations. For an appropriate phase choice, the total electric field is given by:

$$\vec{\mathcal{E}}(z,t) \;=\; \vec{\mathcal{E}}^+(z)e^{-i\omega_L t} + \text{c.c} \tag{7}$$

$$\vec{\mathcal{E}}^+(z) \;=\; \mathcal{E}_0\left[\vec{e}_x e^{ikz} - i\vec{e}_y e^{-ikz}\right] \tag{8}$$

where $\vec{e}_x$, $\vec{e}_y$ are unit vectors along x and y. One can check that the polarization of $\vec{\mathcal{E}}(z,t)$ depends on z, with a periodicity $\lambda_{\rm opt}/2$, where $\lambda_{\rm opt} = 2\pi/k$ is the optical wavelength. For example, in $z = 0, \lambda_{\rm opt}/2, ...$, $\vec{\mathcal{E}}(z,t)$ is σ_- circularly polarized along z; it is σ_+ polarized in $z = \lambda_{\rm opt}/4, 3\lambda_{\rm opt}/4, ...$; it is linearly polarized along $(\vec{e}_x - \vec{e}_y)/\sqrt{2}$ in $z = \lambda_{\rm opt}/8, 5\lambda_{\rm opt}/8, ...$ and along $(\vec{e}_x + \vec{e}_y)/\sqrt{2}$ in $z = 3\lambda_{\rm opt}/8, 7\lambda_{\rm opt}/8$.

Consider an atom with an angular momentum $j_g = 1/2$ in the ground state and an angular momentum $j_e = 3/2$ in the excited state. This atom is illuminated by the previous field configuration, with a laser frequency ω_L below the resonance frequency ω_A:

$$\delta = \omega_L - \omega_A < 0 \tag{9}$$

The transition amplitudes between the atomic energy levels of given angular momentum along z, $\{|g,m\rangle, m = +1/2\}$ and $\{|e,m'\rangle, m' = \pm 3/2, \pm 1/2\}$, by absorption or stimulated emission of laser photons are now weighted by the appropriate Clebsch-Gordan coefficients (see Figure 3a) and thus depend on the polarization of the electric field.

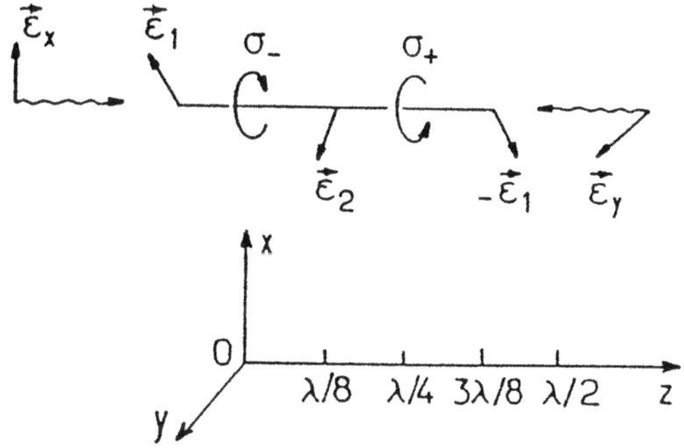

Figure 2: Resulting electric field in the $x - y$ laser configuration.

In a place where the light is purely σ_- polarized, the $|g, -1/2\rangle$ sublevel is coupled to the electric field with an amplitude $\sqrt{3}$ larger than the $|g, +1/2\rangle$ sublevel. It therefore experiences a light shift 3 times larger, $\hbar\delta s$ instead of $\hbar\delta s/3$ for $|g, +1/2\rangle$, with a saturation parameter s given by (3) [2]. In the presence of pure σ_- light, the sublevel $|g, -1/2\rangle$ has not only a well defined lightshift, but is also a trap for the internal atomic state. It is coupled by laser only to the $|e, -3/2\rangle$ excited state sublevel, which by spontaneous emission can only decay to $|g, -1/2\rangle$. If the atom is initially in the $|e, +3/2\rangle$ excited sublevel, it is eventually put in the $|g, +1/2\rangle$ ground sublevel by spontaneous emission. If the atom is initially in the $\{|g, +1/2\rangle, |e, -1/2\rangle\}$ manifold, it is put back in $|g, +1/2\rangle$ after a spontaneous emission with a probability $1/3$ (σ_- emission) and it is trapped into $|g, -1/2\rangle$ with a probability $2/3$ (linearly polarized emission

---

[2] At first sight, there seems to be a discrepancy of a factor 2 between the lightshift (and also the excitation rate) derived here and the one corresponding to (2). In fact, in $z = 0$, the electric field amplitude is $\sqrt{2}\mathcal{E}_0$, instead of $\mathcal{E}_0$ in (2).

along z). The population in $|g, +1/2\rangle$ therefore tends exponentially to zero for an increasing number of fluorescence cycles. The corresponding transition rate from $|g, +1/2\rangle$ to $|g, -1/2\rangle$ by the optical pumping mechanism is found to be $\Gamma s \cdot 1/3 \cdot 2/3 = 2\Gamma s/9$.

In a place where the light is purely σ_+ polarized, the situation is reversed between the sublevels. The atoms are now put by optical pumping in the $|g, +1/2\rangle$ sublevel, which undergoes the largest light shift. The present configuration exhibits therefore a strong correlation between the spatial modulation of lightshifts and the spatial modulation of optical pumping rates. For a negative atom-laser detuning δ, the optical pumping rates are everywhere the largest from the highest energy ground state sublevel to the lowest one.

Sisyphus cooling mechanism

We now explain how cooling occurs in the previously described atom-laser configuration. The key point is that the position dependent lightshifts of the sublevels $|g, \pm 1/2\rangle$ appear as effective potentials $U_\pm(z)$ for a moving atom. It will be thus convenient to introduce the mechanical energies $E_\pm = M v^2/2 + U_\pm(z)$, where v is the atomic velocity along z.

A typical trajectory in $(z, E_\pm)$ space is shown in Figure 3b. It starts in the sublevel $|g, -1/2\rangle$, in a minimum of $U_-(z)$, with a kinetic energy larger than the modulation depth:

$$U_0 = -\frac{2}{3}\hbar\delta s \tag{10}$$

of the potential. If the optical pumping time τ_p from $|g, -1/2\rangle$ to $|g, +1/2\rangle$ is long enough, the atom can climb the potential hill and reach the top of $U_-(z)$ before changing its internal state. During this part of the motion, the mechanical energy E_- remains constant and there is a conversion of kinetic energy into potential energy. In the vicinity of the maxima of $U_-(z)$, the laser is σ_+ polarized and the atom has there the maximal probability to be optically pumped into the $|g, +1/2\rangle$ sublevel, by absorption of a laser photon followed by a spontaneous emission. If such a process takes place, the atom is put back into a valley, but for the $U_+(z)$ potential this time. The resulting decrease of potential energy leads to a corresponding excess of energy for the spontaneously emitted photon with respect to $\hbar\omega_L$. Dissipation of energy thus originates in this cooling scheme from the spontaneous emission of anti-Stokes Raman photons.

The atoms therefore do not stop climbing potential hills, as Sisyphus did in the greek mythology, until their kinetic energy becomes on the order of the modulation depth U_0. This intuitive reasoning is confirmed by a theoretical analysis, close to the one used already for Doppler cooling, which introduces a friction coefficient α and a momentum diffusion coefficient D, as in brownian motion theory. The derivation of the coefficient D and α leads to the equilibrium temperature given in [21]:

$$k_B T = \frac{\langle p^2 \rangle}{M} \propto U_0 \sim -\hbar\delta s \tag{11}$$

Limit of Sisyphus cooling

The equilibrium temperature (11) derived from a brownian motion approach is proportional to the laser intensity, for a given atom-laser detuning δ. Such a scaling law for Sisyphus cooling therefore predicts a vanishing equilibrium temperature in the limiting of vanishing laser intensity. In fact the law (11) is not valid for arbitrary values of the parameters, but only in the regime where the internal atomic variables are evolving faster than the external atomic variables. This requires that $\Omega_{\text{osc}}\tau_p \ll 1$ where $\Omega_{\text{osc}} \sim k\sqrt{U_0/M}$ is the oscillation frequency of the atoms in the bottom of the potential wells, and $\tau_p \sim 1/\Gamma s$ is the optical pumping time. Note that this condition is violated for too low laser intensities I since Ω_{osc} scales as $\sqrt{I}$ whereas τ_p is proportional to $1/I$.

In order to have a more complete understanding of Sisyphus cooling, one has to take into account carefully the heating of the atoms due to random recoil after spontaneous emission.

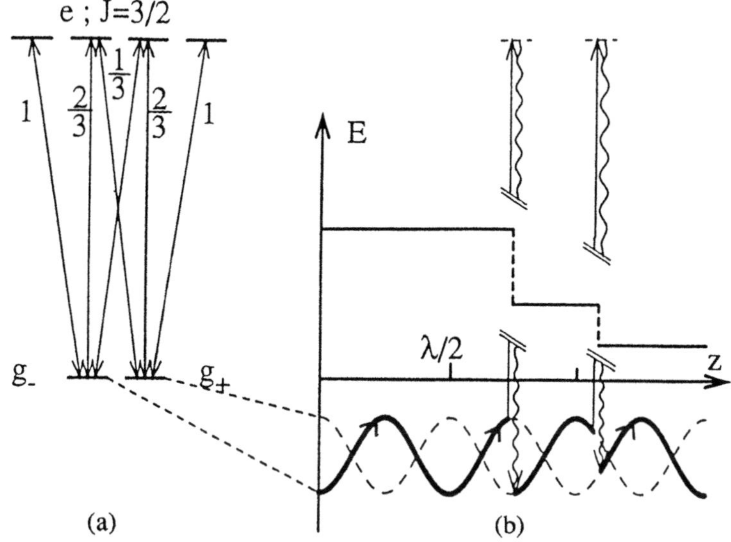

Figure 3: a) Atomic-level scheme and intensity factors (square of the Clebsch-Gordan coefficients) for a $j_g = 1/2 \rightarrow j_e = 3/2$ atomic transition. b) In the Sisyphus cooling configuration, the position dependent lightshifts of the ground state sublevels and a typical atomic trajectory in the position-energy space.

The Sisyphus effect, corresponding to a decrease of the potential energy on the order of U_0 after an optical pumping cycle, is then counterbalanced by an increase of the kinetic energy on the order of the recoil energy $E_R = \hbar^2 k^2 / 2M$. Cooling works only when U_0 is above a threshold of a few E_R, and the minimal achievable temperature is expected from (11) to scale as the recoil energy E_R.

The precise value of the Sisyphus cooling limit has been derived from a quantum treatment of both the internal atomic state and the atomic motion in laser, for the presented simple 1D model [23]. For a given atom-laser detuning $\delta = -2\Gamma$, the dependence of the steady state kinetic energy on the modulation depth U_0 of equation (10) is shown in Figure 4. The scaling law $k_B T \sim U_0$ represented by a straight line in Figure 4 gives only the asymptotic direction of the temperature in the limit of large U_0/E_R. The dependence of the temperature for large U_0 is better fitted by the law $k_B T = a \cdot \hbar |\delta| s + b E_R$, where a and b are dimensionless coefficients depending only on the detuning δ/Γ. The existence of a threshold on U_0 corresponds to a rapid increase of the average kinetic energy, when U_0/E_R becomes too small.

Another result of the quantum treatment concerns the limit of large detunings $|\delta| \gg \Gamma$. In this limit the temperature depends on the potential depth U_0 only (so that the coefficients a and b have a finite limit for $|\delta| \gg \Gamma$), and the coldest distributions are obtained, with a minimum root mean square atomic momentum of $5.5\hbar k$ for $U_0 \simeq 100 E_R$. This regime is discussed in more details in the next section.

Experimentally in 3D measurements of the temperature have been performed on a wide range of detunings and laser intensities, and for several atomic species [19, 20]. For large values of the lightshift $\hbar |\delta| s$, the experimental results are perfectly consistent with a scaling law $k_B T = a \cdot \hbar |\delta| s + b$, with coefficients independent on the detuning δ for $|\delta| \gg \Gamma$.[3] In the regime of very small lightshifts, experimental observations show that the cooling mechanism no longer works [19], and this indicates the existence of a threshold.

---

[3] Note that the values of a and b for the 3D experiments cannot of course be deduced from the simple 1D model that we described.

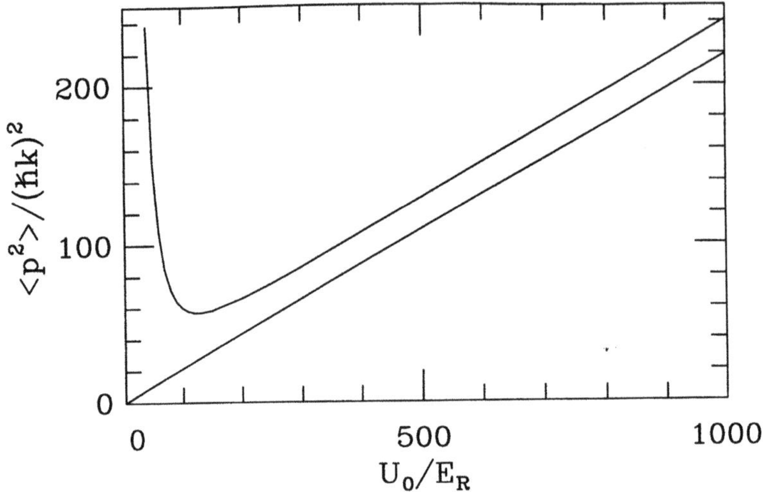

Figure 4: In a full quantum treatment of Sisyphus cooling, mean kinetic energy $\langle p^2\rangle/2M$ in units of recoil energy $E_R = \hbar^2 k^2/2M$ as a function of the modulation depth U_0 of the optical potential wells in units of E_R. The atom-laser detuning is $\delta = -2\Gamma$. The straight line is the prediction of the semi-classical treatment mentioned in the text.

Quantization of atomic motion in the optical potential wells

It has been found fruitful [24] to try to get more insight into the coldest regime of Sisyphus cooling. We have seen that this regime corresponds to the limit of a large detuning $|\delta| \gg \Gamma$ for given values of the potential depth U_0. In this limit the optical pumping time becomes much longer than the oscillation period of the atoms in the optical potential wells:

$$\Omega_{\mathrm{osc}}\tau_p \gg 1 \tag{12}$$

In a quantum picture of the atomic motion this corresponds to a spacing $\sim \hbar\Omega_{\mathrm{osc}}$ of the quasi harmonic levels in $U_\pm(z)$ much larger than the radiative width $\hbar/\tau_p$ of these levels due to optical pumping.

Therefore a good basis for the analysis of the cooling is provided by the stationary solutions of the Schrödinger equation for the atomic motion in $U_\pm(z)$. Since the potentials $U_\pm(z)$ are periodic, of period $\lambda_{\mathrm{opt}}/2$, the corresponding energy spectrum has a band structure. The lowest energy bands, corresponding to quasi-harmonic motion in the bottom of the potential wells (see Figure 5 taken from [25]), have a negligible tunnel width. On the contrary, the energy gaps become very small for the quasi-free motion well above the potential hills. The effect of optical pumping in this basis can be described simply by transition rates between the various energy levels, when inequality (12) holds. The corresponding rate equations allow one to calculate the steady state populations of the energy bands. The maximal population in the most populated band $v = 0$ is found to be 0.34 in [24], for $U_0 \simeq 60E_R$.

The existence of well resolved external quantum levels in 1D Sisyphus cooling has been investigated experimentally.

A first type of experiments is a Raman spectroscopy of the laser cooled atomic sample. A probe laser beam, with a very small intensity, is sent through the atomic cloud along the direction of the laser cooling beams. Two side-bands are clearly observed on the probe absorption spectrum, for a probe frequency $\omega_p \simeq \omega_L \pm \Omega_{\mathrm{osc}}$ (see Figure 6 taken from [26]). They correspond to stimulated Raman transitions between two successive vibrational levels in the potential wells. Since the lowest level is the most populated one, Raman cycles starting

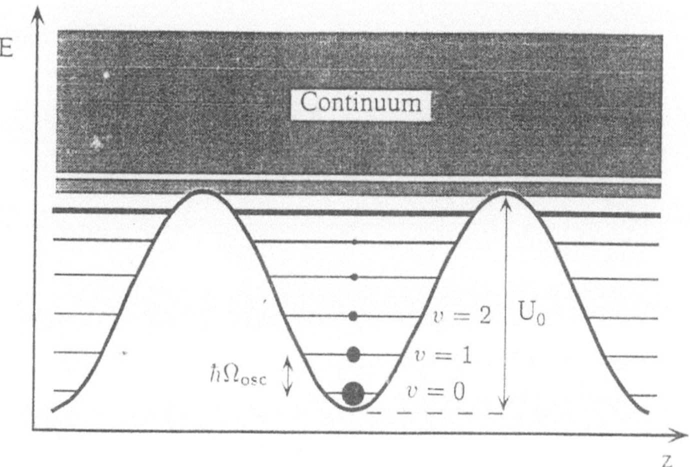

Figure 5: Band structure of the energy spectrum for the quantum motion in $U_+(z)$ optical potential wells. The vibrational levels are labelled $v = 0, 1, \ldots$ The dark circular areas are proportional to the steady state populations of the vibrational levels.

with atoms in $v = 0$ (and transferring them in $v = 1$) are expected to be dominant. When ω_p is below ω_L ($\omega_p \simeq \omega_L - \Omega_{\mathrm{osc}}$), the dominant process is therefore the absorption of one photon in the cooling beams followed by a stimulated emission of one photon in the probe beam, and the probe beam is amplified. When ω_p is above ω_L ($\omega_p \simeq \omega_L + \Omega_{\mathrm{osc}}$), the Raman processes are mainly the absorption of one photon in the probe beam followed by the stimulated emission of one photon in the cooling beams, and the probe beam is absorbed. The so-called "overtones" (Raman transitions with $\delta v = 2$, $\delta v = 3, \ldots$) can also be identified. A stimulated Rayleigh line is observed at $\omega_p = \omega_L$; it can be shown to give evidence for a spatial antiferromagnetic order of the atoms [26, 27].

These observations have been confirmed by studies of the fluorescence spectra of 1D optical molasses [28]. In the vicinity of the elastically scattered component ($\omega_S = \omega_L$), two sidebands are observed, with a frequency $\omega_S \simeq \omega_L \pm \Omega_{\mathrm{osc}}$. On Figure 7 taken from [28], the weights of the two sidebands are clearly different, and this can be interpreted in the simple following way. The "red" sideband, with a frequency $\omega_S = \omega_L - \Omega_{\mathrm{osc}}$, is a signature of the transfer of atoms from v to $v+1$ vibrational levels by fluorescence cycles (i.e. absorption of a laser photon followed by a spontaneous emission). Let us assume for simplicity that its weight is mainly proportional to the number of atoms π_0 in the ground vibrational level $v = 0$. The "blue" sideband, with a frequency $\omega_L + \Omega_{\mathrm{osc}}$, is due to fluorescence cycles from $v+1$ to v vibrational levels, so that its weight, in the same simplified model, is mainly proportional to the number of atoms π_1 in the first excited vibrational level $v = 1$. Since π_0 is larger than π_1, the "red" sideband has a larger weight than the "blue" one, in agreement with the experimental result. The ratio of the two weigths has been used to estimate the temperature of the atomic cloud, assuming the thermal distribution law $\pi_1/\pi_0 = e^{-\hbar\Omega_{\mathrm{osc}}/k_B T}$ (for $\hbar\Omega_{\mathrm{osc}} < k_B T$).[4]

---

[4]Strictly speaking the distribution law of the atoms among the vibrational levels is not thermal [29]. The effective temperature deduced from the ratio π_1/π_0 may differ from the effective temperature deduced from the mean kinetic energy [30].

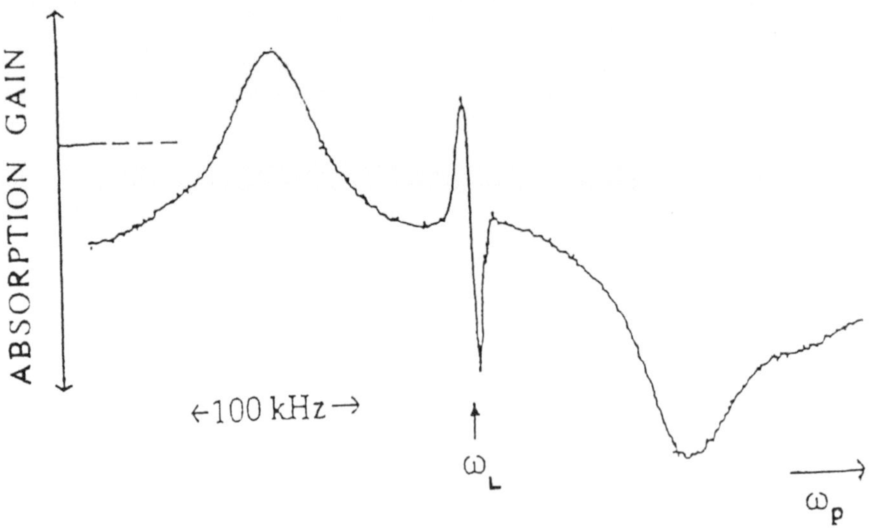

Figure 6: Probe absorption spectrum of a 1D optical molasses. The laser configuration is the one of Figure 2, and the atomic transition is the $j_g = 4 \to j_e = 5$ transition in cesium. The probe beam of frequency ω_p and the copropagating laser cooling beam have the same polarization. One can check than the two Raman sidebands are at the positions $\omega_p = \omega_L \pm \Omega_{osc}$, with a gain on the probe for $\omega_p = \omega_L - \Omega_{osc}$ and an absorption of the probe for $\omega_p = \omega_L + \Omega_{osc}$, as expected from the reasoning given in the text.

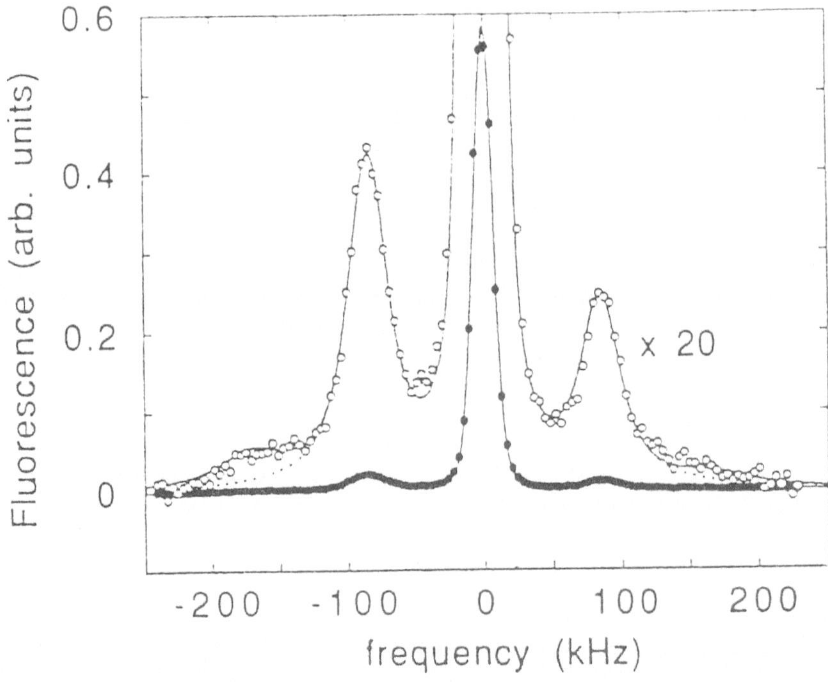

Figure 7: Fluorescence spectrum of a 1D laser cooled atomic sample. The number of scattered photons with a frequency $\nu_S = \omega_S/2\pi$ is given as function of the difference $\nu_S - \nu_L$, where $\nu_L = \omega_L/2\pi$ is the frequency of the laser. The central peak corresponds to elastic scattering of photons. The asymmetry of the two sidebands is cleared visible on the magnified curve.

Optical lattices in 2D and 3D

We have seen that 1D Sisyphus cooling in a laser field can lead to a strong accumulation of the atoms around the points where the field is circularly polarized. Such a property can be generalized to 2D and 3D.

Theoretical results are available in 2D [31]. The quantum approach is similar to the one used in 1D. The population of the energy levels in steady state are found to be strongly reduced: the maximum population of the deepest band shifts from 0.34 in 1D to 0.09 in 2D. This does not mean that cooling is less efficient in 2D, but it is simply a signature of degeneracy effects among the vibrational levels. If the motion along x and the motion along y are considered as decoupled, at least in the bottom of the optical potential wells, the probability of finding the atoms in the ground state of their motion is indeed the product of the probabilities of finding them in the ground states for their motion along x and y respectively. This simple reasoning leads to the estimate $\Pi_0^{2D} \simeq (\Pi_0^{1D})^2$, not far from the exact results.

To get experimental evidence for vibrational levels in $\sigma_\pm$ optical potential wells, one has to get rid of phase fluctuations between the orthogonal running waves, in order to avoid distorsion of the electric field polarization away from $\sigma_\pm$ in the minima of the wells. This has been realized in Münich, with 2 and 3 orthogonal standing waves with well controlled phases [32, 33]. In Paris, an alternative to this approach has been developed. The idea is to use in nD the minimal number of running waves leading to cooling, which is $n + 1$. In this case, as it is evident for the particular case $n = 1$, the periodic pattern of potential wells is only shifted by phase fluctuations. A hexagonal lattice in 2D, and a cubic body centered lattice in 3D have been obtained, and have been analyzed by a probe beam [34].

All these results lead to a completely different physical picture of the laser cooled atomic samples. The concept of optical molasses is replaced by the one of optical lattices, which consist of periodic arrays of microtraps, with vibrational levels for the atomic motion. Note that these optical lattices are for the moment quite lacunar, only a few percent of the sites being occupied.

BEATING THE RECOIL LIMIT

As we have seen in the previous section, the achievable kinetic energies in optical molasses remain larger than the so-called recoil energy $E_R = (\hbar k)^2/2M$, i.e. the kinetic energy of an atom initially at rest after emission of a photon. Indeed, in the optical molasses, spontaneous emission of fluorescence photons, which is essential for carrying away the atomic mechanical energy, takes place at a rate more or less uniform in phase space, i.e. at the optical pumping rate. The atoms thus continuously experience random recoils of size $\hbar k$ in momentum space, so that the mean atomic kinetic energy in steady state cannot be much smaller than the recoil energy.

It is important to know if the recoil energy is a fundamental limit in laser cooling. Atoms with a mean kinetic energy below the recoil energy are indeed in an interesting regime, where the coherence length for the atomic wavepacket, $\lambda_c \simeq \hbar/\Delta p$, is larger than the optical wavelength λ_{opt}. This subrecoil regime is also interesting in the search for quantum statistical effects (such as the Bose-Einstein condensation) with ultracold atoms. In this case, the parameter of quantum degeneracy $n\lambda_c^3$ (where n is the atomic density) can be on the order of unity, with still $n\lambda_{opt}^3$ being very small, so that collisional effects between atoms should remain small.

We see in this section that it is in fact possible to beat the recoil limit. Several subrecoil mechanisms have been proposed. They all rely on the same general idea, which is to accumulate the atoms into states very weakly coupled to the laser and with a very narrow width in momentum space. A simple 1D configuration of the so-called "velocity selective coherent population trapping" (VSCPT) [35, 36] is first analyzed. We then present experimental evidences of subrecoil cooling, based on VSCPT and on another subrecoil scheme, called "Raman cooling" [37]. Other proposals of subrecoil cooling can be found in [38, 39].

1D velocity selective coherent population trapping

Consider the transition between a ground atomic state of angular momentum $j_g = 1$ and an excited atomic state of angular momentum $j_e = 1$. It is illuminated by two mutually coherent laser waves, of opposite wave vectors $k\vec{e_z}$ and $-k\vec{e_z}$ along z and of same frequency ω_L, but with orthogonal circular polarizations, respectively σ_+ and σ_-.

To describe the internal atomic dynamics in this configuration, we introduce the sublevels of angular momentum $m\hbar$ around z, $\{|g, m\rangle_z, |e, m\rangle_z, m = 0, \pm 1\}$. An important point is that the atoms in the Λ system, formed by $\{|g, -1\rangle_z, |e, 0\rangle_z, |g, 1\rangle_z\}$, can never escape from it: this system is stable under the action of the laser, and the excited sublevel $|e, 0\rangle_z$ cannot decay by spontaneous emission into the ground sublevel $|g, 0\rangle_z$ of the V system, because the corresponding $\Delta m = 0$ Clebsch-Gordan coefficient is vanishing. After a few fluorescence cycles, the atoms initially in the V system $\{|e, -1\rangle_z, |g, 0\rangle_z, |e, 1\rangle_z\}$ are all pumped in the Λ system, and we can restrict the discussion of the dynamics to the Λ system (see Figure 8).

To describe in a quantum manner the atomic center of mass motion along z, we introduce the external states $|p\rangle$ corresponding to the plane waves of momentum p. The general atomic state is then expanded on products of internal and external state vectors, and the whole atomic dynamics can be reduced as follows [36].

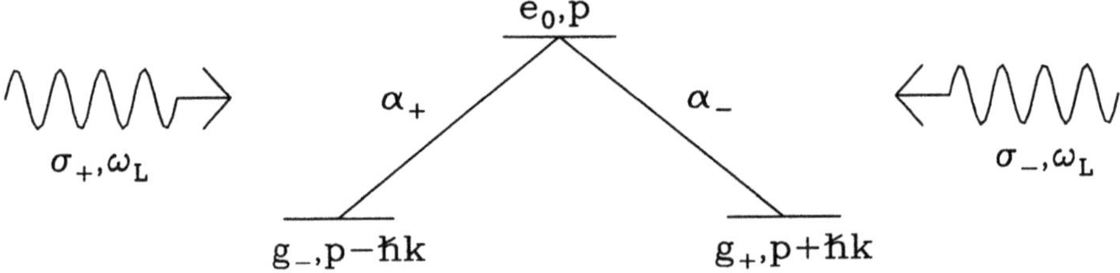

Figure 8: The Λ atomic system in the $\sigma_+ - \sigma_-$ laser configuration. In the absence of spontaneous emission, note the existence of closed families of atomic states labelled by p, atomic momentum along z in the excited state.

In the absence of spontaneous emission, an atom initially in the state $|e_0, p\rangle$ is coupled only to $|g_-, p - \hbar k\rangle$ (by stimulated emission of a photon in the σ_+ wave), with an amplitude α_+^*, and to $|g_+, p + \hbar k\rangle$ (by stimulated emission of a photon in the σ_- wave), with an amplitude α_-^*. The space of atomic states can then be split into a direct sum of an infinite number of closed families $\mathcal{F}(p) = \{|e_0, p\rangle, |g_-, p - \hbar k\rangle, |g_+, p + \hbar k\rangle\}$ labelled by p. Since there is only one internal excited state $|e_0\rangle$ for two ground levels $|g_\pm\rangle$, we can form, inside each family $\mathcal{F}(p)$, a particular linear combination $|\psi_{NC}(p)\rangle$ non coupled to the laser:

$$|\psi_{NC}(p)\rangle = \frac{1}{\sqrt{|\alpha_+|^2 + |\alpha_-|^2}}[\alpha_-|g_-, p - \hbar k\rangle - \alpha_+|g_+, p + \hbar k\rangle] \tag{13}$$

Note that this combination is not a stationary state, because it has no well defined kinetic energy:

$$(p - \hbar k)^2 \neq (p + \hbar k)^2 \tag{14}$$

except for the particular case $p = 0$, for which $|\psi_{NC}\rangle$ is a perfectly trapping state, called "dark state".

Let us take into account spontaneous emission. Starting from $|e_0, p\rangle$, the atom now eventually decays into $|g_\pm, p - p_z^S\rangle$, where p_z^S is the momentum of the fluorescence photon along

z and is randomly distributed between $-\hbar k$ and $+\hbar k$. Spontaneous emission thus couples the family $\mathcal{F}(p)$ to the "neighbouring" families $\mathcal{F}(p')$, with a label $p' = p \pm \hbar k - p_z^S$ in the range $[p - 2\hbar k, p + 2\hbar k]$. This random change of p after a fluorescence cycle may transfer the atoms into the dark state $|\psi_{NC}(0)\rangle$ or into a quasi dark state for p very close to 0, where they remain trapped and where they pile up. The corresponding effect on the atomic momentum distribution is to produce two peaks at $p = \pm\hbar k$, because $|\psi_{NC}(0)\rangle$ is a superposition of states of momenta $+\hbar k$ and $-\hbar k$.

An important question concerns the existence of a fundamental limit on the efficiency of this trapping process. This problem is difficult to address by the standard theoretical techniques using the time evolution of the internal+external atomic density matrix $\sigma(t)$, in particular because $\sigma(t)$ does not reach a steady state. An original statistical approach, reformulating the equation of motion of $\sigma(t)$ in terms of the evolution of stochastic atomic state vectors $|\psi(t)\rangle$, has been developed. For a given momentum interval around $p = 0$, one can then calculate the probability distributions of the escape time from this interval and of the return time to this interval [40]. These probability distributions are found to be the so-called Lévy flights; they are "broad" and have infinite mean values. A first result in [40] is that the width of the peaks in momentum space around $p = \pm\hbar k$ is not limited by the recoil momentum $\hbar k$, but scales as $1/\sqrt{t}$, where t is the atom-laser interaction time; this law was already observed numerically in [36]. A second important result of [40] concerns the height of the peaks: the atomic population in the peaks is proved to go to 0 in the very long time limit; indeed, atoms can in their diffusive motion reach high momenta, where their excitation rate by the laser is very small, because of the atom-laser detuning due to Doppler effect. This disadvantage disappears in recently proposed 1D VSCPT configurations, where a Sisyphus-effect type restoring force confines the atoms in a small region in momentum space around $p = 0$ [41].

Experimental results below the recoil limit

The 1D cooling scheme introduced in the previous section has been first demonstrated experimentally on a beam of metastable helium in [35]. For a maximum interaction time of 350 Γ^{-1}, where Γ^{-1} is the lifetime of the excited state, the half width at half maximum δp of the two peaks in momentum space leads to an effective temperature [5] T_{eff} on the order of 0.5 recoil temperature $T_R = \hbar^2 k^2 / M k_B$.

The velocity selective coherent population trapping on a $j_g = 1 \rightarrow j_e = 1$ transition can be directly generalized to 2D and 3D cooling [36, 42, 43]. One can show in a general fashion that a dark state exists for any monochromatic laser field, and that its components in momentum space are the same than the spatial Fourier components of the laser electric field, i.e. than the momenta of the laser photons. For example, consider a 2D laser configuration realized with four traveling waves, so that the electric field has four Fourier components in the horizontal xy plane, $\pm k\vec{e}_x$ and $\pm k\vec{e}_y$, where $\vec{e}_{x/y}$ are unit vectors along x/y. In this case VSCPT accumulates the atoms in dark states with a well defined momentum dependence in the xy plane:

$$|\psi_{xy}\rangle = |g_{x,+}\rangle|\hbar k\vec{e}_x\rangle + |g_{x,-}\rangle|-\hbar k\vec{e}_x\rangle + |g_{y,+}\rangle|\hbar k\vec{e}_y\rangle + |g_{y,-}\rangle|-\hbar k\vec{e}_y\rangle \tag{15}$$

but with an arbitrary momentum along the vertical axis z. In equation (15) the atomic internal states $|g_{x,\pm}\rangle$ and $|g_{y,\pm}\rangle$ depend on the polarization and the amplitude of the laser waves $\pm k\vec{e}_x$ and $\pm k\vec{e}_y$.

Recent 2D experimental results have been obtained in Paris on metastable helium. The atoms are first cooled and trapped in a magneto-optical trap. Then the magneto-optical trap is switched off and the previous 2D laser configuration, with four traveling waves, is switched on in the horizontal plane. The atoms are doing a free flight along the vertical z direction, until they reach a detector, where their spatial position is measured. During the interaction with the laser, there is a strong accumulation of the atoms close to the dark state (15), for the atomic

---

[5]More precisely, δp is defined as the half width of the peaks at the relative height $e^{-1/2}$, so that the equation $T_{\text{eff}} = \delta p^2 / M$ defining T_{eff} as a function of δp is exact for a Maxwell-Boltzmann distribution.

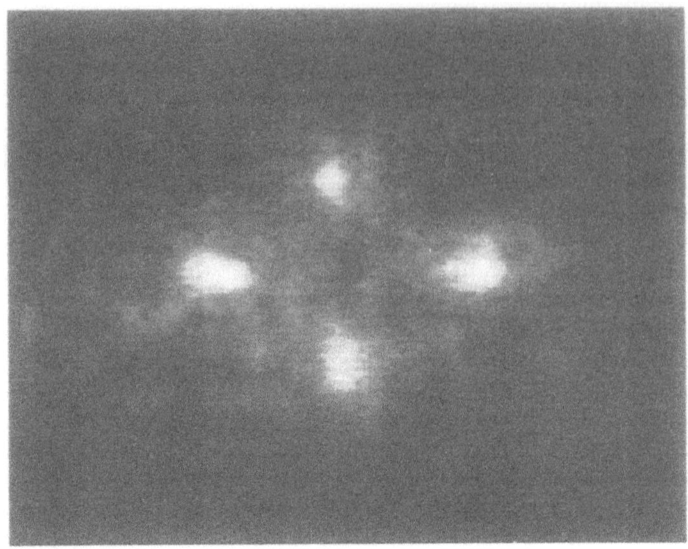

Figure 9: Experimental results for 2D subrecoil cooling. The atoms are released from a trap and interact with the VSCPT laser field during 500 μs. Atoms with a time of free flight between 45 and 65 ms are detected and give rise to four peaks on the detector.

motion in the horizontal plane. In momentum space we get peaks around the four momenta of the laser photons, with a half-width δp. In real space, we expect a separation of four spatial components in the horizontal plane, each one having a mean velocity corresponding to one of the peaks in momentum space. The half-width δr of these spatial components is initially determined by the size δr_0 of the atomic cloud in the magneto-optical trap and is expanding as function of time t as $\sqrt{\delta r_0^2 + (\delta p t/M)^2}$, with $\delta p \ll \hbar k$. If the time of free flight is long enough, the distance $\sim \hbar k t/M$ between the spatial components becomes larger than δr and the four components no longer overlap.[6] It is then possible to deduce the momentum distribution of atoms from their positions on the detector in each spatial component. Experimentally, the four peaks have been clearly resolved (see Figure 9 taken from [44]). Values of δp as low as $\hbar k/4$ have been observed; the corresponding effective temperature, 16 times below the recoil temperature, is 250 nK.

As mentioned in the introductory remarks of this section, another subrecoil cooling scheme has been demonstrated experimentally [37]. Stimulated Raman transitions, between two hyperfine ground states $|1\rangle$ and $|2\rangle$, are produced by two counterpropagating synchronized laser pulses and provide a high selectivity in atomic velocity. The atoms whose velocity v meets the Raman resonance condition $v = v_0$ are transferred into $|2\rangle$ by the pulses and are then "pushed" towards zero velocity back into $|1\rangle$ through a pumping resonant traveling wave. Successive cooling cycles of this type are then performed, with a scanning of v_0 from high velocities towards low velocities, in order to collect most of the atoms. Since v_0 is always different from 0, there is a small region around $v = 0$ where the atoms never meet the Raman resonance condition, and where they pile up. Very sharp momentum distributions around $p = 0$ have thus been obtained and observed in 1D, with effective temperatures down to $T_R/10$.

---

[6] At this stage, it is important to realize that the four spatial components are coherent and could lead to an interference effect if they were recombined.

CONCLUSION

In this lecture we have presented two laser cooling mechanisms.

The first one relies on the spatial correlation between the optical pumping rates and the lightshifts, which leads to a Sisyphus effect. The achievable atomic kinetic energies remain larger than the recoil energy $(\hbar k)^2/2M$. Quantum properties of the atomic motion in the periodic array formed by the light ("optical lattice") have been investigated experimentally. We have mentioned the spectroscopic measurements of vibrational levels in the microtraps where the cooling accumulates the atoms. Also, in the regime of very shallow optical potential wells, a signature of a band structure for the atomic energy spectrum has been reported recently [45].

The second laser cooling mechanism discussed in this lecture is not limited by the recoil energy. It accumulates the atoms in states very weakly coupled to the light and with a well defined velocity dependence. We have reported recent experimental results in 2D. In 3D no subrecoil temperatures have been obtained yet.

Finally, let us mention briefly new directions opened by laser cooling. A cloud of ultra-cold atoms can be a starting point for a number of physics experiments, such as atomic fountain clocks, atomic interferometry, collision physics and non linear optics [46]. It may be a candidate for the search of quantum statistical effects, depending on the bosonic or fermionic nature of the atoms. For example one could hope to put several atoms in the same mode of an atomic cavity, such as the one demonstrated in [47].

References

[1] J. Dalibard and C. Cohen-Tannoudji, J. Opt. Soc. Am. **B2** 1707 (1985)

[2] S. Chu, J. Bjorkholm, A. Ashkin and A. Cable, Phys. Rev. Lett. **57** 314 (1986)

[3] P. Gould, P. Lett, P. Julienne, W.D. Phillips, W. Thorsheim and J. Weiner, Phys. Rev. Lett. **60** 788 (1988)

[4] T. Hänsch and A. Schawlow, Opt. Commun. **13** 68 (1975)

[5] D. Wineland and H. Dehmelt, Bull. Am. Phys. Soc. **20** 637 (1975)

[6] S. Chu, L. Hollberg, J. Bjorkholm, A. Cable and A. Ashkin, Phys. Rev. Lett. **55** 48 (1985)

[7] D.J. Wineland and W.M. Itano, Phys. Rev. **A20** 1521 (1979)

[8] J.P. Gordon and A. Ashkin, Phys. Rev. **A21** 1606 (1980)

[9] E.L. Raab, M. Prentiss, A. Cable, S. Chu and D.E. Pritchard, Phys. Rev. Lett. **59** 2631 (1987)

[10] A.M. Steane and C.J. Foot, Europhys. Lett. **14** 231 (1991)

[11] A.M. Steane, M. Chowdhury and C.J. Foot, J. Opt. Soc. Am. **B9** 2142 (1992)

[12] A. Clairon, P. Laurent, A. Nadir, M. Drewsen, D. Grison, B. Lounis and C. Salomon, Proceedings of the 6th European Frequency and Time Forum, held at ESTEC, Noordwijk (ESA SP-340, June 1992)

[13] C.D. Wallace, T.P. Dinneen, K.Y.N. Tan, A. Kumarakrishnan, P.L. Gould and J. Javanainen, to be published (1994)

[14] P. Lett, R. Watts, C. Westbrook, W.D. Phillips, P. Gould and H. Metcalf, Phys. Rev. Lett. **61** 169 (1988)

[15] P. Lett, W.D. Phillips, S. Rolston, C. Tanner, R. Watts and C. Westbrook, J. Opt. Soc. Am. **B6** 2084 (1989)

[16] J. Dalibard, C. Salomon, A. Aspect, E. Arimondo, R. Kaiser, N. Vansteenkiste and C. Cohen-Tannoudji, Atomic Physics 11, ed. S Haroche, J.C Gay and G. Grynberg (World Scientific, Singapour) p.199 (1989)

[17] Y. Shevy, D.S. Weiss, P.J. Ungar and S. Chu, Phys. Rev. Lett. **62** 1118 (1989)

[18] D.S. Weiss, E. Riis, Y. Shevy, P.J. Ungar and S. Chu, J. Opt. Soc. Am. **B6** 2072 (1989)

[19] C. Salomon, J. Dalibard, W.D. Phillips, A. Clairon and S. Guellati, Europhys. Lett. **12** 683 (1990)

[20] C. Gerz, T.W. Hodapp, P. Jessen, K.M. Jones, W.D. Phillips, C.J. Westbrook and K. Mølmer, Europhys. Lett. **21** 661 (1993)

[21] J. Dalibard and C. Cohen-Tannoudji, J. Opt. Soc. Am. **B6** 2023 (1989)

[22] P.J. Ungar, D.S. Weiss, E. Riis and S. Chu, J. Opt. Soc. Am. **B6** 2058 (1989)

[23] Y. Castin, J. Dalibard, C. Cohen-Tannoudji, Proceedings of Light Induced Kinetic Effects, eds L. Moi, S. Gozzini, C. Gabbanini, E. Arimondo and F. Strumia (ETS Editrice, Pisa, 1991)

[24] Y. Castin and J. Dalibard, Europhys. Lett. **14** 761 (1991)

[25] J.Y. Courtois, thèse de doctorat de l'École Polytechnique, Paris, France (1993)

[26] P. Verkerk, B. Lounis, C. Salomon, C. Cohen-Tannoudji, J.Y. Courtois, G. Grynberg, Phys. Rev. Lett. **68** 3861 (1992)

[27] J.Y. Courtois and G. Grynberg, Phys. Rev. **A46** 7060 (1992)

[28] P.S. Jessen, C. Gerz, P.D. Lett, W.D. Phillips, S.L. Rolston, R.J.C. Spreeuw and C.I. Westbrook, Phys. Rev. Lett. **69** 49 (1992)

[29] Y. Castin, thèse de doctorat de l'Université Paris 6, France (1992)

[30] P. Marte, R. Dum, R. Taïeb, P.D. Lett and P. Zoller, Phys. Rev. Lett. **71** 1335 (1993)

[31] K. Berg-Sørensen, Y. Castin, K. Mølmer and J. Dalibard, Europhys. Lett. **22** 663 (1993)

[32] A. Hemmerich, Zimmerman and T.W. Hänsch, Europhys. Lett. **22** 89 (1993)

[33] A. Hemmerich and T.W. Hänsch, Phys. Rev. Lett. **70** 410 (1993)

[34] G. Grynberg, B. Lounis, P. Verkerk, J.Y. Courtois and C. Salomon, Phys. Rev. Lett. **70** 2249 (1993)

[35] A. Aspect, E. Arimondo, R. Kaiser, N. Vansteenkiste and C. Cohen-Tannoudji, Phys. Rev. Lett. **61** 826 (1988)

[36] A. Aspect, E. Arimondo, R. Kaiser, N. Vansteenkiste and C. Cohen-Tannoudji, J. Opt. Soc. Am. **B6** 2112 (1989)

[37] M. Kasevich and S. Chu, Phys. Rev. Lett. **69** 1741 (1992)

[38] H. Wallis and W. Ertmer, J. Opt. Soc. Am. **B6** 2211 (1989)

[39] K. Mølmer, Phys. Rev. Lett. **66** 2301 (1991)

[40] F. Bardou, J.P. Bouchaud, O. Emile, A. Aspect and C. Cohen-Tannoudji, submitted to Phys. Rev. Lett. (1993)

[41] M.S. Shahriar, P.R. Hemmer, M.G. Prentiss, P. Marte, J. Mervis, D.P. Katz, N.P. Bigelow and T. Cai, Phys. Rev. **A48** R4035 (1993)

[42] F. Mauri, F. Papoff and E. Arimondo, Proceedings of Light Induced Kinetic Effects, eds L. Moi, S. Gozzini, C. Gabbanini, E. Arimondo and F. Strumia (ETS Editrice, Pisa, 1991)

[43] M.A. Ol'shanii and V.G. Minogin, Proceedings of Light Induced Kinetic Effects, eds L. Moi, S. Gozzini, C. Gabbanini, E. Arimondo and F. Strumia (ETS Editrice, Pisa, 1991)

[44] J. Lawall, F. Bardou, B. Saubamea, K. Shimizu, M. Leduc, A. Aspect and C. Cohen-Tannoudji, submitted to Phys. Rev. Lett. (1994)

[45] M. Doery, M. Widmer, J. Bellanca, E. Vredenbregt, T. Bergeman and H. Metcalf, Phys. Rev. Lett. **72** 2546 (1994)

[46] See, for instance, "Laser manipulation of atoms", Proc. of the Int. School of Physics Enrico Fermi, ed. E. Arimondo, W.D. Phillips and F. Strumia, North-Holland (1992)

[47] C. Aminoff, A. Steane, P. Bouyer, P. Desbiolles, J. Dalibard and C. Cohen-Tannoudji, Phys. Rev. Lett. **71** 3083 (1993)

TRANSFER OF SINGLE ELECTRONS AND SINGLE COOPER PAIRS IN METALLIC NANOSTRUCTURES

Michel H. Devoret, Daniel Esteve and Cristian Urbina

Service de Physique de l'Etat Condensé
CEA-Saclay
91191 Gif-sur-Yvette cedex, France

INTRODUCTION

One individual atom, a purely theoretical entity a hundred years ago, can now be imaged and manipulated at the surface of bulk matter [1] or, free-standing, in vacuum [2]. Is the electron, the simplest and most thoroughly studied particle, amenable to such ultimate control? In vacuum, the detection of single electrons is now routine. A spectacular example of the control of individual electrons travelling in a vacuum chamber is the experiment in which Dehmelt *et al.* [3] were able to prove during three months a single electron kept in an electromagnetic trap, thereby measuring to unprecedented accuracy the anomalous part of its magnetic moment. In matter, the manipulation of individual electrons is a very different game, because the separation between electrons is of the same order as their quantum mechanical wavelength. Here we focus on the most basic type of such manipulation. We explain how it is possible to take, at a precise instant, exactly one electron from a first electrode and transfer it with certainty to a second electrode. By making these electrodes part of an electrical circuit and by continuously repeating this transfer process we can achieve a perfectly controlled current source. In particular, for a sequence of single-electron transfers clocked by a radiofrequency signal at frequency f, the current I will be given simply by $I = ef$ where e is the quantum of charge, a fundamental constant. We will also explain how, when at least one of the electrodes is in the superconducting state, electron pairing favors charge transfer by units of $2e$.

Advances in Quantum Phenomena. Edited by E.G. Beltrametti
and J.-M. Lévy-Leblond. Plenum Press, New York, 1995

BASIC PRINCIPLE OF SINGLE ELECTRON TRANSFER

Although the charge of the electron was measured as early as 1911 [4], the granularity of electricity does not usually show up in the macroscopic quantities such as current and voltage, which describe the state of an electric circuit. This is not just a matter of the number of electrons being very large in typical devices. Charge flow in a metal or a semiconductor is a continuous process because conduction electrons are not localized at specific positions. They form a quantum fluid which can be shifted by an arbitrary small amount. The variations of the charge Q on a capacitor C and of the associated potential difference $U = Q/C$ illustrate this property. The charge Q can be any fraction ε of the charge quantum e: if ρ denotes the electron density in the metallic plates of the capacitor and S their surface area, it is easy to see that a bodily displacement $\delta = \varepsilon/(\rho S)$ of the electronic fluid with respect to the ionic background, in the direction perpendicular to the plates, produces the charge $Q = \varepsilon e$.

There exists, however, a solid-state device in which electric charge flows in a discrete manner. It consists of two metallic electrodes separated by an insulating layer so thin that electrons can traverse it by the tunnel effect [5] (Fig. 1). Tunnelling can be considered as an all-or-nothing process because electrons spend a negligible amount of time under the potential barrier corresponding to the insulating layer [6,7]. If one applies a voltage V to such a tunnel junction, electrons will randomly tunnel across the insulator at a rate given by V/eR_t, where the tunnel resistance R_t is a macroscopic parameter of the junction which depends on the area and thickness of the insulating barrier. Apart from allowing the tunnel effect, the two facing electrodes behave as a capacitor whose capacitance C is the other macroscopic parameter of the junction. It is important to stress that the transport of electrons in a tunnel junction and in a metallic resistor are fundamentally different, even though the current-voltage characteristic is linear in both cases. Charge flows continuously along the resistor whereas it flows across the junction in packets of e. Obviously, a tunnel junction provides the means to extract electrons one at a time from an electrode. With a single voltage-biased tunnel junction, however, it is not possible to control the instants at which electrons pass from the upstream electrode to the downstream electrode, because of the stochastic nature of tunnelling. A further ingredient is needed.

Suppose that instead of applying directly a voltage source to the junction one biases it with a voltage source U in series with a capacitor C_s (we reserve the letter symbol V for transport voltage sources that have to deliver a static current). A metallic electrode entirely surrounded by insulating material is formed between the junction and the capacitor (see Fig. 2a). We will call such an isolated electrode, which electrons can enter and leave only by tunnelling, an "island". The island is coupled electrostatically to the rest of the circuit by the capacitances C and C_s whose charges are denoted by Q and Q_s respectively. Although, as we have seen, Q and Q_s are both continuous variables, their difference is the total excess charge of the island. Because charge can enter the island only by tunnelling, this total charge is a multiple of the electron charge: $Q - Q_s = ne$. Suppose furthermore that the island dimensions are

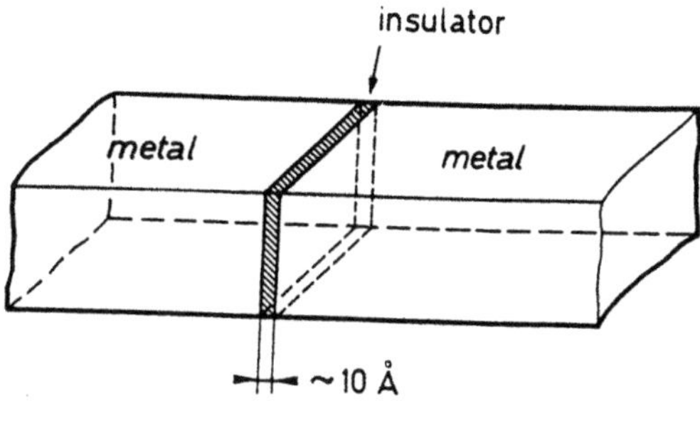

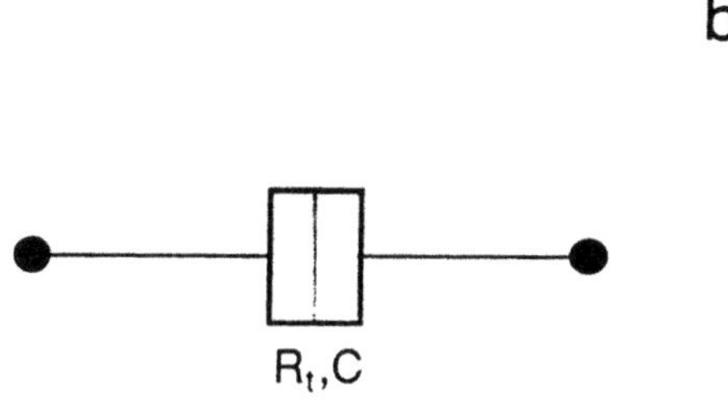

Fig. 1 a) Tunnel junction traversed by a current $I(t)$ which consists, when a fixed voltage is imposed to the junction, of uncorrelated charge packets corresponding to individual electrons. Electrons tunnel through the thin layer of insulator sandwiched between the metal electrodes. The junction is represented in circuit schematics by a double box symbol (b) and is characterized by the tunnel resistance R_t and capacitance C. It is worth noting that although R_t is called a "resistance", it characterizes a purely elastic process. At the insulating barrier, the electron wavefunction is partially transmitted and reflected. Its energy does not change. The tunnel resistance is inversely proportional to the barrier transmission coefficient which decreases exponentially with the thickness of the insulating layer. In practice, measurable tunnel resistances can be achieved only with insulating layers a few nanometres thick.

small enough that the electrostatic energy $E_c = e^2/2C_\Sigma$ of one excess electron on the island is much larger than the characteristic energy $k_B T$ of thermal fluctuations. Here, $C_\Sigma = C + C_s$, and k_B and T denote the total capacitance of the island, the Boltzmann constant and the temperature, respectively. This Coulomb energy E_c is the other ingredient of controlled electron transfer.

When $U = 0$, n will stay identically zero because the entrance or exit of an electron would raise the electrostatic energy of the island to a level much higher than permitted by thermal fluctuations. As U increases from zero, however, the total energy difference between the $n = 0$ and $n = 1$ state of the whole circuit

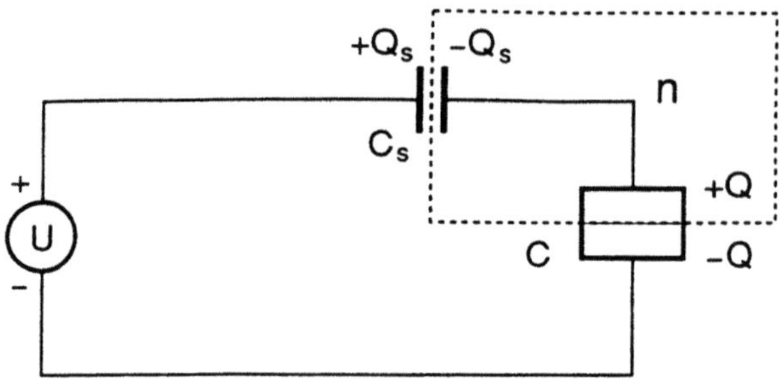

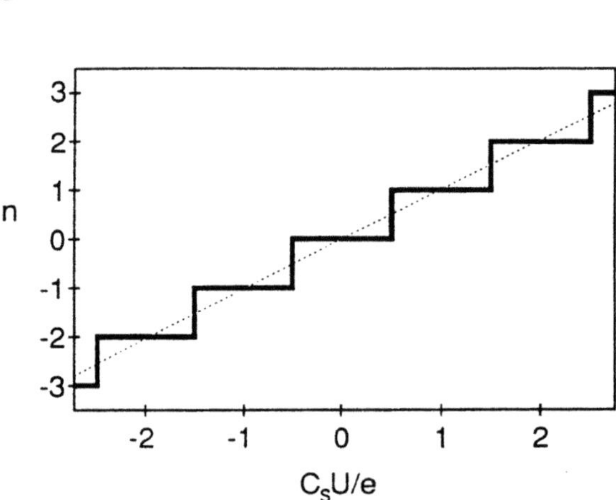

Fig. 2 a) Junction biased by a voltage source U in series with a capacitance C_s. The metal electrode between the junction and the capacitance forms an isolated "island" (box in dashed line) which contains n excess electrons. b) Variation of $\bar{n}$, the average of n as a function of U, when $k_BT \ll E_c$ (full line) and $k_BT \gg E_c$ (dashed line).

decreases, because when an electron tunnels to the island the potential drop C_sU/C_Σ partly compensates the electrostatic energy of the island. In fact, a straightforward calculation of the total energy of the circuit yields $E = E_c(n - C_sU/e)^2 +$ terms independent of n. Thus, when $U = e/2C_s$, the $n = 0$ and $n = 1$ states will have the same energy and an electron can tunnel in and out freely. As U is increased further, the $n = 1$ state becomes the lowest energy state. The maximum stability of the $n = 1$ state against fluctuations is reached at $U = e/C_s$ where, as in the case $U = 0$, and $n = 0$, the charge Q vanishes. It is now easy to see that each time the voltage U is increased by e/C_s, the number n of excess electrons of the island is increased by one. If one plots $\bar{n}$, the average of n, as a function of U, one gets the staircase function shown on Fig. 2b. It is therefore possible to control exactly the number of excess electrons of the island by adjusting the voltage U.

As the temperature is increased, the staircase becomes rounded and for temperatures $k_B T \gg E_c$ it approaches the straight dotted line of Fig. 2b. In practise, one can reliably cool tunnel junctions down to 50 mK but not much below. To satisfy $E_c \gg k_B T$, C_Σ must be of the order of or smaller than one femtofarad. This requires the fabrication of junctions with typical areas of 50 nm $\times$ 50 nm and hence the use of nanofabrication techniques. With such low values of capacitance, the typical voltage corresponding to the addition of an electron is of the order of 100 μV - 1 mV, a value which can be easily controlled electronically. To summarize, tunnelling breaks the continuity of the electron fluid into charge packets corresponding to single electrons. The Coulomb energy of excess charges on an island provides a feedback mechanism that regulates the number of electrons tunnelling in and out the island. At sufficiently low temperature, the exact number of excess electrons on the island does not fluctuate and can be entirely determined by an externally applied voltage. The quenching of the island charge fluctuations for the "single electron box" (the circuit of Fig. 2a), has been demonstrated experimentally by Lafarge *et al.* [8].

We have considered so far only thermal fluctuations of the number n. This variable is also subject to quantum fluctuations. In our analysis of the circuit of Fig. 2a, we have neglected the delocalization energy associated with tunnelling. This energy is very small compared with the Coulomb energy. Perturbative calculations [9,10] show that the quantum fluctuations of n become negligible in the limit $R_t \gg R_K = h/e^2$, where h is Planck's constant. The constant $R_K \approx 26\mathrm{k\Omega}$ is the resistance quantum. In the next three sections we will consider tunnel barriers sufficiently opaque that this latter condition is fulfilled.

SINGLE ELECTRON EFFECTS: A BRIEF HISTORY

A large class of phenomena exist that combine the partial localization of electrons due to tunnelling and the Coulomb charging energy, and may be called "single electron effects". Decades ago, it was proposed that the variation of the island potential due to the presence of only one excess electron could be large enough to react back on the probability of subsequent tunnelling events [11-15]. At that time, the effect could only be observed in granular metallic materials. It was realized that the hopping of electrons from grain to grain could be inhibited at small voltages if the electrostatic energy of a single electron on a grain was much larger than the energy of thermal fluctuations. The interpretation of these pionneering experiments, in which there is an interplay between single-electron effects and random media properties, was complicated by the limited control over the structure of the sample. With modern nanofabrication techniques, it is possible to design metallic islands of known geometry separated by well-controlled tunnel barriers [16]. This led Fulton and Dolan to perform the first unambiguous demonstration of single-electron effects in an island formed by two junctions [17]. Meanwhile, Likharev and coworkers [18,19] had produced detailed predictions of single electron effects in a nanoscale current-biased single junction (this system was also considered in refs. [20] and [21], but only for junctions in the superconducting state) and proposed various applications of the new effects. This current-biased scheme is analogous in some ways to the circuit of Fig. 2a, but with the capacitance replaced by a large resistance. In that case there is no island enforcing charge quantization, because an arbitrarily small amount of charge can flow through the resistance. Only the charge on the junction capacitance would provide the feedback of Coulomb energy on tunnelling.

It was later understood that, in general, the quantum electromagnetic fluctuations due to the finite value of the resistance wash out single electron effects in this single-junction no-island system. Only if the value of resistance is made much larger than the resistance quantum R_K up to frequencies of the order of $e^2/(\hbar C)$ [22-24] can tunnelling be Coulomb blocked in the current biased junction system. In spite of the experimental difficulties involved in fabricating the resistance with adequate characteristics, the competition between single electron effects and quantum electromagnetic fluctuations has been observed [25,26]. The single-junction no-island system is certainly of interest as an illustration of the foundations of the field, but it is not suited for practical applications because getting rid of quantum electromagnetic fluctuations is so difficult experimentally. In what follows, we will resume the discussion of systems that contain at least one island and are thus immune to quantum electromagnetic fluctuations. We will focus mainly on the controlled transfer of single electrons. For general introductions to single-electron effects in normal and superconducting junction systems, see refs. [27-30], for recent snapshots of the state of current research, see refs. [31,32].

THE SINGLE ELECTRON TRANSISTOR

The one-junction one-island circuit of Fig. 2a is the simplest in which single-electron transfer can occur. On the other hand, it cannot produce an externally measurable static current, as the island is a cul-de-sac for electrons. Let us consider the next order of complexity, the two-junction one-island circuit of Fig. 3a [17]. The state of the circuit is now characterized by the two numbers N and N' of electrons having passed through the two junctions. (The sign of N and N' is positive if during tunnelling the electron flows in the direction of increasing voltage, and negative otherwise).

It is convenient to introduce the number $n = N - N'$ of excess electrons on the island and the charge flow index $p = (N + N')/2$. The state $(n, p + 1)$ only differs from the state (n, p) in that one electron has been transferred from one terminal of the transport voltage source V to the other. The electrostatic energies of the various

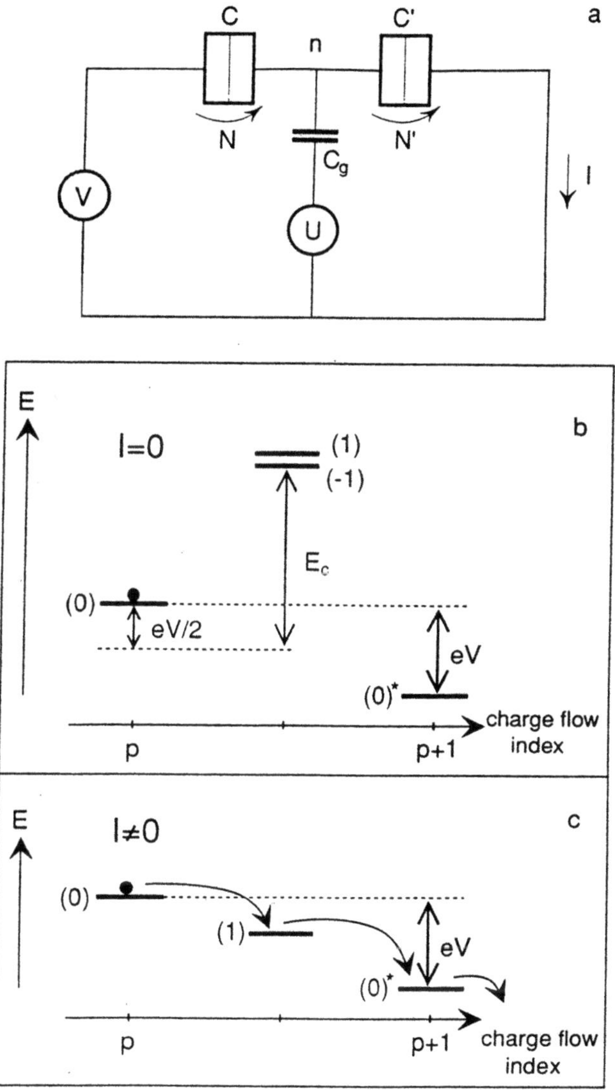

Fig. 3 a) Schematic of single-electron transistor (SET). Energy of the states of the circuit when b) $U = 0$ and c) $U = e/C_g$. The numbers in parenthesis are the values of the number n of excess electrons on the SET island. The charge flow index is half the sum of the numbers N and N' of electrons that have traversed the junctions. In b) no current can flow through the device: this is the Coulomb blockade.

capacitances of the circuit are the same. As the precise value of p does not matter here, we will condense the notations (n, p) and $(n, p+1)$ into (n) and $(n)^*$.

With regard to the total energy of the circuit, which includes the work of the transport voltage, state $(n)^*$ is lower by eV than state (n) and, hence, the circuit has no absolutely stable states. In principle, a steady current I could flow around the loop formed by the two junctions and the transport voltage V. To go from state (n) to state $(n)^*$, however, the circuit must go through state $(n+1)$ or state $(n-1)$, because tunnel events occur one at a time. States $(n+1)$ and $(n-1)$ differ by state (n) by an electron having tunnelled through the first and second junction, respectively. This is where the single-electron Coulomb energy $E_c = e^2/2C_\Sigma$ comes into play (C_Σ is, as before, the island total capacitance given now by $C_\Sigma = C + C' + C_g$, where C and C' are the two junction capacitances). To simplify the discussion, suppose that $eV \ll E_c$.

When the control "gate" voltage is set at $U = 0$, the energy of states (-1) and (1) will be $E_c - eV/2 \approx E_c$ above the energy of state (0) (see Fig. 3c). At low temperature, this will provide a Coulomb barrier for the transport of electrons around the circuit. In this case the current I should be strictly zero. This situation is called the Coulomb blockade. On the other hand, when the control voltage is such that $C_g U \approx e/2$, states (0) and (1) have nearly the same energy (Fig. 3c). As soon as the energy of the (1) state is lowered below that of the (0) state, the $(0) \rightarrow (1)$ transition becomes possible and an electron enters the island through the first junction. If U is such that the energy of the (1) state, although below that of the (0) state, is still above the energy of the $(0)^*$ state, the transition $(1) \rightarrow (0)^*$ takes place and the electron leaves the island through the second junction. Apart from an electron having gone through the device, one is now back to the initial electrostatic state and the cycle can start over again. This cascade of transitions produces a current of order $V/(R_t + R_t')$ through the device (R_t and R_t' are the tunnel resistances of the two junctions). When U is increased further, the energy of the (1) state goes below the energy of the $(0)^*$ state and one enters a new Coulomb blocked state with one excess electron on the island. The domains of the Coulomb blocked states in the set of U values are in a one-to-one correspondance with the flat portions of the staircase of the electron box (Fig. 2b) and its is easy to show that, at voltages low compared with the Coulomb voltage E_c/e, the current I is maximum when $C_g U = ne/2$.

In practise, a current of the order of 10^9 electrons per second can be switched on and off by the presence or absence of half the electron charge on the gate capacitor, hence the name "single electron transistor" (SET) given to this device. The remarkable charge sensitivity of the SET is unrivalled by other devices: it is six orders of magnitude better than conventional FET electrometers [28]. A possible application is the detection of individual photoinduced electron-hole pairs in semiconductors [33]. But the input capacitance of the SET is, by construction, so tiny that its voltage sensitivity is not high. In this respect, it does not compare favourably with the field effect transistor (FET), the semiconductor device on which most of today's applications of solid-state electronics are based. Furthermore, in the SET the modulation of electron flow by the gate ceases as soon as the bias voltage becomes of the order of the Coulomb gap voltage E_c/e, whereas in the FETs used in digital circuits the modulation of the source-drain current by the gate only saturates at large bias voltages

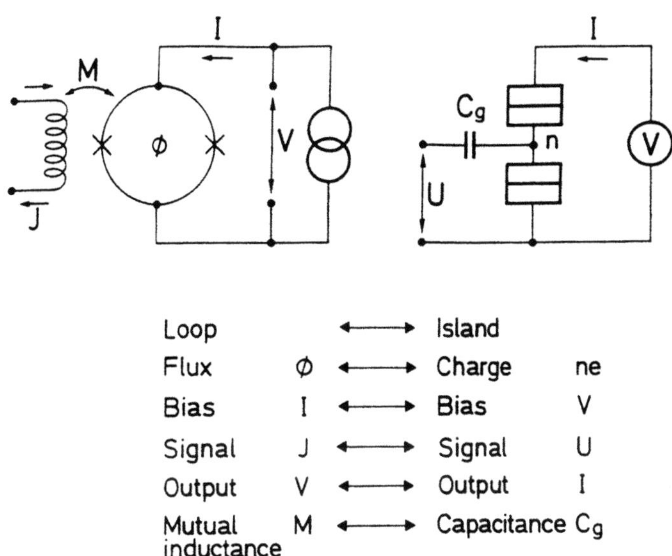

Loop	⟷	Island	
Flux	ϕ ⟷	Charge	ne
Bias	I ⟷	Bias	V
Signal	J ⟷	Signal	U
Output	V ⟷	Output	I
Mutual inductance	M ⟷	Capacitance	C_g

Fig. 4 Comparison between the d.c. SQUID (left) and the SET electrometer (right).

[34]. It is this latter feature which ensures enough voltage gain to compensate for the dispersion in device parameters and which make robust integrated digital circuit design possible with FETs.

An analogy [28] can be drawn between the SET and the dc SQUID with an input coil [35]. The d.c. SQUID (superconducting quantum interference device) consists of two Josephson junctions in parallel biased by a static current. In the d.c. SQUID, the output voltage is a periodic function of the current in the input coil, whereas in the SET, the output current is a periodic function of the voltage on the input capacitor. For the d.c. SQUID the period is set by the flux quantum $h/2e$ whereas for the SET the period is set by the charge quantum e. It is tempting to speculate that the SET will play the same role for ultra-sensitive electrometry that the d.c. SQUID plays for ultra-sensitive magnetometry. However, the fundamental impossibility of building the charge analogue of the superconducting flux transformer which is so crucial to the use of d.c. SQUIDs may severely limit the use of SETs.

The junctions that have been described so far consist in practise of two overlapping metallic films. It is also possible, instead of the three-dimensional gases that conduction electrons form in a metal, to use two-dimensional electron gases which are found in semiconductor heterostructures such as GaAs/GaAlAs. The detailed manifestations of Coulomb blockade have been thoroughly studied in these systems where single-electron effects may coexist with the quantum Hall effect [36].

Finally, Coulomb blockade has been observed with a scanning tunnelling microscope (STM) placed over a tiny metallic droplet [37]. The role of the island is played by the droplet. Unfortunately, it has so far been impossible to modulate the gate voltage independently in the droplet-STM systems. On the other hand, very small island dimensions (a few nanometer) can be achieved in this manner, and Coulomb blockade at room temperature has been reported [38]. In principle the island could even be reduced to a single molecule [39].

It is important to note at this point that we describe the state of a circuit such as the SET by discrete variables like n and p, and not by continuous variables like currents and voltages as in classical electronics. What makes nanojunction circuits of fundamental interest is that we must treat them as single, atom-like, quantum systems. Although we use macrocopic concepts like capacitance, we analyze charge flow in terms of quantum transitions between discrete energy levels of the whole circuit.

CONTROLLED TRANSFER OF CHARGE FLOWING IN AN EXTERNAL CIRCUIT

Although the principle of the SET involves the electrostatic energy of a single electron on the SET island, the charge flow through this device is not controlled at the single-electron level. The voltage U controls only the average value of the current. The instants at which electrons pass through the device are random, as in a single junction. A control of the charge flow electron by electron would mean that, using the control voltage U, one would make a single electron enter the island from the left junction, hold it in the island for an arbitrary time and finally make it leave the island through the right junction. One could then go continuously from a Coulomb-blocked state with $n = 0$ to a Coulomb-blocked state with $n = 1$. This is not possible with only one island. When the energy of the (1) state dips below the energy of the (0) state, it is necessarily above the energy of the (0)* state to which it can decay (see Fig. 3c). An electron cannot be made to enter the island through one junction without setting the electrostatic energies so that it is energetically favourable for another electron to leave the island through the other junction.

The control of charge flow at the single-electron level requires at least three junctions [40]. Let us consider the three-junction two-island circuit of Fig. 5a. As in the case of the SET, the state of the circuit can be described using the numbers n_1 and n_2 of excess electrons on each island and the charge flow index given by the third of the algebraic sum of the number of electrons having tunnelled through each junction. Using the condensed notation defined above, (n_1, n_2) and $(n_1, n_2)^*$ denote two states whose charge flow index differ by one, that is, states differing by an electron which has lost energy eV by passing through the entire device.

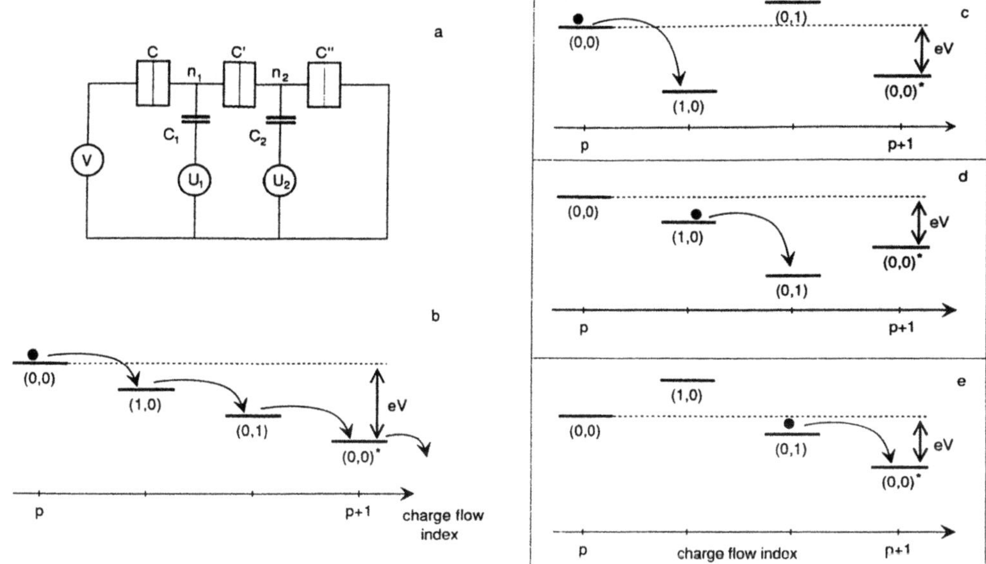

Fig. 5 a) Schematic of the single electron pump. b) Energy states of the circuit when the control voltages U_1 and U_2 are set so that Coulomb blockade is suppressed. c-e) Pumping cycle which transfers one electron around the circuit of a). It is obtained by superposing two phase shifted modulation signals on the values of U_1 and U_2 corresponding to b).

We suppose $V \ll \min(e/C_{\Sigma 1}, e/C_{\Sigma 2})$ where $C_{\Sigma 1}$ and $C_{\Sigma 2}$ denote the total capacitances of the two islands. The two control voltages U_1 and U_2, applied to the two gate capacitances C_1 and C_2, allow us to change the relative energy of the various states of this circuit. If we set U_1 and U_2 to $e/2C_1$ and $e/2C_2$ respectively, the energies of states $(0,0)$, $(1,0)$, $(0,1)$ and $(0,0)^*$ form a cascade (Fig. 5b). We are in a situation equivalent to the suppression of Coulomb blockade depicted in Fig. 3c, and a stochastic current flows through the device. Because there are three junctions instead of two, there are now two intermediate states $(0,1)$ and $(1,0)$ in the cascade. Each one of these states is coupled to $(0,0)$ or to $(0,0)^*$ but not to both. The lowering of either $(0,1)$ or $(1,0)$ below $(0,0)$ and $(0,0)^*$ stops the stochastic current and puts the circuit in a blocked state.

By modulating U_1 and U_2 with dephased periodic signals, the energy of these intermediate states can be cyclically lowered below that of the $(0,0)$ and $(0,0)^*$ states while avoiding the cascade configuration of Fig. 5b (Fig. 5c-e). One starts from the situation where both $(1,0)$ and $(0,1)$ are above $(0,0)$ and $(0,0)^*$. The circuit is in a blocked state with no excess electrons on the islands. At first, an increase of U_1 lowers $(1,0)$ below $(0,0)$ and $(0,1)$. An electron goes through the left-most junction and the circuit adopts a new blocked state with an extra electron on the first island (Fig. 5c). Then U_2 increases while U_1 decreases: this lowers $(0,1)$ below $(1,0)$ and $(0,0)^*$. A tunnel event consequently takes place through the middle junction and the circuit now adopts a blocked state with an extra electron on the second island (Fig. 5d). Finally U_2 is decreased to its initial value, making $(0,1)$ pass above $(0,0)^*$. An electron goes through the right-most junction and, apart for a charge e having crossed the entire device, the circuit returns to its initial blocked state (Fig. 5e). If the transport voltage V is reversed, the same modulation cycle will still carry electrons in the same direction, provided the energy difference eV between $(0,0)$ and $(0,0)^*$ stays small compared with the energy excursions of $(0,1)$ and $(1,0)$. The charge now flows in a direction opposite to that imposed by V. Energy conservation is of course not violated. The work done to "charge" the transport voltage source is provided by the control voltage sources. We have therefore nicknamed this three-junction device the single-electron "pump". The pump is reversible: a time-reversed modulation cycle will transfer electrons from right to left.

The actual operation of a physical device is shown in Fig. 6. We first set U_1 and U_2 to the static values $U_1^{dc} = e/C_1$ and $U_2^{dc} = e/C_2$ corresponding to a maximum zero-voltage conductance (center curve marked "no r.f."). Two periodic signals with the same frequency f but dephased by $\Phi \simeq \pi/2$ are then superimposed on the static components U_1^{dc} and U_2^{dc}. This implements the cycle shown in Fig. 5c-e and a current plateau is observed (see Fig. 6a). One can easily reverse the cycle, leaving all other conditions the same, by changing Φ to $\Phi+\pi$. A current plateau is again observed, with the same absolute value at $V = 0$ but with opposite sign. The height of the plateau is plotted against frequency on Fig. 6b. The relation $I = ef$ is well verified, providing further confirmation that our device does indeed implement the pump principle.

We have seen how two control voltages can transfer electrons one by one in a three-junction device. The transfer of single electron using only one control voltage is possible, but needs at least four junctions. In Fig. 7a we show the schematics of a four-junction three-island circuit which we have nicknamed the "turnstile" [41]. A gate capacitance, with roughly half the value of the capacitance of the junctions, is connected to the central island. Because the gate voltages of the side islands have only to be set to a constant value of zero (in practice, the external gate voltage must be adjusted to compensate for random offset charges [28]), no gate lines have been represented in the figure. The turnstile can be described as a SET with two junctions in the entrance and exit channels. The intermediate islands create energy barriers whose effect is to suppress the stochastic conduction that takes place in the SET for small V, and for U such that $C_g U = e/2$ (see Fig. 7b-d). For these conditions, the circuit can exist in two states characterized by the presence or absence of an extra electron on the central island. Suppose one starts with no electron in the

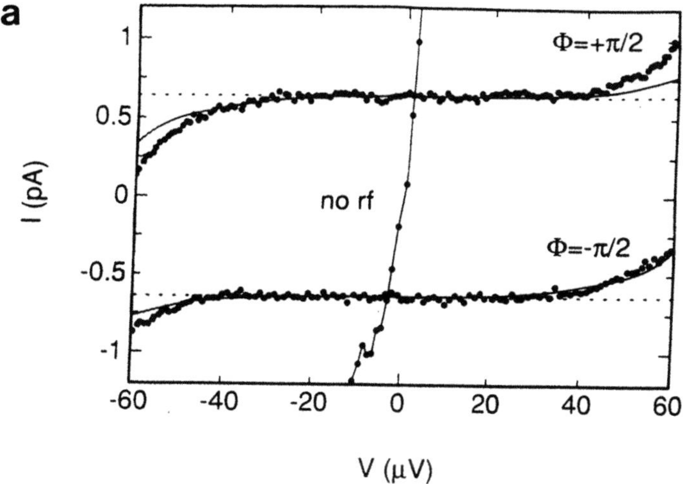

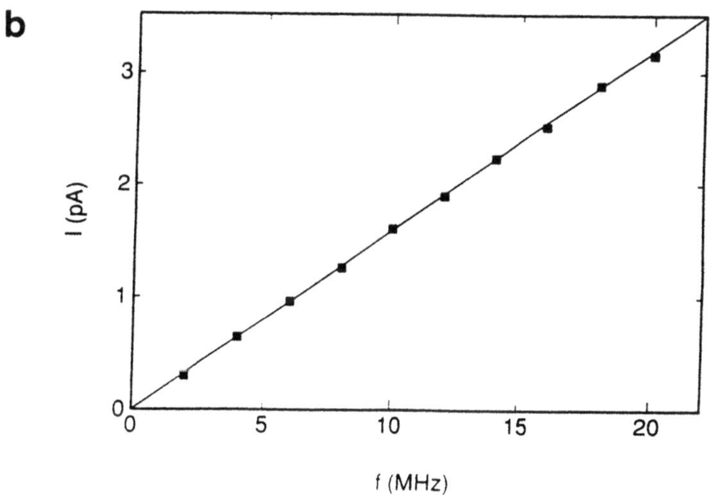

Fig. 6 a) Current-voltage characteristic of the pump with and without a $f = 4\text{MHz}$ control voltage modulation. The two modulation signals were phase-shifted by Φ. Dashed lines indicate $I = \pm ef$. Full lines are the result of numerical simulations taking into account quantum fluctuations of the island electron number. b) Current measured at the inflexion point of the current plateau as a function of the frequency f. Full line is $I = ef$.

central island. As U is increased, all the energies of the intermediate states decrease, although the state with an electron on the central island remains the lowest of the intermediate states (see Fig. 7b). Consequently, one electron enters the central island. If now one decreases U, all the energies of the intermediate states increase and at one

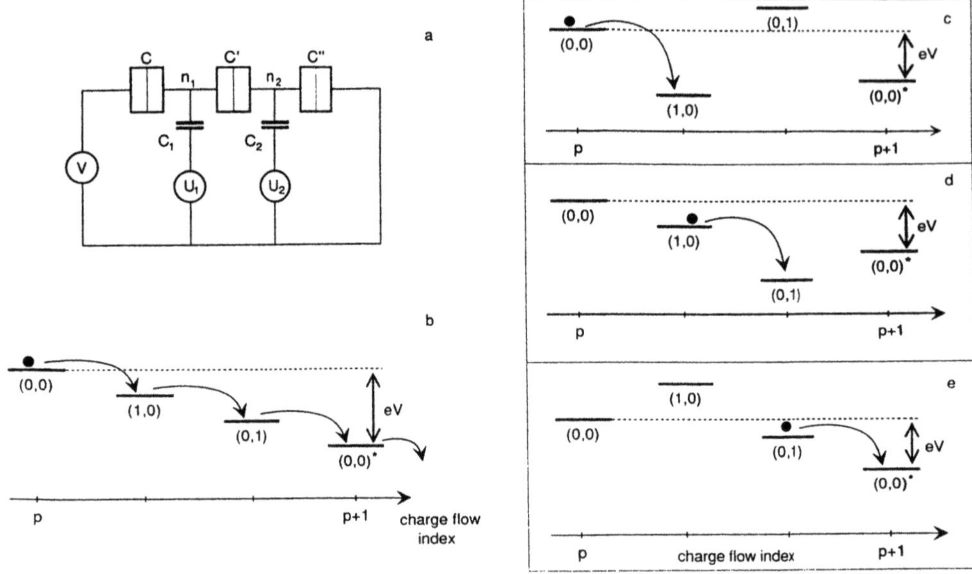

Fig. 7 a) Schematics of single-electron turnstile. b-d) Turnstile cycle which is obtained by modulating the control voltage U and which transfers one electron around the circuit of a).

point the state with an extra electron on the central island is no longer the lowest of the intermediate states. An electron then leaves the central island (Fig. 7d). It is easy to see that after one cycle of modulation of U, a charge of one electron has passed through the whole device. Like the pump, the turnstile produces a current $I = ef$, where f is the modulation frequency. Unlike the pump, however, the turnstile is an irreversible device, the sign of the current being imposed by the sign of the bias voltage V.

METROLOGICAL APPLICATIONS

We have seen that the pump and the turnstile can produce a current determined only by the frequency f and the quantum of charge e. Because frequencies can be accurately determined, these devices would provide in principle a standard of current. The standard is obtained at present by the combination of the Josephson effect [35], which relates a frequency to a voltage through the flux quantum $\Phi_0 = h/2e$, and the quantum Hall effect discovered by von Klitzing [42], which relates current to voltage through the resistance quantum $R_K = h/e^2$. It is important for metrologists

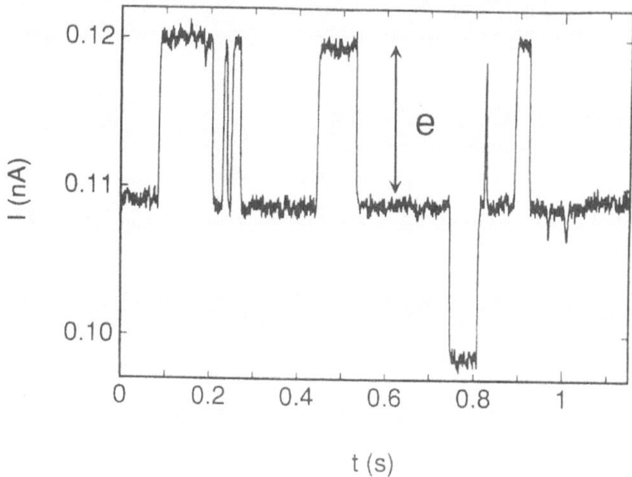

Fig. 8 Time variations of the current through a SET electrometer measuring the charge on an island linked to a charge reservoir through a series of four tunnel junctions. Each jump corresponds to an electron tunnelling into or out of the island.

to check if a direct definition of the ampere using the charge quantum e provided by single electron devices is be compatible with the "Josephson/Klitzing" definition which combines Φ_0 and R_K. The value of the fine-structure constant $\alpha = e^2/(2h\epsilon_0 c)$, where c and ϵ_0 denote the speed of light and the electrical permittivity in vacuum, is another important metrological issue that would benefit from the new access to the charge quantum provided by single-electron devices [43]. This latter application would not necessitate to measure directly the very low current produced by single-electron devices; one would simply charge a calibrated capacitor with a known number of electrons, and compare its voltage with the Josephson volt.

The first experiments carried out to test the precision of the pump and the turnstile were chiefly limited by the precision of current measurements, and it is important to investigate the intrinsic limitations of the devices. One problem is to ensure that the devices are sufficiently cold while passing current. In that respect the pump principle is better than the turnstile, as the pump is reversible and can operate at zero bias voltage. Theoretical analyses show that the fundamental limitation on the accuracy of the devices is due to co-tunnelling events [44] during which several tunnel events take place simultaneously on different junctions. These higher-order processes are a manifestation of the quantum fluctuations of island electron number discussed above. Fortunately, it can be demonstrated that the rate of co-tunnelling events decreases exponentially with the number of junctions in a device. Detailed calculations have shown that an accuracy better than 10^{-8} in the number of transferred electrons is achievable with a pump with five junctions operating at temperatures of 100 mK or less [45,46]. An important step towards the practical realization of high-accuracy transfer devices is to show experimentally that the number of electrons on an island is well determined when this island is connected to a charge reservoir through four junctions that block the quantum fluctuations of electron number. We have made a direct measurement of the charge of such an island by using a SET electrometer [47]. In Fig. 8 we show single tunnelling events in and out the island occurring on a time scale of a tenth of a second. Although this time scale is still shorter than expected theoretically, we believe that if more and smaller junctions were used, the spontaneous tunnel rate could be lowered by two orders of magnitude and thus permit metrological experiments. An important step in this direction has recently been made by Martinis et al. who have operated a five-junction pump with a 10^{-6} accuracy [48].

SINGLE COOPER PAIR TRANSFER

Up to now we have considered metallic nanojunction circuits in the normal state. For circuits in the superconducting state, one could naively expect that single electron transfer can be transposed into single Cooper pair transfer, e being simply replaced by $2e$. Several features of the superconducting state complicate this direct transposition and early experiments on Cooper pair transfer in superconducting nanojunction circuits showed unexpected results [49-51,8] which we begin to understand in detail only now. Let us go back to our basic circuit, the electron box of section 2, and examine the simplest case where only the island is in the superconducting state. The energy of the circuit as a function of the number n of electrons in the island is now $E = E_c(n - C_sU/e)^2 + (n \bmod 2)\tilde{\Delta} +$ terms independent of n. The first term is the same electrostatic energy as in the normal state, i.e. the electrostatic energy of C and C_s and the work of the voltage source U. The superconducting nature of the island manifests itself in the second term which is the internal energy of the island which we suppose for the moment at $T = 0$. This internal energy depends on n only through its parity, the parameter $\tilde{\Delta}$ denoting the minimum energy of a quasiparticle excitation. Such an odd-even difference is expected for a superconductor, since for an odd number of electrons, one of them cannot be paired and must remain as a quasiparticle excitation [52]. However, it is crucial to realize that the energy cost of this remaining quasiparticle excitation coincides with the superconducting energy gap Δ only for an ideal BCS superconductor in zero magnetic field. From this circuit energy we can predict the ensemble average $< n >$ which we suppose equal to the temporal average $\bar{n}$ measured in the experiment.

In Fig. 9a we show as a function of U the energy of the different n states, for the non-superconducting case $\tilde{\Delta} = 0$. As we have shown in section 2, n will adopt the value of the integer closest to C_sU/e, which corresponds to the lowest energy state. We thus get the staircase pattern of Fig. 9b which is identical to the full line curve of Fig. 2b. In Fig. 9c we show the case of a superconducting island such that $\tilde{\Delta} < E_c$.

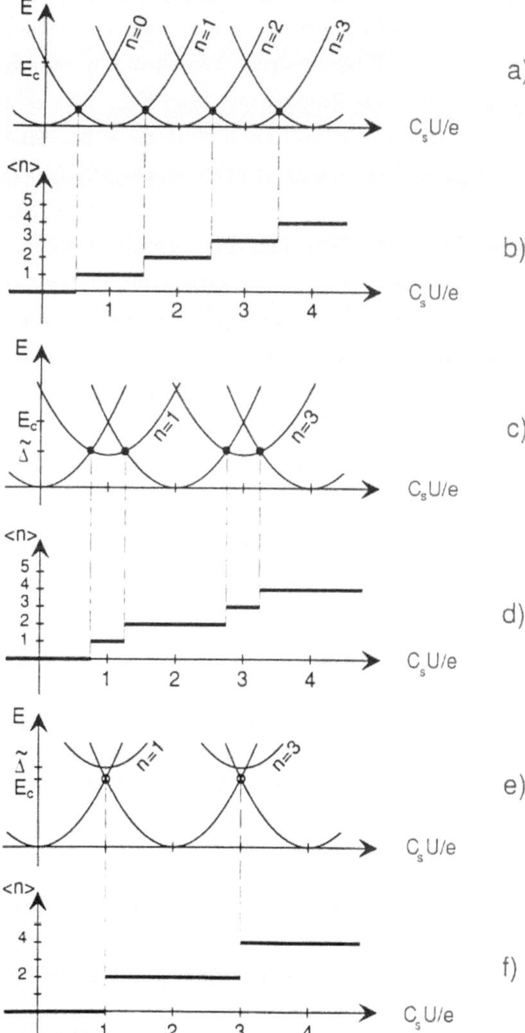

Fig. 9 Total energy of the electron box (Fig. 2a) as a function of the polarization C_sU/e, for several values of the excess number n of electrons in the island, in the non-superconducting state (a) and superconducting state (c, e). E_c is the electrostatic energy of one excess electron on the island for $U = 0$. The minimum energy for odd n is $\tilde{\Delta}$ above the minimum energy for even n. Panels c and e differ by the relative magnitude of $\tilde{\Delta}$ and E_c. The black dots correspond to level crossings where a single electron tunnel into and out of the island. The hollow circles correspond to level crossings where the only allowed process is the simultaneous tunnelling of two electrons into the island to form a pair (Andreev process). The equilibrium value $< n >$ versus C_sU/e is shown in the non-superconducting (b) and superconducting (d, f) states, at $T = 0$. The Andreev process is shown in f) by a vertical dashed line to distinguish it from the single electron tunneling process shown in b) and d) by a vertical continuous line.

The effect of the odd-even difference is simply to reduce the span of U over which the system will adopt an odd n ground state and, conversely to increase the span of U where an even n state will be favored. We thus get an asymmetric staircase which again has e-steps but which is $2e$-periodic (see Fig. 9d). Finally, in Fig. 9e we show the case of a superconducting island such that $\tilde{\Delta} > E_c$. In that case, for every value of U, the ground state of the circuit always correspond to an even n, which explains the doubling in Fig. 9f of the height and length of the steps with respect to Fig. 9b.

These theoretical predictions can be extended at temperatures T such that $k_B T \ll E_c$, provided one replaces the odd-even energy difference $\tilde{\Delta}$ by the odd-even <u>free</u> energy difference $\tilde{\Delta}(T) = \tilde{\Delta} - k_B T \ln \mathcal{N} + \mathcal{O}(T^2)$ [53,54], where $\mathcal{N} \sim 10^4$, the total number of electron states in the island participating in the superconductivity, is a measure of the degeneracy of the odd ground state with one unpaired electron. In Fig. 10 we show our experimental results for a superconducting aluminium island at $T = 28$ mK. In this experiment we vary $\tilde{\Delta}$ by means of a magnetic field applied to the sample. The evolution of the staircase as the field is varied provides a complete confirmation of the predictions of Fig. 9.

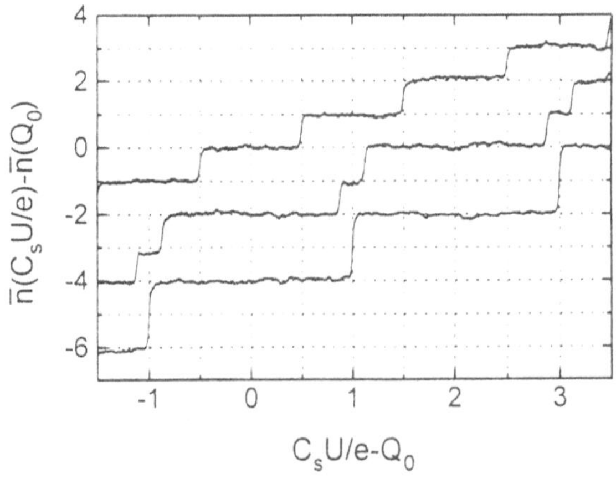

Fig. 10 Variations of the average charge of the island, in units of e, with the polarization $C_s U/e$, at $T = 28$ mK, for 3 values of the magnetic field H applied to the sample. For the top trace ($H = 0.2$ T) the island is non-superconducting. For the middle ($H = 0.05$ T) and bottom ($H = 0$) traces the island is superconducting. For clarity, the middle and bottom traces have been offset vertically by 2 and 4 units, respectively. The symbol Q_0 refers to the offset charge on the island, an uncontrolled parameter which drifts on the time scale of hours.

A remarkable point which is not completely understood is why the odd-even symmetry breaking which manifests itself in trace b) and c) of Fig. 10 corresponds to a $\tilde{\Delta}(T \to 0) \simeq \Delta_{BCS}$. The superconducting islands are far from ideal: they contain many defects like impurities, grain boundaries and surface states. Apparently, none of these defects provides available states near the Fermi energy for an unpaired electron, thus ruining the ideal BCS behavior. However, at the time of this writing, the superconducting box experiment has been unsuccessful when both sides of the junctions are superconducting. A possible explanation is that in the expression for the odd-even free energy difference $\tilde{\Delta}(T)$, the temperature that enters is the temperature of the quasiparticles, not the phonon temperature. In an all-superconducting circuit at low temperature, quasiparticles can have a very long lifetime and thus, it is possible that the remaining out-of-equilibrium quasiparticles completely suppress the odd-even asymmetry. Having normal metal on one side of the junction provides an efficient way of relaxing out-of-equilibrium quasiparticles, since they can diffuse into the normal metal but not out.

It is also important to note that when both sides of the junction are superconducting, the Andreev process of Fig. 9e, by which two electrons from the normal side tunnel coherently to form a pair in the island, is replaced by Josephson tunnelling. In contrast with single electron tunnelling or Andreev two-electron tunnelling which are irreversible processes characterized by a rate, Josephson tunnelling is a reversible process characterized by a macroscopic coupling energy $E_J \sim \Delta(R_K/R_t)$. When $k_B T \ll E_J$, the superconducting electron box polarized at $C_s U/e = 1$ should be in the macroscopic coherent superposition of charge states $| \, n = 0 > + \, | \, n = 2 >$ [55]. This has been indirectly observed in recent experiments in which the critical current of the superconducting SET has been measured as a function of gate voltage [56,57]. In particular, the experiment of Joyez et al. [57] demonstrates that the naive picture of single Cooper pair transfer in nanojunction circuits can only hold if the characteristic energies of the superconductors are set properly, i.e. $\tilde{\Delta}(T) > E_c \gg E_J \gg k_B T$.

FUTURE PROSPECTS

It has been suggested [27,28] that single-electron devices might find applications in digital electronics. A single electron would code for one bit, obviously the most economical way to store information. In fact, the electron pump is already very similar to the shift registers found in computers. The SET would be the building block of this "single electronics". A problem, however, is that metallic SETs made using today's technology have no "engineer gain": one transistor can barely feed one other transistor in the chain of signal processing, once the dispersions on parameters are accounted for. An no one understands how to get rid of random offset charges [28] which at present ruin any attempt to have more than a few transistors one one chip. In semiconductor devices, single-electron effects may even appear to be a nuisance because they imply that the electrons go through the dots or channels one at a time, a slowing down of the conventional FET operation. The main benefit of understanding single-electron effects in semiconductor nanotechnology may be just to provide the knowledge to fight them efficiently.

The real virtue of single-electron devices, as far as industrial applications are concerned, is that they teach us how to produce digital functions using only tunnelling and the Coulomb interaction, basic ingredients that are available down to the molecular level. In the as yet undeveloped "molecular electronics" technology [58], basic time constants are very short, there is no dispersion in the parameters of individual components and there are few electrons to work with anyway. There, the principles underlying the devices that have been discussed in this article may be fruitfully implemented. The ultimate computer imagined by Feynman [59], in which elementary information is carried by a single electron on a single atom, would then cease to be a mere theoretical construction and become a reality.

Acknowledgements

We thank P. Joyez, P. Lafarge, P.F. Orfila and H. Pothier, with whom the experimental results described in this article have been obtained, for helpful discussions and help with the figures. We are grateful to G.-L. Ingold for the preparation of the camera-ready manuscript. This work has been partly supported by the Bureau National de la Métrologie.

REFERENCES

[1] Eigler, D.M. and Schweizer, E.K., Nature, **344**, 524 (1990).

[2] Wineland, D.J., Itano, W.M. and Van Dyck, R.S. Jr., Adv. At. Mol. Phys. **19**, 135 (1983).

[3] Van Dyck R.S. Jr., Schwinberg, P.B. and Dehmelt, H.G., Phys. Rev. **D 34**, 722 (1986).

[4] Millikan, R.A., Phys. Rev. **32**, 349 (1911).

[5] Solymar, L., "Superconductive Tunneling" Chap. 2 (Chapman and Hall, London, 1972).

[6] Büttiker, M. and Landauer, R., Phys. Rev. Lett. **49**, 1739 (1982).

[7] Persomn, B.N.J. and Baratoff A., Phys. Rev. **B 38**, 9616 (1988).

[8] Lafarge, P., Pothier, H., Williams, E.R., Esteve, D., Urbina, C. and Devoret, M.H., Z. Phys. B **85**, 327 (1991).

[9] Matveev, K.A., Zh. Eksp. Teor. Fiz. **99**, 1598 (1991) [Sov. Phys. JETP **72**, 892, (1991)].

[10] Grabert, H., Physica B **194-196**, 1011 (1994).

[11] Gorter, C.J., Physica **17**, 777 (1951).

[12] Neugebauer, C.A. and Webb, M.B., J. Appl. Phys. **33,** 74 (1962).

[13] Giaver I., and Zeller, H.R., Phys. Rev. Lett. **20**, 1504 (1968).

[14] Lambe, J. and Jaklevic, R.C., Phys. Rev. Lett., **22**, 1371 (1969).

[15] Kulik, I.O. and Shekter, R.I., Zh. Eksp. Teor. Fiz. **68**, 623 (1975) [Sov. Phys. JETP **41**, 308 (1075)].

[16] Dolan, G.J. and Dunsmuir, J.H., Physica B **152**, 7 (1988).

[17] Fulton, T.A. and Dolan G.J., Phys. Rev. Lett. **59**, 109 (1987).

[18] Likharev, K.K. and Zorin, A.B. J. Low. Temp. Phys. **59**, 347 (1985)

[19] Averin, D.V. and Likharev, K.K., J. Low Temp. Phys. **62**, 345 (1986).

[20] Widom, A., Megaloudis, G., Clark, T.D., Prance, H. and Prance, R.J., J. Phys. A **15**, 3877 (1982).

[21] Ben-Jacob, E. and Gefen Y. Phys. Lett. A **108**, 289 (1985).

[22] Nazarov, Yu. V., Pis'ma Zh. Eksp. Teor. Fiz. **49**, 105 (1989) [JETP Lett. **49**, 126 (1990)].

[23] Devoret, M.H., Esteve, D., Grabert, H., Ingold, G.-L., Pothier, H. and Urbina, C. Phys. Rev. Lett. **64**, 1824 (1990).

[24] Girvin, S.M., Glazman, L.I., Jonson, M., Penn, D.R. and Stiles, M.D., Phys. Rev. Lett. **64**, 3183 (1990).

[25] Cleland, A.N., Schmidt, J.M. and John Clarke, Phys. Rev. Lett. **64**, 1565 (1990).

[26] Kuzmin, L.S., Nazarov, Yu. V., Haviland, D.B., Delsing, P., and T. Claeson, Phys. Rev. Lett. **67**, 1161 (1991).

[27] Likharev K.K., IBM J. Res. Dev. **32**, 144 (1988).

[28] Averin, D.V. and Likharev, K.K., in "Quantum Effects in Small Disordered Systems", ed. by Altshuler, B.L., Lee P.A. and Webb, R.A. (Elsevier, Amsterdam, 1991).

[29] Schön G. and Zaikin A.D., Phys. Rep. **198**, 237 (1990).

[30] "Single Charge Tunneling", ed. by Grabert, H. and Devoret, M.H. (Plenum, New York, 1992).

[31] "Single Electron Tunneling and Mesoscopic Devices", Proc. 4th Int. Conf. SQUID '91,

ed. by Koch, H. and Lübbig, H. (Springer-Verlag, Berlin, 1992).

[32] "The Physics of Few-Electron Nanostructures", Proc. of the Noordwijk NATO ARW, 1992, ed. by Geerligs, L.J., Harmans, C.J.P.M., and Kouwenhoven, L.P. (North-Holland, Amsterdam, 1993).

[33] Cleland, A.N., Esteve, D., Urbina, C. and Devoret, M.H., Appl. Phys. Lett. **61**, 2820 (1992).

[34] Fraser, D.A., "The Physics of Semiconductor Devices" (Clarendon, Oxford, 1986).

[35] Barone, A. and Paterno, G. "Physics and Applications of the Josephson Effect", (Wiley, New York, 1982).

[36] see Beenakker, C.W.J., "Single Charge Tunneling", Ch.5, ed. by Grabert, H. and Devoret, M.H., (Plenum, New York, 1992).

[37] Wilkins, R., Ben-Jacob, E. and Jaklevic, R.C. Phys. Rev. Lett. **63**, 801 (1989).

[38] Schönenberger, C., van Houten, H. and Beenakker, C.W.J., Physica B **189**, 218 (1993).

[39] Nejoh, H., Nature **353**, 640, (1991).

[40] Pothier, H., Lafarge, P., Urbina, C., Esteve, D. and Devoret, M.H. Physica B **169**, 573 (1991); Europhys. Lett. **17**, 259 (1992).

[41] Geerligs, L.J., Anderegg, V.F., Holweg, P.A.M., Mooij, J.E., Pothier, H., Esteve, D., Urbina, C. and Devoret, M.H., Phys. Rev. Lett. **64**, 2691 (1990).

[42] von Klitzing, K., Rev. Mod. Phys. **58**, 519 (1986).

[43] Williams, E.R., Gosh, R.N. and Martinis J.M., J. Res. Natl. Ins. Stand. and Technol. **97** (1992).

[44] Averin D.V and Odintsov A.A., Phys. Lett. **A 149**, 251 (1989); Averin D.V., Odintsov A.A. and Vyshenskii S.V., J. Appl. Phys. **73**, 1297 (1993).

[45] Jensen, H.D., and Martinis, J.M., Phys. Rev. **B46**, 13407 (1992).

[46] Pothier, H., Lafarge, P., Esteve, D., Urbina, C. and Devoret, M.H., IEEE Trans. Instr. and Meas., **42**, 324 (1993).

[47] Lafarge, P., Joyez, P., Pothier, H., Cleland, A., Holst, T., Esteve, D., Urbina, C. and Devoret, M.H., C. R. Acad. Sci. Paris **314**, 883 (1992).

[48] Martinis J.M., Nahum M. and Dalsgaard Jensen H., Phys. Rev. Lett. **72**, 904 (1994).

[49] Fulton, T.A., Gammel, P.L., Bishop, D.J. and Dunkleberger, L.N., Phys. Rev. Lett. **63**, 1307 (1989).

[50] Geerligs, L.J., Anderegg, V.F., Rommijn, J. and Mooij, J.E., Phys. Rev. Lett. **65**, 377 (1990).

[51] Geerligs, L.J., Verbrugh, S.M., Hadley, P., Mooij, J.E., Pothier, H., Lafarge, P., Urbina, C, Esteve, D., and Devoret, M.H., Z. Phys. B **85**, 349 (1991).

[52] Averin, D.V. and Nazarov, Yu. V., Phys. Rev. Lett. **69**, 1993 (1992).

[53] Tuominen, M.T., Hergenrother, J.M., Tighe, T.S., and Tinkham, M., Phys. Rev. Lett. **69**, 1997 (1992).

[54] Lafarge P., Joyez, P., Esteve, D., Urbina, C. and Devoret, M.H., Phys. Rev. Lett. **70**, 994 (1993)

[55] Lafarge P., Ph. D. thesis, Paris 6, (1993)

[56] Eiles, T.M., and Martinis, J.M., Phys. Rev. B **50**, 627 (1994).

[57] Joyez, P., Lafarge, P., Filipe, A., Esteve, D. and Devoret, M.H., Phys. Rev. Lett. **72**, 2458 (1994).

[58] Aviram, A. and Ratner, M., Chem. Phys. Lett. **29**, 27 (1974); "Molecular Electronic Devices" (ed. Carter, F.L.) (North-Holland, Amsterdam, 1991).

[59] Feynman, R.P., Optics News, 11 (1985).

INTERFEROMETRY WITH PARTICLES OF NON-ZERO REST MASS: TOPOLOGICAL EXPERIMENTS

Geoffrey I. Opat

School of Physics
The University of Melbourne
Parkville, Victoria 3052
Australia

These lectures deal with two main topics. Firstly, interferometry as a space-time process is described, together with its topology in space-time. Starting from this viewpoint, a convenient unified formalism for the phase shifts which arise in particle interferometry is developed. This formalism is based on a covariant form of Hamilton's action principle and Lagrange's equations of motion. It will be shown that this Lorentz invariant formalism yields a simple perturbation theoretic expression for the general phase shift that arises in matter-wave interferometry. The Lagrangian formalism is compared with the more usual formalism based on the wave propagation vector and frequency. The resulting formalism will be used to analyse the Sagnac affect, gravitational field measurements, and several Aharonov-Bohm-like topological phase shifts.

Secondly, several topological interferometric experiments using particles of non-zero rest mass are discussed. These experiments involve the use of electrons, neutrons and neutral atoms. Neutron experiments will be emphasised as interferometry with electrons and atoms will be discussed by other lecturers.

1. INTERFEROMETRY IN SPACE-TIME

In this paper we describe interferometry from a unified viewpoint, and present a unified fomalism for the calculation of phases for the matter waves of particles of non-zero rest mass. Such a discussion has become timely because, in addition to experiments with electrons and neutrons, experiments with neutral atoms are becoming increasingly important owing to the development of laser cooling.

In a typical interferometry experiment, radiation is emitted as waves from a source region, and travels via two (or more) paths to a region of superposition, where the radiation is detected.

Advances in Quantum Phenomena. Edited by E.G. Beltrametti
and J.-M. Lévy-Leblond, Plenum Press, New York, 1995

The behaviour of an interferometer may be understood by considering the pattern of the 3-dimensional surfaces of constant phase, Φ, in 4-dimensional space-time. Such surfaces are given by

$$\Phi(x^0,x^1,x^2,x^3) = \text{constant} \qquad (1)$$

In general, the Φ is not single-valued but is determined up to integral multiples of 2π. A point in space-time is labelled by the coordinates x^α, $\alpha =$ 0,1,2,3. (Usually x^0 will be reserved for the time coordinate.) It is important to realize that phase surfaces only exist in regions of non-vanishing wave function.

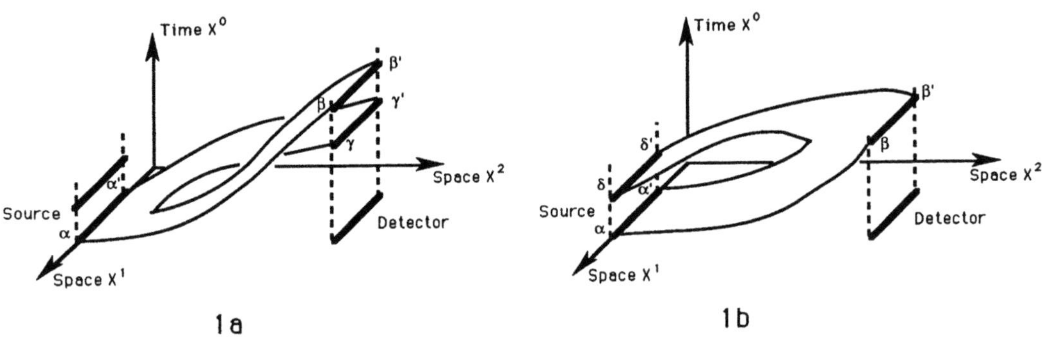

Fig. 1. Surfaces of constant phase. Fig. 1a. Shows the development forward in time of the phase surface emerging from the source. Fig. 1b. Shows the phase surface developing backward in time from the detector.

Some typical constant phase surfaces are shown in figure 1. Figure 1a depicts the phase surface of radiation emitted from the source region $\alpha\alpha'$, which divides and encircles a spatial region, arriving at the region of the detector, $\beta\beta'$ and $\gamma\gamma'$ at different times. Figure 1b depicts the constant phase surface obtained by retracing the waves backward from the detector to the source region, where it arrives at $\alpha\alpha'$ and $\delta\delta'$, again at two different times. Figure 1b emphasises among other things, that the source needs to have adequate phase coherence over a number of periods of oscillation for constructive interference to be possible. We also see that by joining the region of the phase surface $\beta\beta'$ on figures 1a and 1b a surface of a helical nature may be generated. Thus the total phase surface in an interferometer has a topologically interesting shape in space-time.

To calculate the phase shift, $\Delta\Phi$, which determines the degree of constructive or destructive interference in a given experimental configuration, we consider the phase changes $\Delta\Phi_I$ and $\Delta\Phi_{II}$ that occur along the two space-time paths I and II (see fig. 2a or more simply fig. 2b) which begin at the space-time point S on the source and end at the space-time point D on the detector. The paths I and II are chosen to be in the two regions of non-vanishing wave function which encircle the interferometer along different paths. By considering figures 1a, 1b, 2a and 2b together we see that the phase must change along at least one of the paths I, II to produce a phase shift. On figure 2b some constant phase surfaces are depicted.

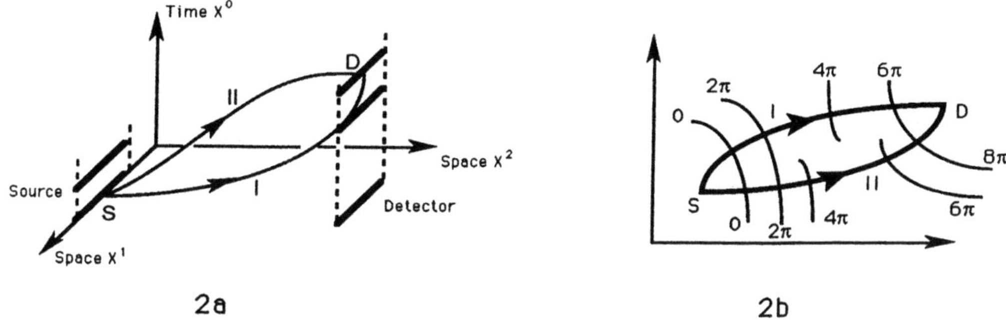

Fig 2.   Paths in space-time along which the phase shift is calculated. Fig. 2a should be viewed in conjunction with Figs. 1a and 1b. Fig. 2b is a more conventional diagram, which shows the paths cutting phase surfaces.

The phase shift is given by

$$\Delta\Phi = \Delta\Phi_I - \Delta\Phi_{II} \tag{2}$$

In fact, $\Delta\Phi$ is essentially a topological quantity, being the algebraic difference of the number of phase surfaces cut in going around the closed space-time loop, path I - path II. Figure 2b depicts this in a formalised way.

2 THE RAY APPROXIMATION

In most interferometric experiments the wavelength of the radiation is short in comparison to the dimensions of the apparatus. Under these conditions the propagation of the radiation is well approximated by the equations of ray optics. For the case of particles with mass we may use Hamilton's action principle to compute the phases. [1]

Consider the action functional S given by

$$S\{x^\alpha(\lambda); x_1^\alpha, x_2^\alpha\} \equiv \int_1^2 L(x^\alpha(\lambda), dx^\alpha(\lambda)/d\lambda)d\lambda \tag{3}$$

where the path $x^\alpha(\lambda)$ leads from $x_1^\alpha = x^\alpha(\lambda_1)$ to $x_2^\alpha = x^\alpha(\lambda_2)$. We have used a more general parameter λ along the path (instead of the usual time) with consequential modifications to the Lagrangian, L.

Hamilton's principle [1] states that the classical trajectories are those paths between x_1^α and x_2^α for which the functional $S\{x^\alpha(\lambda); x_1^\alpha, x_2^\alpha\}$ is stationary with respect to the path variations (see fig. 3a). In short, if

$$\delta S = 0 \tag{4}$$

for coterminous path variations then

$$\frac{\partial L}{\partial x^{\alpha}} - \frac{d}{d\lambda}\left\{\frac{\partial L}{\partial(dx^{\alpha}/d\lambda)}\right\} = 0 \tag{5}$$

a differential equation which determines the actual path of the system.

We now consider the function

$$S(x_2^{\alpha}, x_1^{\alpha}) = \int_P L(x^{\alpha}, dx^{\alpha}/d\lambda)d\lambda \tag{6}$$

The integration in equation 6 is over the actual path P of the system.

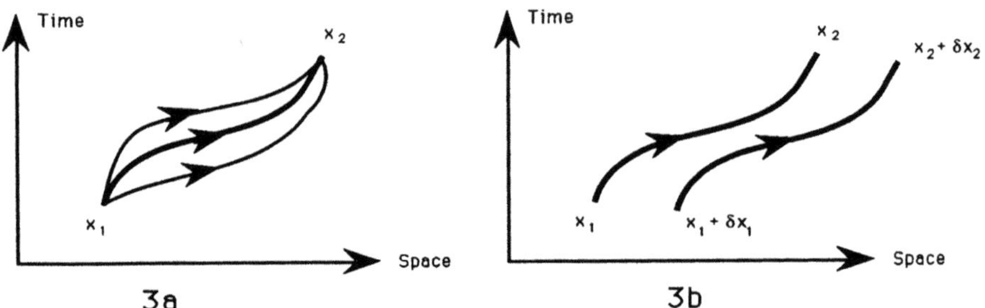

Fig. 3. Fig. 3a shows several adjacent paths leading from initial to the final space-time point. The solid curve is the path actually followed by the particle. Fig. 3b shows two actual paths which differ by end point variations.

Again, following standard theoretical arguments [1] it may be shown that end point variations (depicted in fig. 3b) result in

$$\delta S = -p_{\alpha}\delta x^{\alpha}\ \Big|_1^2 \tag{7}$$

where the p_{α} are the covariant components of the canonical 4-momentum. For the simple case of free propagation in Minkowski space with a metric tensor, $\eta = \text{diag}(+1, -1, -1, -1)$, we note that

$$p_{\alpha}\delta x^{\alpha} = E\delta t - \mathbf{p}\cdot\delta\mathbf{x} \tag{8a}$$

$$= \hbar(\omega\delta t - \mathbf{k}\cdot\delta\mathbf{x}) \tag{8b}$$

$$= -\hbar\delta\Phi \tag{8c}$$

The second line embodies deBroglie's hypothesis relating energy and momentum to frequency and wavelength. The third line identifies the expression with the phase, Φ.

In general, we see that S, the action function, defined by equation 6 is related to the quantum mechanical phase by

$$S = \hbar\Phi \tag{9}$$

This was the origin of Feynman's [2,3] approach to quantum mechanics.

Combining equations 2, 6, and 9 yields

$$\hbar\Delta\Phi = \oint_P L\,d\lambda \tag{10}$$

The path P is composed of two segments which are the actual paths which together encircle the interferometer.

We note that under canonical (gauge) transformations L may change according to

$$L \rightarrow L' = L + d\chi/d\lambda \ , \tag{11}$$

where χ is an arbitrary single valued function.

Thus we see that although the canonical momenta may change under canonical transformation, and with that change the phase differences along path segments, the phase shift does not, as the insertion of equation 11 into 10 will show. This result implies that the phase surfaces depicted in figs. 1a and 1b may be moved in space-time by canonical transformations but their basic topology is left unaltered.

We now consider how $\Delta\Phi$ changes with changes in the Lagrangian. Let ε be a (small) parameter.

$$L = L_0 + \varepsilon L_1 + \varepsilon^2 L_2 \tag{12}$$

$$\hbar\Delta\Phi = \oint_P L\,d\lambda = \oint_{P_0} L\,d\lambda + O(\varepsilon)^2 \tag{13}$$

where P_0 is the loop in space-time similarly composed of the two paths evaluated between the same points on the source and detector but calculated from the unperturbed Lagrangian L_0 (not L). The error, being of order ε^2, (not ε), is a consequence of equation 4 which ensures the stationarity of the action between adjacent coterminous paths. If we insert equation 12 into equation 13 we find

$$\hbar\Delta\Phi = \oint_{P_0} L_0\,d\lambda + \varepsilon \oint_{P_0} L_1\,d\lambda + O(\varepsilon^2)$$

$$\equiv \hbar(\Delta\Phi_0 + \varepsilon\Delta\Phi_1 + \varepsilon^2\Delta\Phi_2 \ldots) \tag{14}$$

We thus see that the first order change in phase shift due to the perturbation is given by

$$\hbar \Delta \Phi_1 = \oint_{P_0} L_1 d\lambda \tag{15}$$

i.e., the first order change in phase is due to the perturbation to the Lagrangian integrated over the orbits determined by the unperturbed Lagrangian. This is a principal result of the present paper, and a convenient starting point for calculations. I.e., equation 15 is a convenient starting point for the experiments analyzed below.

3. DISCUSSION

In the previous section we have deduced a way to calculate phase shifts based on the formalism of equation 15, i.e. the integration of the Lagrangian perturbation around the unperturbed closed space-time loop. In practice this is usually rather simple, as is illustrated in section 4 below.

Another commonly used method is to integrate equation 8. To use equation 8, the actual space-time trajectory must be determined, and $\omega, \mathbf{k}$ (or equivalently $E, \mathbf{p}$) evaluated on that trajectory. Such calculations can be quite difficult. Further, from time to time these equations are misused, in one of two ways.

(i)   In the case of a variable potential V (or refractive index) along the path, $E \equiv p_0^2/2m = p^2/2m + V$ is misidentified with the kinetic energy, (not the total energy), which then varies from point to point. The time differential dt is also equated to ds/v, where v is the particle speed, and ds is a length of path. This procedure is erroneous. Nevertheless, the integral $+\oint E dt$ so mis-evaluated happens to agree with $-\oint \mathbf{p} \cdot \mathbf{dx}$ to lowest order in V. Thus there is a feeling that somehow $\oint E dt$ and $-\oint \mathbf{p} \cdot \mathbf{dx}$ are equally responsible for the same phase shift, and either one may be calculated. On rare occasions, both are used resulting in an erroneous doubling of the calculated phase.

(ii) Sometimes only the 3-space expression $\oint \mathbf{p} \cdot \mathbf{dx}$ is used. In fact this is a correct procedure for all time independent processes as these have constant energy. For the particular case of a time independent process, for which E is constant

$$\oint E dt \equiv 0 \tag{16a}$$

and so

$$\Delta \Phi = \oint \mathbf{p} \cdot \mathbf{dx}/\hbar = \oint \mathbf{k} \cdot \mathbf{dx} \tag{16b}$$

Figure 4, depicts a Bonse-Hart single silicon crystal neutron interferometer [4,5].   Such an interferometer is hewn from a single perfect crystal of silicon, leaving three "ears" which through Bragg diffraction act as beam splitters. Figure 5 depicts the same interferometer in space-time. Equation 16b shows that for stationary processes in that interferometer the trajectories shown projected on to the XY plane suffice for the calculation of phases, as in common practice. For non time independent processes the $\oint E dt$ term must be used as well.

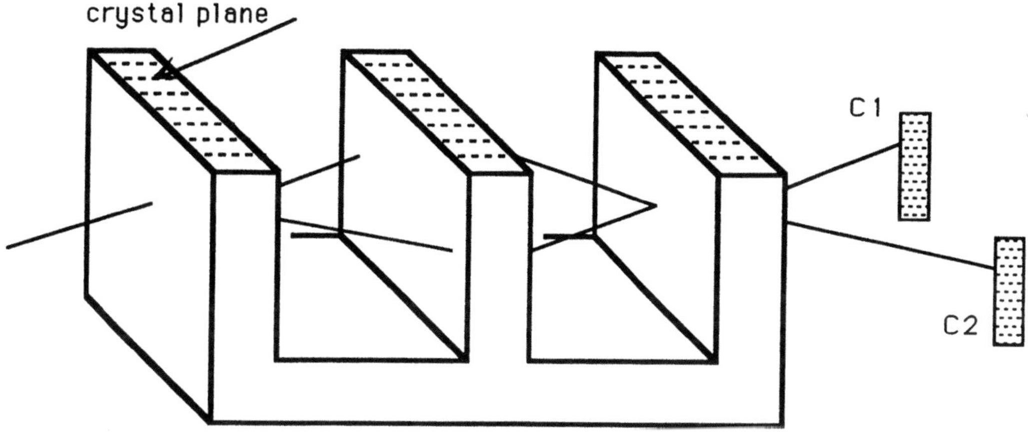

Fig. 4. A Bonse-Hart single crystal neutron interferometer. The three "ears" act as beam splitters by making use of Bragg diffraction from the crystal planes. Because the planes in each ear originally belonged to a single perfect crystal, they are very parallel from ear to ear.

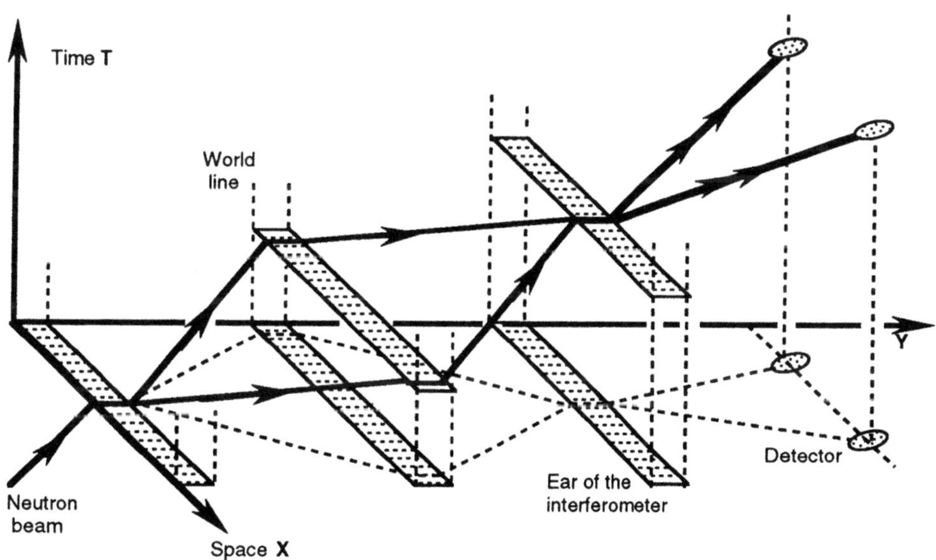

Fig. 5. A space-time representation of a Bonse-Hart neutron interferometer. For convenience, the Z-axis is not shown. The paths shown projected into the XY plane are those normally employed calculations of time stationary problems.

These last comments are important for the calculation of the Sagnac phase in an interferometer which is rotating with respect to an inertial frame. For an observer in the non-inertial frame, co-stationary with respect to the interferometer, the diffraction process is stationary in time, so that equation 16b may be used. For such an observer, the trajectories are not straight, and **p** is position dependent. For an inertial observer, the interferometer rotates, and the process is non-stationary. The particles do move in straight lines, but with different energies in different parts of the interferometer. The full expression must be used in this case.

4. SPECIFIC APPLICATIONS

(a) A Potential Hump or Refractive Block

Suppose that in arm I of an interferometer the particles travelling along the X-axis, encounter an additional potential hump of height V and spatial width D. The unperturbed trajectory through this region is given by

$$x = x_0 + v_0 t \; . \tag{17a}$$

Note that

$$L dt = L_0 dt - V dt \tag{17b}$$

We thus see that,

$$\hbar \Delta \Phi_1 = \int (-V) dt = \int -V dx/v_0$$
$$= -VD/v_0 \tag{17c}$$

Conventional methods would yield

$$\Delta \Phi_1 = (k - k_0)D = (n-1)k_0 D \tag{17d}$$

where n is the refractive index.

Comparison of 17c and 17d yields the usual result for weak potentials [5]

$$n - 1 = -V/2E \tag{17e}$$

(b) Static Gravitational Field (Newtonian)

A constant field in the negative Z-direction yields a potential $V = mgz$. If the interferometer path is initially a horizontal parallelogram of sidelength D and width W, which is tilted by an angle Θ about a line parallel to the side of length D thereby raising path I, the phase shift is given by a similar integration to that of section (a) to be

$$\Delta \Phi = -mgA \sin\Theta / \hbar v_0 \tag{18}$$

where $A = DW$ is the area of the interferometer loop. This result agrees with the work of Colella et al. [6] and Staudenmann et al. [7].

(c) Electric and Magnetic Perturbations; Aharonov-Bohm Effects

If a particle of charge q is placed in an electromagnetic field the Lagrangian is changed by the addition of terms dependent on the scalar potential ϕ and the vector potential $\mathbf{A}$

$$L_1 dt = -q(\phi - \mathbf{A} \cdot \mathbf{v}/c)dt$$

$$= -q(\phi c dt - \mathbf{A} \cdot d\mathbf{x})/c$$

$$= -q A_\alpha dx^\alpha/c \ . \tag{19}$$

The effects of electric and magnetic fields on the motion of the particle are included. Note that under gauge transformation

$$A_\alpha \rightarrow A_\alpha + \partial_\alpha \chi \tag{20}$$

which results in a change to the Lagrangian of $-qd\chi/c$, a total differential and hence of no physical consequence. For a pure magnetic field the magnetic flux contained in a closed loop is given by
$F = \oint \mathbf{A} \cdot d\mathbf{x}$, as Stokes' theorem shows. Applying this to the motion of a charged particle yields the "vector" or magnetic Aharonov-Bohm result [8],

$$\Delta\Phi_1 = qF/\hbar c \ . \tag{21}$$

As pointed out by Aharonov and Bohm this result is surprisingly valid, even when the magnetic field is zero along the orbit of the particle.

The phase shift due to pulsing the electric potential along one path of an interferometer, in such a way as to produce no electric field at the particle, yields the "scalar" or electric Aharonov-Bohm [8] result,

$$\Delta\Phi_1 = -q \int \phi dt/\hbar \ . \tag{22}$$

(d) Metric Perturbations

The space-time metric tensor $g_{\alpha\beta}$ depends on the gravitational field, and the frame of reference. The Lagrangian for a particle moving under gravity and inertia alone is given by [9]

$$L d\lambda = -mc\sqrt{g_{\alpha\beta}(dx^\alpha/d\lambda)(dx^\beta/d\lambda)} \ d\lambda \tag{23}$$

The independence on the choice of the parameter λ is manifest. Further, if λ is chosen as the time, the non-relativistic weak static field limit of equation (23) becomes

$$L d\lambda = \left(-mc^2 + \frac{1}{2} mv^2 - m\phi\right) dt \tag{24}$$

as expected.

If we write

$$g_{\alpha\beta} = g^0_{\alpha\beta} + h_{\alpha\beta} \tag{25}$$

we find

$$L_1 d\lambda = -\frac{1}{2} mc\, h_{\alpha\beta}(dx^\alpha/d\tau)(dx^\beta/d\tau)d\tau \tag{26}$$

where τ/c is the proper time along the unperturbed orbit.

For slow particles in a weak gravitational field in a frame which rotates with angular velocity Ω with respect to an inertial frame (such as the laboratory frame on earth, the metric may be approximated by

$$(ds)^2 = \left(1+[2\phi-(\omega\times\mathbf{r})^2]/c^2\right)(cdt)^2 - (d\mathbf{r})^2 - 2\boldsymbol{\omega}\times\mathbf{r}\cdot d\mathbf{r}dt \tag{27}$$

we find therefore, that

$$L_1 d\lambda = -m\left\{(\phi - \frac{1}{2}(\omega\times\mathbf{r})^2)dt - \boldsymbol{\omega}\cdot\mathbf{r}\times d\mathbf{r}\right\} \tag{28}$$

The first term contains the effective potential

$$\phi_{\text{eff}} = \phi - \frac{1}{2}(\boldsymbol{\omega}\times\mathbf{r})^2 \tag{29}$$

which allows for the centrifugal reduction of true gravity. This first term may be treated as in Eq. 17c and yields Eq. 18, with g changed to the as-measured effective gravitational acceleration.

The terms linear in Ω in equations 26 and 27 give rise to the Coriolis force and Foucault pendulum precession when applied to the classical mechanics of point particles, and to the Sagnac effect in wave interferometry. Thus, the Foucault pendulum is the classical manifestation of the Sagnac phase shift.

If we integrate the terms linear in Ω of Eq. 27 around a closed loop using the expression for the vector area of the loop as **S.**

$$2\mathbf{S} = \oint \mathbf{r}\times d\mathbf{r} \tag{30}$$

we find

$$\Delta\Phi_{\text{Sagnac}} = 2m\boldsymbol{\omega}\cdot\mathbf{S}/\hbar \tag{31}$$

This expression agrees with the Sagnac phase shifts calculated by many others [10, 11, 12, 13]. There are two approaches to deriving the expression for the Sagnac effect. One (as at present) is based on using the physical methods appropriate to a laboratory based non-inertial frame, in which case the physical process is stationary in time. The other is based on the use of an inertial observer in space, for whom the entire process is not stationary in time. For such an observer, the world lines of the collimator, monochromator, interferometer and detector are all helical in space-time. Further, each optical element Doppler shifts or changes ω and $\mathbf{k}$ in some way. Also, both the time and space parts of the expression 8b contribute to the phase. Both methods have been used to calculate the Sagnac phase. Hybridizing the two approaches is fraught with pitfalls.

(e) The Inclusion of Spin

Spin effects may be included by adding a term

$$L_s d\lambda \; = \; \tfrac{1}{2}\gamma \hbar s_{\alpha\beta} F^{\alpha\beta} dt \; = \; \gamma \hbar \mathbf{s} \bullet (\mathbf{B} - \mathbf{v}/c \times \mathbf{E}) dt \tag{32}$$

where γ is the gyromagnetic ratio and $F^{\alpha\beta}$ is the antisymmetric electromagnetic field tensor, and $s_{\alpha\beta}$ is the antisymmetric spin tensor. The spin tensor obeys the constraint

$$s_{\alpha\beta} dx^\beta / d\lambda = 0 \tag{33}$$

This last equation leads to the second line of equation 32.

(i)   If the vector $(\mathbf{B} - \mathbf{v}/c \times \mathbf{E})$ is fixed in direction, then using this direction as the quantisation axis yields separate equations for the spin "up" and "down" states. Note that $(\mathbf{B} - \mathbf{v}/c \times \mathbf{E})$ may be time (or space) dependent, however its direction must not change.

The extra Lagrangian is simply the negative of an extra potential.

$$L_\alpha d\lambda \; = \; \gamma \hbar s \, | \mathbf{B} - \mathbf{v}/c \times \mathbf{E} | \, dt \tag{34}$$

where $s = \pm \tfrac{1}{2}$.

Integration of equation 34 yields two different phases. The magnetic term

$$\Delta\Phi_B \; = \; \frac{s\gamma}{v_0} \oint B ds \; , \tag{35}$$

is the additional phase shift due to the presence of a magnetic field which creates an additional potential for the particle. Such a term is completely analogous to the ordinary electric potential acting on a charged particle. With a suitable time structure added to B it can yield results analogous to the electric Aharonov-Bohm effect [14, 15, 16, 17, 18].

The electric term for the case of a conducting charged rod about which the particles circulate, yields a phase shift

$$\Delta\Phi_E = s\gamma\,\Lambda/c \tag{36}$$

where Λ is the charge per unit length on the rod. This term has been discussed by Aharonov and Casher and Anandan [19, 20] and demonstrated experimentally by Cimmino et al [21].

(ii)   If the vector $(\mathbf{B} - \mathbf{v}/c \times \mathbf{E})$ changes direction, then the interferometer output intrinsically depends on two phases, whose values are coupled by the equations of motion. Provided the field direction changes slowly when compared with the phase, we may write down the separate equations for the phases.

5. GENERAL INTERFEROMETRY - A SUMMARY

The calculation of phase shift perturbations in interferometry may be based on the integration of equation 8 or, alternatively, integration of equation 15. To use equation 8, the actual ray paths must be found, and in general the four components of the propagation 4-vector
(ω, k^x, k^y, k^z) computed along the paths. For many situations this is arduous work, although for time stationary interferometry the frequency terms may be ignored.

By returning to the foundations of classical and quantum mechanics, we arrive at equation 15, the main result of this paper. To use this equation to compute the phase shift, the perturbation to the Lagrangian (a scalar) need only be integrated over the unperturbed paths. The utility and simplicity of this Lagrangian approach was illustrated in section 4 above, in which the results of many papers were calculated in a few lines.

Although the formalism is covariant, the applications have been mainly to non-relativistic problems. Relativistic problems, often based on Hamiltonians have been considered in the literature, by Anandan [20], Fabri and Picasso [22] and Cai and Papini [23].

A general formulation of interferometry for the multicomponent deBroglie waves of particles with several internal states (e.g. spin) is yet to be considered. In spite of the absence of such a general theory, a significant number of situations may be treated using the above formalism because the internal state remains unchanged throughout the motion, e.g. as in the Aharanov-Casher effect [19].

6. INTERFEROMETRIC EXPERIMENTS

We now survey and discuss several interferometric experiments. As has been pointed out in section 1, in some sense all interferometric experiments are topological in space-time. Nevertheless the word "topological" will be reserved for experiments in which the phase shift does not depend on the detailed shape of the path. The Aharonov-Bohm [8] and Aharonov-Casher [19] effects are of this type.

As there are now many interferometric experiments with electrons, neutrons, and more recently neutral atoms, it is not possible to do justice to all of them. I shall mainly discuss the experiments in which I was involved, not because they are necessarily the most important, but simply because I know them the best.

(a) Spinor Rotations

The rotation matrix for the internal state of a spin 1/2 particle is given by

$$d(\mathbf{a}) = \mathbf{1}\cos(a/2) - i\sin(a/2)\hat{\mathbf{a}} \cdot \sigma \tag{37}$$

The rotation is specified by a vector $\mathbf{a}$; $a = |\mathbf{a}|$ = angle of rotation, and $\hat{\mathbf{a}}$ is the axis of rotation. In normal geometry, a full rotation which has $a = n2\pi$, n = integer, is an identity operation, as all objects are restored to their former disposition by the rotation. The rotation matrix

$$d(n2\pi\hat{\mathbf{a}}) = (-1)^n\mathbf{1} \tag{38}$$

so that for n odd, a phase factor -1 occurs, i.e. the spinor wave-function Ψ changes sign

$$\Psi \rightarrow \Psi_r = (-1)^n\Psi \tag{39}$$

As observables are quadratic in the wave function it was felt that full rotations would not be observable in any way. However, Bernstein [24] and Aharonov and Susskind [25] argued that this was not necessarily the case under all circumstances. In fact the observation of the spinorial phase factor -1 was the subject of experiments by three groups, Rauch et al. [26] and Werner et al [27] using Bragg diffraction in a single crystal interferometer, and by Klein and Opat [28] using Fresnel diffraction of neutrons through a slit.

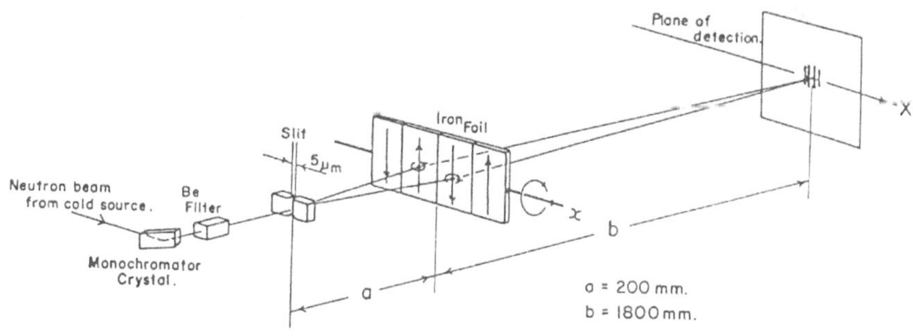

Fig. 6. Schematic layout of the apparatus (from ref. 28).

Figure 6 shows a schematic layout of the apparatus. Neutrons which are given a high degree of transverse coherence by the slit, diffract through an iron foil containing large ferromagnetic domains and long straight domain boundaries. A neutron passing to the left and right of the domain boundary suffers equal but opposite precessions of its spin, irrespective of the original spin direction. Thus in two separate regions of the wavefunction of the same neutron, the neutron suffers different precessions, a totally non-classical situation. Downstream these two parts of the wave function interfere, making manifest any relative phase shifts. The angle of rotation is determined by the saturation field of iron and time spent in the foil. The time spent in the foil may be changed by changing the thickness of the foil traversed by the neutron. This is done by rotating the foil about axis x.

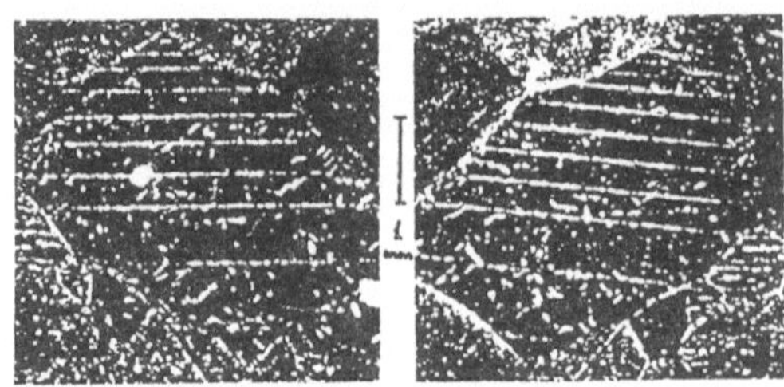

Fig. 7. Bitter patterns on front and back surfaces of cubic-textured Fe-3% Si foil. Straight domain walls several millimeters in length are seen to intersect both faces of the (100) crystal (From ref.28).

The domains in the foil are shown in Fig. 7. The diffraction pattern on one screen may be regarded as the superposition of two straight edge patterns. The overlap of the two patterns in the "forbidden" region may be constructive or destructive depending on the relative rotation of the two halves of the wave-function. The results shown in fig.8, confirm the negative sign on odd 2π rotation.

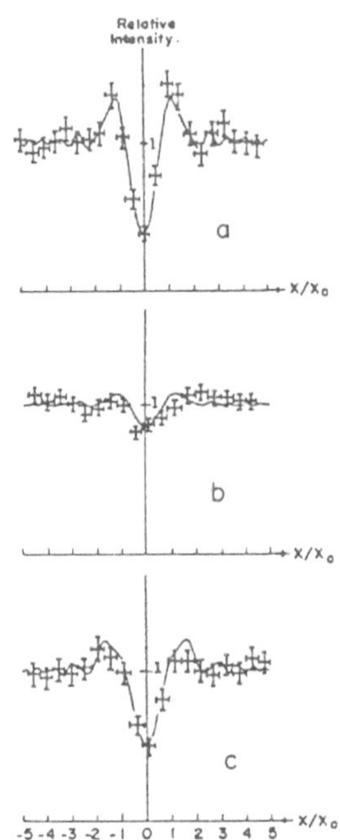

Fig.8. Fresnel-diffraction patterns showing interference of neutrons whose spins were precessed by relative angles of (a) $9.2\times2\pi$ rad, and (b) $10.2\times2\pi$ rad, and (c) $11.2\times2\pi$ rad. (From ref.28).

(b) The Aharonov-Bohm Effects with Charged Particles.

Aharonov and Bohm [8] pointed out that the diffraction pattern of charged particles could be changed by the presence of potentials under circumstances in which the charged particle never experiences a force, having never encountered an electric or magnetic field. This is surprising from a classical standpoint. There are two Aharonov-Bohm effects, one magnetic or vector, and another, electric or scalar. Peshkin and Tonomura [29] have extensively reviewed these effects.

The magnetic Aharonov-Bohm effect is depicted in figure 9a. The wave of a charged particle divides and encircles a tube of magnetic flux, without actually encountering the flux directly. Of necessity a non-vanishing vector potential must exist outside the region of flux, a region in which the electron moves. A phase shift given by eq. 21 is predicted.

The experiments of Chambers [30], Möllenstedt and Bayh [31], and more recently Tonomura and co-workers [32] have verified the prediction of eq. 21 in detail. The experiments of Tonomura are particularly elegant.

The electric Aharonov-Bohm effect is depicted in figure 11a. Whilst the charged particle is totally within the metal flight tubes, the potential of at least one of the tubes is changed. The charged particle has had its electric potential changed without ever having encountered a force-producing electric field. The phase shift given by eq. 22 is predicted. Technical difficulties have prevented a full realisation of this experiment, however, Mateucci and Pozzi [33] have carried out a related experiment.

(c) The Aharonov-Casher Experiment

Aharonov and Casher [19] considered a situation dual to the magnetic Aharonov-Bohm experiment. This may be understood in terms of fig. 9. The tube of magnetic flux of fig. 9a is equivalent to a line of magnetic dipoles, as Ampères circuital law shows. Thus the magnetic Aharonov-Bohm effect is equivalent to the diffraction of a charged particle around a line of magnetic dipoles as is shown in fig. 9b. The Aharonov-Casher effect is the diffraction of a magnetic dipole around a line of charges, as is shown in fig. 9c. The predicted phase shift is given by eq. 36. This phase shift depends on the charge per unit length on the charged "cylinder". It does not depend on the detailed path around that cylinder or the velocity of the particle.

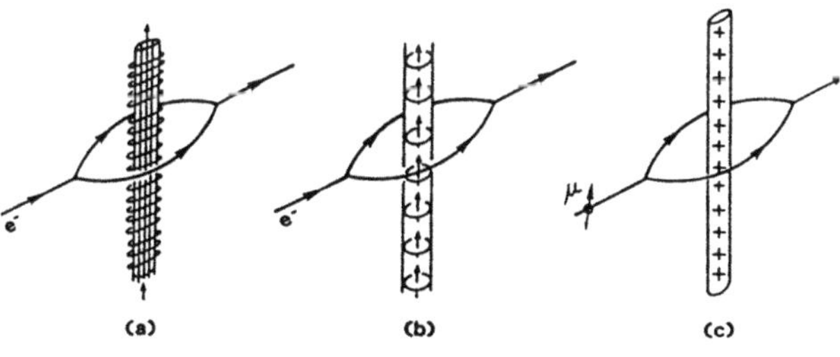

Fig. 9. Schematic diagram of (a) the magnetic Aharonov-Bohm experiment, (b) equivalent Aharonov-Bohm experiment with a line of magnetic dipoles, (c) the duality between the Aharonov-Bohm and Aharonov-Casher topology. (From ref.34)

The Aharonov-Casher experiment was carried out by a University of Melbourne-University of Missouri(Columbia) group [34] in which neutrons in a Bonse-Hart interferometer circulated around a charged triangular cylinder. The apparatus is shown in figure 10.

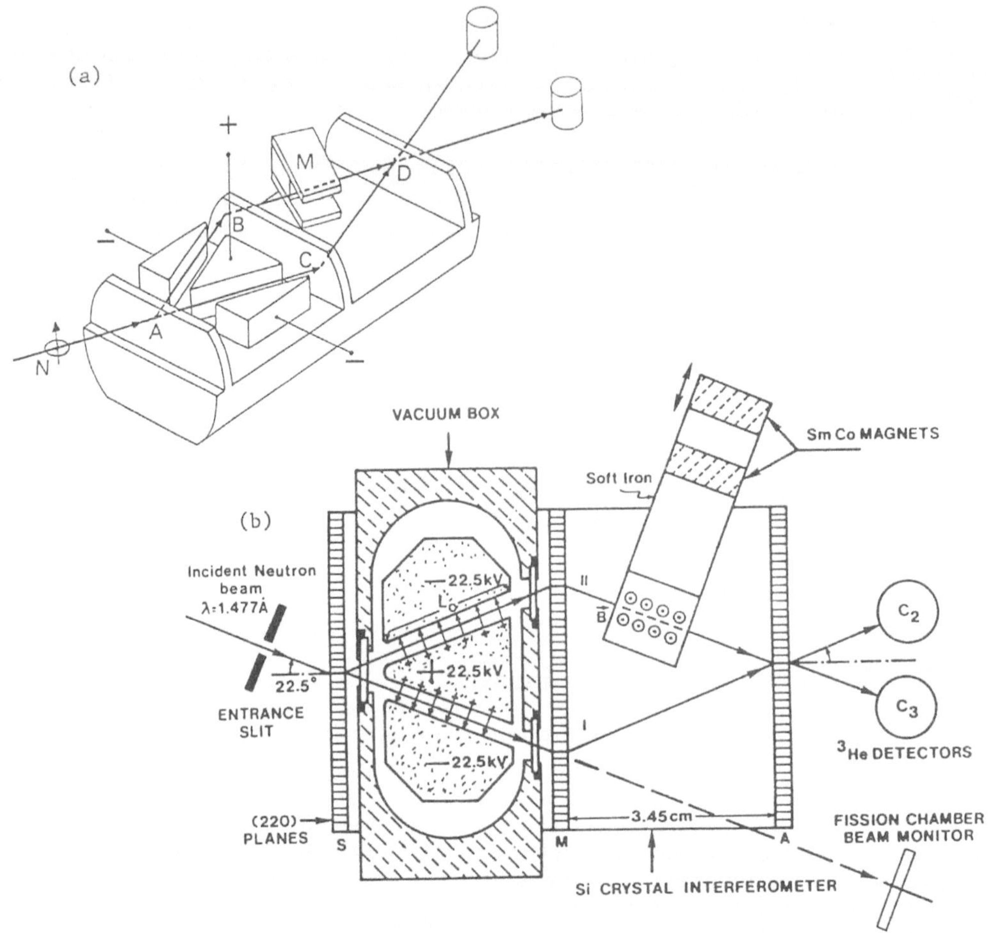

Fig. 10. The apparatus is shown in perspective in (a) and in plan view in (b). The plan view depicts the Bonse-Hart perfect-Si-crystal interferometer. An unpolarised neutron beam (λ = 1.477 Å) is used. The neutron on path II passes through an electric field **E** and then through a vertical magnetic bias field **B**. The neutron on path I passes on the opposite side of the electrode. The interferometer as a whole could be tilted about the incident beam direction to adjust the gravitational phase shift $\Delta\Phi_g$. The ^{3}He proportional detectors C_2 and C_3 count the exit beams. (From ref. 34)

The number of neutrons of entering counters 2 and 3, is given by

$$N_2 = N_o(a_2 + b_2 \cos\Delta\Phi) \tag{40a}$$

$$N_3 = N_o(a_3 + b_3 \cos\Delta\Phi) \tag{40b}$$

N_o is the number of neutrons entering the apparatus, and a_2 a_3 b_2 b_3 are constants of the interferometer. We find experimentally, $b_2 = -b_3$ as is required by the conservation of neutrons. The phase shift $\Delta\Phi$ is the sum of four terms

$$\Delta\Phi = \Delta\Phi_n + \Delta\Phi_g + \sigma(\Delta\Phi_m + \Delta\Phi_e) \qquad (41)$$

$\Delta\Phi_n$ is the offest phase of the instrument of nuclear origin.

$\Delta\Phi_g$ is a gravitational phase shift which is changed by tilting the interferometer (of order $1°$) about the incident neutron direction. (See Colella et al. [33]).

$\Delta\Phi_m$ is a magnetic phase shift produced by the magnets

$\Delta\Phi_e$ is the   (electric) phase i.e. The Aharonov-Casher phase to be determined.

As the magnetic and electric phase shifts depend on whether the spin is "up" or "down" with respect to the normal to the plane of the interferometer a factor, $\sigma = \pm 1$ is made explicit.

The experiment does not require that polarised neutrons be used provided the phases are correctly chosen. Averaging eq. 40b over neutron spins, yields the equation,

$$N_3 = N_0(a_3 + b_3 \cos(\Delta\Phi_n + \Delta\Phi_g)\cos(\Delta\Phi_m + \Delta\Phi_e)) \qquad (42)$$

with a similar expression for N_2. The apparatus is tilted to change $\Delta\Phi_g$ until $\cos(\Delta\Phi_n + \Delta\Phi_g) = 1$. The magnetic field is adjusted to change $\Delta\Phi_m$ to ensure that $\cos(\Delta\Phi_m + \Delta\Phi_e) = \sin\Delta\Phi_e \approx \Delta\Phi_e$.

As a result of several months of running, the detection of 2.5×10^7 neutrons, and many electric polarity reversals, we obtained the experimental value

$$\Delta\Phi_e \equiv \Delta\Phi_{AC} = 2.19 \pm 0.52 \text{ millirad.}$$

This value is 1.46 ± 0.35 times the theoretical prediction. Given the neutron counting statistics of this experiment the agreement is satisfactory.

Recently another Aharonov-Casher experiment was carried out by Sangster et al [36] using a beam of TlF molecules. The agreement with theory in this experiment is at the 4% level. In addition, the phase shift was shown to be independent of the molecular speed and proportional to the applied electric field strength. This experiment does not have the conventional topology [see however Casella [36]], and uses the Ramsey method to obtain high phase precision.

Another observation of the Aharonov-Casher effect of a very different kind was made by Elion et al [37] using vortices in superconducting networks. This work followed a theoretical work of Reznick and Aharonov [38] and Wees [39].

(d)   The Analog Scalar Aharonov-Bohm Effect with Neutrons.

As has been pointed out above, the electric (scalar) Aharonov-Bohm has not been performed as it is technically too difficult. However, it had been pointed out by Zeilinger [40] and Anandan [41, 42] that instead of using charged particles in electric potentials (as in fig. 11a), one could use instead magnetic dipole bearing neutral particles (neutrons) in a magnetic field (as in fig. 11b).

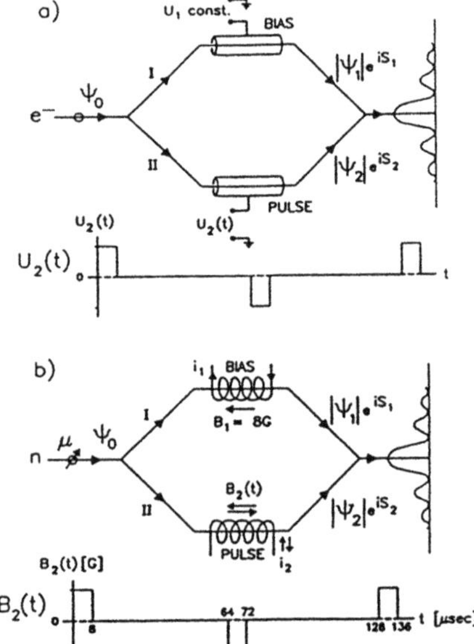

Fig.11. Schematic diagram of (a) the scalar Aharonov-Bohm experiment for electrons and (b) the scalar Aharonov-Bohm experiment with neutrons. The wave forms of applied pulses are also shown. (From ref.44)

Let us suppose that a neutron polarised along the axis of the solenoid enters the solenoid with the field off. When it is fully inside the solenoid, a current pulse produces a spatially uniform magnetic field which disappears before the neutron emerges from the far end solenoid. As the neutron only ever encounters an uniform magnetic field, no force is applied to it, and its momentum is constant. Its potential energy is changed during the magnetic pulse by

$$\Delta E = - \sigma\mu B_2 \qquad [43]$$

where μ is the neutron's magnetic moment, B is the magnetic field, and $\sigma = \pm 1$ depending on whether the neutron spin is quantised along or against the magnetic field $\mathbf{B_2}$. This energy change results in a phase shift

$$\Delta\Phi_{AB} = (\sigma\mu/\hbar)\int B_2 dt \qquad [44]$$

As expected such a phase shift is readily detectable by neutron interferometry.

The experiment was carried out by a University of Melbourne-University of Missouri(Columbia) collaboration (Allman et al. [43,44]). A skew-symmetric Si single crystal interferometer was used as it provided more working space. (See Fig. 12). The interferometer was surrounded by four bar magnets. These provided a uniform background magnetic field over the apparatus in the manner of a Helmholtz coil, except that magnets, unlike Helmholtz coils, do not heat the apparatus, (thanks to quantum mechanics). This background field was an order of magnitude greater than the Earth's field. Two points of technique are worthy of mention:

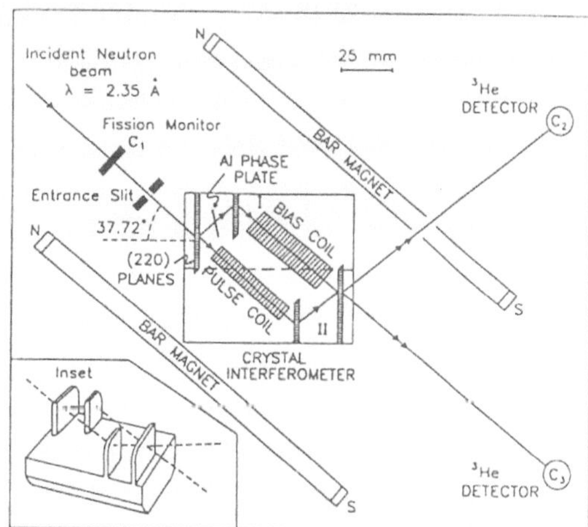

Fig.12. Layout of AB experiment using a skew-symmetric single-Si-crystal neutron interferometer. Inset: An isometric view of the interferometer crystal. (From ref.44)

(i)   Instead of chopping the neutron beam, the magnetic field was pulsed periodically by a "clock". Each neutron had its time of detection recorded. From the known velocity, the range of times for which that neutron was within the central portion of the solenoid could be found, and from that time information, where it was in the magnetic pulse cycle could be determined. Alternate magnetic pulses had opposite polarity, so in reality two interleaved experiments were done together. Fig. 13 shows the neutron counts in detectors 2 and 3 for a particular given magnetic field pulse height. It will be noted that the plateau near 12 µs and 76 µs correspond to neutrons within the uniform field central section of the solenoid at the time of the pulse. From the dependence of these plateau heights on magnetic field pulse heights fig. 14 may be constructed. (See eq. 47 below).

(ii)   By adjusting the magnetic field in auxiliary bias coil it is possible to carry out this experiment with unpolarised neutrons. Again the overall phase shift $\Delta\Phi$ is the sum of four terms

$$\Delta\Phi = \Delta\Phi_n - \Delta\Phi_{A\ell} + \sigma(\Delta\Phi_{AB} - \Delta\Phi_m) \tag{45}$$

$\Delta\Phi_n$ is an instrumental offset phase

$\Delta\Phi_{A\ell}$ is a phase due to the adjustable aluminium phase plate, which
      refracts neutrons.

$\Delta\Phi_{AB}$ is the Aharonov-Bohm phase (eq.44) and

$\Delta\Phi_m$ is the phase which is adjustable by the field in the bias coil.

The counts in detectors 2 and 3 are given by expressions similar to eq. 40. Averaging over the neutron spin yields for detector 3.

The aluminium phase plate was adjusted to ensure $\cos(\Delta\Phi_n - \Delta\Phi_{A\ell}) = 1$, and the bias magnetic field to ensure $\cos(\Delta\Phi_{AB} - \Delta\Phi_m) = \pm \sin\Delta\Phi_{AB}$. We find

$$N_3 = N_0(a_3 \pm b_3 \sin\Delta\Phi_{AB}) \tag{47}$$

The form and magnitude of the graphs in fig. 14 agree with this expression, thereby verifying the analog scalar Aharonov-Bohm effect.

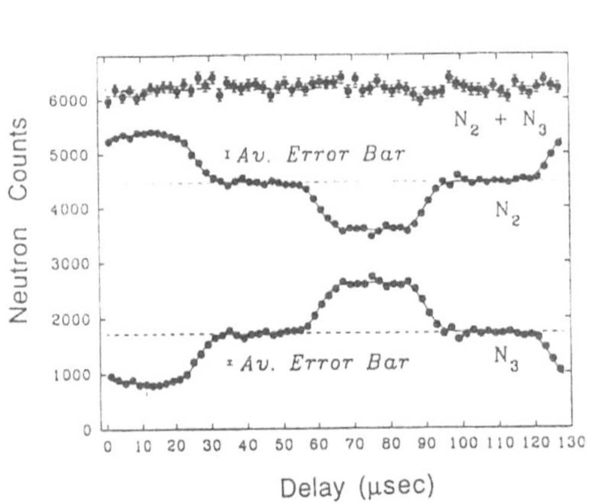

Fig. 13. Scan of counts per channel in the two detectors N_2 , N_3 , and their sum, plotted against delay time, for the particular case of pulsed field amplitudes of ±19G, which corresponds to $\Delta\Phi_{AB} = 1$ rad. The solid lines are theoretical fits, as described in the text. The total data collection time for this scan was about 10 h. (From ref. 44).

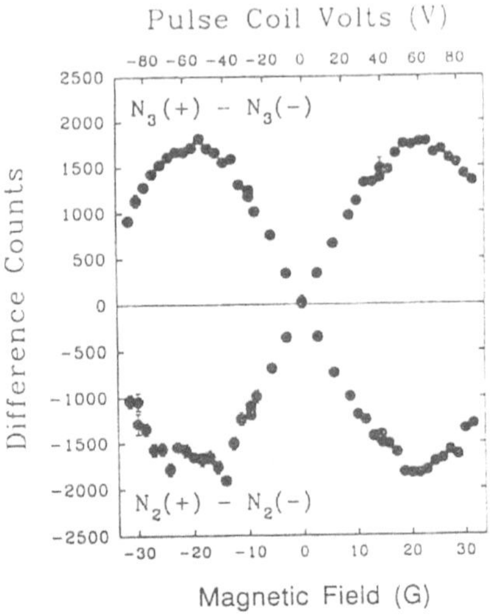

Fig. 14. Interferometer output signals as a function of pulse coil field strengths. This obtained from the average of the central four points in each plateau region of data sets such as the one shown in fig. 13. (Certain kinds of systematic error are avoided by taking the difference between the counts for positive and negative pulses.) (From ref. 44)

7. DISCUSSION

There continues to be analysis of Aharonov-Bohm like effects in the literature. Some of this discussion is semantic, such as an inadequate definition of energy or momentum. It seems that the essence of the matter is that a true Aharonov-Bohm effect should be force free and have a topological aspect. The words force-free are often taken to mean non-dispersive. On this last point Badurek et al. [45] have carried out an elegant experiment. They show explicitly that if neutrons enter, traverse, and leave a region of static magnetic field, that forces are applied to the neutron on entering and leaving the region through the magnetic gradient. The sense of these forces is opposite for spin-up and spin-down. This results in the spin-up wave-packet being shifted in space with respect of the spin down wave packet. If the coherence length is very short, all coherence between the spin-up and spin-down cases is lost. (See fig. 15a). On the other hand, if a uniform magnetic field rises and falls within a region already occupied

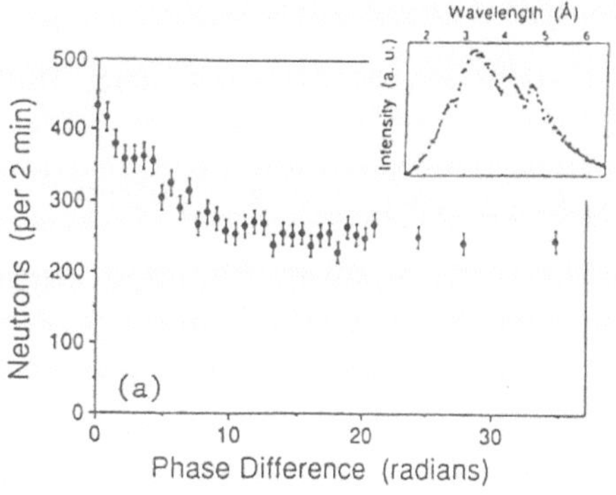
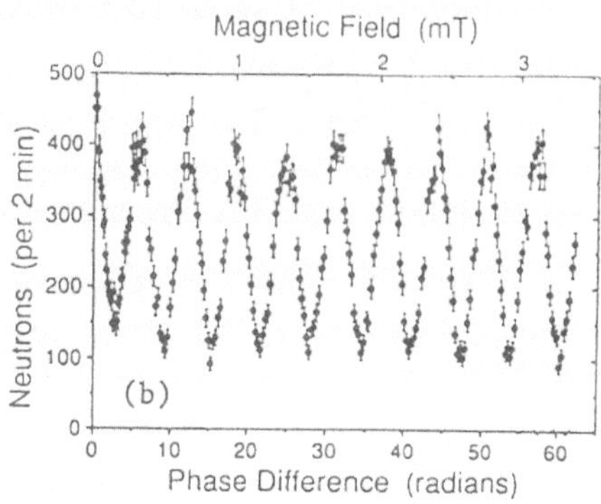

Fig. 15. In (a), neutrons enter, traverse and leave a region of static magnetic field. In (b) the field is swtiched on and off in a region of space already containing the neutrons. Coherence is not lost as the beautiful interference pattern shows. (From ref. 45).

by the neutron, no force is applied to the neutron, and in spite of a short coherence length, the spin-up and spin-down wave packets are not displaced one from the other, and coherence is not lost. The phase shift (given by eq. 44) which applied equally to the force and force-free situations, is clearly manifest in the pulsed results of fig. 15b.

As atom interferometry becomes more developed, we can look forward to many more tests of basic quantum mechanics by these methods, almost certainly with higher sensitivity. In addition, mesoscopic electronics will no doubt offer additional opportunities for interferometric experiments.

Acknowledgements

The author is indebted to S.A.Werner for many interesting discussions, and to the Physics Department at the University of Missouri(Columbia) for its hospitality where this work began. The author also wishes to thank B.Mashhoon, and A.G.Klein. The author is indebted to the Australian Research Council for support.

References

[1]   H. Goldstein Classical Mechanics (2nd Edit. Addison-Wesley) (1980)

[2]   R.P. Feynman, R.B. Leighton, and M. Sands, *The Feynman Lectures in Physics*, Vol. 3, Quantum Mechanics, (Addison-Wesley) (1965).

[3]   R.P. Feynman, and A.R. Hibbs, Quantum Mechanics and Path Integrals, (McGraw-Hill) (1965).

[4]   A.G. Klein and S.A. Werner, Rep.Prog.Phys. **46** pp259-335 (1983)

[5]   S.A. Werner and A.G. Klein in *Methods of Experimental Physics* edited by K.Skold and D.L.Price (Academic New York 1986) **Vol.23** pp259-337.

[6]   R. Colella, A.W. Overhauser, and S.A. Werner, Phys. Rev. Lett. <u>34</u>, 1472 (1975).

[7]   J-L. Staudenmann, S.A. Werner, R. Colella and A.W. Overhauser, Phys. Rev. **A21** 1419-38 (1980).

[8]   Y. Aharonov and D. Bohm, Phys. Rev., **115,** 485 (1959).

[9]   S. Weinberg, *Gravitation and Cosmology* (Wiley New York 1972)

[10]   S.A. Werner, J-L. Staudenmann and R. Colella, Phys. Rev. Lett. <u>42</u>, 1103 (1979).

[11]   J.J. Sakurai, Phys. Rev. D **21** 2993 (1980).

[12]   B. Mashhoon, Phys. Rev. Letts. **61** 2639 (1988).

[13]   M. Dresden and C.N. Yang, Phys. Rev. **20D** 1846 (1979).

[14]   A. Zeilinger, in *Fundamental Aspects of Quantum Theory*, Vol. 144, *NATO ASI Series B: Physics*, edited by V.Sorini and A. Frigerio (Plenum, New York, 1985).

[15]   J. Anandan, Phys. Lett. **A138**, 347 (1989); **152**, 504 (1991)

[16]   J. Anandan, in *Proceedings of the 3rd International Symposium on the Foundations of Quantum Mechanics*, Tokyo, 1989 (Physical Society of Japan, Tokyo, 1990), pp 98-106.

[17]   B.E. Allman, A. Cimmino, A.G. Klein, G.I. Opat, H. Kaiser, and S.A. Werner, Phys. Rev. Lett. **68**, 2409 (1992).

[18]   B.E. Allman, A. Cimmino, A.G. Klein, G.I. Opat, H. Kaiser, and S.A. Werner, Phys. Rev. **A48** 1799-1807 (1993).

[19]   Y. Aharonov, and A. Casher,.Phys. Rev. Lett. **53** 319 (1984) .

[20] J. Anandan, Phys. Rev. Lett. **48**, 1660 (1982).

[21] A. Cimmino, G.I. Opat, A.G. Klein, H. Kaiser, S.A. Werner, M. Arif, and R. Clothier, Phys. Rev. Lett. **63**, 380 (1989).

[22] E. Fabri and L.E. Picasso, Phys. Rev. **39A** 4641-4650 (1989).

[23] Y.Q. Cai and G. Papini, Class. Quantum Grav. **6** 407-418 (1989).

[24] H.C. Bernstein, Phys.Rev.Lett. **18** 1102 (1967)

[25] Y. Aharonov and L. Susskind, Phys.Rev. **158** 1237 (1967)

[26] H. Rauch, A. Zeilinger, G. Badurek, A. Wilfing, W. Bauspiess and U. Bonse, Phys.Lett. **54A**, 425 (1975)

[27] S.A. Werner, R. Colella, A.W.Overhauser and C.F.Eagen, Phys.Rev.Lett. **35**, 1053 (1975)

[28] A.G. Klein and G.I. Opat, Phys.Rev.Letts. **37**, 238 (1976)

[29] M. Peshkin and A. Tonomura, *The Aharonov-Bohm Effect*, Lecture Notes in Phjysics Vol. 3450 (Springer-Verlag, Berlin, 1989).

[30] R.G. Chambers, Phys.Rev.Lett. **5**, 3(1960)

[31] G.Möllenstedt and W. Bayh, Physica B **18**, 299 (1962).

[32] A. Tonomura, T. Matsuda, R. Suzuki, A. Fukuhara, N. Osakabe, H. Umezaki, J. Endo, K. Shinagawa, Y. Sugita and H. Fujiwara, Phys.Rev.Lett. **48**, 1443 (1982); A. Tonomura, Rev. Mod. Phys. **59**, 639 (1987)

[33] G. Mateucci and G. Pozzi, Phys.Rev.Lett. **54**, 2469 (1985).

[34] A. Cimmino, G.I. Opat, A.G. Klein, H. Kaiser, S.A. Werner, M. Arif and R. Clothier, Phys.Rev.Lett. **63**, 380 (1989)

[35] K. Sangster and E.A. Hinds, Phys. Rev. Letts. **71**, 3641 (1993).

[36] R. Casella, Phys.Rev.Lett. **65**, 2217 (1990)

[37] W.J. Elion, J.J. Wachters, L.L. Sohn and J.E. Mooij, Phys.Rev. Lett. **71**, 2311 (1993)

[38] B. Reznik and Y. Aharonov, Phys.Rev. D **43**, 8717 (1991)

[39] B.J. van Wees, Phys.Rev.Lett. **65**, 255 (1990)

[40] A. Zeilinger, in *Fundamental Aspects of Quantum Theory*, Vol. 144 of *NATO ASI Series B: Physics*, edited by V. Sorini and A. Frigerio (Plenum, New York, 1985).

[41]   J. Anandan, Phys. Lett. A **138**, 347 (1989); **152**, 504 (1991)

[42]   J. Anandan, in *Proc. of the 3rd International Symposium on the Foundations of Quantum Mechanics*, Tokyo 1989 (Physical Society of Japan, Tokyo, 1990), pp.98-106.

[43]   B.E. Allman, A. Cimmino, A.G.Klein, G.I.Opat, H. Kaiser and S.A. Werner, Phys.Rev.Lett. **68**, 2409 (1992)

[44]   B.E. Allman, A. Cimmino, A.G.Klein, G.I.Opat, H. Kaiser and S.A. Werner, Phys.Rev. A **48**, 1799 (1993)

[45]   G. Badurek, H. Weinfurter, R. Gähler, A. Kollmar, S. Wehinger and A. Zeilinger, Phys. Rev. Letts. **71**, 307 (1993).

ACHIEVEMENTS IN NEUTRON INTERFEROMETRY

Helmut Rauch

Atominstitut der Österreichischen Universitäten
Schüttelstrasse 115, A-1020 Wien, Austria

INTRODUCTION

Different kinds of neutron interferometers have been tested in the past. The slit interferometer is based on wavefront division and provideslong beam paths but only a very small beam separation[1,2]. The perfect crystal interferometer[3,4] is based on amplitude division. The interferometer based on grating diffraction is a more recent development and has its main application for very slow neutrons[5]. A schematical comparison is shown in Fig.1. The perfect crystal interferometer provides highest intensity and highest flexibility for beam handling and is now most frequently used due to its wide beam separation and its universal availability for fundamental-, nuclear- and solid-state physics research.

The perfect crystal interferometer represents a macroscopic quantum device with characteristic dimensions of several centimeters. The basis for this kind of neutron interferometry is provided by the undisturbed arrangement of atoms in a monolithic perfect silicon crystal[3,6]. An incident beam is split coherently at the first crystal plate, reflected at the middle plate and coherently superposed at the third plate (Fig.1, middle).

From general symmetry considerations follows immediately that the wave functions in both beam paths, which compose the beam in the forward direction behind the interferometer, are equal ($\psi_0^I = \psi_0^{II}$), because they are transmitted-reflected-reflected (TRR) and reflected-reflected-transmitted (RRT), respectively. The de Broglie wavelength of the neutrons diffracted from such crystals is about 1.8 A and their energy is about 0.025 eV. The theoretical treatment of the diffraction process from the perfect crystal is described by the dynamical diffraction theory, which can also be found in the literature for the neutron case[7-10]. Inside the perfect crystal two wave fields are excited when the incident beam fulfills the Bragg condition, one of them having its nodes at the position of the atoms and the other in between them. Therefore, their wave vectors are slighly different ($k_1 - k_2 = 10^{-5}\ k_0$) and due to mutual interference processes, a rather complicated interference pattern is built up, which changes substantially over a caracteristic length Δ_0 - the so-called Pendellösung length, which is of the order of 50 μm for an ordinary silicon reflection. To preserve the interference properties over the length of the interferometer, the dimensions of the monolithic system have to be accurate on a scale comparable to this

Advances in Quantum Phenomena, Edited by E.G. Beltrametti
and J.-M. Lévy-Leblond, Plenum Press, New York, 1995

quantity. The whole interferometer crystal has to be placed on a stable goniometer table under conditions avoiding temperature gradients and vibrations. A phase shift between the two coherent beams can be produced by nuclear, magnetic or gravitational interactions. In the first case, the phase shift is most easily calculated using the index of refraction[11,12]:

$$n = \frac{k_{in}}{k_0} = 1 - \frac{\lambda^2 N}{2\pi} \sqrt{b_c^2 - \frac{\sigma_r}{2\lambda}} + i\frac{\sigma_r N\lambda}{4\pi} \tag{1}$$

which simplifies for weakly absorbing materials $(\sigma_r \to 0)$ to

$$n = 1 - \lambda^2 \frac{Nb_c}{2\pi} \tag{2}$$

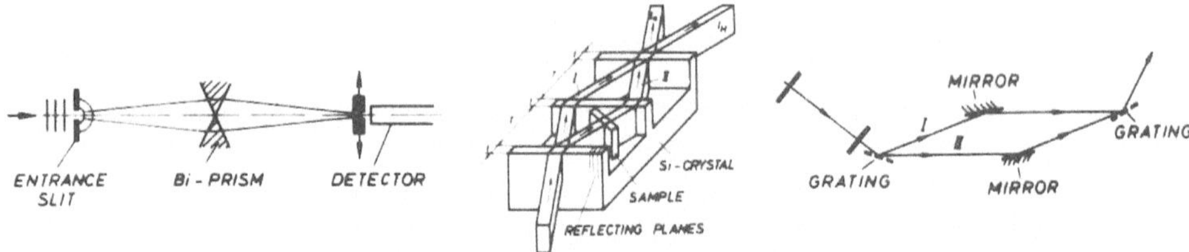

Figure 1: Scheme of a slit-, a perfect-crystal and a grating interferometer

where b_c is the coherent scattering length and N is the particle density of the phase shifting material. The different k-vector inside the phase shifter causes a spatial shift $\vec{\Delta}$ of the wave packet which depends on the orientation of the sample surface $\hat{s}$

$$\vec{\Delta} = \frac{\vec{k} - k_0}{k} D_0 \quad with \quad \vec{k} - \vec{k}_0 = \frac{(1-n)k\hat{s}}{(\hat{k}\cdot\hat{s})} \tag{3}$$

As in ordinary light optics the change of the wave function is obtained as follows:

$$\psi \to \psi_0 e^{i\vec{\Delta}\cdot\vec{k}} = \psi_0 e^{-iNb_c\lambda D} = \psi_0 e^{i\chi} \tag{4}$$

Therefore, the intensity behind the interferometer becomes

$$I_0 \propto \left|\psi_0^I + \psi_0^{II}\right|^2 \propto (1+\cos\chi) \tag{5}$$

The intensity of the beam in the deviated direction follows from particle conservation:

$$I_0 + I_H = const \tag{6}$$

Thus, the intensities behind the interferometer vary as a function of the thickness D of the phase shifter, the particle density N or the neutron wavelength λ. A wave-packet description has to be used to define coherence properties in real experiments.

$$\psi(\vec{r}) \propto \int a(\vec{k}) e^{i\vec{k}\cdot\vec{r}} d^3k \tag{7}$$

Standard quantum mechanics defines the momentum distribution of the beam by

$$g(\vec{k}) = \left|\psi(\vec{k})\right|^2 = \left|a(\vec{k})\right|^2 \tag{8}$$

and, therefore, one gets the real part of the coherence function as the Fourier-transform of the momentum distribution

$$\left|\Gamma(\vec{\Delta})\right| \propto \left|\int g(\vec{k}) e^{i\vec{k}\cdot\vec{\Delta}} d^3\vec{k}\right|, \tag{9}$$

which simplifies for Gaussian momentum distributions

$$g(\vec{k}) \propto \exp\left[-(\vec{k} - \vec{k}_0)^2 / 2\delta\vec{k}^2\right], \tag{10}$$

with characteristic widths δk_i to

$$\left|\Gamma(\vec{\Delta}_0)\right| = \prod_{i=x,y,z} \exp\left[-(\Delta_i \delta k_i)^2 / 2\right]. \tag{11}$$

The mean square distance related to $\left|\Gamma(\vec{\Delta})\right|$ defines the coherence length D which is for Gaussian distribution functions directly related to the minimum uncertainty relation $\left(\Delta_c^{\,i} - 1/(2\delta k_i)\right)$.

Any experimental device deviates from the idealized assumptions made by the theory: the perfect crystal can have slight deviations from its perfectness, and its dimensions may vary slightly; the phase shifter contributes to imperfections by variations in its thickness and inhomogeneities; and even the neutron beam itself contributes to a deviation from the idealized situation because of its momentum spread δk. Therefore, the experimental interference patterns have to be descibed by a generalized relation

$$I \propto A + B\cos(\chi + \Phi_0) \tag{12}$$

where A, B and Φ are characteristic parameters of a certain set-up. It should be mentioned, however, that the idealized behaviour described by equation (5) can nearly be approached by a well balanced set-up[13]. The reduction of the contrast at high order results from the longitudinal coherence length which is determined by the momentum spread of the neutron beam ($\Delta_x = (2\delta k_x)^{-1}$). This causes a change in the amplitude factor $\hat{B}$ of equ.(12) as $\left(B \to B \exp\left[-(\Delta_i \delta k_i)^2 / 2\right] \right)$. The wavelength dependence of χ in equ.(4) disappears in a special sample position where the surface of the sample is oriented parallel to the reflecting planes and the path length through the interferometer becomes $D_0/\sin\Theta_B$ and, therefore, the phase shift $\chi = -2d_{hkl} N b_c D_0$ becomes independent of the wavelength and the damping at high interference orders is strongly reduced. not appear as in the standard position. Related results of a recent experiment where the interference pattern in the 256th interference order have been measured in the dispersive and the nondispersive sample position are shown in Fig.2[14]. The much higher visibility of the interferences in the nondispersive sample arrangement is visible and is caused by the much smaller momentum spread perpendicular to the reflecting planes.

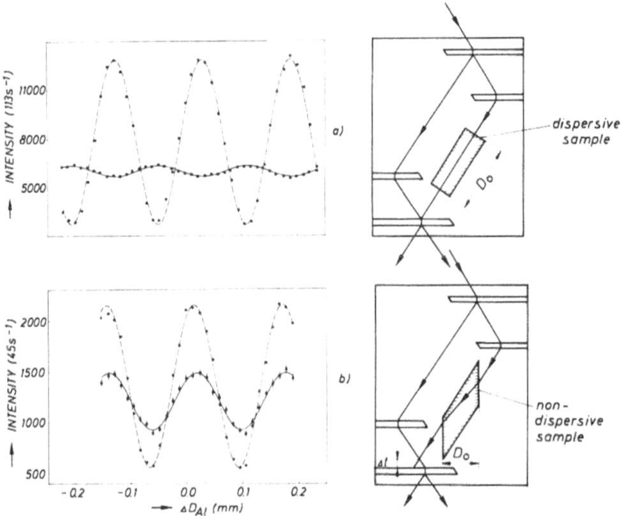

Figure 2: Interference pattern observed at high order (m = 256) with a dispersively (above) and a nondispersively arranged sample[14]. (Dashed lines correspond to measurements at low order.)

Various post-selection measurements in neutron interferometry have shown that interference fringes can be restored even in cases when the overall beam does not exhibit any interference fringes due to spatial phase shifts larger than the coherence length of the interfering beams[15-18]. This indicates that the simple picture which predicts interference only when wave packets spatially overlap is untrue. Interference actually occurs no matter how large the optical path difference may be. From classical optics it has been known for many years that the coherence properties manifest themselves in a spatial intensity variation for phase shifts smaller than the coherence length and in a spectral intensity variation for large phase shifts[19-23]. This phenomenon becomes more apparent for less monochromatic beams and can cause overall spectral shifts[24,25] and even squeezing phenomena[26,27]. The related phenomena for matter waves have been discussed recently [28,29] and will be elucidated in more detail in this paper.

All the results of interferometric measurements, obtained up until now can be explained well in terms of the wave picture of quantum mechanics and the complementarity principle of standard quantum mechanics. Nevertheless, one should bear in mind that the neutron also carries well defined particle properties, which have to be transferred through the interferometer. These properties are summarized in Table 1 together with a formulation in the wave picture. Both particle and wave properties are well established and, therefore, neutrons seem to be a proper tool for testing quantum mechanics with massive particles, where the wave-particle dualism becomes very obvious.

All neutron interferometric experiments pertain to the case of self-interference, where during a certain time interval, only one neutron is inside the interferometer, if at all. Usually, at that time the next neutron has not yet been born and is still contained in the uranium nuclei of the reactor fuel. Although there is no interaction between different neutrons, they have a certain common history within predetermined limits which are defined, e.g., by the neutron moderation process, by their movement along the neutron guide tubes, by the monochromator crystal and by the special interferometer set-up. Therefore, any real interferometer pattern contains single particle and ensemble properties together.

Table 1: Properties of the neutrons

Particle properties:

mass	:	$m = 1.6748220(25).10^{-24}$ g
spin	:	$s = 1/2\, h$
magnetic moment	:	$\mu = -1.91304308(54)\ \mu_K$
lifetime	:	$\tau = 882.6(2.7)$ s
electric charge	:	$q < 2.2.10^{-20}$ e
electric dipol moment	:	$d < 4.8.10^{-25}$ e.cm
electric polarizability	:	$\alpha = 12.0(2.5).10^{-4}$ fm^3
confinement radius	:	$R = 0.7$ fm
quark structure	:	$n = u - d - d$

Wave properties:

Compton wavelength	:	$\lambda = h/mc = 1.32.10^{-13}$ cm
de Broglie wavelength	:	$\lambda_B = h/mv \cong 1.10^{-8}$ cm *)
coherence length	:	$\Delta_c = \lambda^2/\Delta\lambda = 1.10^{-6}$ cm *)
packet length	:	$\Delta_p = v.\Delta t = 1$ cm *)
decay length	:	$\Delta_d = v.T_{1/2} = 2.10^8$ cm *)
phase difference	:	$0 \leq \chi \leq 2\pi$

*) values belong to thermal neutrons ($\lambda_B = 1.8$ A, v = 2 200 m/s)

PARTIAL BEAM PATH DETECTION EXPERIMENTS

A certain beam attenuation can be achieved either by a semi-transparent material or by a proper chopper system. The transmission probalility in the first case is defined by the absorption cross section σ_a of the material [a = I/I$_0$ = exp(-σ_aN D)] and the change of the wave function is obtained directly from the complex index of refraction (equ.1):

$$\psi \rightarrow \psi_0 e^{i(n-1)kD} = \psi_0 e^{i\chi} e^{-\sigma_a ND/2} = e^{i\chi} \sqrt{a}\, \psi \tag{13}$$

Therefore, the beam modulation behind the interferometer is obtained in the following form

$$I_0 \propto \left|\psi_0{}^I + \psi_0{}^{II}\right|^2 \propto \left[(1+a) + 2\sqrt{a}\cos\chi\right] \qquad (14)$$

On the other hand, the transmission probability of a chopper wheel or another shutter system is given by the open to closed ratio, $a = t_{open}/(t_{open} + t_{closed})$, and one obtains after straightforward calculations

$$I \propto \left[(1-a)\left|\psi_0{}^{II}\right|^2 + a\left|\psi_0{}^I + \psi_0{}^{II}\right|^2\right] \propto \left[(1+a) + 2a\cos\chi\right] \qquad (15)$$

i.e. the contrast of the interference pattern is proportional to $\sqrt{a}$, in the first case, and proportional to a in the second case, although the same number of neutrons are observed in both cases. The absorption represents a measuring process in both cases because a compound nucleus is produced with an excitation energy of several MeV, which is usually deexcited by capture gamma rays. These can easily be detected by different means.

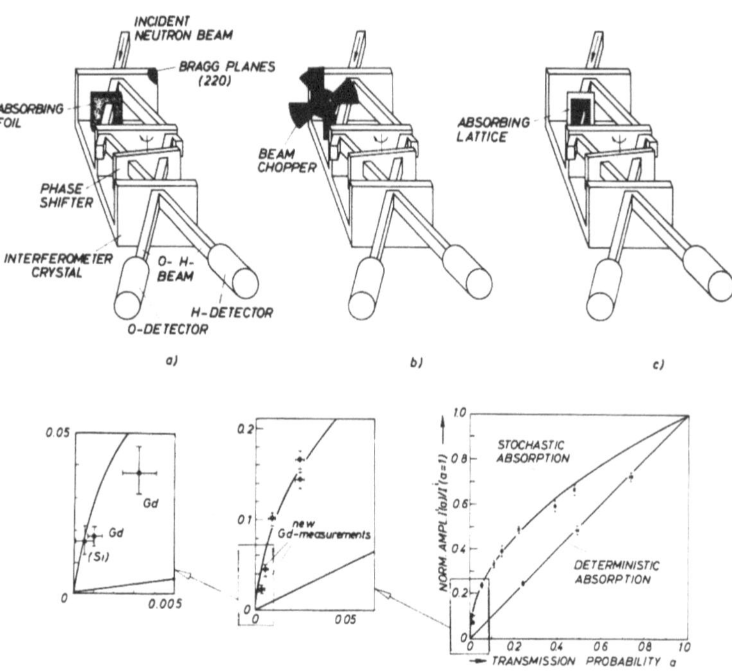

Figure 3: Sketch of the experimental arrangement for absorber measurements (above). (a) stochastic absorption, (b) deterministic absorption, (c) attenuation by a transmission grating. Reduction of the contrast as a function of beam attenuation for different attenuation methods (below)[31,32].

Figure 3 shows the dependence of the normalized contrast of the measured interference pattern on the transmission probability[30-32]. The different contrast becomes especially obvious for low transmission probabilities where the interfering part of the interference pattern is distinctly larger than the transmission probability through the semi-transparent absorber sheet. The discrepancy diverges for $a \rightarrow 0$ but it has been shown that in this regime the variations of the transmission due to variations of the thickness or of the density of the absorber plate have to be taken into ccount which shifts the points below the $\sqrt{a}$-curve[33]. This can most easily be understood if the variation of the beam attenuation due to variations of the thickness or density fluctuations is included $a = \bar{a} + \Delta a$, which yields after averaging

$$\sqrt{a} < \sqrt{\bar{a}} \tag{16}$$

indicating that the points fall below the $\sqrt{a}$-curve.

The region between the linear and the square root behaviour can be reached by very narrow chopper slits or by narrow transmission lattices, where one starts to loose information of through which individual slit the neutron went. This is exactly the region which shows the transition between a deterministic and a stochastic situation and, therefore, it can be formulated by a Bell-like[34,35] inequality

$$\sqrt{a} > x > a \tag{17}$$

The stochastic limit corresponds to the quantum limit when one does not know anymore through which individual slit the neutron went. Which situation exists depends how the slit widths l compares to the coherence lengths $\left(\Delta_i \approx (2\Delta k_i)^{-1} \right)$ in the related direction. In case that the slit widths become smaller than the coherence lengths, the wave function behind the slits show distinct diffraction peaks which correspond to new quantum states ($n \neq 0$), which now do not overlap with the undisturbed reference beam. The creation of the new quantum states means that those labbeled neutrons carry information about the chosen beam path and, therefore, do not contribute to the interference amplitude[36]. A related experiment has been carried out by rotating an absorption lattice around the beam axis where one changes from $l << \Delta_x$ (vertical slits) to $l >> \Delta_y$ (horizontal slits), Fig.4, because

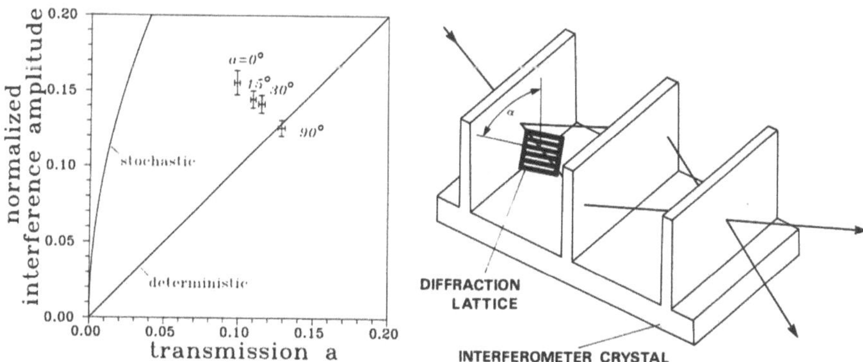

Figure 4: Lattice absorber in the interferometer approaching the classical limit when the slits are oriented horizontally and the quantum limit when they are oriented vertically [31,36].

the coherence length parallel to the reflecting lattice vector is much larger than in any other directions. Thus, the attenuation factor a has to be generalized including not only nuclear absorption and scattering processes but also lattice diffraction effects if they remove neutrons from the original phase space.

A very similar situation exists if a very fast chopper produces beam bursts (packet lengths) shorter than the coherence time $\Delta t_c = \Delta/v$. In this case, difraction in time occurs which also removes neutrons out of the original phase space. This limit is very difficult to reach with a mechanical chopper but it can probably be tackled with a high frequency spin flipper.

NEUTRON JOSEPHSON EFFECT

This phenomenon is based on the dipol coupling of the magnetic moment $\vec{\mu}$ of the neutron to a magnetic field $\vec{B}\left(H = -\vec{\mu}\vec{B}\right)$ which causes the famous 4π-symmetry of spinor wave functions, as measured in early neutron interferometer experiments[37,38]. The change of the wave function reads as:

$$\psi \rightarrow \psi_0 e^{-i\left(Ht/h\right)} = \psi_0 e^{-i\left(\vec{\mu}\vec{B}t/\hbar\right)} = \psi_0 e^{-i\vec{\sigma}\vec{\alpha}/2} = \psi(\alpha) \tag{18}$$

where $\vec{\alpha}$ represents a formal description of the Larmor rotation angle around the field $\vec{B}\left(\alpha = (2\mu/\hbar)\int Bdt = (2\mu/\hbar v)\int Bds\right)$. This enabled also the realization of the spin superposition experiments where spin-up $(|\uparrow>)$ and spin-down states $(|\downarrow>)$ are superposed producing a final state perpendicular to both initial states[39,40]. It is interested to mention that in the case of spin reversal by means of a resonance flipper the spin term is accompanied by an energy exchange equal the Zeeman energy $\hbar\omega_r = 2\mu B$. This provided the basis for the observation of a new quantum beat effect; the magnetic Josephson analog.

A double coil arrangement can be used for the observation of a new quantum beat effect. If the frequencies of the two coils are chosen to be slightly different, the energy transfer becomes different too $\left(\Delta E = \hbar(\omega_{r1} - \omega_{r2})\right)$. The frequency difference can be made very small, if high quality frequency generators are used for the field generation. The flipping efficiencies for both coils are always very close to unity (better than 99%). Now, the wave functions change according to

$$\psi \rightarrow e^{i\left(\omega - \omega_{r1}\right)t}\left|\downarrow> + e^{i\chi} e^{i\left(\omega - \omega_{r2}\right)t}\right|\downarrow> \tag{19}$$

Therefore, the intensity behind the interferometer exhibit a typical quantum beat effect, given by

$$I \propto 1 + \cos\left[\chi + (\omega_{r1} - \omega_{r2})t\right] \tag{20}$$

Thus, the intensity behind the interferometer oscillates between the forward and deviated beam without any apparent change inside the interferometer[41]. The time constant of this modulation can reach a macroscopic scale which is correlated to an uncertainty relation $\Delta E \, \Delta t \leq \hbar/2$. Figure 5 shows the result of an experiment, where the periodicity of the intensity modulation, $T = 2\pi/(\omega_{r1} - \omega_{r2})$, amounts to $T = (47.90 \pm 0.15)$s caused by a frequency difference of about 0.02 Hz. This corresponds to a mean difference ΔE of the energy transfer between the two beams, $\Delta E = 8.6 \cdot 10^{-17}$ eV, and to an energy sensitivity of $2.7 \cdot 10^{-19}$ eV, which is by many orders of magnitudes higher than that of other advanced spectroscopic methods. This high resolution is strongly decoupled from the monochromaticity of the neutron beam, which was $\Delta E = 5.5 \cdot 10^{-4}$ eV around a mean energy of the beam $E_B = 0.023$ eV in this case. It should be mentioned, that the result can also be interpreted as being the effect of a slowly varying phase $\Delta(t)$ between the two flipper fields, but the more physical description is based on the argument of a different energy transfer. The extremely high resolution may be used for fundamental, nuclear and solid state physics applications.

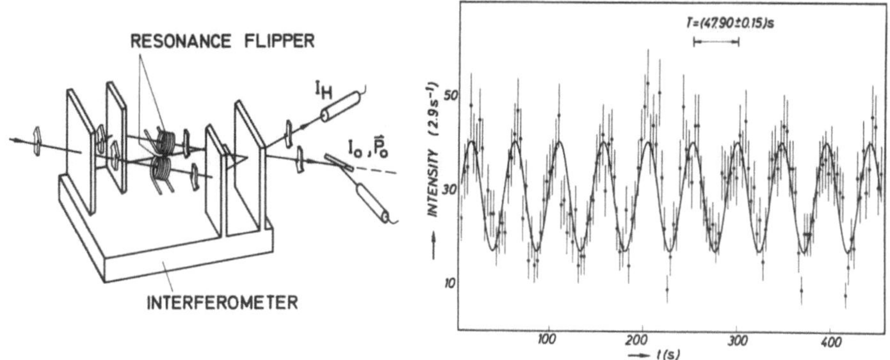

Figure 5: Quantum beat effect observed when the frequencies of the two flipper coils differ by about 0.02 Hz around 71.89979 kHz[41].

The quantum heat effect can also be interpreted as a magnetic Josephson effect analog. In this case, the phase difference is driven by the magnetic energy $\left(\Delta B_0 = B_{02} - B_{01}\right)$

$$\frac{\delta}{\delta t}\left(\Delta_2 - \Delta_1\right) = \omega_{r2} - \omega_{r1} = \frac{2\mu\Delta B_0}{\hbar} \tag{21}$$

and, therefore,

$$\Delta(t) = \left(\omega_{r2} - \omega_{r1}\right)t = \frac{2\mu\,\Delta B_0}{\hbar}\cdot t \tag{22}$$

This yields the observed modulation (compare equ.(20))

$$I \propto \left(1 + \cos\Delta(t)\right) \tag{23}$$

This is analogous to the well-known Josephson-effect in superconducting tunnel junctions [42], where the phase of the Cooper-pairs in both superconductors is related according to

$$\frac{\delta}{\delta t}\left(\Phi_2 - \Phi_1\right) = \frac{1}{\hbar}\left(E_2 - E_1\right) = \frac{1}{\hbar}2eV \tag{24}$$

which is driven by the electrical potential V between both superconductors. This gives

$$\Phi(t) = \frac{2eV}{\hbar}\cdot t \tag{25}$$

and a superconducting Josephson current

$$I_S = I_{sMax} \sin\Phi(t) \qquad (26)$$

POSTSELECTION OF STATES

In the course of several neutron interferometer experiments[15-18] it has been established that smoothed out interference properties at high interference order can be restored even behind the interferometer when a proper spectral filtering is applied. The experimental arrangement with an indication of the wave packets at different parts of the interference experiment is shown in Figure 6. An additional monochromatization is applied behind the interferometer by means of various single crystals brought into Bragg-position.

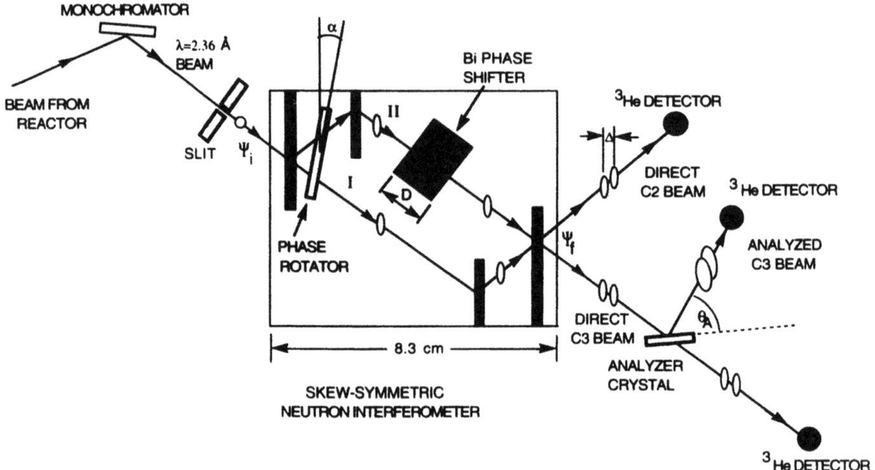

Figure 6: Scheme of the experimental arrangement with a skew-symmetrically cut perfect crystal interferometer and a post-selection analyzer crystal[18,28].

Using Eqs.(5) and (7), the momentum-dependent intensity reads as:

$$I_0(\vec{r},\vec{k}) = \left|\psi_0^I(\vec{r},\vec{k}) + \psi_0^{II}(\vec{r}+\Delta,\vec{k})\right|^2 \propto \left|a(\vec{k})\right|^2 \left(1 + \cos(\vec{\Delta}(\vec{k}).\vec{k})\right) \qquad (27)$$

whereas the overall beam reads as

$$I_0(\vec{\Delta}_0) \propto 1 + \left|\Gamma(\vec{\Delta}_0)\right| \cos\vec{\Delta}_0.\vec{k}_0 \qquad (28)$$

where $\vec{\Delta}_0$ represents the spatial phase shift for the $\vec{k}_0$-component of the packet. Equ.(28) describes the interference fringes when $\vec{\Delta}_0$ is varied. The formula also shows that the overall interference fringes disappear for spatial phase shifts larger than the coherence lengths $\left[\Delta_i \geq \Delta_c^i = 1/(2\delta k_i)\right]$ (see equ. (11)). This behavior is shown in Fig.7 and has been verified experimentally by several investigations for Gaussian and non-Gaussian neutron beams[43-45].

In our experiment we deal with the coherence properties along the interferometer axis (x), where the (tangential) components of the momentum vectors (and coherence length) do not change due to Bragg diffraction. According to basic quantum mechanical laws, the related momentum distribution follows from equ.(27) and for Gaussian packets it can be rewritten in the form

$$I_0(k) = \exp\left[-(k-k_0)^2 / 2\delta k^2\right]\left\{1+\cos\left(\chi_0 \frac{k_0}{k}\right)\right\} \qquad (29)$$

where the mean phase shift is introduced $\left(\chi_0 = k_0\Delta_0 = Nb_c\lambda_0 D_{eff}\right)$. The surprising feature is that $I_0(k)$ becomes oscillatory for large phase shifts where the interference fringes described by equ.(28) disappear (see Fig.7). This indicates that interference in phase space has to be considered[46 47] rather than the simple wave function overlap criterion described by the coherence function (equ.(10)). The second beam behind the interferometer (H) just shows the complementary modulation $I_H = I_{total} - I_0$.

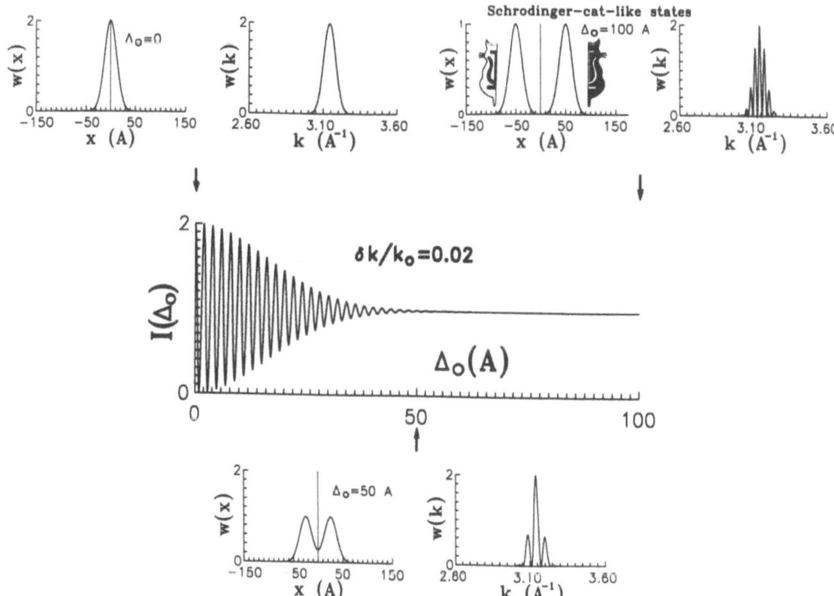

Figure 7: Interference pattern as a function of the relative phase shift (middle) and related wave packets and momentum spectra behind the interferometer for different values of the phase shift [18].

The amplitude function[48] of the packets arising from beam paths I and II determines the spatial shape of the packets behind the interferometer

$$I_0(x) = \left|\psi(x)+\psi(x+\Delta)\right|^2 \qquad (30)$$

which separates for large phase shifts into two peaks (Fig.7). For Gaussian packets, having a spatial width, δx, which corresponds to the coherence length Δ_c the minimum uncertainty relation $\delta x \delta k = 1/2$ is fulfilled. For an appropriately large displacement $(\Delta \gg \Delta_c)$, the related state can be interpreted as a superposition state of two macroscopically distinguishable states, that is a stationary Schrödinger-cat like state[26,49,50], but here first for massive particles. These states - separated in ordinary space and oscillating in momentum space - seem to be notoriously fragile and sensitive to dephasing effects[51-54].

Measurements of the wavelength spectrum were made with a silicon crystal with a rather narrow mosaic spread which reflects in the parallel position a very narrow band of neutrons only ($\delta k'/k_0 \approx 0.0003$) causing an enhanced visibility at large phase shifts (Fig.8). This feature shows that an interference pattern can be restored even behind the interferometer by means of a proper postselection procedure. In this case the overall beam does not show interference fringes anymore and the wave packets originating from the two different beam paths do not overlap.

The momentum distribution has been measured by scanning the analyzer crystal through the Bragg-position. The related results are shown in Fig.8 for different phase shifts. These results clearly demonstrate that the predicted spectral modulation (equ.(29)) appears when the interference fringes of the overall beam disappear. The modulation is somehow smeared out due to averaging processes across the beam due to various imperfections, unavoidably existing in any experimental arrangement. The contrast of the empty interferometer was 60%.

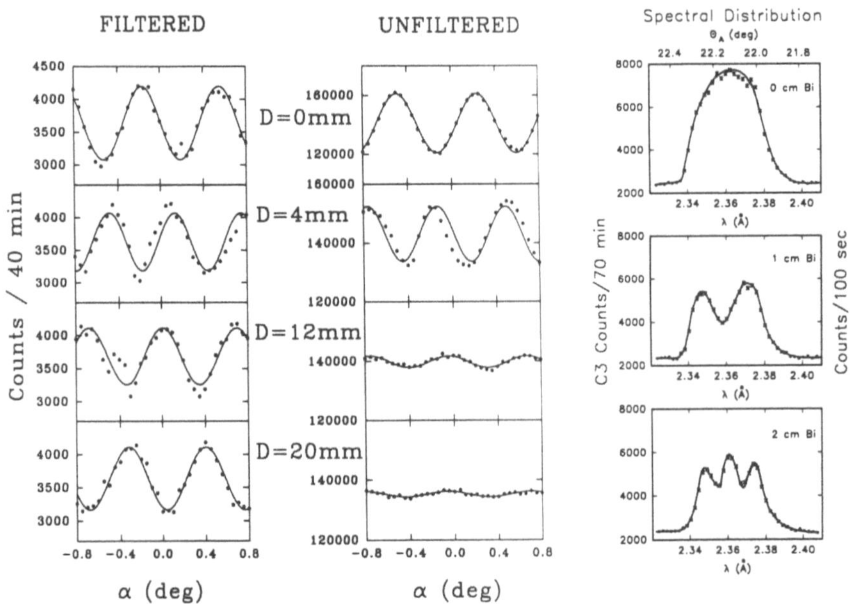

Figure 8: Interference pattern of the unfiltered overall beam ($\delta k/k_0 = 0.012$, middle) and the filtered beam reflected from a nearly perfect crystal analyzer in the antiparallel position ($\delta k'/k_0 = 0.0003$, left) and the observed spectral modulation (right) of the outgoing beam for different phase shifter thicknesses[18].

Each peak in the momentum distribution shown in Fig.8 corresponds to a different number of phase shifts experienced by the neutrons of that wavelength band during its passage through the interferometer. In that sense, the minimum quantum entity of the incident wave packet becomes a new diverse entity respresenting different quantum states with distinguishable properties. This kind of labelling shows that constructive interference is restricted to a certain wavelength band only; a situation similar to that where new states have been created due to lattice diffraction inside the interferometer (Fig.4)[36].

The new quantum states created behind the interferometer can be analyzed with regard to their uncertainty properties. Analogies between a coherent state behavior and a free but coherently coupled particle motion inside the interferometer have been addressed previously[32]. In such cases, the dynamical conjugate variables x and p minimize the uncertainty product with identical uncertainties $(\Delta x)^2 = (\Delta k)^2 = 1/2$ (in dimensionless units). Using $I_0(k)$ and $I_0(x)$ (equs. (29) and (30)) as distribution function we get in our case

$$\Gamma(\bar{\Delta}) = <\psi^*(0)\,\psi(\bar{\Delta})> \tag{31}$$

and (for $\delta k / k_0 << 1$)

$$<(\Delta k)^2> = <k^2> - <k>^2$$

$$= (\delta k)^2 \left\{ 1 - \left(\frac{\Delta_0}{2\delta x}\right)^2 \frac{e^{-(\Delta_0/2\delta x)^2/2}\cos(\Delta_0 k_0) + e^{-(\Delta_0/2\delta x)^2}}{\left[1 + e^{-(\Delta_0/2\delta x)^2/2}\cos(\Delta_0 k_0)\right]^2} \right\} \tag{32}$$

These relations are shown in Fig.9 indicating that for $(\Delta k)^2$ a value below the coherent state value can be achieved, which in quantum optic terminology means squeezing[55-58,26]. One emphasizes that a single coherent state does not exhibit squeezing, but a state created by superposition of two coherent states can exhibit a considerable amount of squeezing. Thus highly nonclassical states is made by the power of the quantum mechanical superposition principle.

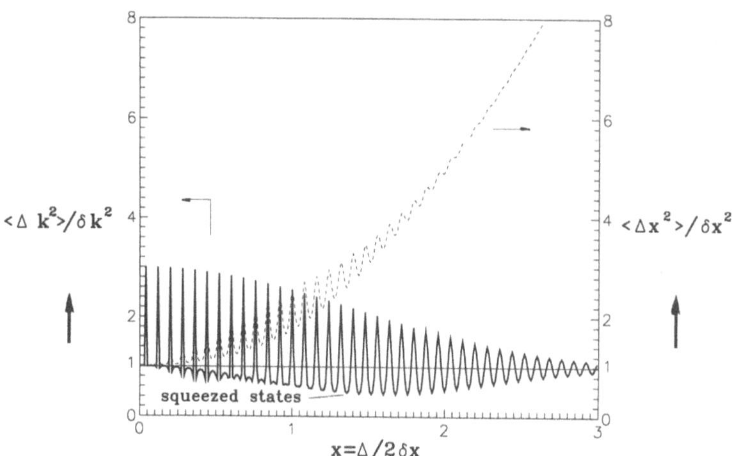

Figure 9: Spatial and momentum uncertainties of the outgoing beams with the indication of squeezing in the momentum domain[18].

CONTRAST RETRIEVAL BY PHASE ECHO

Phase echo is a similar technique to spin echo, which is routinely used in neutron spectroscopy[59]. A large phase shift ($\Delta > \Delta^c$) can be applied in one arm of the interferometer, which can be compensated by a negative phase shift acting in the same arm or by the same phase shift applied to the second beam path[44,45,60]. According to equ.(4), the phase shift is additative and the coherence function depends on the net phase shift only. Thus, the interference pattern can be restored as it is shown schematically and in form of an experimental example in Fig.10. The phase-echo method can also be applied behind the interferometer loop when multiplate interferometers are used[26]. In this case, the situation becomes even more similar to the situation discussed in the previous section.

These results tell us that information first appearing in a spatial phase shift becomes transferred into a momentum modulationwhich can be retrievaled to ordinary space modulation effects again. A comment has to be made that it becomes intrinsically more difficult to restore the original contrast the wider the separation of the wave packets in ordinary space happened.

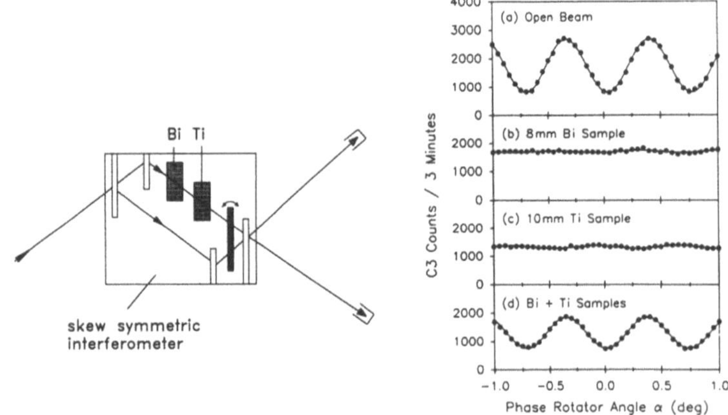

Figure 10: Loss of contrast at high interference and its retrieval by an opposite phase shifter inserted into the same beam[45].

REQUEST FOR POST-SELECTION IN EPR-EXPERIMENTS

The previously discussed neutron experiments have shown us that phase space coupling persists even if the overlap in one parameter space does not exist anymore. The stored information becomes exchanged between parameter spaces and can be measured by a proper experimental method. This has consequences for EPR-experiments too. The entangled states (e.g.[62]) of two photons produced by an atomic decay cascade (Fig.11)

$$\psi \propto \left|-k>_1\right|k>_2 + \left|-k>_2\right|k>_1$$

(33)

are correlated due to the energy conservation of the transition

$$k_1 + k_2 = k_{01} + k_{02} = const.$$

(34)

This produces a momentum and space dependent intensity distribution when the packet structure of the related wave functions is taken into account[28]

$$I(k_1, k_2, \vec{r}) = |\psi|^2 = 2|a(k_1)|^2 |a(k_2)|^2 \cdot \left(1 + \cos\left[2(k_2 - k_1)r\right]\right) \qquad (35)$$

This shows a characteristic intensity modulation for each photon pair (Fig.11) and indicates that individual |k>-states remain interacting even at arbitrarily large spatial separation of the wave packets. For large distances $(r > (2\delta k)^{-1})$ the appearance of a momentum distribution modulation follows from equ.(35) too[28]. If one of these photons are registered on one side its wave-function collapses which instantaneously changes the wave function on the other side to $\left\| k >_2 \right|^2$. This shows again that much more information can be gained than it is usually extracted. Therefore, it is recommended to repeat this experiment with a proper momentum resolution which would show that the right and the left wave fields of the related momentum band (i.e. the partner photons) remain coupled even at arbitrarily large spatial separation of the overall wave-packets.

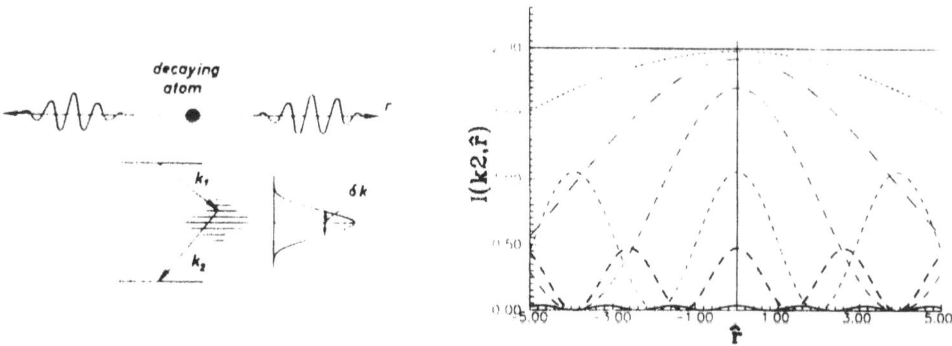

Figure 11: Scheme of a typical EPR-transition and expected intensity distribution for individual momentum pairs for $|k_1\text{-}k_2|/\delta k$ = 0, 0.1, 0.2, 0.4, 0.8, 1.2 and 2 (from above to below[28].

Related experiments will show that this coupling of the partner pairs of photons persists independently from its overall spatial separation. That indicates that locality should be treated in phase space rather than in ordinary space only. Here, too the required momentum resolution becomes more stringent when the packets become wider separated in ordinary space.

IRREVERSIBILITY AND MEASURING PROCESS

In the previous sections it has been shown that more information about a quantum system can be extracted when more experimentally accessible parameters are measured. It becomes obvious that a system remains coupled in phase space even when it becomes separated in any parameter space. Thus, interference properties can be shifted from one parameter space to another one and back again. Related bands of plane wave components which compose the wave packets (equ.7) may be considered as a responsible factor for the understanding of the non-locality phenomenon in quantum mechanics.

REVERSIBILITY - IRREVERSIBILITY

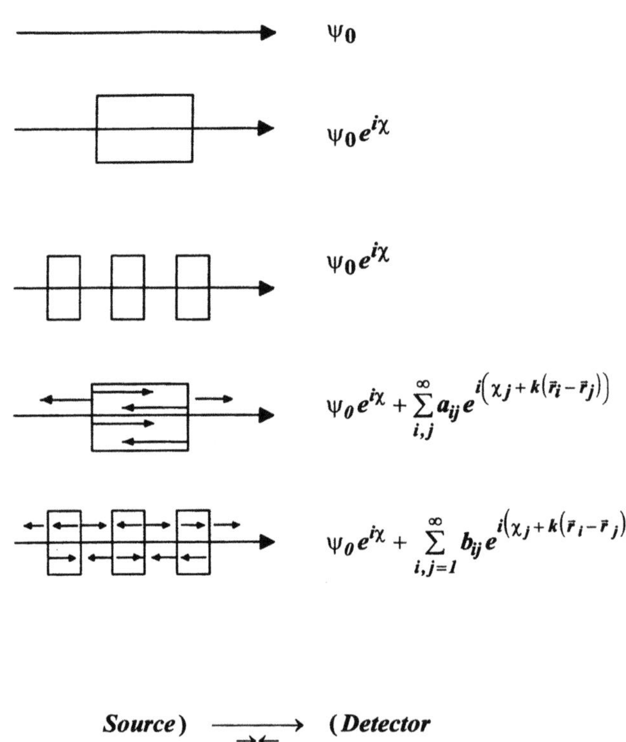

Figure 12: Approximative and complete wave functions behind a compact and a split interaction region.

The summaries drawn for the different experimental situations discussed in this article are followed by a statement that the retrieval of the interference properties by several post-selection procedures become increasingly more difficult the wider the separation of the quantum system happened before. A more detailed view even show that a complete retrieval is impossible, in principle, as it is shown in Fig.12 for the case of a phase echo system. In a more complete and more accurate measurement, more and more parts of the complete (not approximative) wave functions become visible which contain more and more of the detailed history the quantum system has experienced between the source and the detector. This indicates a basic irreversiblity process not caused by parasitic effects like absorption or incoherent scattering processes but by the appearance of an infinite number of additional terms in the wave function which indicates that, by no means, the original state can be restored completely. Unavoidable fluctuations (even zero-point fluctuations) cause an irreversibility effect which becomes more influential for widely separated Schrödinger-cat like states. All these effects can be described by an increasing entropy inherently associated with any kind of interaction[63]. This also supports the idea that irreversibility is a fundamental property of nature and reversibility, an approximation only, as stated by several authors (e.g.[64-66]).

This shows that irreversibility and, therefore, the measurement process starts with the first interaction the quantum system experiences in the experimental set-up. The assignment of a source and a detector region define the direction of increasing entropy.

DISCUSSION

All the results of the neutron interferometric experiments are well described by the formalism of quantum mechanics. According to the complementarity principle of the Copenhagen interpretation, the wave picture has to be used to describe the observed phenomena. The question how the well-defined particle properties of the neutron are transferred through the interferometer, is not a meaningful one within this interpretation, but from the physical point of view it should be an allowed one.

The newly discovered persisting phase space coupling in cases of large spatial shifts of the wave packets may bring some attenuation to the action of plane wave components outside the packet. The shown results clearly demonstrate that a spectral modulation can be observed in neutron interference experiments at high interference order and that interference has to be treated in phase space rather than in ordinary space. It looks like that the plane wave components of the wave packets, i.e. narrow band width components, interact over a much larger distance than the size of the packets. This interaction guides neutrons of certain momentum bands to the 0- or H-beam, respectively. These phenomena throw a new light on the discussion on Schrödinger-cat-like situations in quantum mechanics and, therefore, on the discussion about the EPR-experiments too[28,62,67,68]. Spatially separated packets remain entangled in phase space and nonlocality appears as a result of this entanglement. The analogy with optical experiments performed in the time-frequency domain is striking[23]. An analogue situation exists in neutron spin-echo systems where multiple spin rotation plays an equivalent role as high order interferences discussed here[59,69].

More complete quantum experiments show that a complete retrieval of all wave components behind an interaction the quantum system experienced becomes impossible, in principle. This implies on a high accuracy level a basic non-commutivity of operators A.B $|\psi> \neq$ B.A$|\psi>$ and indicates that the irreversible quantum measuring process starts with the first interaction of the quantum system with the experimental set-up.

ACKNOWLEDGEMENT

Most of the experimental results discussed in detail have been obtained by our Dortmund-Grenoble-Vienna interferometer group working at the high flux reactor in Grenoble, and some recent ones stem from our cooperation with the Columbia-Missouri group working at the MURR-reactor. The cooperations with these groups and especially the cooperation with colleagues from our Institute, which are cited in the references, are gratefully acknowledged.

REFERENCES

1. H. Maier-Leibnitz, T. Springer, Z. Physik 167 (1962).
2. R. Gaehler, J. Kalus, W. Mampe, J. Phys. E13: 546 (1980).
3. H. Rauch, W. Treimer, U. Bonse, Phys. Lett. A47: 369 (1974).
4. W. Bauspiess, U. Bonse, H. Rauch, W. Treimer, Z. Physik 271: 177 (1974).
5. A.I. Ioffe, V.S. Zabiyankan, G.M. Drabkin, Phys. Lett. 111: 373 (1985).
6. U. Bonse, M. Hart, Appl. Phys. Lett. 6: 155 (1965).
7. H. Rauch, D. Petrascheck, "Neutron Diffraction" H. Dachs, ed., Springer Verlag, Berlin 1978, Chap.9.
8. V.F. Sears, Can. J. Phys. 56: 1261 (1978).
9. W. Bauspiess, U. Bonse, W. Graeff, J. Appl. Cryst. 9: 68 (1976).
10. D. Petrascheck, Acta Phys. Austr. 45: 217 (1976).
11. M.L. Goldberger, F. Seitz, Phys. Rev. 71: 294 (1947).
12. V.F. Sears, Phys. Rep. 82: 1 (1982).
13. U. Bonse, H. Rauch (Eds.), "Neutron Interferometry" (Clarendon Press, Oxford 1979).
14. H. Rauch, E. Seidl, D. Tuppinger, D. Petrascheck, R. Scherm, Z. Physik B69: 69 (1987).
15. S.A. Werner, R. Clothier, H. Kaiser, H. Rauch, H. Wölwitsch, Phys. Rev. Lett. 67: 683 (1991).
16. H. Kaiser, R. Clothier, S.A. Werner, H. Rauch, H. Wölwitsch, Phys. Rev. A45: 31 (1992).
17. H. Rauch, H. Wölwitsch, R. Clothier, H. Kaiser, S.A. Werner, Phys. Rev. A46: 49 (1992).
18. D.L. Jacobson, S.A. Werner, H. Rauch, Phys. Rev. A49: 3196 (1994).
19. L. Mandel, J. Opt. Soc. Am. 52: 1335 (1962).
20. L Mandel, E. Wolf, Rev. Mod. Phys. 37: 231 (1965).
21. F. Heineger, A. Herden, T. Tschudi, Opt. Comm. 48: 237 (1983).
22. D.F.V. James, E. Wolf, Phys. Lett. A157: 6 (1991).
23. X.Y. Zou, T.P. Grayson, L. Mandel, Phys. Rev. Lett. 69: 3041 (1992).
24. E. Wolf, Phys. Rev. Lett. 63: 2220 (1989).
25. D. Faktis, G.M. Morris , Opt. Lett. 13: 4 (1988).
26. W. Schleich, M. Pernigo, Fam Le Kien, Phys. Rev. A44: 2172 (1991).
27. J. Janski, A.V. Vinogradov, Phys. Rev. Lett. 64: 2771 (1990).
28. H. Rauch, Phys. Lett. A173 240 (1993).
29. H. Rauch, Proc. Quantum Measurement & Control, E.Ezawa, Y.Murayama, eds., North Holland, 1993, p.223.
30. H. Rauch, J. Summhammer, Phys. Lett. 104A: 44 (1984).
31. J. Summhammer, H. Rauch, D. Tuppinger, Phys. Rev. A36: 4447 (1987).
32. H. Rauch, J. Summhammer, M. Zawisky, E. Jericha, Phys. Rev. A42: 3726 (1990).
33. M. Namiki, S. Pascazio, Phys. Lett. 147A: 430 (1990).
34. J. Bell, Physics 1: 195 (1965).
35. D. Home, F. Selleri, Revista del Nuovo Cim. 14: 1 (1991).
36. H. Rauch, J. Summhammer, Phys. Rev. 46: 7284 (1992).
37. H. Rauch, A. Zeilinger, G. Badurek, A. Wilfing, W. Bauspiess, U. Bonse, Phys. Lett. A54: 425 (1975).
38. S.A. Werner, R. Colella, A.W. Overhauser, C.F. Eagen, Phys. Rev. Lett. 35: 1053 (1975).

39. J. Summhammer, G. Badurek, H. Rauch, U. Kischko, A. Zeilinger, Phys. Rev. A27: 2523 (1983).

40. G. Badurek, H. Rauch, J. Summhammer, Phys. Rev. Lett. 51: 1015 (1983).

41. G. Badurek, H. Rauch, D. Tuppinger, Phys. Rev. A34: 2600 (1986).

42. B.D. Josephson, Rev. Mod. Phys. 46 251 (1974).

43. H. Rauch, in: "Neutron Interferometry",U.Bonse, H.Rauch, eds., Clarendon Press, 1979, p.161.

44. H. Kaiser, S.A. Werner, E.A. George, Phys. Rev. Lett. 50: 563 (1983).

45. R. Clothier, H. Kaiser, S.A. Werner, H. Rauch, H. Wölwitsch, Phys. Rev. A44: 5357 (1991).

46. W. Schleich, J.A. Wheeler, Nature 326: 574 (1987).

47. W. Schleich, D.F. Walls, J.A. Wheeler, Phys. Rev. A38: 1177 (1988).

48. J.-M. Levy-Leblond, F. Balibar, "Quantics", North-Holland, 1990.

49. A. Legett, Proc. Found. Quantum Mechanics, S.Kamefuchi, ed., Phys. Soc. Japan, 1984, p.74.

50. B. Yurke, W. Schleich, D.F. Walls, Phys. Rev. A42: 1703 (1990).

51. D.F. Walls, G.J. Milburn, Phys. Rev. A31: 2403 (1985).

52. R.J. Glauber, "New Techniques and Ideas in: Quantum Measurement Theory", D.M.Greenberger, ed., N.Y. Acad. Sci. 1986, p.336.

53. M. Namiki, S. Pascazio, Phys. Rev. A44: 39 (1991).

54. H. Zurek, Physics Today, Oct. 1991, p.36.

55. D.F. Walls, Nature 306: 141 (1983).

56. S.L. Braunstein, R.I. McLachlan, Phys. Rev. A35: 1659 (1987).

57. R. Loudon, P.L. Knight, J. Mod. Opt. 34: 709 (1987).

58. J. Jansky, A.V. Vinogradov, Phys. Rev. Lett. 64: 2771 (1990).

59. F.Mezei, ed. "Neutron Spin Echo", Lect. Notes in Physics 128, Springer (1980).

60. G.Badurek, H.Rauch, A.Zeilinger, in: "Neutron Spin Echo", F.Mezei, ed., Lect. Notes in Physics 128, Springer (1980), p.136.

61. M.Heinrich, D.Petrascheck, H.Rauch: Z.Physik B72: 357 (1988).

62. N.D. Mermin, Phys. Rev. Lett. 65: 1838 (1990).

63. H.A.Lorentz: Theorie der Strahlung, Akad. Verlagsges. Leipzig (1927).

64. I.Prigogine: Proc. Ecol. Phys. Chem., p. 8, Siena, Elsevier, Amsterdam 1991.

65. F.Haag: Comm. Math. Phys. 123: 245 (1990).

66. P.Blanchard, A.Jadczyk: Phys. Lett. A175: 157 (1993).

67. A. Einstein, B. Podolsky, N. Rosen, Phys. Rev. 47: 777 (1935).

68. D.M. Greenberger, M.A. Horne, A. Zeilinger, in: "Bell's Theorem, Quantum Theory and Conceptions of the Universe"; M.Kafatos, ed., Kluwer Publ. 1989, p.69.

69. G. Badurek, H. Weinfurter, R. Gähler, A. Kollmar, S. Wehinger, A. Zeilinger, Phys. Rev. Lett. 71: 307 (1993).

ELECTRON INTERFEROMETRY
AND HOLOGRAPHY

Akira Tonomura

Advanced Research Laboratory, Hitachi, Ltd.
&
Tonomura Electron Wavefront Project, ERATO, JRDC
Hatoyama, Saitama 350-03, Japan

INTRODUCTION

The wavelength of an electron beam can be as short as on the order of 1/100 Å. Therefore, microscopic objects and fields can be measured or observed with this small unit of length by electron interferometry. In fact, extremely interesting electron-interference experiments, though limited in number, were carried out in the 1950s to 1970s to measure inner and contact potentials,[1] magnetic fluxons,[2] and others.[3]

Electron interference experiments required highly skillful techniques in those days, since only thermal electron beams of low coherence were available. Difficulties in electron-interference experiments thus can be compared to those of optical interference experiments using high-pressure mercury-arc lamps as a light source. Therefore, these experiments were done in only a few laboratories, such as at Tübingen University in West Germany,[1] CNRS Toulouse[4] in France, and later Berlin University,[2] Bologna University,[3] Tohoku University[5] and others.

The advent of a "coherent" field-emission electron beam[6] in 1979 changed the situation. The maximum number of observable interference fringes increased by an order of magnitude, and interference fringes became observable directly on a fluorescent screen when their number was 50 or less.

This coherent beam improved the performance of electron holography[7] to the extent that it can be used for practical applications. Since electron wavefronts are faithfully transformed into optical wavefronts through electron holography, versatile optical techniques can be used for electron optics. One example is that electron holography has made it possible to obtain phase contour maps, which are inaccessible with an electron microscope equipped with an electron biprism.[8] Furthermore, the precision in phase measurement was increased to $2\pi/100$ by using a special technique peculiar to holography.[9]

The recent development of such electron interferometry has made it possible to carry out fundamental experiments in physics that were previously not feasible. In addition, it has engendered new methods of measurement and observation such as quantitative

Advances in Quantum Phenomena. Edited by E.G. Beltrametti
and J.-M. Lévy-Leblond. Plenum Press. New York, 1995

measurement of the thickness distribution of a uniform material[9] and the magnetic field distribution inside a ferromagnetic film,[10] and the observation of flux lines[11] in a superconductor.[11]

HOLOGRAPHIC INTERFERENCE MICROSCOPY

Electron holography[7] is a two-step imaging process that uses electrons and light (Fig. 1). In the first step, an interference pattern produced by the electron beam transmitted through an object and a reference beam is formed in a field-emission transmission electron microscope equipped with an electron biprism[8] and is recorded on film as a hologram.

The hologram is subsequently illuminated by a collimated laser beam, and the exact image, or the wavefront, is optically reproduced in one of the two diffracted beams. Another image, called a "conjugate image," is produced in the other diffracted beam. Its amplitude is a complex conjugate of the reconstructed image, with the phase value reversed in sign.

An interference micrograph that displays the phase distribution can be obtained by overlapping an optical plane wave with the reconstructed image as a comparison beam. If a conjugate image rather than a plane wave overlaps this image, the phase difference doubles, as if the phase distribution were amplified by a factor of two. By repeating this technique, a phase shift as small as 1/100th of the electron wavelength can be detected.[9]

INTERACTION OF ELECTRONS WITH ELECTROMAGNETIC FIELDS

To interpret electron interference micrographs, we have to solve the Schrödinger equation to determine the interaction of an electron beam with an object. This problem was essentially solved by Y. Aharonov and D. Bohm[12] in 1959. They concluded that an electron wave was influenced in the form of a phase shift, $\Delta\Phi$, by electromagnetic potentials, ϕ and A, and that if the relative change in the potentials is much smaller than 1, in the distance λ, electron wavelength, then $\Delta\Phi$ is given by

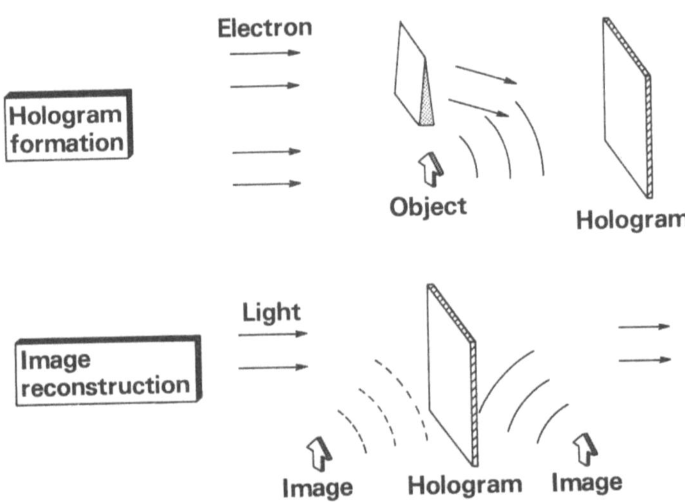

Fig. 1. Principle of electron holography.

$$\Delta\Phi = -\frac{e}{\hbar}\int(A\,ds - \phi\,dt). \qquad\qquad .....(1)$$

Their analysis revealed the exciting discovery that two electron waves passing through only field-free regions on both sides of an infinite solenoid can be physically influenced by inaccessible electromagnetic fields, E and B, to produce the observable effect as displacements of the interference fringes.

To be more specific, in a magnetic case, two electron waves enclosing a magnetic flux produce a relative phase shift proportional to the flux, even when the waves never touch the flux. Aharonov and Bohm attributed this effect to the vector potential. The vector potential cannot vanish even in a field-free region surrounding the magnetic flux, since the circulation integral of the vector potential around any loop is equal to the magnetic flux flowing through the loop.

The significance of this effect has recently taken on weight,[13] since it has been regarded as a direct experimental manifestation of the validity of the gauge principle, a guiding principle in the search for a unified theory of all fundamental interactions in nature. However, several theoreticians[14] here questioned the existence of the AB effect, even asserting that the AB effect was purely a mathematical concoction. Previous experimental results were also attributed to flux leakage from the finite solenoids used in these experiments. Therefore, we were convinced that a firm experimental foundation should be established.

EXPERIMENTAL CONFIRMATION OF THE AB EFFECT

The crucial experimental point discussed during the AB effect controversy concerned the effect of magnetic flux leakage from finite solenoids on an electron wave. The infinite solenoid assumed in the AB effect theory is experimentally unattainable. However, an ideal geometry can be achieved by using a finite toroidal solenoid.[15] Although toroidal geometry had often been proposed during the controversy, we had to wait for the development of microlithography to be able to fabricate the necessary tiny and complicated samples.

Since no overlap should exist between the electron wave and the magnetic field, the following tiny toroidal samples were fabricated. The toroidal magnet was covered with a metal layer to prevent electron penetration into the magnet. To avoid even a small amount of flux leakage from the magnet, the metal layer is made of superconducting material. Magnetic fields cannot pass through a superconducting layer because of the Meissner effect, so there are no magnetic-field leaks from the sample.

A scanning electron micrograph of a fabricated sample is shown in Fig. 2. Since magnetic flux cannot be varied as in case of a solenoid, many toroidal samples with different magnetic flux values ranging from $4(h/e)$ to $6(h/e)$ were fabricated.

An electron wave was incident on a toroidal sample cooled to 4.5 K and, the relative phase shift between two electron waves, one passing inside and one outside the toroidal sample, was measured by means of an interferogram formed by electron holography.[16] Although many samples with various magnetic flux values were measured, the relative phase shifts detected were either 0 or π (Fig. 3). The conclusion is now obvious. A relative phase shift of π (Fig. 3(b)) is produced even when the magnetic fields are confined within the superconductor and shielded from the electron wave. This conclusively proves that the AB effect exists.

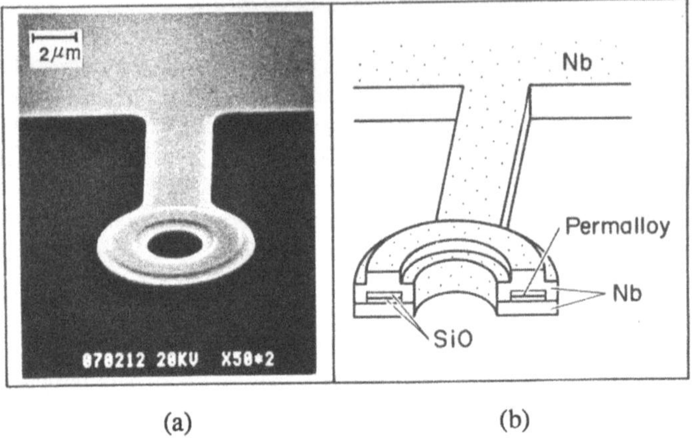

(a)   (b)

Fig. 2. Toroidal permalloy covered with niobium layer.
(a) Scanning electron micrograph. (b) Cross-sectional diagram.

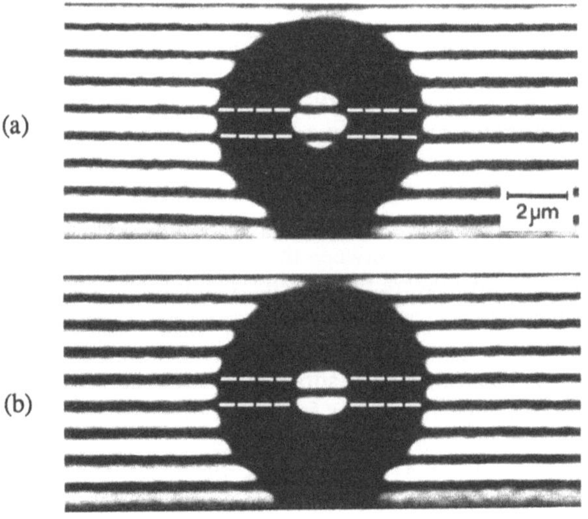

Fig. 3. Electron interferograms of toroidal samples at 4.5 K.
(a) $\Delta\phi = 0$. (b) $\Delta\phi = \pi$.

APPLICATIONS OF INTERFERENCE MICROSCOPY

It can be seen from Eq. (1) that electromagnetic fields can be observed quantitatively in electron waves as phase shifts. In fact, the development of a "coherent" field-emission electron beam has opened the way to the observation of microscopic electromagnetic fields.

Magnetic Domain Structure

The interpretation of an interference micrograph of a ferromagnetic thin film is straightforward. The contour fringes follow the projected magnetic lines of force in h/e flux units. This is easily derived from Eq. (1). The relative phase difference $\Delta\Phi$ between two points P_1 and P_2 on the transmitted electron beam (see Fig. 4) is defined as the phase difference between two beams starting from a single point, passing through two points P_1 and P_2 in the object plane, and recombining at another point. Theus $\Delta\Phi$ is given by the magnetic flux enclosed by the two electron tranjectories. Therefore, when the two points, P_1 and P_2, are along a single magnetic line of force, $\Delta\Phi$ vanishes, i. e., the phase contour lines lie along the magnetic lines. When two electron beams passing through P_1 and P_2 enclose a flux of h/e, then $\Delta\Phi = 2\pi$. Therefore, a constant magnetic flux of h/e flows between two adjacent contour lines in an interference micrograph.

An example for a hexagonal cobalt particle is shown in Fig. 5. It is difficult to determine the magnetic domain structure in a fine particle by Lorentz microscopy, since Fresnel fringes are also produced inside the particle image by image defocusing to hide the magnetic contrast. In interference microscopy, there is no need to defocus the image, and magnetic lines of force can be seen as contour fringes on the image when the particle thickness is uniform. Whether the magnetization direction is clockwise or counter-clockwise corresponds to whether the wavefront protrudes like a mountain, or is hollowed like a valley. This difference does not appear in the contour map of the wavefront. However, this can be decided from the interferogram obtained by slightly tilting the two interfering beams in the optical reconstruction stage. In this case, the wavefront is advanced in the particle center, meaning that the magnetization direction is clockwise.

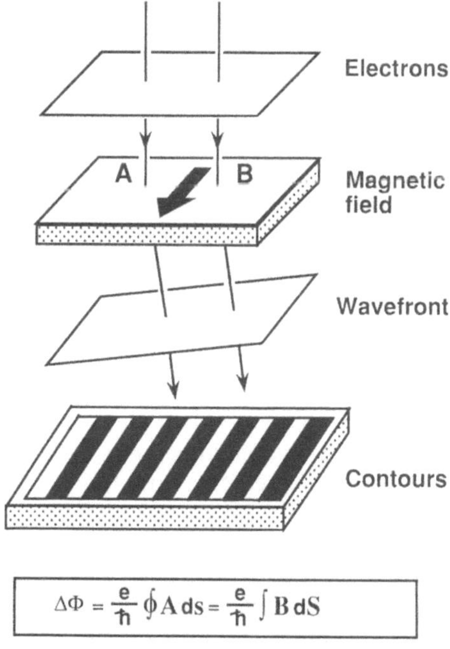

$$\Delta\Phi = \frac{e}{\hbar} \oint A\,ds = \frac{e}{\hbar} \int B\,dS$$

Fig. 4. Principle of magnetic line observation by interference electron microscopy.

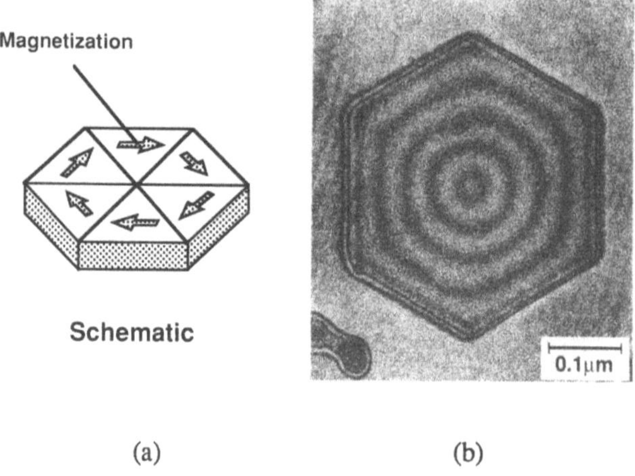

(a)   (b)

Fig. 5. Interference micrograph of hexagonal cobalt particle.
(a) Schematic. (b) Interference micrograph.

Flux Quantization Process

The process of magnetic flux quantization can be observed directly using the toroidal ferromagnet (Fig. 2) used in the AB effect experiment. Holographic interference microscopy was used to measure at various temperatures the relative phase shift between two electron beams, one passing through the hole and the other passing outside the toroidal sample.

An example of this result is shown in Fig. 6. The phase shift at room temperature was 0.3 π, as shown in Fig. 6(a). However, when the sample temperature was reduced, the phase shift gradually increased to 0.8 π at $T = 15$ K and then jumped to π at Tc (= 9.2 K).

This behavior can be interpreted as follows. Above Tc, the phase shift is determined by the magnetic flux flowing inside the magnet. The temperature dependence of the phase shift from 300 K to 15 K arises from the fact that magnetization in the permalloy increases by 5% due to the decreasing thermal fluctuations of the spins.

When T decreases below Tc, supercurrent begins to flow in the inner surface layer of the hollow superconducting torus so that the total magnetic flux is an integral multiple of $h/2e$. The phase shift becomes π, since the number of fluxons is odd in this sample.

Fluxons Penetrating Superconductors

In the previous experiments, changes in hidden fluxons were observable. However, fluxons penetrating a superconductor, which play an important role in practical superconductivity applications, have evaded direct observation in spite of several attempts. This is because in addition to being extremely small, ($h/2e = 2 \times 10^{-15}$Wb), their flux is in the shape of a very thin thread.

We therefore observed such fluxons by measuring the magnetic fields leaking out from a superconductor surface.[11] The experimental arrangement is shown in Fig. 7. A magnetic field of a few gauss or less was applied perpendicularly to an evaporated lead film. By applying an electron beam to the magnetic lines of force penetrating the superconductor from above, we can observe fluxons through electron holography.

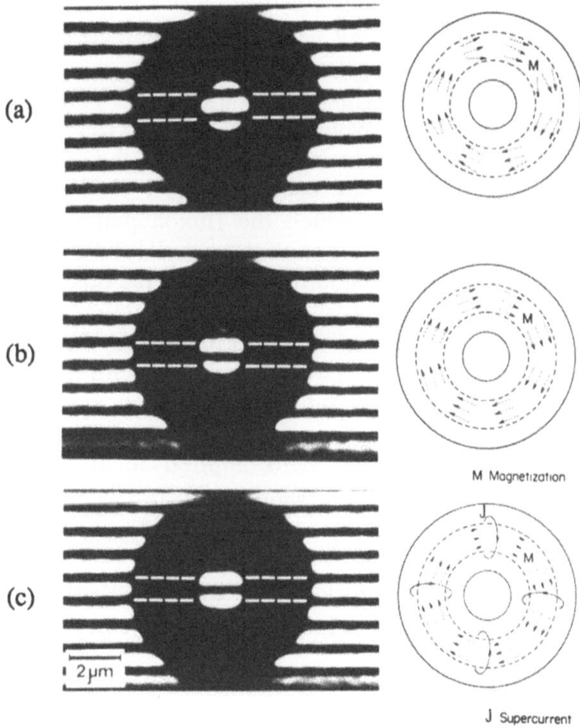

Fig. 6. Temperature dependence of electron interferogram of toroidal samples.
(a) $\Delta\Phi = 0.3\,\pi$ at $T = 300$ K. (b) $\Delta\Phi = 0.8\,\pi$ at $T = 15$ K.
(c) $\Delta\Phi = \pi$ at $T = 4.5$ K.

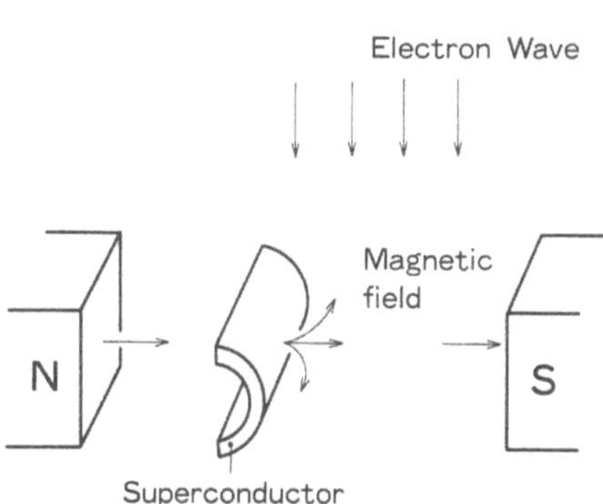

Fig. 7. Experimental arrangement for observing fluxons.

Figure 8(a) shows single fluxons penetrating a 0.2μm - thick film. In this figure, the phase difference is amplified by a factor of two. Therefore, one interference fringe corresponds to one fluxon. A single fluxon is captured at the right side of this photograph. The magnetic line of force is produced by an extremely small area of the lead surface and spreads out into free space.

In addition to observing isolated fluxons, we observed a pair of fluxons oriented in opposite directions and connected by magnetic lines of force (left side in Fig. 8(a)). One possible reason for this pair production is as follows. When the specimen is cooled below the critical temperature, the lead becomes superconductive. During the cooling, however, the specimen experiences a state in which the fluxon pair appears and disappears repeatedly due to thermal excitation in a two-dimensional system, and is pinned by some imperfection in the superconductor, eventually resulting in the flux being frozen.

What happens when the thickness of the superconducting thin film is increased? Figure 8 (b) shows the state of the magnetic lines of force when the thickness is 1 μm. We can see that the state changes completely. Magnetic flux penetrates the superconductor not as individual fluxons but in bundles. The figure does not show any fluxon pairs.

An explanation for this phenomenon is as follows. Because lead is a type-I superconductor, the strong magnetic field applied to it partially destroys the superconductive state in some parts of the specimen (intermediate state). Figure 8 (b) is a photograph that shows that the magnetic lines of force penetrate the parts of the specimen where superconductivity has been destroyed. However, since the surrounding parts are still superconductive, the total amount of penetrating magnetic flux is an integral multiple of the flux quantum, $h/2e$. Thin superconducting films (Fig. 7(a)) are an exception. The lead behaves like a type-II superconductor and the flux penetrates the superconductor in the form of individual fluxons.

In the method described so far an electron wave is passed near a superconductor surface so that we can observe fluxons extending from the surface. With this method, however, a two-dimensional array of fluxons cannot be observed. Furthermore, we cannot observe the inside of the superconductor.

Recent enhancements in our 350-kV holography electron microscope,[16] which provides a more "coherent" and more "penetrating" electron wave, have made it possible to observe both the static image of a fluxon array by holographic interference microscopy[17] and their dynamic behavior by Lorentz microscopy.[18]

Fig. 8. Interference micrographs of fluxons leaking from Pb film (Phase amplification: × 2). (a) Thickness = 0.2 μm. (b) Thicknes = 1 μm.

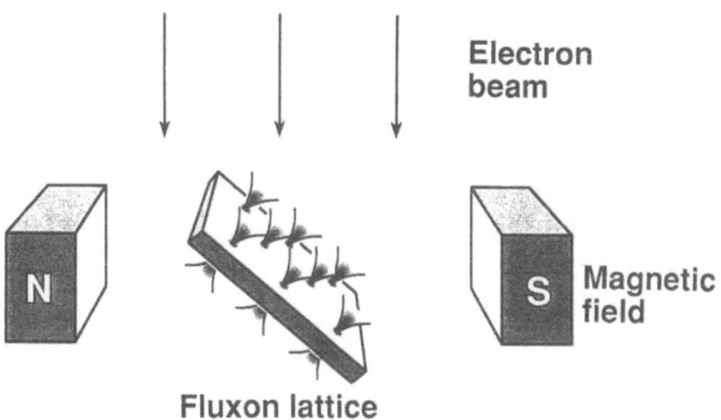

Fig. 9. Schematic diagram for fluxon lattice observation.

The experimental arrangement is shown in Fig. 9. A Nb thin film, set on a low-temperature stage, was tilted 45° to an incident beam of 300 kV electrons so that the electrons could be affected by the fluxon's magnetic fields. An external magnetic field of up to 150 gauss was applied horizontally. An example of a fluxon array in a single-crystalline Nb thin film[17] is shown in Fig. 10. In this interference micrograph, projected magnetic lines of force can be observed. They become dense in the localized regions indicated by the circles in the photograph, corresponding to individual flux lines.

Lorentz microscopy is useful for observing the dynamic behavior of fluxons. In this experiment, the sample was first cooled down to 4.5 K and the applied magnetic field, B, was gradually increased. When B was increased to 32 gauss, fluxons suddenly began to penetrate the film. Their number increased as B increased and their dynamic behavior was quite interesting. At first, only a few fluxons appeared here and there in the 15×10 μm^2 field of view. They oscillated around their own pinning centers and occasionally hopped from one center to another. These movements continued for as long as the fluxons remained closely packed ($B \leq 100$ gauss).

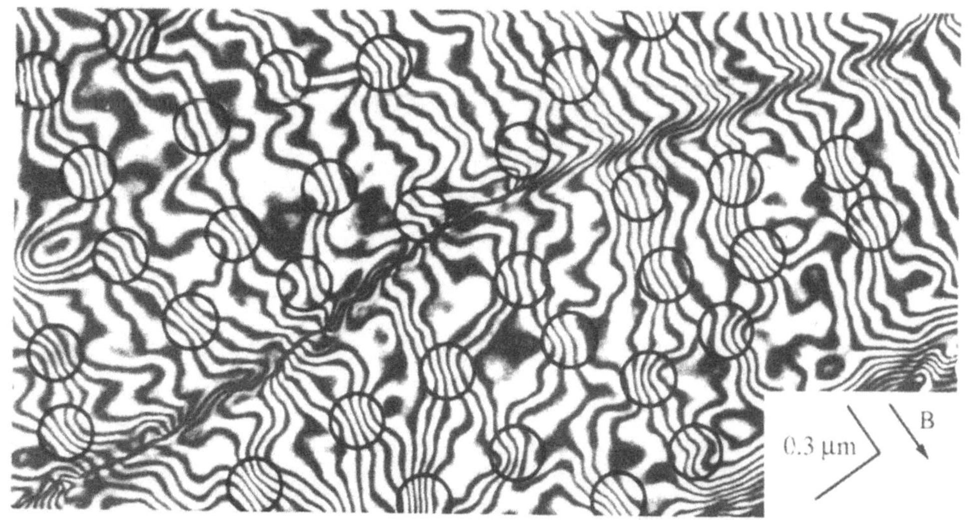

Fig. 10. Interference micrograph of a superconducting Nb film at $B = 100$ gauss (Phase amplification: $\times$ 16).

An equilibrium Lorentz micrograph[18] at $B = 100$ gauss is shown in Fig. 11. The film has a fairly uniform thickness in the region shown, but is bent along the black curves, called *bend contours*, which are caused by Bragg reflections at the atomic plane. Each spot showing black and white contrast is the image of a single fluxon. This contrast reversed, as expected, when the applied magnetic field was reversed. The tilt direction of the sample can be discerned from the line dividing the black and white parts of the spots. Since the black part is on the same side of all the spots, the polarities of all the fluxons seen in the region are the same.

CONCLUSIONS

The performance of electron interferometry has been improved thanks to the development of a coherent field-emission electron beam and electron holography. This technique can be applied to measure the phase distribution of an electron wave interacting with an object to a precision of $2\pi/100$, which opens up a new window for the direct observation of magnetic lines of force in both ferromagnetic and superconductive samples. This technique has a wide range of potential application fields which will be developed together with an even more coherent electron beam.

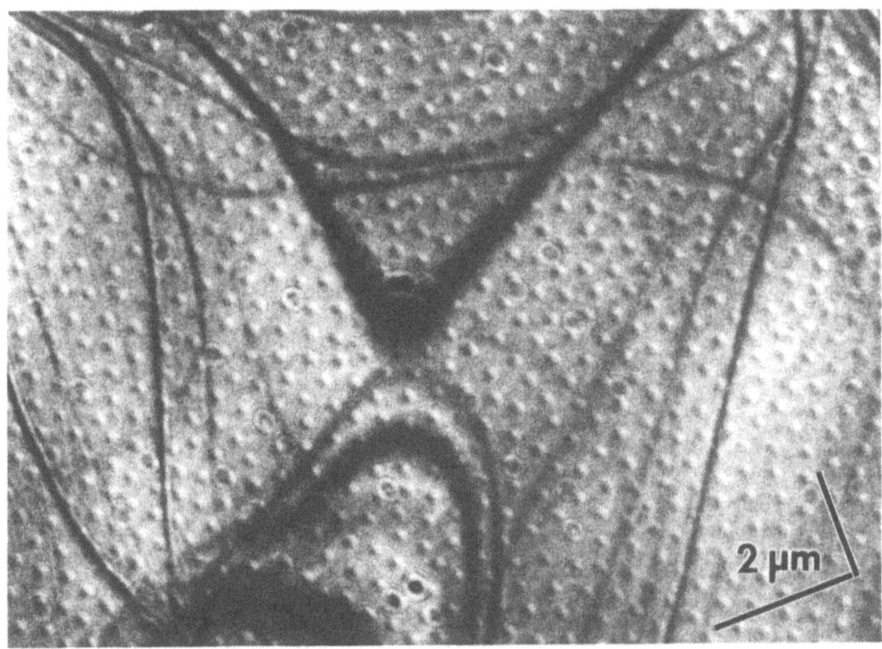

Fig. 11. Lorentz micrograph of a two-dimensional array of fluxons in a superconducting Nb film.

REFERENCES

1. G. Möllenstedt and M. Keller, Elektroneninterferometrische Messung des inneren Potentials, Z. Physik 148: 34 (1957).
2. R. Lischke, Direct Observation of Quantized Magnetic Flux in a Superconducting Hollow Cylinder with an Electron Interferometer, Phys. Rev. Lett. 22: 1366.
3. G. F. Missiroli, G. Pozzi and U. Valdrè, J. Phys. E: Sci. Instrum. 14: 649 (1981).
4. J. Faget and C. Fert, Diffraction et Interféreces en Optique Électronique, Cah. Phys. 83: 285 (1957).
5. T. Hibi and S. Takahashi, Electron Interference Microscope, J. Electron Microsc. 12: 129 (1963).
6. A. Tonomura, T. Matsuda, J. Endo, H. Todokoro and T. Komoda, Development of a Field Emission Electron Microscope, J. Electron Microsc. 28: 1 (1979).
7. D. Gabor, Microscopy by Reconstructed Wave-fronts, Proc. R. Soc. A197: 454 (1949).
8. G. Möllenstedt and H. Düker, Beobachtungen und Messungen an Biprisma-Interferenzen mit Elektronenwellen, Z. Phys. 145: 377 (1956).
9. A. Tonomura, T. Matsuda, T. Kawasaki, J. Endo and N. Osakabe, Sensitivity-Enhanced Electron-Holographic Interferometry and Thickness-Measurement Applications at Atomic Scale, Phys. Rev. Lett. 54: 60 (1985).
10. A. Tonomura, T. Matsuda, J. Endo, T. Arii and K. Mihama, Direct Observation of Fine Structure of Magnetic Domain Walls by Electron Holography, Phys. Rev. Lett. 44: 1430 (1980).
11. T. Matsuda, S. Hasegawa, M. Igarashi, T. Kobayashi, M. Naito, H. Kajiyama, J. Endo, N. Osakabe, A. Tonomura and R. Aoki, Magnetic Field Observation of Single Flux Quantum by Electron-Holographic Interferometry, Phys. Rev. Lett. 62: 2519 (1989).
12. Y. Aharonov and D. Bohm, Significance of Electromagnetic Potentials in Quantum Theory, Phys. Rev. 115: 485 (1959).
13. T. T. Wu and C. N. Yang, Concept Nonintegrable Phase Factors and Global Formulation of Gauge Fields, Phys. Rev. D 12: 3845 (1975).
14. P. Bocchieri and A. Loinger, Nonexistence of the Aharonov-Bohm Effect, Nuovo Cimento 47A: 475 (1978).
15. C. G. Kuper, Electromagnetic Potential in Quantum Mechanics: A Proposed Test of the Aharonov-Bohm Effect, Phys. Lett. 79A: 413 (1980).
16. T. Kawasaki, T. Matsuda, J. Endo and A. Tonomura, Observation of a 0.055 nm Spacing Lattice Image in Gold Using a Field Emission Electron Microscope, Jpn. J. Appl. Phys. 29: L508 (1980).
17. J. E. Bonevich, K. Harada, T. Matsuda, H. Kasai, T. Yoshida, G. Pozzi and A. Tonomura, Electron Holography Observation of Vortex Lattices in a Sperconductor, Phys. Rev. Lett. 70: 2952 (1993).
18. K. Harada, T. Matsuda, J. Bonevich, M. Igarashi, S. Kondo, G. Pozzi, U. Kawabe and A. Tonomura, Real-time Observation of Vortex Lattices in a Superconductor by Electron Microscopy, Nature, 360: 51 (5 November 1992).

QUANTUM PHENOMENA AND THEIR APPLICATIONS IN SEMICONDUCTOR MICROSTRUCTURES

Federico Capasso

AT&T Bell Laboratories
600 Mountain Avenue
Murray Hill, NJ 07974

INTRODUCTION

During the last decade a powerful new approach for designing semiconductor structures with tailored electronic and optical properties, bandgap engineering, has spawned a new generation of semiconductor materials and of electronic and photonic devices.[1] Central to bandgap engineering is the notion that by spatially varying the composition and the doping of a semiconductor over distances ranging from a few microns down to ≈ 2.5 Å (≈ 1 monolayer), one can tailor the band structure of a material in a nearly arbitrary and continuous way.[1] Thus semiconductor structures with new electronic and optical properties can be custom-designed for specific applications.

The enabling technology which has made band-structure engineering a reality is Molecular Beam Epitaxy pioneered at Bell Labs by Cho and Arthur.[2] This epitaxial growth technique allows multilayer heterojunction structures to be grown with atomically abrupt interfaces and precisely controlled compositional and doping profiles over distances as short as a few tens of angstroms. Such structures include quantum wells, which are a key building block of bandgap engineering. These potential energy wells are formed by sandwiching an ultrathin lower gap layer (of thickness comparable or smaller than the carrier thermal de Broglie wavelength, which is ≈ 250 Å for electrons in GaAs at 300 K) between two wide-gap semiconductors (for example, AlGaAs). Electrons can form standing wave patterns in the direction perpendicular to the layer so that the corresponding kinetic energy is quantized similar to the particle in a box problem in elementary quantum mechanics (quantum size effect). Of course, the motion along the plane is free so that the net effect of size quantization is that the conduction band is divided up in many subbands whose bottom is represented by energy levels (as, for example, in Fig. 1(a)). The spacing and position of the discrete energy levels in the well depend on the well thickness and depth. These quantum confined states were first

Advances in Quantum Phenomena. Edited by E.G. Beltrametti
and J.-M. Lévy-Leblond. Plenum Press. New York. 1995

observed in pioneering optical and transport experiments performed at Bell Laboratories and IBM, respectively, in 1974.[3]

If many quantum wells are grown on top of one another and the barriers are made so thin (typically < 50 Å) that tunneling between the coupled wells becomes important, a superlattice is formed. Superlattices are materials with novel optical and transport properties introduced by the artificial periodicity.[3] In this paper we give a brief overview of recent developments in the area of quantum phenomena in quantum wells and superlattice structures and of their applications.

RESONANT TUNNELING STRUCTURES

One can tunnel through the quantized states of the well if the perpendicular kinetic energy of electrons matches that of the energy levels of the well and the barriers are made sufficiently thin (Fig. 1(a)). In the limit of negligible scattering in the double barrier region this resonant tunneling process is conceptually similar to that giving rise to transmission resonances in an optical cavity at well defined wavelengths. In practice scattering processes tend to destroy the coherent interference so that the process in many practical situations is better described in terms of sequential tunneling into and out of the well.

It is clear from the above discussion that double barriers with heavily doped regions as contact layers behave as negative differential resistance diodes (Fig. 1(a)), as first shown by Chang et al.[3] The current exhibits a sharp drop as the bias is increased corresponding to the state of the well being lowered below the conduction band edge in the emitter layer. In this situation resonant tunneling of electrons from the emitter into the well is inhibited for lack of lateral momentum conservation. In a series of pioneering experiments Sollner, and coworkers at Lincoln Laboratory operated AlGaAs/GaAs resonant tunneling diodes as mixers up to 1.8 Terahertz and as oscillators up to 420 Gigahertz. More recently using AlSb for the barriers and InAs for the well they have obtained oscillation frequencies of 675 GHz.[3]

Although resonant-tunneling diodes may find application in oscillators, logic circuits require devices with input-output isolation in order to achieve gain and avoid loading of the input stages. This isolation is best accomplished with a transistor.

In general terms, a resonant-tunneling transistor is a device that uses an applied control voltage to modulate the difference between the energy levels of the quantum well and the energy of the incident electrons. Thus the resonant-tunneling current through the double barrier can be made to peak at one or more values of the control voltage, corresponding to the different energy levels.[4,5] Note that a transistor with such a current-voltage characteristic has multiple on and off states, corresponding respectively to the peaks and valleys of the I-V curve. Although multiple-valued logic has been the subject of considerable theoretical investigation, all proposed and demonstrated circuit architectures have until now relied on conventional two-state devices or tunnel diodes. Multiple-valued logic may one day reduce the interconnection complexity of integrated circuits.

Conceptually, the simplest way to build a resonant-tunneling transistor is to form a contact with the heavily doped quantum well of a double barrier. This low-resistance

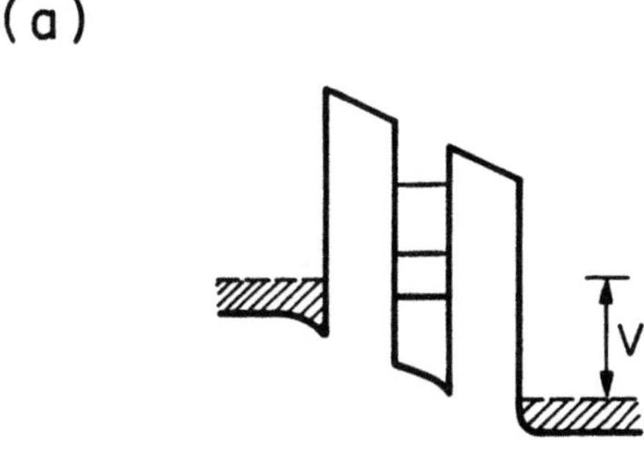

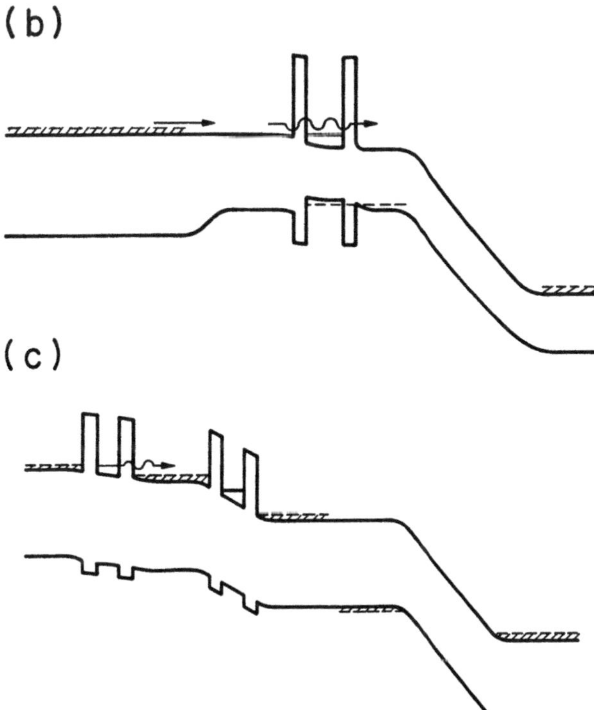

Fig. 1   Band diagrams of various resonant tunneling devices. The quantum wells
barrier thicknesses are typically in the range $50-100$ Å and $20-50$ Å
respectively.

    a.   Biased double barrier diode showing the states of quantum well, each
corresponding to a transmission resonance.

    b.   Resonant tunneling bipolar transistor under operating conditions.

    c.   Multistate resonant tunneling transistor.

contact would serve as the control terminal. This approach is, however, fraught with major technical difficulties, and attempts in this direction have not yet succeeded. A conceptually similar but technologically much easier approach is to incorporate a double barrier into the base of a bipolar transistor.[4,5]

Fig. 1(b) shows the energy-band diagram for a resonant-tunneling bipolar transistor. As the base-emitter voltage is increased, resonant tunneling through each subband first reaches a maximum and is then suppressed as the bottom the subband is lowered below the conduction-band edge in the emitter. This should produce multiple peaks in the collector current, i.e. multiple negative transconductance.

The first operation of a resonant-tunneling bipolar transistor was reported in 1986.[4,5] Room temperature operation with a single peak in the I-V curve and a 3:1 peak-to-valley ratio was demonstrated. With the band structure of Fig. 1(b) it is difficult, however, to achieve multiple peaks of comparable height - a useful feature for circuit applications.

To solve this problem the new structure shown in Fig. 1(c) was introduced.[4,5] In this new device, two or more double barriers are placed in the emitter rather than in the base. To achieve multiple peaks in the dependence of collector current on base-emitter voltage space charge buildup in the quantum wells is exploited. The attendant electrostatic screening effect makes the field across the two double barriers spatially nonuniform, and higher in the well closer to the base. Thus as the bias is increased, resonant tunneling through the two wells is suppressed at two different voltages, yielding two peaks in the I-V characteristic. (When resonant tunneling is suppressed at only one well, current continuity is insured by inelastic scattering).

We have fabricated such a structure with $Al_{0.48}In_{0.52}As$ and $Ga_{0.47}In_{0.53}As$ by MBE, putting two double barriers in the emitter. At room temperature the device has demonstrated two current peaks with excellent peak-to-valley ratios, current gains in excess of 60 and a cutoff frequency $f_T \geq 24$ GHz.[5]

The kind of transfer characteristic discussed above makes possible a class of circuits with greatly reduced complexity - requiring far fewer transistors per function than do conventional circuits. For example, a parity-bit checker for four-bit words has been built with a *single* multistate transistor.[5] Such circuits are normally used in digital communication systems to detect errors by checking if there is an odd or even number of ones (or zeros) in the binary word. Conventional 4-bit checkers require 24 transistors. Multistate transistors are also attractive for frequency multipliers. With the device shown in Fig. 6(b), an input frequency of 350 MHz has been multiplied by five.[5] Other possible applications include fast analog-to-digital converters and memory cells.[5]

SUPERLATTICE STRUCTURES

Consider now a periodic structure with many quantum wells. The barrier layers should be thin enough to allow significant tunneling. The perpendicular kinetic energy of electrons in such a structure breaks up into allowed bands separated by forbidden gaps, just as we would expect in any periodic structure. This is of course analogous to the formation of energy bands in solids in consequence of their periodic crystal lattices.

There are, however, important differences between the natural crystal lattice and the man-made superlattice - as these structures are called.[3] First, the period of the superlattice is typically $30-100$ Å, whereas the period of the crystal lattice is only a few angstroms. Therefore the energy bands of the superlattice, which arise from the coupling between the energy levels of the potential wells, are typically much narrower - on the order of 10 meV; they are called minibands. Second, the superlattice is generally one dimensional, unlike the three-dimensional crystal lattice. (Two-dimensional superlattices with lateral structures and patterned metallic electrodes, have also been studied.)

The relation between energy and wavenumber in a miniband of with Δ can be written[3]

$$ E = \frac{\Delta}{2} (1 - \cos ka) \tag{1} $$

reflecting the periodicity of the superlattice with period a. Consequently, the electron group velocity v(k) oscillates with the wavenumber k according to

$$ v(k) = \frac{1}{\hbar} \frac{dE}{dk} = \frac{a\Delta}{2\hbar} \sin ka \tag{2} $$

This is in sharp contrast to the velocity of a free electron, which increases linearly with k. This band picture of transport in a superlattice is valid only if Δ is significantly larger than the collisional broadening $\hbar/\tau$ (where τ is the scattering time) or, equivalently, if the mean free path is substantially greater than the superlattice period.

Applying an electric field F causes the wavevector k to increase linearly with time according to the quasi-Newtonian expression $\hbar dk/dt = eF$. For free electrons this would lead to a continuous increase in the velocity. But in a superlattice the velocity decreases once the wavevector k crosses $\pi/2a$, as we see from Eq. 2. Esaki and Tsu, in their seminal 1970 paper,[3] showed that the steady-state drift velocity also decreases in the presence of collisions, once the field F exceeds the threshold value $\hbar/ea\tau$. For such fields the electron distribution transfers to regions of the band structure where the group velocity decreases with increasing k according to Eq. 2, because of Bragg reflections from the miniband boundary. For typical superlattice periods, on the order of 100 Å and collision times τ of about 3×10^{-12} seconds, one gets a threshold field of 3×10^3 V/cm for the onset of negative differential resistance. This effect has proved elusive until recently. One difficulty in the observation of this phenomenon arises from the space charge injection from the contacts which gives rise to a nonuniform electric field and high field domains in the superlattice. Another problem is the presence of negative differential resistance due to the electron transfer effect into the satellite valleys, which in most III-V semiconductors occurs for values of the electric field comparable to those predicted by Esaki and Tsu in a superlattice.

The above difficulties can be overcome using the structure of Fig. 2(a).[6] Here electrons can be injected at arbitrarily low density into the superlattice by controlling the base emitter bias. The superlattice is placed in the collector of a bipolar transistor, which is very low doped. Under these conditions the electric field across the superlattice remains uniform. In addition, in our bipolar transistor structure one cannot observe NDR by intervalley transfer in conditions of constant emitter current injection. In these conditions, in fact, the decrease in velocity caused by the higher effective mass of the

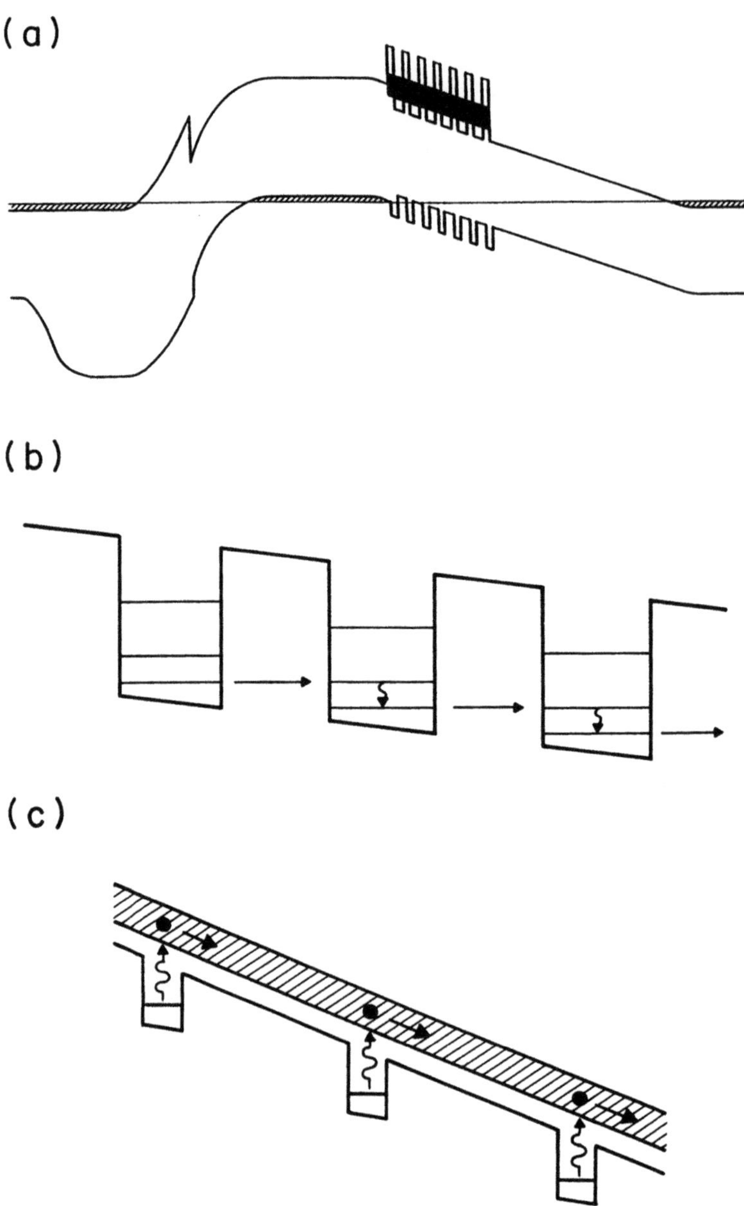

Fig. 2   Superlattice devices.

  a. Heterojunction transistor used to inject electrons into the miniband of the superlattice placed in the collector layer. The shaded region in the superlattice indicates the first miniband.

  b. Sequential resonant tunneling between coupled wells.

  c. Quantum well infrared photodetector based on intersubband absorption between bound and continuum states.

satellite valleys is compensated by an increase in the carrier density; the collector current is therefore not altered. On the other hand, the Esaki and Tsu mechanism can be observed since the Bragg-reflected electrons in the negative-mass region of the miniband give rise to an opposite flux so that the collector current decreases while the base current increases to maintain a constant emitter bias. We recently implemented this structure. It had an AlInAs emitter, a quaternary AlInGaAs base and an AlInAs (17 Å)/GaInAs (37 Å) 14 periods undoped superlattice followed by an InGaAs undoped layer in the collector. Negative differential conductance was observed in the collector current measured as a function of collector base bias, at a constant injected emitter current.[6] This represents the first clear evidence of the negative differential conductance by Bragg reflection predicted by Esaki and Tsu.

If the field is much greater than $\hbar/ea\tau$, one might expect the electron's wavenumber to increase to a value several times $\pi/2a$ before it is scattered. Consequently the velocity should display oscillatory behavior in real space as well as velocity space. This suggests the possibility for extracting coherent radiation with an appropriately designed device (Bloch oscillator).[3]

In a recent landmark experiment, coherent submillimeter-wave emission from optically excited electrons undergoing Bloch oscillations has been observed in an AlGaAs/GaAs superlattice. The oscillation frequency was tuned with the applied electric field from 0.5 THz to 2 THz, in accordance with the expression for the Bloch frequency

$$\omega_B = \frac{eFa}{\hbar}.[7]$$

In superlattices with weak coupling between wells (typically ≥ 100 Å), transport is best described in terms of tunneling between the localized states of the individual wells (Fig. 2(b)). This sequential resonant tunneling process has been clearly observed via photocurrent measurements in a 35 period AlInAs/GaInAs superlattice with wells and barriers about 140 Å thick.[8]

Quantum states are also formed in the classical continuum above the barriers, due to quantum reflections. Barry Levine and coworkers at Bell Labs have recently used these states to demonstrate a new infrared detector (Fig. 2(c)) sensitive in the $8-12$ μm wavelength region.[1] High quality thermal images, with a noise equivalent temperature difference of 10 mK have been obtained using 128×128 arrays of these detectors multiplexed to Si C-MOS circuitry. The potential advantage of such detectors over conventional HgCdTe devices lies in the virtues of GaAs-AlGaAs technology. This material system, which includes the substrate, provides superior overall quality and reliability. Furthermore, the readout electronics might in future be monolithically integrated on the same GaAs substrate.

COUPLED QUANTUM WELL MOLECULES WITH GIANT NONLINEAR OPTICAL PROPERTIES

The structures discussed in this section can be viewed as "quasi-molecules" with giant dipole matrix elements and nearly equally spaced energy levels (Fig. 3). These characteristics are responsible for their very large nonlinear optical susceptibilities.[9] Both were grown in the AlInAs/GaInAs system lattice matched to InP, and only the thickest well is doped n-type. The choice of this material system facilitates the tunnel

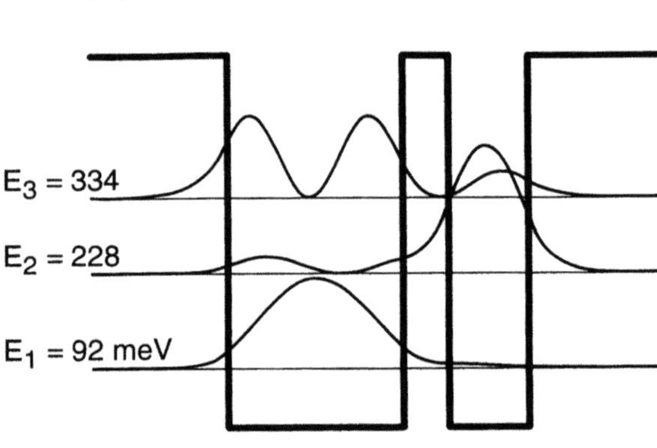

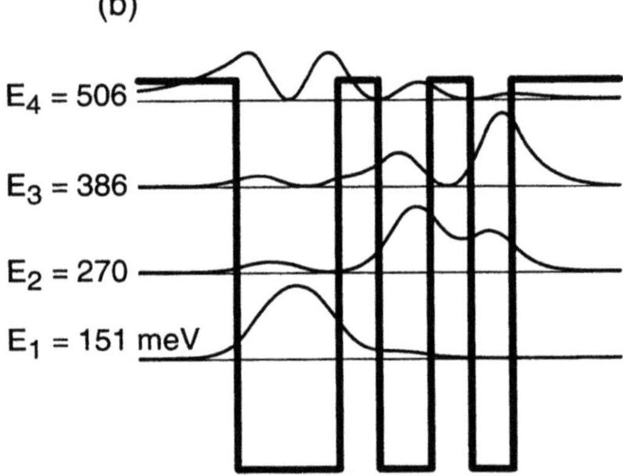

Fig. 3 Conduction band energy diagrams of a single period of the AlInAs/GaInAs coupled quantum well nonlinear optical structures. Shown are the positions of the calculated energy subbands and the corresponding modulus squared of the wave functions. (a) Structure used for resonant second harmonic generation. The GaInAs wells have thicknesses of 64 Å and 28 Å and are separated by a 16 Å AlInAs barrier; (b) Structure used for triply resonant third harmonic generation. The GaInAs wells have thicknesses of 42, 20 and 18 Å, respectively and are separated by 16 Å AlInAs barriers.

coupling between the layers due to the low effective mass of the barrier region compared to AlGaAs and provides a large band discontinuity essential for confining four states in the three-well structure separated by an energy corresponding to that of the photon from a CO_2 laser (≈ 130 meV). To optically excite the quantized motion normal to the layer interfaces, one must use light with a component of the polarization normal to the layers. In our experiments this was done using linearly polarized light in a multipass waveguide structure wedged at 45 degrees. In our second harmonic generation (SHG) experiment a coherent polarization is created at double the frequency ω of the pump wave (a CO_2 laser beam) due to the lack of reflection symmetry of our two-well structure (Fig. 8(a)). This coherent polarization radiates a wave of frequency 2ω colinear with the pump. The vicinity of the pump photon energy to $E_2 - E_1$ and of the second harmonic photon to $E_3 - E_1$ produces a strong resonant enhancement of the nonlinear susceptibility $\chi^{(2)}_{2\omega}$ associated with SHG.[9] The maximum susceptibility $(\chi^{(2)}_{2\omega})$ corresponds to exact matching, i.e. $\hbar\omega = E_2 - E_1 = E_3 - E_2$. This can be achieved using the large linear Stark effect typical of this structure by applying an electric field of suitable polarity. In these conditions a $|\chi^{(2)}_{2\omega}| = 10^{-7}$ m/V was measured, approximately three hundred times the value of $|\chi^{(2)}_{2\omega}|$ in bulk GaAs at $\lambda = 10$ μm.[9]

The three-well structure (Fig. 3(b)) with the near equal separation of its four energy levels is suitable for triply resonant third harmonic generation (THG). In this process a pump wave at frequency ω sets up a nonlinear polarization at 3ω which coherently radiates a wave at this frequency.[9] The nonlinear susceptibility $\chi^{(3)}_{3\omega}$ which enters the expression for the polarization is strongly enhanced when the condition $\hbar\omega = E_2 - E_1 = E_3 - E_2 = E_4 - E_3$ is met. It is important to note that a parabolic well (i.e. an harmonic oscillator potential) is unsuitable for this purpose since in such a system the electron oscillations are linear. THG experiments in the structure of Fig. 3(b) have found a $|\chi^{(3)}_{3\omega}| = 10^{-14}$ (m/V)^2 at 300 K.[9] At cryogenic temperatures $|\chi^{(3)}_{3\omega}|$ is four times larger. These are the highest third order susceptibility of any known material system. The three-coupled well structure was also used to study multiphoton electron escape from a well, the analog of multiphoton ionization of an atom.[9] In this process electrons in the biased well are photoexcited into the continuum via a CO_2 laser using a three photon transition giving rise to a photocurrent. The cross section for this process is found to be many orders of magnitude larger than in atomic and molecules.

FABRY-PEROT ELECTRON FILTERS AND BOUND STATES IN THE CONTINUUM

In the previous section we have seen how confined states of quantum wells are central in number of tunneling and optical phenomena. Highly localized states can also be created at energies above the barrier height in a potential well using constructive interference phenomena.[10,11]

Consider first a conventional rectangular well (Fig. 4(a)). At energies greater than the barrier height one has a continuum of scattering states. For discrete energies corresponding to a semi-integer number of electron wavelengths across the well one finds transmission resonances. Although at these energies the electron amplitude in the well layer is enhanced, the wave functions do no decay exponentially in the barrier, unlike the

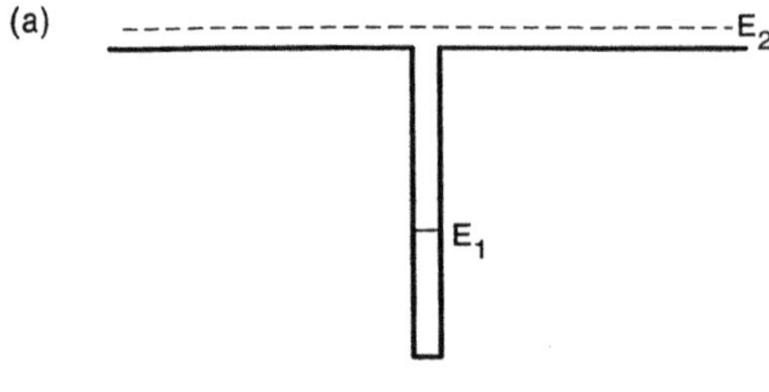
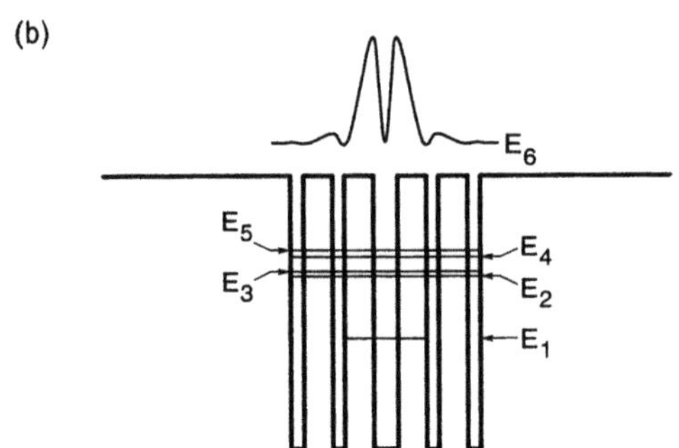
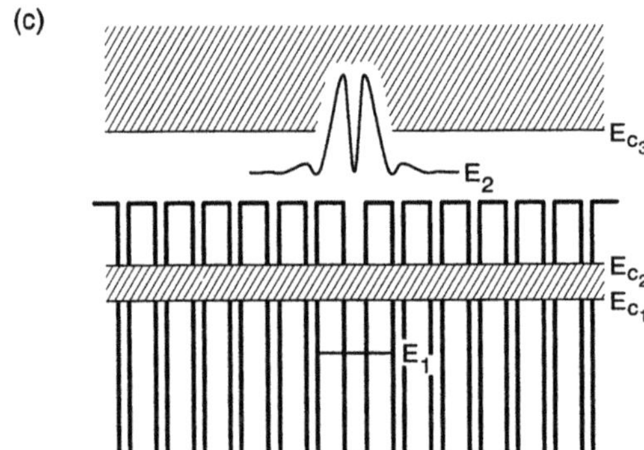

Fig. 4   Conduction band diagrams of AlInAs/GaInAs heterostructures used in the study of continuum states: (a) reference sample. Shown are the ground state of the well (E_1 = 204 meV) and the position (dashed line) of the first transmission resonance in the continuum (E_2 = 560 meV); (b) quantum well cladded by two-period quarter-wave stacks (Fabry-Perot electronic filter). Shown is $|\psi|^2$ of the localized quasi-bound state (E_6 = 560 meV) formed in correspondence to the transmission resonance and the positions of new states created at lower energies; (c) in the superlattice limit the $\lambda/4$ stacks behave as Bragg reflectors. The state above the well now becomes a bound state localized by the superlattice minigap (= 266 meV).

confined states of the well, but are plane-wave-like. These states can be localized in the well using as barriers stacks of layers of thickness $\lambda/4$ each where λ is the de Broglie wavelength in the layer (at the energy of the selected transmission resonance) (Figs. 4(b)-(c)). Constructive interference between the waves partially reflected by the heterointerfaces of the $\lambda/4$ stacks leads to the formation of a quasi-bound state above the center well (Fig. 4(b)). This strongly narrows the transmission resonance in analogy with a Fabry-Perot optical filter where sharp optical resonances are produced using as high reflectivity mirrors quarter-wave stacks. The degree of localization increases with the number of periods; in the structure with just two-period stacks, the wave function is already highly confined (Fig. 4(b)). In the superlattice limit and at low temperatures the stacks become Bragg reflectors; a minigap opens up (Fig. 4(c)) and the localized state becomes a bound state at energies greater than the barrier height. The prediction that certain oscillatory potentials support bound states in the continuum, due to quantum interference, was first put forth by von Neumann and Wigner in 1929.

The reference sample (Fig. 4(a)) had twenty 32 Å InGaAs quantum wells n-type doped separated by 150 Å undoped AlInAs barriers. In the other three structures the 32 Å wells, doped to the same level, were cladded, respectively, by one-period, two-period (Fig. 4(b)) and six-period (Fig. 4(c)) $\lambda/4$ stacks consisting of 39 Å AlInAs barriers and 16 Å GaInAs wells, designed as discussed above. The phase coherence length in the superlattice structure of Fig. 4(c) is estimated to be ~300 Å at 10 K.

The absorption spectra of the reference sample is broad with a long wavelength cutoff determined by the height of the barrier. In the structure with one $\lambda/4$ period the peak is considerably narrower and centered at an energy corresponding to the transition between the ground state of the well and the localized resonant state at the energy E_6. As the number of quarter-wave stacks is doubled the absorption peak does not shift and considerably narrows, precisely the behavior expected for a Fabry-Perot. In fact the observed narrowing (16 meV) can be quantitatively explained in terms of the reflectivity increase of the $\lambda/4$ stacks.[11] In the structure with six periods at cryogenic temperatures, the highly localized state becomes effectively a bound state confined by Bragg reflectors from the superlattice. The absorption spectrum shows an isolated peak at 360 meV of width ~10 meV corresponding to the transition from the state E_1 to the state E_2 in Fig. 4(c).[10] It is worth noting that the width of the transition to the confined state above the well in the two- and six-period structure is identical to that of the bound-to-bound state transition measured in a conventional 55 Å thick GaInAs well with 300 Å thick barriers, thus demonstrating the bound nature of the state above the well.

QUANTUM CASCADE LASER: A UNIPOLAR INTERSUBBAND SEMICONDUCTOR LASER

A new semiconductor injection laser (Quantum Cascade Laser) which differs in a fundamental way from diode lasers has been recently demonstrated.[12,13] It relies on only one type of carrier (it is a unipolar semiconductor laser), and on electronic transitions between conduction band energy levels of coupled quantum wells.

Intersubband lasers were originally proposed 25 years ago,[14] but despite considerable effort this is the first structure to achieve laser action. The present device operates at a wavelength of 4.26 microns, but since the wavelength is entirely determined by quantum confinement, it can be tailored from the mid-infrared to the submillimeter region using the same heterostructure material. Electrons streaming down a potential staircase sequentially emit photons at the steps. The latter consists of coupled quantum wells in which population inversion between discrete conduction band excited states is achieved in a 4-level atomic-like laser scheme using tunneling injection (Fig.5). The AlInAs/GaInAs structure grown by MBE comprises 25 stages, each consisting of a graded gap n-type injection layer and a three coupled-well active region, cladded by waveguiding layers. The undoped active region includes 8 Å and 3.5 Å thick GaInAs wells separated by 3.5 Å AlIAs barriers. The reduced spatial overlap between the states of the laser transition and the strong tunnel-coupling to a nearby 2.8 Å GaInAs well ensure population inversion. The samples were processed into mesa etched waveguides and the laser facets were obtained by cleaving. Powers $\approx$ 20 mW in pulsed operation (20 ns pulses with a 10^{-3} duty cycle) have been obtained at 80 K. Operating temperatures up to 125 K have been achieved with 5 mW of power. A dramatic narrowing of the emission spectrum provides direct evidence of laser action.[12,13] An outstanding feature of this laser is that the gain is much less sensitive to temperature than conventional semiconductor lasers. In addition, the intrinsic linewidth of these lasers in cw single mode operation is expected to be Schawlow-Townes limited, similar to atomic lasers, without the linewidth enhancement factor typical of diode lasers. Well defined longitudinal modes are observed in the high resolution spectrum at two different currents. The mode spacing is in good agreement with the calculated one.[12,13] The linewidth of the dominant mode is presently limited by heating effects and mode hopping during the pulse.

ACKNOWLEDGEMENTS

It is a pleasure to acknowledge the many collaborators and colleagues who have significantly contributed to this work: F. Beltram, A. Y. Cho, S. N. G. Chu, J. Faist, A. R. A. Kiehl, L. M. Lunardi, R. J. Malik S. S. Sen, D. L. Sivco, C. Sirtori, P. R. Smith,

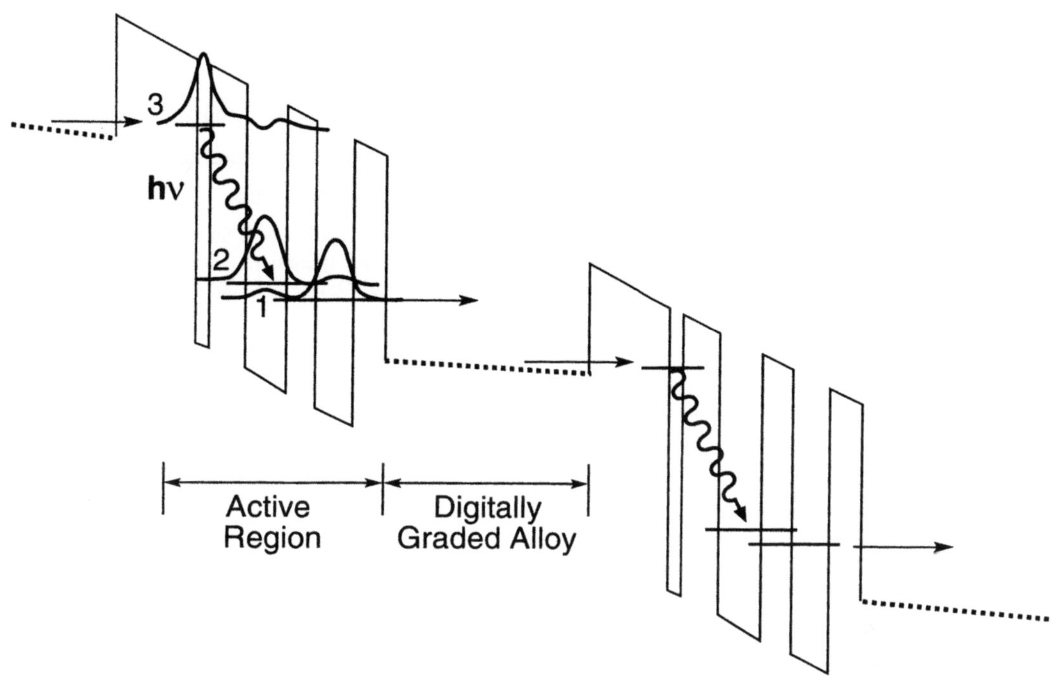

Fig. 5 Energy diagram of quantum cascade laser under operating conditions. The energy level separations are: $E_3 - E_2 = 295$ meV (laser transition); $E_2 - E_1 \simeq 30$ meV. The latter separation, close to the optical phonon energy, ensures ultrafast electron relaxation ($\tau_{21} \approx 0.6$ ps) between the two states. The scattering time from subband n = 3 to subband n = 2 is $\simeq 4.6$ ps due to the large energy separation and the reduced spatial overlap of the two states. Thus population inversion is ensured since $\tau_{32} > \tau_{21}$. Electrons tunnel readily out of state n = 1.

REFERENCES

1. F. Capasso and A.Y. Cho, Bandgap engineering of semiconductor heterostructures by molecular beam epitaxy: physics and applications, *Surf. Sci.* 299/300:878 (1994).

2. A.Y. Cho, Advances in molecular beam epitaxy (MBE), *J. Cryst. Growth* 111:1 (1991).

3. C. Weisbuch and B. Vinter. "Quantum Semiconductor Structures," Academic Press, San Diego (1991).

4. F. Capasso and S. Datta, Quantum electron devices, *Physics Today* 43:74 (1990).

5. F. Capasso, S. Sen, F. Beltram, L. Lunardi, A. Vengurlekar, P.R. Smith, N.J. Shah, R.J. Malik, and A.Y. Cho, Quantum functional devices: resonant tunneling transistors, circuits with reduced complexity and multiple-valued logic, *IEEE Trans. Electron. Devices* 36:2065 (1989).

6. F. Beltram, F. Capasso, D.L. Sivco, A.L. Hutchinson, S.N.G. Chu, and A.Y. Cho, Scattering controlled transmission resonances and negative differential conductance by field induced localization in semiconductor superlattices, *Phys. Rev. Lett.* 64:3167 (1990).

7. C. Waschke, H.G. Roskos, R. Schwedler, K. Leo, H. Kurz, and K. Köhler, Coherent submillimeter-wave emission from Bloch oscillations in a semiconductor superlattice, *Phys. Rev. Lett.* 70:3319 (1993).

8. F. Capasso, K. Mohammed, and A.Y. Cho, Sequential resonant tunneling through a multiquantum well superlattice, *Appl. Phys. Lett.* 48:478 (1986).

9. F. Capasso, C. Sirtori, and A.Y. Cho, Coupled quantum well semiconductors with giant electric field tunable nonlinear optical properties in the infrared, *IEEE J. Quantum Electron.* 30:133 (1994).

10. F. Capasso, C. Sirtori, J. Faist, D.L. Sivco, S.N.G. Chu, and A.Y. Cho, Observation of an electronic bound state above a potential well, *Nature* 358:565 (1992).

11. C. Sirtori, F. Capasso, J. Faist, D.L. Sivco, S.N.G. Chu, and A.Y. Cho, Quantum wells with localized states at energies above the barrier height: a Fabry-Perot electron filter, *Appl. Phys. Lett.* 61:898 (1992).

12. J. Faist, F. Capasso, D.L. Sivco, C. Sirtori, A.L. Hutchinson, and A.Y. Cho, Quantum cascade laser, *Science* 264:553 (1994).

13. J. Faist, F. Capasso, D.L. Sivco, C. Sirtori, A.L. Hutchinson, and A.Y. Cho, Quantum cascade laser: an intersubband semiconductor laser operating above liquid nitrogen temperatures, *Electron. Lett.* 30:865 (1994).

14. R.F. Kazarinov and R.A. Suris, Possibility of the amplification of electromagnetic waves in a semiconductor with a superlattice, *Sov. Phys. Semicond.* 5:707 (1971).

ELECTRON CONDUCTION AND QUANTUM PHENOMENA IN 2D HETEROSTRUCTURES

Bernard Etienne

Laboratoire de Microstructures et de Microélectronique (L2M)
C.N.R.S.
196 av. H. Ravera, B.P. 107, 92225 Bagneux Cedex
France

ABSTRACT

A quasi two-dimensional electron gas confined at the interface between two semiconductors is a nearly ideal low disordered system. Its low temperature transport properties are dominated by a large variety of quantum phenomena. Some recent theoretical and experimental studies on the correlation energy of the plasma, on the spin-splitting of the conduction band, on the observation of weak antilocalization and on a unified understanding of the Integer and Fractional Quantum Hall effects are discussed in order to illustrate the very rich physics of these artificial structures.

INTRODUCTION

In contrast with most other lectures of this course, we deal here with a quantum system exhibiting a very large number of degrees of freedom : namely, a two-dimensional (2D) degenerate plasma of electrons in Coulomb interaction[1]. Such a plasma is confined in the vicinity of a potential barrier by a strong transverse built-in electrostatic field ($> 10^4$ V/cm). This potential step results from the conduction band offset ($\sim$300 meV) at the interface between two different semiconductors (fig. 1a). The field is created by the interaction

Advances in Quantum Phenomena, Edited by E.G. Beltrametti
and J.-M. Lévy-Leblond, Plenum Press, New York, 1995

between the electrons in the channel and the ionized dopant impurities in the barrier. The electrons are free to move along the plane of the interface but the excitations in the transverse direction are not possible at low temperature. Indeed the inter-subbands transition energy is large (>10 meV) because of the quantization of the transverse motion in the narrow potential well (fig. 1b). The system is referred to therefore as a quasi two-dimensional electron gas (2DEG) but one has to keep in mind that the Coulomb interactions remain three-dimensional.

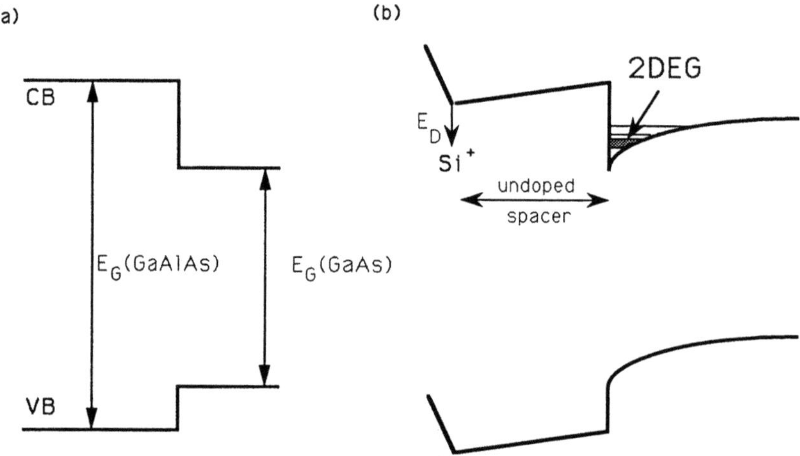

Fig. 1   Sketch of the band structure at the interface between two semiconductors. Left part (a) before and right part (b) after the transfer of electrons from the barrier to the channel.

The most widely used material system is made of AlGaAs/GaAs. The heterojunctions are grown by Molecular Beam Epitaxy (MBE) : the interface is nearly abrupt at atomic scale and the GaAs channel can be obtained with an extremely low content of background impurities ($\sim$1 impurity/10^8 atoms). A calculation of the conduction band profile in the self consistent Hartree approximation[1] is shown in Fig.2. The electrons, which are spatially separated from the ionized donors by an undoped spacer, can have a very high mobility at low temperature (from 10^5 to 10^7 cm^2 V^{-1} s^{-1}). The 2DEG density can be varied between 2×10^{12} and 10^{10} cm^{-2} according to the spacer thickness (0 to 2000 Å). However above 7×10^{11} cm^{-2} , a second subband might be filled, which limits the electron mean free path because of inter subbands scattering.

The physics of these 2D electron gas, obtained in such selectively doped heterostructures, is very rich for many different reasons :

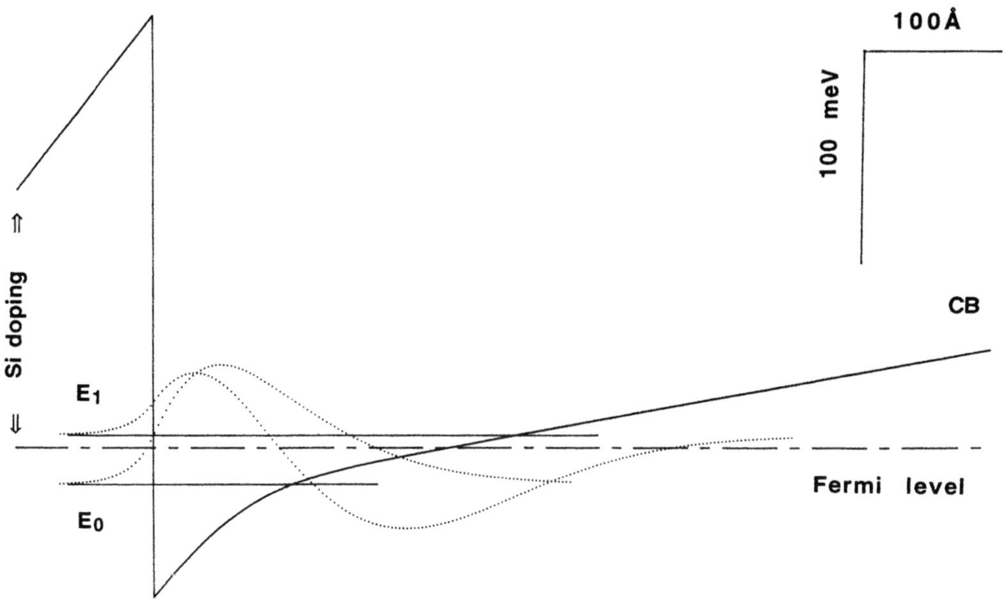

Fig.2   Calculation of the conduction band (CB) profile along the growth axis with the envelope wave function corresponding to the two first subbands (E_0, E_1). The electrical equilibrium in the structure is defined by a constant Fermi level. The scale is as indicated.

i) the electron elastic mean free path l_e is quite large at low temperature (several μm up to several 10 μm, depending on the mobility) because of the spatial separation with the ionized donors. Using small samples, it is possible to reach the regime of ballistic transport (instead of being in the diffusive regime)[2]. The electrical conduction becomes similar to the propagation of a radiation in a wave guide. Ohm's law do not apply.

In addition, the Fermi wave length λ_F being large, as compared to the case of metals, because the carrier density is much weaker (here $\lambda_F = 40$nm for a 2D charge density of 4×10^{11} cm^{-2}), the modes of the wave guide are not so many (typically their number is equal to the ratio between the width W of the sample and λ_F). The nice thing is that, because we deal with charged particles instead of photons, the width of the waveguide can be continuously varied by a gate voltage or the electron trajectories can be deflected by a magnetic field. This leads to the observation of phenomena such as the conductance quantization in quantum point contacts and coherent electron focussing[2].

ii) the phase coherence length L_ϕ of the electrons is large at low temperature (larger then l_e).
The wave like nature of the carriers, which modify the electrical conductivity of metallic systems, is therefore revealed by various quantum effects[3] : weak localization and antilocalization, universal conductance fluctuations (in wires), Aharanov-Bohm effect and persistent current[4] (in rings). These effects have usually been discovered in thin metallic layers but are conveniently studied in heterojunctions because l_e and consequently L_ϕ can be varied by the choice of the structure.

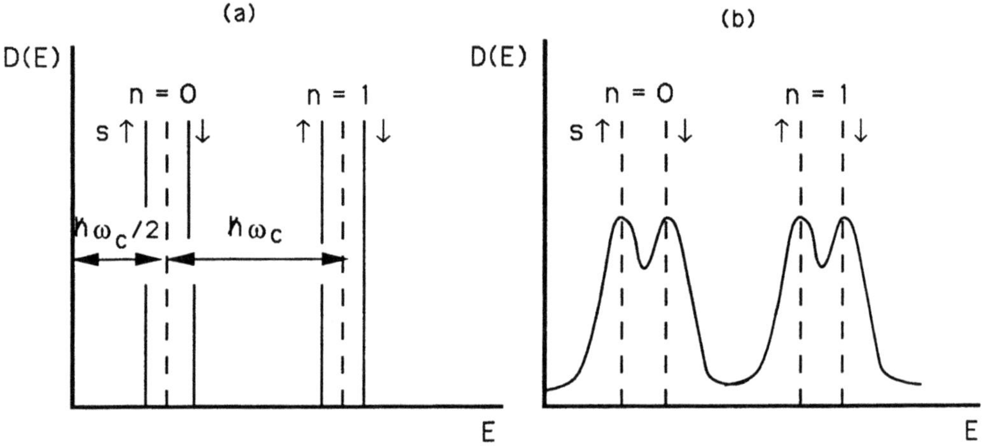

Fig.3   Density of electron states in the regime of Landau quantization. Left (a) in an ideal (not disordered) structure, right (b) in a real (disordered) structure.

iii) under a strong transverse magnetic field, the electron states are fully quantized : in the transverse direction by the potential well and in the plane by Landau quantization (Fig. 3). At integer or fractional ratio of the number of electrons N_e to the number of quantum fluxes $N_\Phi = B/(h/e)$, it has been discovered that the Hall resistance is quantized and equal to the value of h/e^2 divided by this ratio. A plateau is observed over a finite range of the magnetic field (Fig. 4) : this is the Integer Quantum Hall Effect (IQHE)[5] and the Fractional Quantum Hall Effect (FQHE)[6]. The remarkable fact is that the quantized resistance is only determined by fundamental constants. These unexpected discoveries have been the onset of very active theoretical and experimental research on the quantum states of matter.

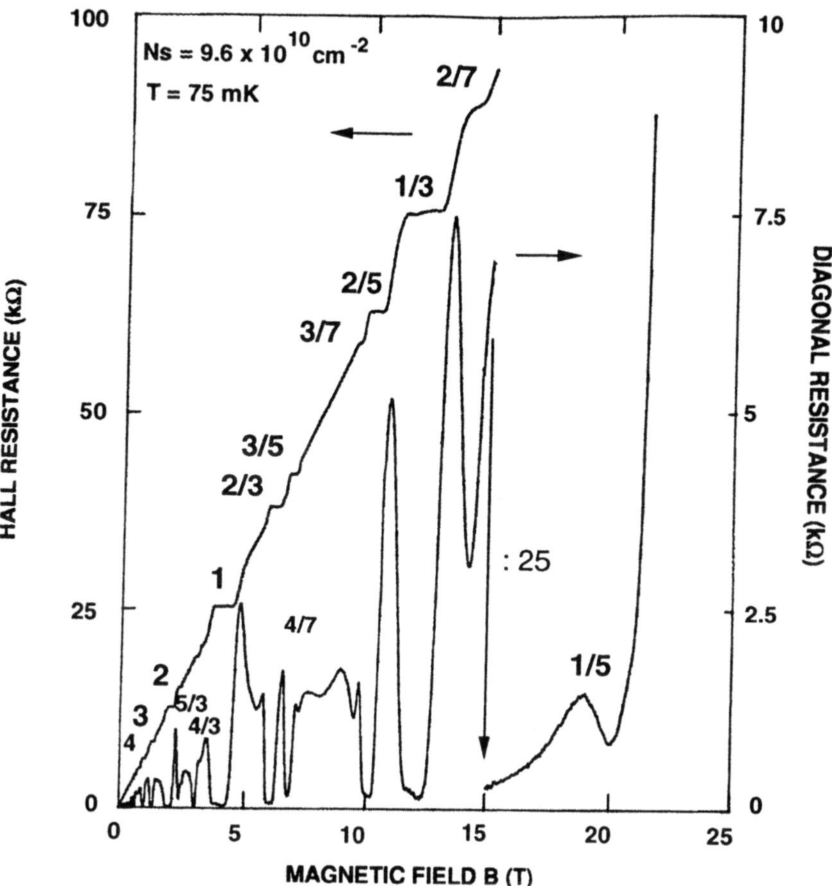

Fig.4 Hall resistance (left scale) and diagonal resistance (Shubnikov-de Haas oscillations) (right scale) as a function of magnetic field in a high quality heterojunction. Hall plateaus and S.-de H. minima are observed for the integer or fractional values of ν indicated by bold numbers (E. Paris, PhD thesis, U. of Paris VI, unpublished).

At still higher magnetic field, it is possible to reach the regime of Wigner crystallization : this is a quantum transition to an ordered state (in real space) of the electron plasma (Fig.5). The existence of this magnetically induced Wigner solid (MIWS), predicted long ago, is only possible in 2DEG in which the correlations between electrons dominate over the single particle localisation resulting from potential fluctuations[7]. This requires very low disordered heterojunctions. A more complete understanding of the competition between the Wigner transition and the FQHE is limited by the quality of the samples.

The aim of this article is necessary limited and detailed reviews can be found on the electronic properties of two-dimensional systems[1], the quantum transport in semiconductor nanostructures[2], the integer[8] or the fractional[9] quantum Hall effect. I have chosen to discuss some problems, which are becoming important or some new developments which are not discussed extensively in these reviews. The aim is to be informative to people working in other fields. The discussion is limited to the case of translationally invariant 2D systems for sake of simplicity, even if some points can be of great interest for the currently very active research on modulated 2DEG and on quantum wires[2].

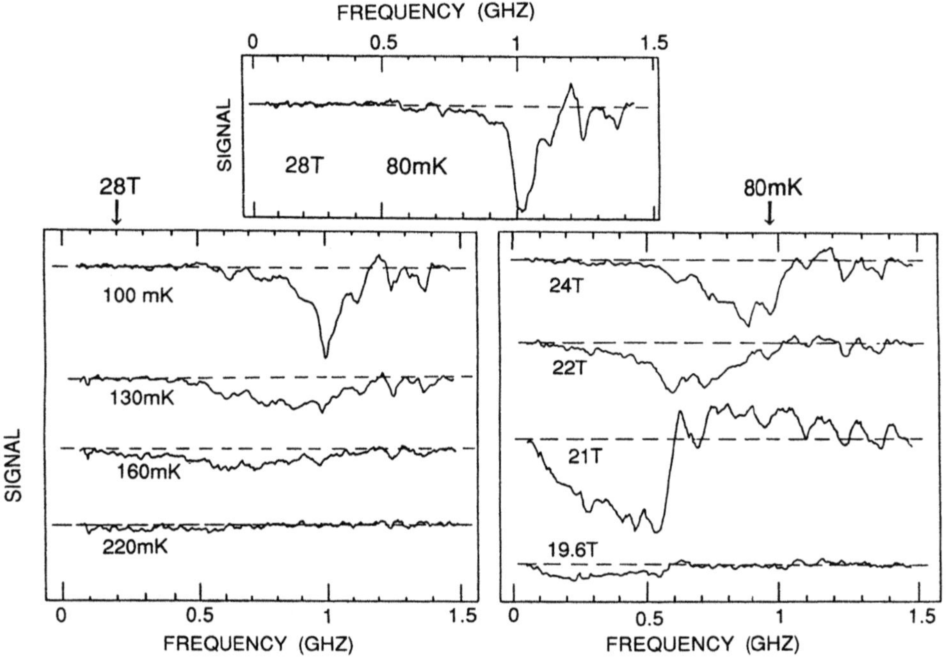

Fig.5   The Wigner transition is detected by the rf absorption due to shear excitations in an electron solid. This signature disappears either on increasing the temperature (bottom left) or decreasing the magnetic field (bottom right) (E.Y. Andrei et al., ref.7).

COHESIVE ENERGY OF THE 2D ELECTRON GAS

In the electron systems considered here, the plasma parameter r_s (this is the mean interelectron distance in unit of the 3D effective Bohr radius, $a_B = 100\text{Å}$ in GaAs) varies between 1 and 2. But it is well known that, unless r_s is much smaller than 1 (dense plasma), the contribution to the cohesive energy coming from correlations cannot be neglected with respect to the kinetic and exchange energies[10]. The Hartree-Fock approximation gives a mean energy per electron equal in 2D to $1.0/r_s^2 - 1.2/r_s$ in unit of the effective Rydberg ($R_Y^* = 4$ meV in GaAs). It is easy to guess however that the correlations should be even more important than in 3D so that calculations requiring the use of many-body theories are needed[11].

We have in our case the additional difficulty that we deal with quasi 2D electron systems : there are several electric subbands (Fig.2), and intersubbands transitions should be included in the calculations. But this is too difficult and the only effect usually taken into account is the screening of the Coulomb interaction by a form factor obtained with a realistic transverse density profile[1]. A reduction of the band gap of the host semiconductor in the region occupied by the plasma of the order of 10 meV is obtained[12]. More involved calculations of the correlation energy and of the pair-correlation function (i.e. of the hole exchange) have appeared recently[13].

These calculations are usually done in the case of the so-called jellium model : the electrons are imbedded into a uniform background of positive compensating charges. For quasi 2D systems, the direct term (Hartree term) is non zero, because of the charge separation between the positive ions in the barrier and the electrons in the channel : the neutrality is global but not local. This direct term is usually combined with the electron-ion interaction and the sum has the following expression :

$$V_H(r,z) = - \frac{2\pi e^2}{S} \sum_{ns} N_{ns} \int dz' \ |z\text{-}z'| \left\{ \ |\zeta_{ns}(z')|^2 - \rho_{ion.}(z') \right\} \qquad (1)$$

where $\zeta_{ns}(z)$ is the electron wave function in the subband of order n and spin s, which is occupied by N_{ns} electrons, and $\rho_{ion.}(z)$ is the ionic density. Eq.1 shows that V_H grows with the charge separation.

In the case of charge transfer into a quantum well, the electrons are confined in the structure by the atomic potential : it is possible therefore to calculate first the ground state without electron-electron interaction and then to introduce the Coulomb interactions only in a second step[14]. This can be done using pertubation theory up to second order (divergences appear only at third order in 2D instead of second order in 3D). This is already enough to get some part of the exchange and correlation energies[15] (without neglecting the intersubbands transitions). The result proves clearly that it is not enough to take into account only the Hartree term as it is done most of the time, or that, if exchange and correlation are neglected, a pertubation treatment of this term is sufficient[15].

In the case of a single interface heterojunction, it is not possible to neglect the Coulomb interactions at the beginning, in order to get a bound state. Therefore the Hartree term (Eq. 1) is included in a one-electron hamiltonian :

$$H = T + V_C(z) + V_H(z) \tag{2}$$

in order to get the ground state of the system in a self-consistent Hartree approximation (Vc is the potential profile of the heterojunction for the conduction band). The solution is usually obtained solving self-consistently Schrödinger and Poisson equations. This is the easiest way to get a reasonable value of the energy levels and of the charge transfer. It can be also proved that, in spite of the intricacies due to intersubband transitions, no direct contribution remains in the N electron hamiltonian written in the basis of these self-consistent Hartree states. This is therefore a good starting point.

This is not the case with the method of ref. 15, in which all the interaction terms are considered simultaneously : diagrams involving the Hartree term would appear at any order. It is not yet clear to know what is the best compromise between the different methods. This is not so important for 2DEG because the charge density can be easily measured and the quantum effects on the transport and magnetotransport properties do not depend on a detailed knowledge of the band structure. But this is no longer true in the case of modulated structures, with potential modulation on a lateral scale comparable to the Fermi wavelength, because it is then important to predict the Fermi level position (in a miniband or in a minigap) and the charge modulation.

SPIN-SPLITTING OF THE CONDUCTION BAND

In high density 2DEG, the spin dependance of the electron states can be measured and the splitting gives rise to original properties. This is discussed in the Hartree approximation, ignoring all of the correlations discussed above.

The III-V semiconductors, such as GaAs, crystallize in the zinc-blende structure. The atoms of each family (III or V) belong to a face-centered cubic lattice. Each atom is at the center of a regular tetrahedron formed by four atoms of the other type. There is no inversion symmetry in the crystal, contrary to the case of the diamond structure, in which the same atoms (for instance Si) occupy both sublattices. Therefore it can be deduced from group theory that in

III-V compounds, there is no Kramers degeneracy. The energy of state (k, s) is not equal to the energy of state (k, -s) with opposite spin.

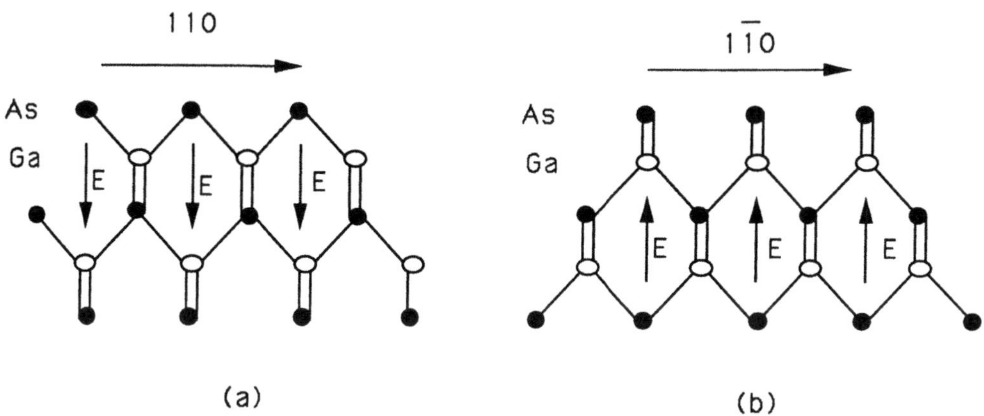

Fig.6   GaAs crystal structure along [110] and [1̄10] projected in the plane of the figure (each line
        represents a bond). The direction of the electric field E in a cell is indicated.

The physical origin of the splitting can be understood as following. The bondings are not purely covalent, as in Si, but have a slight polar character. As a consequence, an electron or a hole moving in a cristallographic direction such as [110] feels an electric field E. In fact this field is averaged to zero in a state of symmetry S, but not for lower symmetry. In the reference frame attached to the particle moving with velocity $\mathbf{v}$, an internal magnetic field results from the Lorentz transformation : $\mathbf{B}_{int} = -(\mathbf{v}/c^2) \wedge \mathbf{E}$. The spin states are therefore split (Zeeman effect). In the conduction band this effect is very small : the states are exactly S type at Brillouin zone center but they have anywhere else some P type component, proportional to the wave vector k, coming from a coupling with other upper and lower bands. The splitting can be calculated by considering a 14-band model[16]. The velocity and the associated Lorentz field being also proportional to k, the matrix element giving the spin splitting is expected to go as k^3 in bulk GaAs. The spin-orbit coupling hamiltonian for electron states can be written :

$$H_{so} = g \mu_B \mathbf{B}_{int}.\sigma \tag{3}$$

where g is the gyromagnetic ratio, μ_B the Bohr magneton and σ (σ_x, σ_y, σ_z) are the Pauli matrices. The directions x, y, z are along the principal axis of the crystal ([100] and equivalent). The components of the field are :

$$\mathbf{B}_{int,x} = \frac{\gamma}{\mu_B} \; k_x \, (k_y^2 - k_z^2) \tag{4}$$

$$\mathbf{B}_{int,y} \text{ and } \mathbf{B}_{int,z} \text{ by circular permutation}$$

It can be easily verified that $\mathbf{B}_{int}$ is orthogonal to the electron velocity as expected from the Lorentz transformation. The magnetic field is zero for electron moving in the directions such as [100] and the sign of the electric field is reversed for electrons moving [110] or [1$\bar{1}$0] directions, as expected from the lattice structure.

In most of the studies in bulk GaAs, this splitting is usually ignored because it is very small. But this is not so in selectively doped heterojunctions because, at the Fermi energy, the wavevector of a high density degenerate electron gas is large. In addition the spin-splitting modify strongly the quantum corrections to the conductivity as will be shown in the next part.

For a quasi 2D electron system in the electric quantum limit, i.e. with one electric sub-band populated, eq. (4) are used replacing k_z and $(k_z)^2$ by their mean value in this state, i.e. respectively 0 (because we have a bounded state) and κ^2. We get :

$$\mathbf{B}_{int,x} = \frac{\gamma}{\mu_B} \; k_x \, (k_y^2 - \kappa^2) \tag{5}$$

$$\mathbf{B}_{int,y} = \frac{\gamma}{\mu_B} \; k_y \, (\kappa^2 - k_x^2)$$

$$\mathbf{B}_{int,z} = 0$$

The dependance with the in-plane wave vector contains now linear terms and cubic terms, which can be of the same order of magnitude at large 2DEG density. The magnetic field is however no longer orthogonal to the electron velocity. This is in fact not so paradoxical because a confined quasi 2D state can be viewed as the coherent superposition of two 3D states of opposite k_z wave vector. Adding the two 3D Lorentz fields, we get again the same kind of expressions.

The relations given above lead to the following expression for the spin splitting :

$$\Delta E(k, \theta) = \pm \gamma \sqrt{\kappa^4 k^2 - \kappa^2 k^4 \sin^2 2\theta + k^6 \, \frac{\sin^2 2\theta}{4}} \tag{6}$$

where k is the module of the wavevector and θ its azimuth with respect to the in-plane [100] axis. In a first approach this formula can be averaged over azimuth.

Spin-flip single-particle excitations (Fig. 7) between the spin-split conduction bands of a high density 2DEG (confined in a selectively doped quantum well of width 180 Å) has been observed by Raman scattering[17]. An electron on the Fermi surface is excited into the band of opposite spin, with a change in wave vector **q** defined by the incident laser energy and the geometry of the scattering experiment. A value of 0.37 meV is obtained for n_s = 1.3 10^{12} cm^{-2} (κ = 1.7 10^6 cm^{-1} and k_F = 2.8 10^6 cm^{-1}).

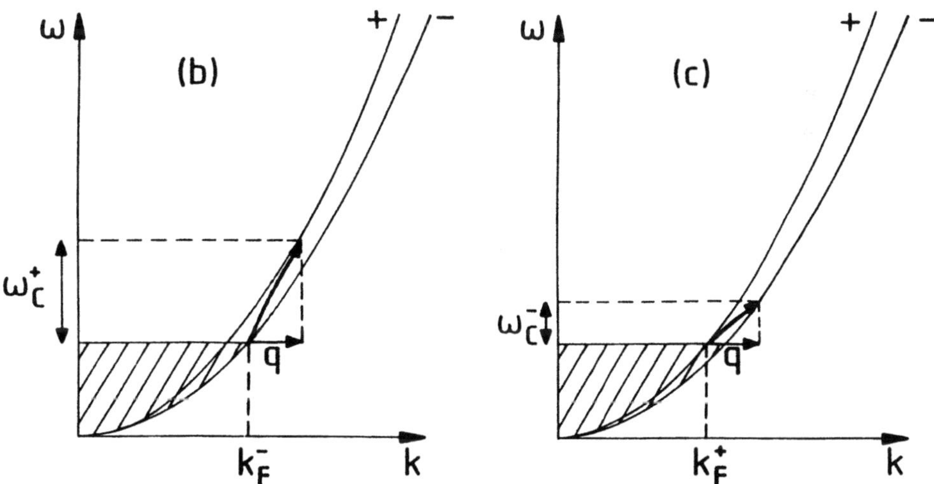

Fig. 7   If the conduction band is spin-split, the single-particle spin-flip excitations of wave-vector q correspond to two different energies (B. Jusserand et al., ref. 17).

In fact this lack of centro-symmetry of the host crystal is not the only origin of the measured spin splitting. Selectively doped structures are not symmetric along the growth axis because of the confinement built-in electric field (Fig.2). This gives an other contribution to the spin splitting, the so-called Rashba term[18] which is isotropic and depends linearly on k. The field associated to the potential step at the barrier has also to be included in this term[19].

When the conduction band Hamiltonian is diagonalized with the band structure and the Rashba contributions, it appears that the four-fold in-plane symmetry, obtained when each contribution is considered separately, is broken : the splitting is no longer the same in the [110] and [1$\bar{1}$0] directions. This can be easily understood keeping in mind the change of sign of the polar electric field between these directions, mentioned previously. The azimuthal dependance can be probed by Raman scattering and the experimental results confirm that [110] and [1$\bar{1}$0] are not equivalent while [100] and [010] are equivalent directions. The measurements gives a value for the coupling parameters (such γ as in Eq. 6) for both types of contribution[19].

WEAK LOCALIZATION AND ANTILOCALIZATION

The spin splitting is small but is revealed by the low-field magneto-resistance. We discuss the differences of behavior expected without or with spin splitting.

a) without spin

The conductivity of the electron gas can be obtained semiclassically using the Boltzmann transport equation. At zero temperature, the conductivity can be expressed using the Drude formula :

$$\sigma_D = \frac{n_s e^2 \tau}{m} \qquad (7)$$

where τ is the elastic collision time.

This expression can be transformed easily into the following form (valid in 2D) in which the quantum of conductance ($\sigma_Q = e^2/h$) appears explicitly :

$$\sigma_D = \frac{e^2}{h} (k_F l_e) \qquad (8)$$

where l_e is the elastic mean free path. The semi-classical calculation is meaningful in the case $k_F l_e \gg 1$ which is well obeyed in 2DEG (the limit $k_F l_e \ll 1$ corresponds to the strong localization regime)[1]. But non negligible corrections, resulting from electron-electron interactions and interference effects, can be evidenced by the study of the conductivity variations as a function of temperature or magnetic field[2].

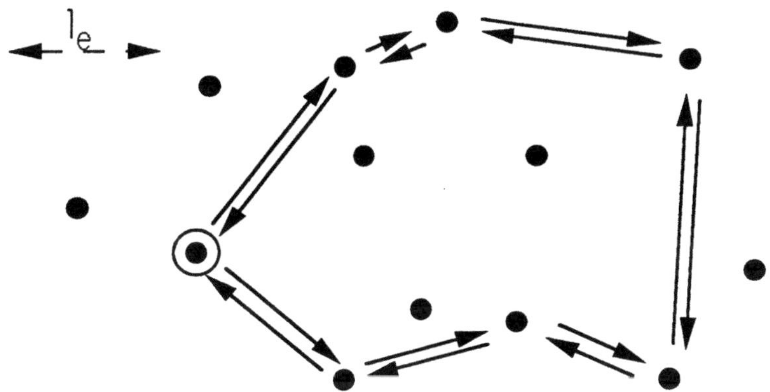

Fig.8   A closed path (plain line) with its time reversed path (dotted line) contributing to the coherent back scattering (weak localization).

The interferences along a closed coherent path between two propagating waves $\Psi_+(R)$ and $\Psi_-(R)$, one being the time-reversed of the other (Fig. 8), increase the back scattering probability by a factor of 2 because these waves come back in phase. The probability P of return to the origin is :

$$P(0) = [\Psi_+(O) + \Psi_-(O)]^2 = 2\left[|\Psi_+(O)|^2 + |\Psi_-(O)|^2\right] \qquad (9)$$

$$= [A\,e^{i\Phi} + A\,e^{-i\Phi'}]^2 = 2A^2[1+\cos(\Phi-\Phi')]$$

$$= 4A^2 \qquad\qquad \text{because } \Phi = \Phi'$$

As a result, the conductivity is decreased : this effect is known as weak localization. The correction to the conductivity can be obtained using classical diffusion (the inelastic length $L\phi$ is supposed to be much larger than the mean free path l_e and is the cut-off length in the problem) :

$$\delta\sigma_{WL} = -\frac{e^2}{\pi^2\hbar}\,\text{Log}\,\frac{L_\phi}{l_e} \qquad (10)$$

This result applies for 2D weak localization (the dimensions of the sample are much larger than Lϕ). A more sophisticated calculation which does not require the diffusion approximation, (Lϕ >>l$_e$), can be performed[20].

When a transverse magnetic field is applied, the time-reversal invariance is broken. The associated potential vector modifies the phase of the two waves around a closed loop in the same way as in an Aharonov-Bohm experiment, i.e. by $\pm 2\pi\Phi/\Phi_0$ (Φ = BS is the enclosed flux and Φ_0 = h/e is the flux quantum). This is equal to $S/(l_B)^2$, where $l_B = (\hbar/eB)^{1/2}$ is the magnetic length. This reduces the backscattering because the loops with area larger than $(l_B)^2$ add classically in average. A negative magnetoresistance is observed (Fig.9). The theoretical analysis can be made in the diffusion approximation up to a field B for which $l_B \sim l_e$. If l_e is large, the coherent backscattering is not completely destroyed at this field. A recent calculation shows that, in the case of isotropic scattering, the quantum correction decreases as $B^{-1/2}$ beyond this limit[21].

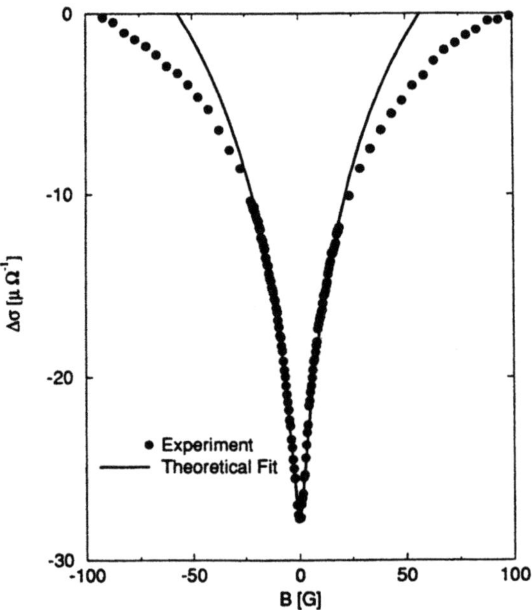

Fig.9   Variation of the conductance versus magnetic field in a 2D electron gas of density n_s = 2.2 x 10^{11} cm^{-2} at T = 0.5 K (P. D. Dresselhaus , PhD Thesis, U.of Yale, 1992, unpublished).

b) with spin

The electronic states are now two component spinors $\Psi = (\Psi\uparrow, \Psi\downarrow)$ and the probability of return to the origin is a sum of the probabilities calculated for each component :

$$P(0) = [\Psi_{\uparrow+}(0) + \Psi_{\uparrow-}(0)]^2 + [\Psi_{\downarrow+}(0) + \Psi_{\downarrow-}(0)]^2 \qquad (11)$$

Let us suppose that we start from an initial spin state $|\uparrow>$ and that at the end of the loop, the spin has been rotated by R. Any rotation can be described as the product of the 3 following rotations : a rotation by angle φ around the z axis, followed by a rotation θ around the x axis and a rotation ϕ around the z axis. The matrix corresponding to the rotation R is easily obtained[22] :

$$R = \begin{pmatrix} \cos(\frac{\theta}{2})\ e^{-i\varphi/2}\ e^{-i\phi/2} & i\ \sin(\frac{\theta}{2})\ e^{i\varphi/2}\ e^{-i\phi/2} \\ i\ \sin(\frac{\theta}{2})\ e^{-i\varphi/2}\ e^{i\phi/2} & \cos(\frac{\theta}{2})\ e^{i\varphi/2}\ e^{i\phi/2} \end{pmatrix}$$

$$(12)$$

using the usual expression for a rotation by an angle α around a unit vector $\mathbf{u}$:

$$R_{\mathbf{u}}(\alpha) = \cos(\frac{\alpha}{2})I - i\sigma.\mathbf{u}\ \sin(\frac{\alpha}{2}) \qquad (13)$$

where I is the unit 2x2 matrix.

The matrix R' associated to the time reversed path is $R^{-1} = \tilde{R}^*$, the matrix R being unitary. In R^{-1}, the product of the reversed rotations is done in a reverse sequence.

There are other contributions to the phase shift of the components of the spinor, namely the dynamical phase and the Berry (or geometrical) phase[23]. They are the same for the time-reversed paths and therefore we can ignore them and write directly the amplitude A as a real number. We get :

$$\Psi_{\uparrow+}(0) = A\ \cos(\frac{\theta}{2})\ e^{-i\varphi/2}\ e^{-i\phi/2} \qquad (14)$$

$$\Psi_{\downarrow+}(0) = i\ A\ \sin(\frac{\theta}{2})\ e^{-i\varphi/2}\ e^{i\phi/2}$$

and

$$\Psi_{\uparrow\text{-}}(O) = A \, \cos(\frac{\theta}{2}) \, e^{i\varphi/2} \, e^{i\phi/2}$$

$$\Psi_{\downarrow\text{-}}(O) = i \, A \, \sin(\frac{\theta}{2}) \, e^{-i\varphi/2} \, e^{i\phi/2}$$

It is important to notice that both components of the spinor are not complex conjugate of each other. A simple calculation gives finally :

$$P(O) = 2A^2 \, [1 + \cos^2(\frac{\theta}{2})\cos(\varphi + \phi) - \sin^2(\frac{\theta}{2})] \tag{15}$$

If we average over all possible rotation, the back scattering probability is :

$$P(O) = A^2 \tag{16}$$

This value is lower than the classical return probability ($2A^2$). So, if there is some spin-orbit coupling which rotates the spin components, the back scattering is reduced instead of being increased! This is related to the fact that upon a rotation by 2π, the radial part comes out in phase whereas the spin part comes out with a phase shift by π. The product is phase shifted by an odd multiple of π. The conductivity is therefore enhanced if compared to the result of a calculation which neglects the coherence effects. The phenomenon is called weak antilocalisation.

Let us now discuss how the spin dephasing along a closed path can be understood. A first source of spin-orbit scattering might be due to the Coulomb field around impurities. This is known as Elliott mechanism[24]. The Hamiltonian describing the interaction between an electron and an impurity is as follows :

$$H_{int} = V(r) + \frac{\hbar}{4m^2c^2} \, (\nabla V(r) \wedge \mathbf{p}).\sigma \tag{17}$$

It follows that the matrix element of transition between the states k and k' is :

$$(H_{int})_{kk'} = V_{k-k'} \, [1 + i \, \frac{\hbar}{4m^2c^2} \, (\mathbf{k} \wedge \mathbf{k'}).\sigma] \tag{18}$$

This appears to be the product of $V_{k\text{-}k'}$ by the first two terms of the development of $e^{i(\sigma.\phi)/2}$. This is a rotation by a small angle $\phi = 2\varepsilon \, (k{\wedge}k')/|k|^2$. In III-V heterojunctions, this is quite inefficient because $\varepsilon \sim 10^{-8}$ and the mean free path is quite large.

A more efficient spin-orbit coupling term is that due to the non centro-symmetry of the structure. We have seen above that there are two contributions to the spin splitting of the conduction band : one due to the absence of inversion center in the crystallographic structures of GaAs, another due to the confinement electric field and the potential step in an heterojunction. In the reference frame of an electron at the Fermi surface, these contributions add to give an internal magnetic field B_{int} in the plane of confinement of the 2DEG. If the elastic scattering time τ is short compared to the inverse of the Larmor frequency $\nu_{lar} = \omega_{lar}/(2\pi) = (g\mu_B B_{int})/h$ (i.e. there is either a low electron mobility $\mu_e = e\tau/m$ or a small spin-splitting ΔE), the spin cannot reach an eigenstate and precesses around the internal field B_{int} between collisions. These conditions corresponds to the Dyakonov-Perel mechanism[25]. Obviously in the state associated by time reversal symmetry, the electron velocity is reversed and so is B_{int}. The precession goes the other way.

After the precession of an initial spin state $S_z(t)$ by a small angle β during time t, around an in-plane magnetic field B_{int}, the final state, calculated with the Larmor relation, is :

$$S_z(t+\tau) = S_z(t) \cos(\beta) = S_z(t) \, (1 - \frac{\beta^2}{2}) \tag{19}$$

We have $\beta = \omega\tau$, so that we get:

$$\frac{\Delta S_z}{\tau} = -\frac{\omega_{lar}^2 \, \tau}{2} \, S_z(t) \tag{20}$$

If the scattering time τ is short, the first term can be taken as equal to the derivative dS_z/dt, which is related to the spin dephasing time τ_{so} by :

$$\frac{dS_z}{dt} = -\frac{1}{\tau_{so}} \, S_z(t) \tag{21}$$

We obtain finally the relation :

$$\frac{1}{\tau_{so}} = \frac{\omega_{lar}^2 \, \tau}{2} \qquad\qquad (22)$$

The spin diffuses as the electron is scattered along the closed path. With the Dyakonov-Perel mechanism, the spin dephasing time is inversely proportional to the transport scattering time. This analysis has been refined recently[26]. On the contrary, if τ is large compared to $(\nu_{lar})^{-1}$ (i.e. there is either a large electron mobility μ_e or a large spin splitting ΔE), a regime which can be reached in selectively doped heterojunctions but which has not yet been analyzed in detail, the spin is going to reach an eigenstate, parallel or antiparallel to B_{int}, between each collision.

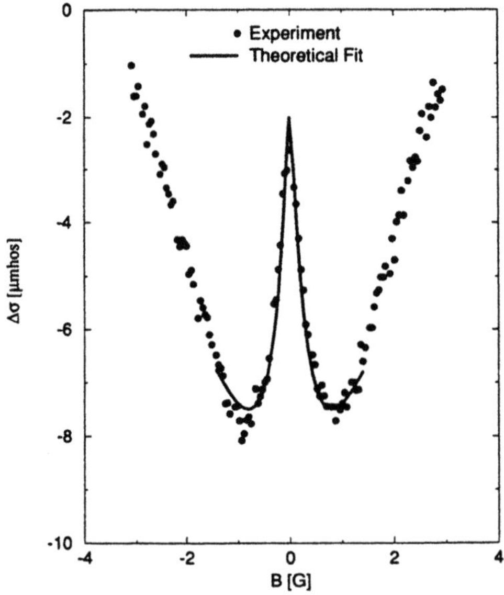

Fig.10   Variation of the conductance with magnetic field in a 2D electron gas of density $n_s = 4.1 \times 10^{11}$ cm^{-2} at T = 1 K (P. D. Dresselhaus et al., ref. 28).

In the same way as it has been explained for weak localisation, a transverse magnetic field is going to destroy the interferences (now negative) giving rise to a positive magnetoresistance. This effect has been observed in metallic thin films[27] (with heavy atoms like Au) and in III-V selectively doped heterojunctions[28]. When the magnetic field is applied, the interferences between the time-reversed closed paths are reduced. It appears that the spin negative interferences are destroyed first and then the positive interferences at higher field. As a consequence, there is a positive magnetoresistance followed by a negative magnetoresistance (Fig. 10). The observation of this result is a clear evidence for spin splitting. A value for τ_{so} is obtained from variation of the magnetoconductance. This dephasing time can be also measured by optical methods[29].

Weak localization is a macroscopic evidence of the wave like nature of the charge carriers. It is striking that in structures of low enough symmetry, the spin part of the wave function can also manifest itself at a macroscopic level (the size of the samples is in the cm range) in the antilocalization effect. In some laterally modulated structures, as those obtained by organized epitaxy on vicinal surfaces, the lack of symmetry of the in-plane potential could also give an additional contribution to this effect.

THE QUANTUM HALL EFFECT

K. von Klitzing discovered in 1980 that the Hall resistance of a quasi-two dimensional electron gas exhibits plateaus and is quantized at low temperature for some ranges of magnetic field (Fig.4) : $R_H = h/(ie^2)$, i being an integer[5]. The diagonal resistance is simultaneously vanishing[8] : $R < 0.1 m\Omega$. This implies that the current is flowing orthogonally to the electric field and that there is no dissipation of energy. The accuracy of the quantized value of R_H, which does not depend on any geometrical or material parameter, is close to 1 part in 10^8. This is the same order of magnitude as the ratio R/R_H. Being conveniently reproduced, the quantized value is now taken as the standard resistance etalon (25 812.904 Ω for i = 1).

This quantization is observed when the ratio $v(B) = n_s/d$, between the electron density n_s and the degeneracy of the Landau levels, $d = eB/h$, is close to an integer value i. It can be immediately checked that R_H, varying as B/en_s in a classical calculation, is indeed equal to quantized value when $v(B) = i$. But the occurence of this condition was not expected to be observable. The essence of the Integer Quantum Hall Effect (IQHE) is that R_H remains strictly constant and equal to the quantized value in a finite range of B.

The width of a plateau is linked to the potential fluctuations, which break the translational invariance of the system. The quantization of the Hall resistance is not related to symmetry (a well-known example of this case is the quantization of kinetic momentum) but is related to topology[8]. Let us explain how this can be understood. The degeneracy d of the Landau levels is lifted by disorder : they are transformed into Landau bands. Nearly all Landau states are then spatially localized except those close to the center of the bands. When the localization radius $\xi(E)$ of a state of energy E is smaller than the phase coherence length Lϕ, the state is localized. This occurs when ν is not too far from an integer value i. The diagonal part σ_{xx} of the magneto-conductivity vanishes (at T = 0 K) because there is a zero density of extended states at the Fermi level energy. The diagonal resistivity is then also zero (in 2D $\rho_{xx} = \sigma_{xx}/(\sigma_{xx}^2+\sigma_{xy}^2)$) and there is no heat dissipation. In the regime of variable range hopping, $\xi(E)$ diverges as $|E-E_i|^{-2.3}$, where E_i is the energy of the center of the Landau band[30]. The observation of an increase of the Hall resistance between plateaus proves that some (at least one!) extended states exist in each Landau band.

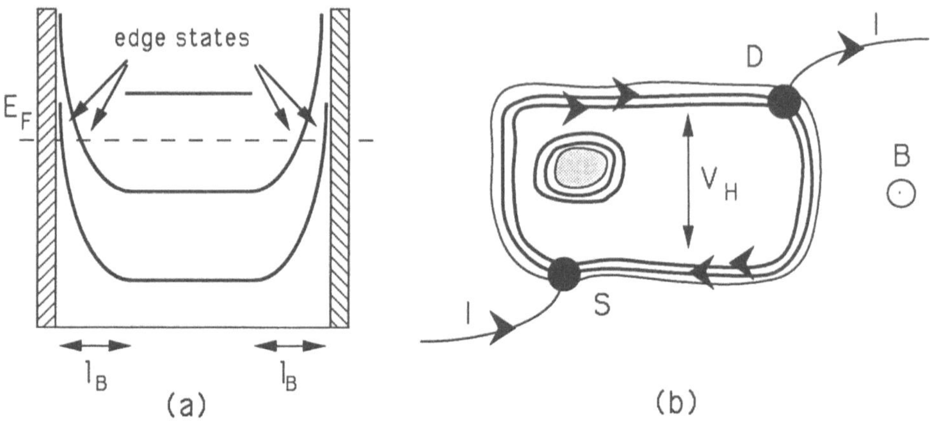

Fig. 11 Landau levels and edge states (a, on the left). Path of the edge states in a sample (crossed by a current I) with disordered boundaries and a defect (b, on the right).

Around a boundary region, the Landau states are bent upward and the extended states cross the Fermi level[8]: these are the edge states (Fig. 11b). It does not matter whether the path along these states (from the current source S to the current drain D) is irregular because of some surface imperfections or impurities (Fig. 11 b) so long as there is a path. If the disorder is so strong that there is no Landau states in some region, edge states do exist however in the vicinity of its border. The IQHE is observed when the edge states on the left side of a line joining S and D are not coupled to the edge states on the right side of this line except in the contact regions. This sets conditions on the electron mobility and on the temperature, and also on the width W of the sample (W >> l_B), but not on its symmetry at any level.

The geometry of the sample is therefore irrelevant and Laughlin's theory[31] considers a ribbon of 2DEG bent into a loop, which is pierced everywhere by a transverse magnetic field B. A potential difference V_H exists between the inner and outer edges. From Faraday's law, it results that the current I around the loop is equal to the derivative of the total electronic energy U of the system with respect to the magnetic flux Φ through the loop. When Φ changes by a quantum flux $\Delta\Phi = \Phi_0 = h/e$, the localized states are not affected (the phase of the wave function is changed by 2π) and the extended states can be rearranged by gauge transformation : this results in a transfer of i electrons (one per filled Landau level) from one edge of the ribbon to the other. The change of energy is $\Delta U = eiV_H$. From $I = \Delta U/\Delta\Phi$ we get a quantized value for the resistance $R_H = V_H/I$. This picture of the IQHE is not modified by electron interactions, though it would be no longer possible to assign a specific Landau state to each electron. This is related to the fact that, when we have a Landau level integer filling i, the ground state of the electron system is not degenerate (there is exactly one electron per state). The nature of the Quantum Hall fluid has lead to a refined description of the edge states[32,33], as the formation of alternating compressible and incompressible strips of electron plasma. The confining potential is screened in the compressible regions.

The Coulomb interactions allow the quantization of the Hall resistance for additional values of ν (see Fig.4), which are fractional numbers with odd denominator (such as 1/3, 2/3, 1/5, ...) : this is the fractional quantum Hall effect (FQHE). The interplay between the degeneracy of the ground state in a one-electron picture ($d=n_s/\nu$) and the Coulomb interactions which favor correlations, leads to an incompressible collective ground state with a band gap for the excitations of fractionally charged quasiparticles obeying to fractional statistics. Laughlin[34] has proposed a wave function, allowing the calculation of the ground state energy, for $\nu = 1/q$ (q odd). States for $\nu = 1-1/q$, are obtained by electron-hole symmetry. Additional observed fractions (such as 2/5, 3/7, ...,n/(2n+1), ...)(see Fig.4) have been explained as a result of a hierarchy[9]. The observation of further states in these series[35] require very low disordered samples because the band gaps are smaller and smaller and are destroyed by the fluctuations. In the same way as for the IQHE, edge states exist at the boundary of the sample in the FQHE regime. The fractional edge states are considered to be the most ideal systems to study the quantum state of 1D electron fluid, because the boundary disorder remains weak and there no localization nor backscattering as one-dimensional metallic wire.

As can be noticed in Fig.4, the experimental observation of the IQHE and of the FQHE effect, for the main fractional states and for the other states of a hierarchy as well, are quite similar : a minimum in the Shubnikov-de Haas resistance and, if not too small, a plateau in the Hall resistance. Jain[36] has proposed a theory unifying the description of the IQHE and FQHE. The FQHE is decribed as a IQHE of composite fermions : the integer filling factor i and the fractional filling factor f are related by the following relation :

$$\frac{1}{f} = \frac{1}{i} + 2m \tag{23}$$

We check that for m = 1, we can associate f = 1/3 to i = 1, f = 2/5 to i = 2, f = 3/7 to i = 3, and so on. Adding 2m quantum fluxes Φ_0 to each electron, the long range Coulomb interaction is renormalized by gauge transformation into weak interactions between the resulting composite fermions[37]. These ideas are related to the theory of anyons[38]. Some consequences can be easily deduced. Indeed, using Eq. (23) with i = ∞ and m = 1, we get f = 1/2 : we have to relate the Fermi gas at B = 0 (i.e. in the $v = \infty$ state) with a composite fermions system at effective magnetic field $\tilde{B} = B - B(v=1/2)$, corresponding to half filling of the first Landau band. We can feel the correspondence between the IQHE and FQHE states converting the magnetic field B at which the fractional states are observed into the field $\tilde{B}$ of an effective integer series. This works perfectly[39] : reminding that the Landau level filling factor is the ratio of the number of electrons to the number of magnetic fluxes, from $B(f) = n_s\Phi_0/f$, we get easily $\tilde{B}(i) = n_s\Phi_0/i$. Using Eq. 23 with other values of m, we would get other FQHE series. The Fermi wave vector of the "metal" of composite fermions at $v = 1/2$ would be of order $1/l_B$, the inverse of the magnetic length. Investigations of the properties of this "Fermi surface" are still lacking.

CONCLUSION

Unanticipated quantum effects such as the Integer Quantum Hall Effect and the Fractional Quantum Hall Effect have been found in two-dimensional electron gas now more than twelve years ago. These discoveries have revealed the very rich physics of 2D systems. The understanding of the nature of these quantum states of matter in extreme conditions (very low temperature, very high magnetic field) is still a very active field of theoretical and experimental research. The possibility of growing higher quality heterojunctions with even lower disorder should allow a deeper understanding of ideas such as the fractional statistics in 2D systems, the nature of 1D metals (Luttinger liquid versus Fermi liquid). At high electron density, the lack of Kramers degeneracy, revealed so far by weak antilocalization, could bring additional features related to the symmetry of the structures.

Acknowledgement

Thanks are due especially to B. Jusserand (Bagneux Lab., CNET) and also to E.Paris, C. Dorin, L. Sfaxi, F. Petit for their active contributions to the results presented here. Stimulating discussions with M. Combescot (G.P.S., Paris) and W. Knap (G.E.S., Montpellier) have been very helpful.

REFERENCES

1. T. Ando, A.B. Fowler and F. Stern, Electronic Properties of Two-dimensional Systems, Review of Modern Physics 54: 437 (1982).
2. C.W.J. Beenakker and H. van Houten, Quantum Transport in Semiconductor Nanostructures, Solid State Physics, ed. H. Ehrenreich and D. Turnbull, Academic Press, 44:1 (1991)
3. Y. Imry, in Directions in Condensed Matter Physics, ed. by G. Grinstein and G. Mazenko, World Scientific Singapore, 1:101 (1986).
4. D. Mailly, C. Chapelier and A. Benoit, Experimental Observation of Persistents Currents in a Single Loop, Phys. Rev. Lett. 70:2020 (1993).
5. K. von Klitzing, G. Dorda and M. Pepper, New Method for High Accuracy Determination of the Fine Structure Constant based on Quantum Hall Resistance, Phys. Rev. Lett. 45:494 (1980).
6. D.C. Tsui, H.L. Stormer and A.C. Gossard, Two-Dimensional Magnetotransport in the Extreme Quantum Limit, Phys. Rev. Lett. 48:1559 (1982).
7. E.Y. Andrei, G. Deville, D.C. Glattli, F.I.B. Williams, E. Paris and B. Etienne, Observation of a Magnetically Induced Wigner Solid, Phys. Rev. Lett. 60:2765 (1988).
8. The Quantum Hall Effect, ed. by R.E. Prange and S.M. Girvin, Graduate Texts in Contemporary Physics, Springer Verlag (1987).
9. T. Chakraborty and P. Piettiläinen, The Fractional Quantum Hall Effect, Springer Series in Solid-State Sciences, vol. 85 (1988).
10. D. Pines, Elementary Excitations in Solids, W.A. Benjamin Inc. (1964).
11. A. Ishihara, Electron Correlations in Two Dimensions, Solid State Physics, ed. H. Ehrenreich and D. Turnbull, Academic Press, 42:271 (1989).

12. G.E. Bauer and T. Ando, Theory of Band Gap Renormalisation in Modulation-Doped Quantum Wells, J. Phys. C 19:1537 (1986).

13. A. Gold and L. Calmels, Correlation in Fermi Liquids : Analytical Results for the Local-Field Correction in Two and Three Dimensions, Phys. Rev. B48:11622 (1993).

14. M. Combescot, O. Betbeder-Matibet, C. Benoit à la Guillaume and K. Boujdaria, Hartree Energy in Semiconductor Quantum Wells, Solid State Comm. 88: 309 (1993).

15. O. Betbeder-Matibet, M. Combescot and C. Tanguy, Coulomb Energy of Quasi-2D Electron Gases in Quantum Wells, Phys. Rev. Lett. 72:4125 (1994).

16. P. Pfeffer and W. Zawadski, Conduction Electrons in GaAs : Five Level k.p Theory and Polaron Effects, Phys. Rev. B41:1561 (1990).

17. B. Jusserand, D. Richards, H. Peric and B. Etienne, Zero-Magnetic Spin Splitting in the GaAs Conduction Band from Raman Scattering on Modulation-Doped Quantum Wells, Phys. Rev. Lett. 69:848 (1992).

18. E.I. Rashba, Properties of Semiconductors with an Extremum Loop I, Fizika Tverdovo Tela, 2:1224 (1960).

19. B. Jusserand, D. Richards, G. Allan, C. Priester and B. Etienne, Spin Orientation at Semiconductor Interfaces, submitted for publication.

20. A. Cassam-Chenai and B. Shapiro, Two-dimensional Weak Localisation beyond Diffusion Approximation, preprint.

21. M.I. Dyakonov, Magnetoconductance due to Weak Localization Beyond the Diffusion Approximation : the High Field Limit, preprint.

22. G.Bergmann, Weak Anti-Localization - An experimental Proof for the Destructive Interference of Rotated Spin 1/2, Solid State Comm. 42:815 (1982).

23. M.V. Berry, Quantal Phase Factors Accompanying Adiabatic Changes, Proc. R. Soc. Lond. A 392:45 (1984).

24. R.J. Elliott, Theory of the Effect of Spin-Orbit Coupling on Magnetic Resonance in Some Semiconductors, Phys. Rev. 96:266 (1954).

25. M.I. Dyakonov and V.Yu. Kachorovskii, Spin Relaxation of Two-Dimensional Electrons in Noncentrosymmetric Semiconductors, Sov. Phys. Semicond. 20:110 (1986).

26. W. Knap, C. Skierbiszewski, E. Litwin-Staszewska, F. Kobbi, J.L. Robert, Weak Antilocalization and Spin Precession in Quantum Wells, preprint.

27. G. Bergmann, Weak Localization in Thin Films, Phys. Rep. 107:1 (1984).

28. P.D. Dresselhaus, C.M.A. Papavassiliou, R.G. Wheeler and R.N. Sacks, Observation of Spin Precession in GaAs Inversion Layers Using Antilocalization, Phys. Rev. Lett. 69:106 (1992).

29. H. Riechert, H.-J. Drouhin and C. Hermann, Phys. Rev. B 38:4136 (1988).

30. D.G. Polyakov and B.I. Shklovskii, Variable Range Hopping as the Mechanism of the Conductivity Peak Broadening in the Quantum Hall Regime, Phys. Rev. Lett. 70:3796 1993).

31. R.B. Laughlin, Quantized Hall Conductivity in Two Dimensions, Phys. Rev. B 23:5632 (1981).

32. D.B. Chklovskii, B.I. Shklovskii and L.I. Glazman, Electrostatics of Edge Channels, Phys. Rev. B 46:4026 (1990).

33. D.B. Chklovskii, K.A. Matveev and B.I. Shklovskii, Ballistic Conductance of Electrons in the Quantum Hall Regime, Phys. Rev. B 47: 12605 (1993).

34. R.B. Laughlin, Anomalous Quantum Hall Effect : An Incompressible Quantum Fluid with Fractionally Charged Excitations, Phys. Rev. Lett. 50:1395 (1983).

35. R. Willett, J.P. Eisenstein, H.L. Stormer, D.C. Tsui, A.C. Gossard and J.H. English, Observation of an Even-Denominator Quantum Number in the Fractional Quantum Hall Effect, Phys. Rev. Lett. 59:1776 (1987).

36. J.K. Jain, Theory of the Fractional Quantum Hall Effect, Phys. Rev. B 41:7653 (1990).

37. J.K. Jain, Strongly Correlated Fractional Quantum Hall States : Description in terms of Weakly Interacting Composite Fermions, Surface Science 263:65 (1992).

38. S.Forte, Quantum Mechanics and Field Theory with Fractional Spin and Statistics, Rev. of Modern Physics 64:193 (1992).

39. R.R. Du, H.L. Stormer, D.C. Tsui, L.N. Pfeiffer, and K.W. West, Experimental Evidence for New Particles in the Fractional Quantum Hall Effect, Phys. Rev. Lett. 70:2944 (1993).

PHOTONIC BANDS AND TUNNELING

Günter Nimtz and Winfried Heitmann

II. Physikalisches Institut
Universität zu Köln
D-50937 Köln

1 INTRODUCTION

Photonics is the electromagnetic analogy to electronics. In photonics the propagation of electromagnetic wave packets is studied in periodic dielectric structures. The whole theory of electrons in a periodic potential applies directly to the problem of electromagnetic waves. As a matter of fact this result is general and independent of the special physical meaning of the waves considered, and it must be the same for elastic, electromagnetic, and Schrödinger waves.

In a periodic dielectric structure there are photonic bands separated by forbidden gaps, as well as defect levels as we are used to from the elctronic states in a semiconductor crystal. Figure 1 illustrates a hypothetical combination of an electronic and a photonic band-gap in one crystal. The electron wave vector is scaled by a factor of 10^{-3} taking into consideration the larger wavelength of a photon and the correspondingly smaller photonic Brillouin zone. In this fictitious crystal the radiative electron-hole recombination would be inhibited, as the photonic band is opened at the relevant photon energy [1]. Such a crystal would extraordinarily improve the specifications of several optoelectronic devices.

Also tunneling which represents another important field of wave mechanics can be studied in photonic structures. The propagation of so-called evanescent electromagnetic waves is analogous to particle tunneling. Evanescent modes are the purely imaginary solutions of the Helmholtz equation, they correspond to the tunnel solutions of the Schrödinger equation. Accordingly, investigations with electromagnetic wave packets do promise to obtain general informations on the tunneling process. Particularly the tunneling time is very difficult to be measured in the case of electrons, this appears to be somewhat easier for evanescent electromagnetic wave packets.

In this lecture we shall introduce in the first part some basic features of one- and three-dimensional photonic bands and defect levels. The second part is devoted to photonic tunneling experiments and the interpretation of observed zero tunneling time.

Advances in Quantum Phenomena, Edited by E.G. Beltrametti
and J.-M. Lévy-Leblond, Plenum Press, New York, 1995

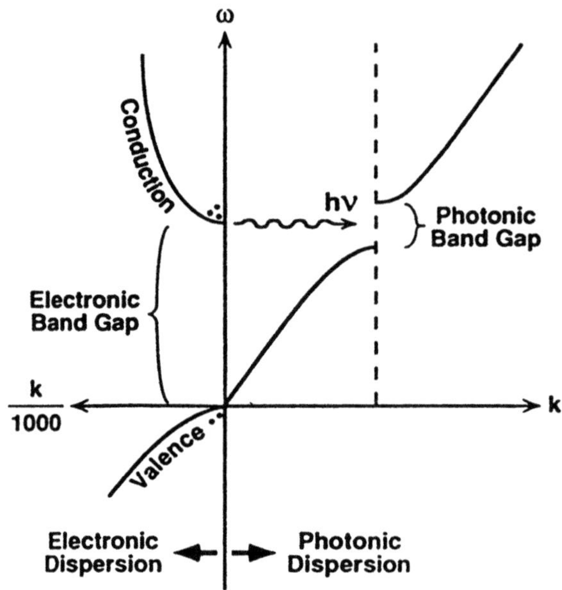

Figure 1. On the right is the electromagnetic dispersion, with a forbidden gap at the wavevector of the dielectric periodicity. On the left is the electron wave dispersion of a semiconductor. Since the photonic band gap straddles the electronic band edge, electron-hole recombination via photon emission is inhibited.

2 BANDSTRUCTURE

2.1 One-dimensional lattice

A simple one-dimensional lattice is obtained by the periodic arrangement of two types of dielectric layers as shown in the insert of Fig. 2. Their refractive indices are given by n_1 and n_2, respectively. The transmission of this structure is shown in the same figure [2]. There is a forbidden photonic band in the displayed frequency range, i.e. a stopping band where the transmission is reduced up to four orders of magnitude and the electromagnetic wave becomes evanescent modes. In principle this band structure could be calculated from the Kronig-Penney model [3].

2.2 Three-dimensional lattice

A gap in the energy dispersion relation is always opened at the Brillouin zone of a crystal. If this gap is narrow and the Brillouin zone deviates strongly from a sphere, there is no overlap of the individual gaps in the various directions of the reciprocal lattice, i.e. in momentum space. Therefore it was suggested, to investigate a priori only face centred cubic lattice structures (FCC), which have a Brillouin zone deviating least from a sphere compared with other common Brillouin zones [1, 4]. The Brillouin zones of the two most important crystal structures are displayed in Fig. 3. The Brillouin zone of the body-centred cubic (BCC) deviates much more from a sphere than the FCC one, thus for instance at the symmetry points L and X the forbidden gaps do not overlap.

However, it was figured out soon, that even the FCC structure does not ensure an overlap of the frequency gaps for all possible directions in reciprocal space. There was still a degeneracy of two bands for electromagnetic waves of different polarization at the

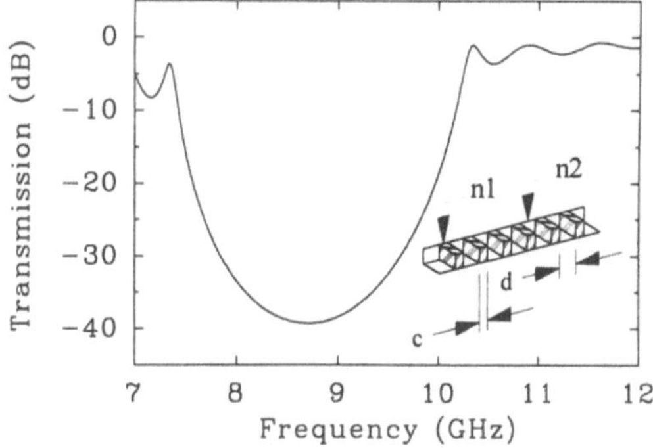

Figure 2. Transmission vs frequency of a one-dimensional dielectric lattice. The periodic structure is shown in the insert. $n_1 = 1.6$, $n_2 = 1$, six layers with $c = 6$ mm separated by $d = 12$ mm.

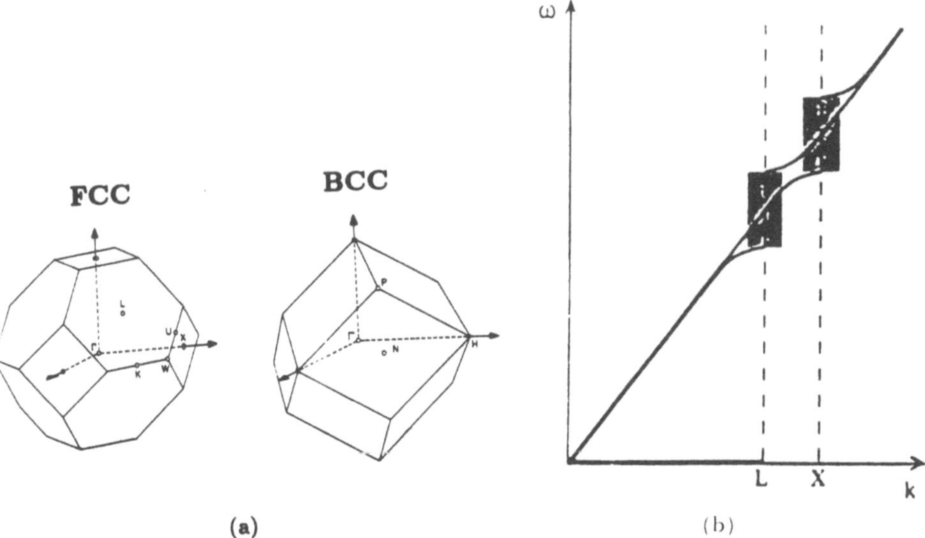

Figure 3. (a) Two common Brillouin zones for body-centred and face-centred cubic crystals. (b) The dispersion relation illustrates the difficulty to create a forbidden frequency gap overlapping all points along the surface of the Brillouin zone

W-point of the Brillouin zone. Eventually, this degeneracy was lifted by destroying the spherical symmetry of the lattice elements, which was achieved by the drilling procedure presented in Fig. 4 [1].

The band structure was measured at microwave frequencies. A gap was found between 12 and 17 GHz. Calculated results are displayed together with a sketch of the Brillouin zone of a FCC lattice in Fig. 5. In consequence of the non-spherical shape of the drilled holes some of the symmetry points have a reduced symmetry now. The experimental data of this photonic band structure shown in Fig. 7(a) are in agreement with several vector wave calculations [1]. There are some deviations between the measured and calculated dispersion data seen in Fig. 5. They are essentially caused by the approximations made in simulating the crystal geometry on the computer.

The introduced photonic crystal was designed for microwave frequencies with

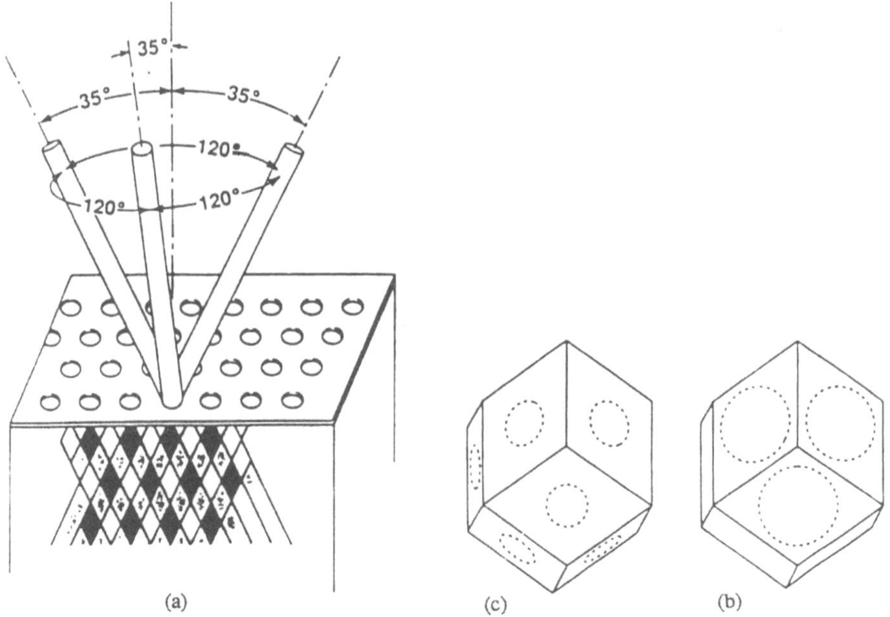

(a)   (c)   (b)

Figure 4. (a) The method of constructing an FCC lattice. A slab of dielectric material is covered by a mask consisting of a triangular array of holes. Each hole is drilled through three times, at an angle $35.26°$ away from normal, and spread out $120°$ on the azimuth. (b) The corresponding Wigner-Seitz real-space unit cell is a rhombic dodecahedron and has no spherical symmetry. (c) Example for a Wigner-Seitz cell with spherical symmetry.

wavelenghts of several cm, however, the results are of general character. By a proper scaling photonic crystals can be made for infrared or even for optical applications.

2.3 Lattice defects

Figure 6 shows a cross-sectional view of the photonic crystal which was introduced in Fig. 4. The unit cell is indicated by the dashed line. Photonic crystal defects can easily be generated either by breaking one of the interconnecting dielectric ribs or by filling an "air atom" with some dielectric material. In the first case the eigen frequency of the unit cube is upwards shifted, in the language of semiconductor physics an acceptor state is created by this procedure. If on the other hand a unit cell is filled with some dielectric material a donor-like state is produced. Such an experiment was performed recently by Yablonovitch et al. [5], the results are shown in Fig. 7. The transmission data reveal an acceptor state as well as two donor states in the forbidden band gap. Presumably, like in a semiconductor, a high density of defects would result in a transmission band eventually. Such an induced defect level or band can be used for design of special filters in solid state laser technology.

3 INHIBITION OF SPONTANEOUS EMISSION

In the introduction we have mentioned, that special photonic environments with a frequency gap may inhibit spontaneous emission in a microelectronic device. Because of this promised ability to control spontaneous emission of light in quantum optics,

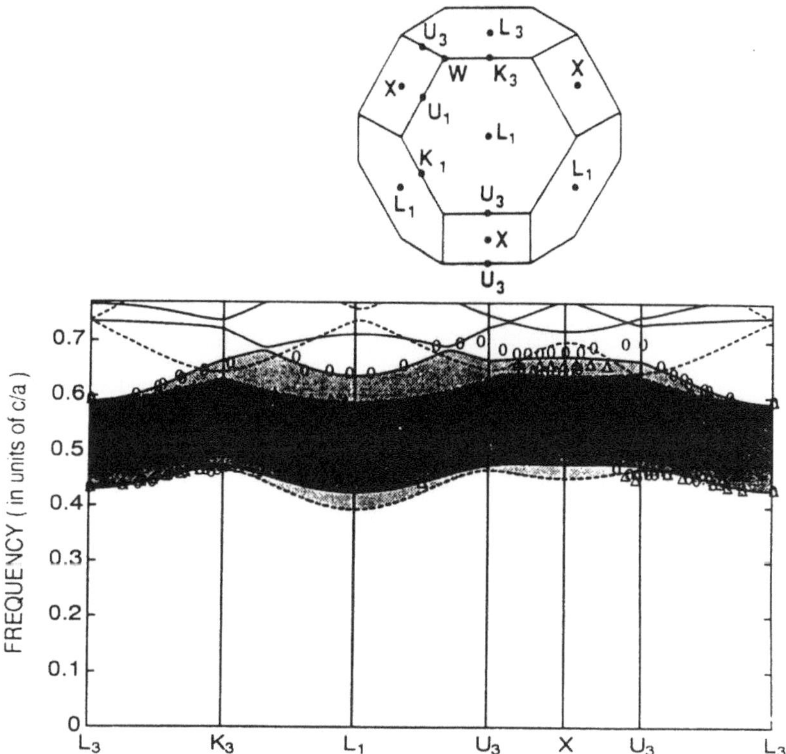

Figure 5. Brillouin zone of a FCC structure and frequency vs wave vector dispersion along the surface of the Brillouin zone. The frequency is given by c/a, where c is the speed of light and a the FCC cube length. The ovals and triangles are the experimental points for s and p polarization respectively. The full and broken curves are the calcutations for s and p polarization respectively.

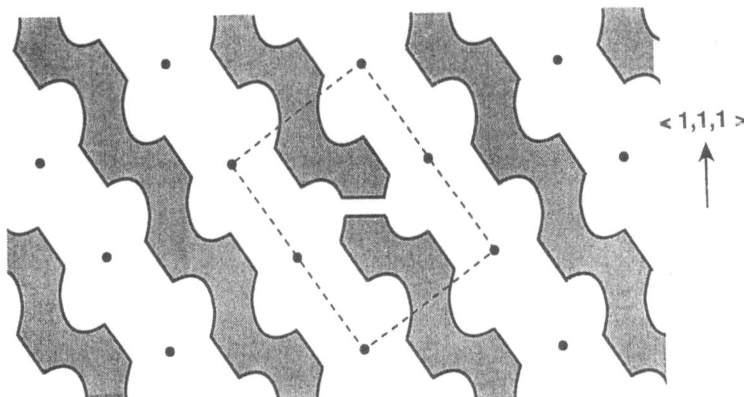

Figure 6. A $< 110 >$ cross-sectional view of the photonic crystal of Fig. 4. Dielectric material is represented by the shaded area. The rectangular dashed line is a face-diagonal cross section of the unit cube. Donor defects consisted of a dielectric sphere centered in an air atom. An acceptor defect is shown as a missing horizontal slice in a single vertical rib.

this has been a major motivation for studying photonic band structures. There is

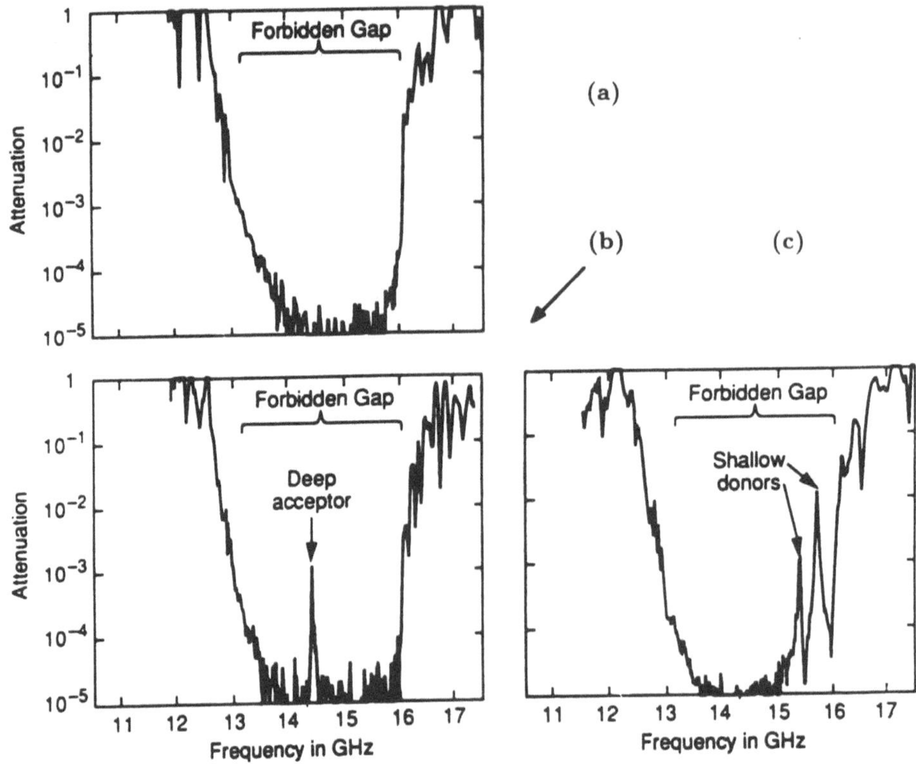

Figure 7. (a) Transmission attenuation through a defect-free photonic crystal as a function of microwave frequency. (b) Attenuation through a photonic crystal with a single acceptor in the center. (c) Attenuation through a photonic crystal with a single donor defect, an uncentered dielectric sphere, leading to two shallow donor modes.

ample motivation since the performance of semiconductor lasers, heterojunction bipolar transistors, and solar cells are all limited by spontaneous emission, but each in a different way.

For a long time spontaneous emission was thought to be a phenomenon which does not depend on any environmental influence. In 1946, Purcell gave the first hint that the confinement of an excited atom in a cavity should alter the spontaneous emission rate from the free space value [6]. In 1985, Hulet et al. [7] could show, that in fact the spontaneous emission rate of an atom is changed in a cavity. The emission was inhibited, if the cavity had no electromagnetic modes at the atom's transition frequency. A photonic crystal with a band gap represents a medium which corresponds to a cavity.

Cavity-induced changes in atomic spontaneous emission rates are often interpreted in terms of quantum electrodynamical zero-point fluctuations. Here we shall introduce a classical analogon: a dipole radiating between parallel mirrors. The calculation was carried out by Dowling [8] recently. The confirmation of the model was presented by Seeley et al. in a simple microwave experiment [9]. The calculated values of the dipole emission in the field of the parallel mirrors are shown together with the measured data in Fig. 8. The agreement is within the experimental accuracy. At a distance $d < \lambda/2$, which corresponds to the cut-off condition of the cavity, the radiation is suppressed.

The dipole study is of pedagogical and applied interest in that it is a simple mockup of several famous cavity QED experiments that investigate the change of atomic spon-

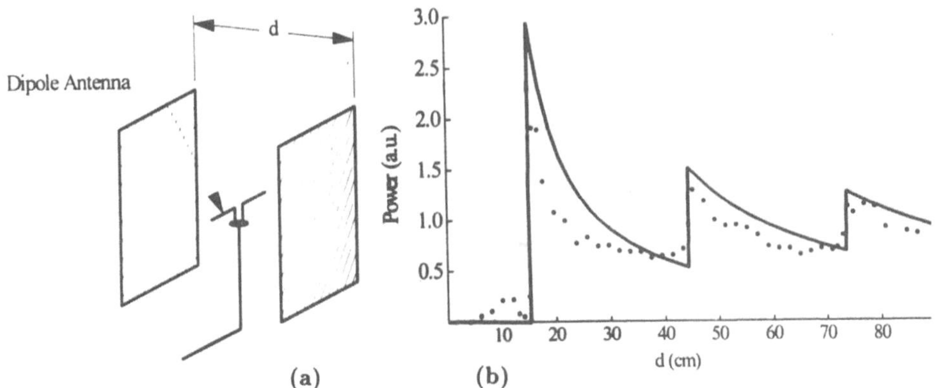

(a)   (b)

Figure 8. (a) Experimental set-up, the dipole is placed in the center between the two mirrors (metal plates), their distance is d. (b) Dipole radiation power vs mirror distance. The solid line represents the calculated data, the dots represent the measured values.

taneous emission rates in such resonators. The suppression of the parasitic spontaneous emission in semiconductor lasers by dielectric crystals has already been achieved.

4 THE PHOTONIC TUNNELING ANALOGY

The analogy between electrons following the Schrödinger equation and electromagnetic wave propagation has been a central ingredient in the development of quantum mechanics [10]. In the following sections we are dealing with the analogy between the tunneling of electrons and evanescent waves in rectangular metal tubes as wave-guides. One guide is operated at a frequency below cut-off (evanescent region), which is put in between other guides where this frequency is above cut-off, as illustrated in Fig. 9 [2, 11].

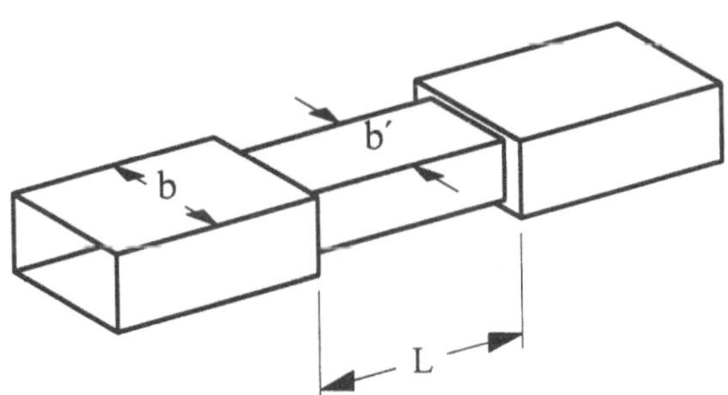

Figure 9. Wave-guide

The dispersion relation of such a rectangular wave-guide is given by

$$k^2 = (2\pi/\lambda_o)^2(\epsilon - (\lambda_o/2b)^2) = (2\pi\nu/c)^2(\epsilon - (\nu/\nu_c)^2)$$

where λ_o is the free space wavelength, c the velocity of light, b the wave guide width, $\nu_c = c/2b$ the cut-off frequency, and ϵ the dielectric constant of the material in the guide.

Instead of changing the size of the guide, an evanescent region can also be obtained by reducing the dielectric constant in the same wave-guide within a finite region. In both cases the wave number is purely imaginary inside the evanescent region; outside that region the wave number is of course purely real.

The electromagnetic wave propagation in a guide, where the dielectric constant was changed at a constant cross-section, has been investigated and compared with the one-dimensional particle tunneling by Martin and Landauer [10]. They have shown that at the barriers both the continuity conditions and their derivatives are equivalent for the guided electromagnetic waves and for the particle tunneling problem. The formal equivalence allows to conclude that results obtained from photonic tunneling experiments should hold also for electronic tunneling. This is particularly of interest in the case of the tunneling time. During the last sixty years many theoretical approaches to the tunneling time were presented yielding quite different results [12]. However, for a test unambigous experimental tunneling data have not been available. Experiments are difficult and the measured time may be governed by various Coulomb interactions of the electron with its electrodynamic environment, either between electrons or with lattice defects. Often the electronic tunneling time was deduced from the high-frequency rectifying limit of a tunnel diode or from the line width in the case of resonant tunneling. We shall demonstrate, that the latter time is not governed by the tunneling process. It is therefore highly desirable to study an experimental system which can be described as an analog to tunneling, but where the mentioned effects can be disregarded.

We shall demonstrate, that in consequence of modern sophisticated electronic equipment it is possible to measure the propagation time of evanescent Gaussian-like electromagnetic wave packets and of amplitude-modulated incident waves. We want to emphasize that in these experiments nearly ideal conditions are realized: **the experiments are non-invasive and asymptotic**. The stationary transmission and reflection coefficients at each frequency point are determined as if source and detector do not interfere with the evanescent region. This is achieved by the process of calibration. From the physical point of view source and detector are ideal and suppress their interference with the cut-off wave-guide, i.e., a perfectly matched device which is equivalent to an infinite long transmission line [14].

It has been found, that the experimental results are in full agreement with the phase time approach for particle tunneling [10, 17, 18, 19]. However, the obervations started a violent dispute about the interpretation of the measured times. We shall discuss this problem in the analysis at the end of this contribution.

5   TUNNELING TIMES

First we shall introduce some of the times frequently used in the literature to describe tunneling of particles [19]. The **dwell time** $\tau_D(E_k)$ is the mean time spent by an incident particle of energy $E_k = \frac{\hbar^2 k^2}{2m}$ in the barrier region regardless of whether it is ultimately transmitted or reflected; the **transmission time** $\tau_T(E_k)$ is the corresponding mean time if the particle is finally transmitted and the **reflection time** $\tau_R(E_k)$ is the mean time if it is finally reflected. Thus τ_T is the time a wave packet spent tunneling through a potential barrier. As regards τ_T some relate to it the **phase time** $\tau_\varphi = d\varphi/d\omega = z\,dk/d\omega = z/v_{gr}$. This phase time was measured. There are transmission times τ_T^{BL} introduced by Büttiker and Landauer obtained by either the time-modulated barrier or Lamor clock method. The differences between τ_φ and τ_T^{BL} can be very significant. For instance the first one is in the opaque barrier regime independent of barrier length, whereas the latter is proportional to it. The dwell time τ_D has been identified with a quantity τ_d defined within the context of a stationary-state scattering problem as the average number entering (or leaving) the barrier per unit time, i.e.

$\tau_d(E_k) = \int_{z_1}^{z_2} |\psi_k(z)|^2 \, dz \big/ \frac{\hbar k}{m}$ with the evanescent region $z_1 < z < z_2$. (This time is related to the **energy velocity** according to $\tau_d = (z_2 - z_1)/v_E$. v_E will be defined in section 5.3.2.) However, it has been claimed that while this definition is perfectly valid classically it is not correct in the quantum-mechanical regime because of interference effects. In general τ_d cannot be identified with τ_D [19]. In the microwave measurements reported here, the **delay time of the center of gravity** of transmitted wave packets or of amplitude-modulated signals were measured. There is much confusion about the definition of signals. Therefore we shall remind the reader to the classical definitions by Sommerfeld and Brillouin of a signal as well as of group, signal, and front velocities in the final section.

At present there are two experimental approaches known to measure the time delay of electromagnetic wave packets when crossing an evanescent region. The first one studies the propagation of evanescent microwave modes in wave guides [2, 11, 13, 15], whereas in a second experiment the delay time of optical photons was measured in a dielectric stopping band structure, a structure similar to that introduced in Fig. 2 [16]. The most extentive material on the tunneling process, however, including the frequency dependence and the barrier length variation, was obtained in microwave studies [2, 11, 13, 14]. We shall focus on it in the following. The center frequency ν_0 of the studied wave packets was choosen that the condition $\nu_c(b) < \nu_0 < \nu_c(b')$ holds, with b the width of the normal wave-guide and b' the width of the undersized wave-guide in between. Effects of the geometrical discontinuities between the wave-guides have been taken into considerations, see Ref. [14].

Two different measuring methods have been applied, one in the frequency domain and the other one in the time domain. In the frequency domain the complex frequency dependent transmission matrix was measured for individual monochromatic waves. A Fourier transform brought this transmission spectrum multiplied by a wave packet-shape (Kaiser-Bessel window [20]) into the time domain. This procedure is performed by the help of a network analyzer and is explained in more details in Refs. [11, 13, 14]. Measurements in the time domain were carried out with a transition analyzer equipment [13, 14]. Here the transmitted signals have been amplitude-modulated waves.

5.1 Resonant tunneling

Quite often the line width $\Delta\omega$ of a resonant transition was assumed to depend on the tunneling time. In photonics this resonant line width can be studied in a double barrier set-up as shown in Fig. 10. Three resonant transition lines are seen in the studied range of frequency. Their line widths according to $\tau = (\Delta\omega)^{-1}$ correspond to **life times** of 40 ns and 80 ns for the analyzed lines III and II, respectively. On the other hand the phase time analysis yields the same values as displayed at the bottom of the figure. Thus the delay times deduced from the real and from the imaginary parts of the transmission coefficient do agree. As we shall see in the next section, these times are very long compared with those in the case of a non-resonant transition measured either in a double-barrier structure or in a single barrier. The **resonant-transition time** corresponds to the localization of the wave packet in the well of the double-barrier structure.

5.2 Tunneling

The first tunneling time experiments were carried out with frequency limited Gaussian wave packets using a network analyzer [11]. In this study wave packets were sent through an undersized wave-guide as shown in Fig. 9 with the barrier length as parameter.

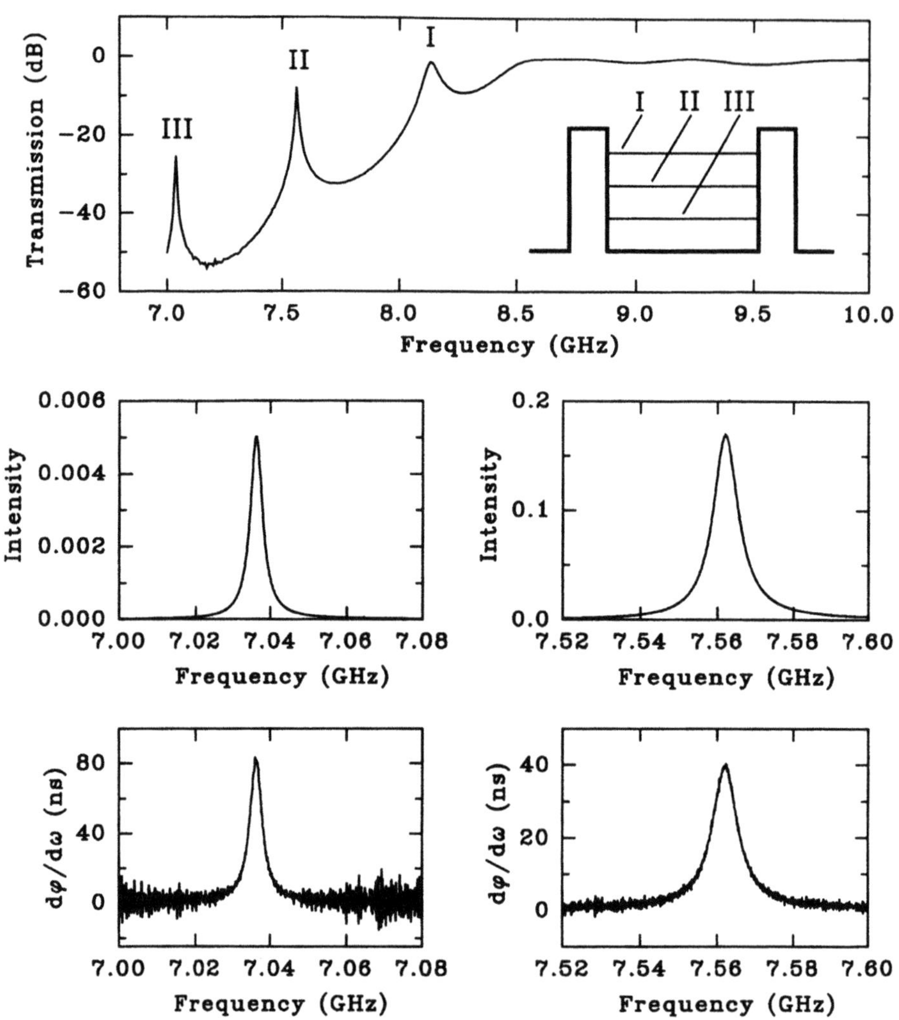

Figure 10. Transmission vs frequency of a double-barrier structure with three resonant transition lines below the barrier's cut-off frequency (insert). The figures present the intensity vs frequency of the resonant lines (center) and the phase time $d\varphi/d\omega$ of lines II and III as a function of frequency (bottom).

The surprising result has been, that the measured delay time of the maxima of the transmitted pulses is independent of the length of opaque barriers; opaque means $|kL| > 1$, with k the imaginary wave number and L the barrier length. The normalized results of the experiment are shown in Fig. 11. A rather short delay time of the maxima in this experiment was found to be 130 ps, that is much less than the corresponding vacuum time of 333 ps, and more than two orders of magnitude shorter than the delay times measured in the resonant tunneling regime. As the result is independent of barrier length, that means that a transmission over an additional distance takes no time at all. So the finite delay time measured must be assigned to the dielectric boundaries only.

In order to compare reference signal and tunneled signal at the same magnitude, and to see distortions due to the propagation in the evanescent region, an amplitude-

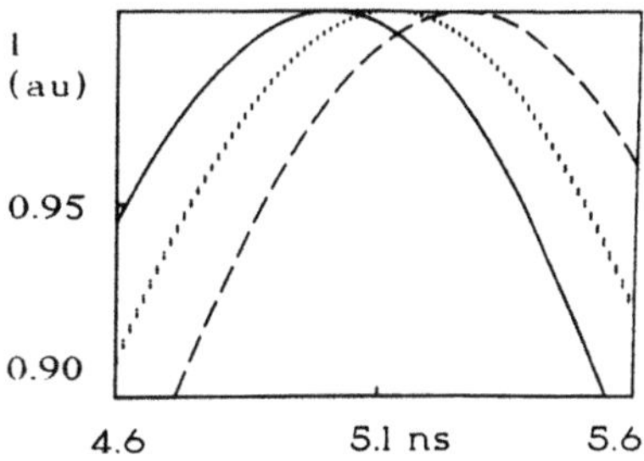

Figure 11. Reference puls at the barriers frontside ($L = 0$, solid line), the dotted line represents the *four* normalized pulses transmitted through barriers of 4, 6, 8, and 10 cm length. All of them had the same delay time of 130 ps, the vacuum time is 333 ps for 10 cm. The dashed line represents the measured puls after a barrier length of 10 cm with an additional large wave-guide of 2.84 cm length, which was operated above cut-off, the latter experiment was performed to test the time resolution of the measurement conditions.

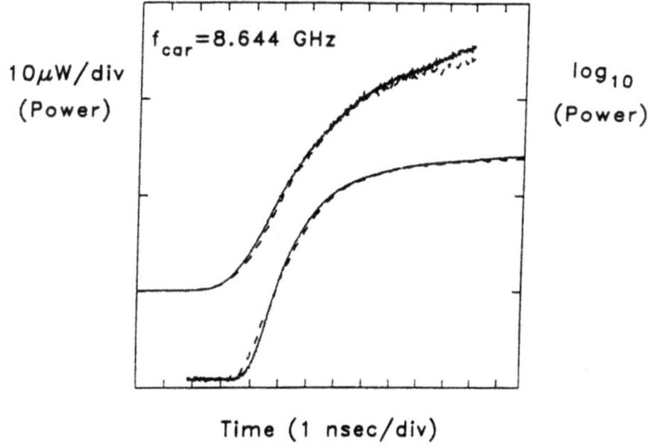

Figure 12. Power envelope of the rising edge of a 8.644 GHz carrier frequency transmitted by a step-attenuator at 40 dB attenuation (solid line). Dotted line the same signal transmitted by the step attenuator now at 0 dB attenuation and transmitted by an evanescent wave-guide of 6 cm length. In spite of the longer distance, there is no significant delay time, some distortion of the signal envelope by the dispersion of the evanescent region is seen.

modulated microwave (having a rising edge) was sent through a classical step attenuator. This experiment was performed in the time domain with a transition analyzer [13, 14]. In the first step a calibrated high-precision attenuator was adjusted to 40 dB and the signal was recorded at the attenuator's output. In a second step, the attenuator was adjusted to 0 dB (changing its delay time less than 25 ps) and an evanescent waveguide of 6 cm length was added, which had at this frequency in first order approximation also 40 dB attenuation loss. Thus we could measure and compare two signals of similar size, however, one of them was transmitted over a 6 cm longer distance. The result is displayed in Fig. 12. Obviously, they arrived at the two different distances at the same time within the experimental accuracy. At the log-scale presentation the tunneled signal even appears to be faster than the classically damped one. This is an apparent non-causality as the dispersion of the evanescent guide favours compared to the step attenuator the high-frequency components at the onset of the signal. This experiment differs from the above one as now the rising edge and not the maximum of a wave packet has been measured. As will be explained in Sec. 5.3.1 the detection of the front of a rising edge, i.e. the very beginning of a signal, may depend on the sensitivity of the detector.

Delay time measurements of wave packets in the stopping band of a one-dimensional dielectric lattice, (an example has been introduced in Fig. 2), revealed also delay times much shorter than the corresponding vacuum time [2, 21]. The advantage of the evanescent stopping band and also of non-resonant tunneling through a double-barrier is the small and symmetrical dispersion around a center frequency. This results in a negligible distortion of a signal with a narrow band width (see also Sec. 5.3.1).

5.3 Tunneling time analysis

The most important results of the tunneling experiments have been: the time for transmitting through opaque evanescent regions is for the peaks of wave packets shorter than the equivalent time in free space and independent of barrier length, of course the signals are exponentially damped with barrier length. The result corresponds to superluminal group and energy velocities of the tunneled part of the initial incident wave packet. The experimental data presented in Fig. 11 are in quantitative agreement with phase time calculations for one-dimensional *quantum-mechanical* tunneling of a wave packet by Hartman [18, 14]. So the investigation has presented the first proof of the analogy between one-dimensional particle tunneling and propagation of evanescent wave packets.

However, there is a controversy about the interpretation of the times in question [10, 12]. Before discussing the experimental data we shall present the classical definitions of a signal and of its velocities.

5.3.1 Signal velocities.

A group of waves or a wave packet is a signal of finite length, comprising only a limited number of frequencies. Whereas the phase velocity $v_{ph} = \omega/k$ (the velocity of a monochromatic wave) and the group velocity $v_{gr} = d\omega/dk$ (the velocity of the center of gravity of a wave packet or the maximum velocitity of propagation of beats) are well established quantities, the signal velocity v_S and the energy velocity v_E represent obviously fuzzy quantities.

The problem of signal velocities has been treated in famous papers by Sommerfeld and by Brillouin, see Ref. [22]. To investigate the wave propagation in a medium they assumed a signal given by the formula $f(t) = 0$ ($t < 0$ and $t > T$) and $f(t) = \sin(\omega_0 t)$ ($t > 0$ and $t < T$) which is terminated at both ends (a classical signal begins at a well defined time). Such a wave form is composed of two unterminated waves, one beginning at $t = 0$ and the second at $t = T$ with opposite phase, so that the two cancel for all time $t > T$.

The Fourier analysis of this signal yields after several transformations [22]

$$f(t) = \frac{1}{2\pi} \Re \int_{-\infty}^{+\infty} \left[e^{i\omega(t-T)} - e^{i\omega t} \right] \frac{d\omega}{\omega - \omega_0}$$

If this signal traverses a distance z each wave ω propagates with its phase velocity $v_{ph}(\omega)$ and the integral becomes

$$f(t, z) = \frac{1}{2\pi} \Re \int_{-\infty}^{+\infty} \left[e^{i\omega(t-T-z/v_{ph})} - e^{i\omega(t-z/v_{ph})} \right] \frac{d\omega}{\omega - \omega_0}$$

In any dispersive medium (forced oscillations of the particles of the material, either electrons or ions) the highest frequency components will arrive at z with c the velocity of light in free space. They do not interact with the medium and their weak oscillations are called forerunners or front. At a lower speed the main signal will arrive, which will be deformed. If it is possible to determine **the exact moment when this main signal arrives, this defines the signal velocity**. One must distinguish between the **wavefront velocity**, which might be determined by a forerunner, and the colloquial **signal velocity**, with which the main part of the wave propagates in a dispersive medium. In general, the signal velocity measured depends on the sensitivity of the detecting apparatus used. With a very sensitive detector, even the forerunners might be detected. *Only* if the detector is restricted to a quarter or to all of the final signal intensity, then an unambiguous definition of the signal velocity can, in general, be given, for instance with a quantum detector. It became obvious from the classical definition that a Gaussian puls having no defined start, does not represent a signal.

If we are not dealing with a signal with a sudden start and a sudden ending (the realistic case), we can suppress frequencies very different from ω_0 and the formula becomes by expansion of the exponents

$$
f(t, z) \;=\; \frac{1}{2\pi} \Re \left\{ e^{i\omega_0(t - z/v_{ph}(\omega_0))} \int_{\omega_0 - \Delta\omega}^{\omega_0 + \Delta\omega} \left[e^{i(\omega - \omega_0)(t - T - z/v_{gr}(\omega_0))} \right. \right.
$$
$$
\left. \left. - \, e^{i(\omega - \omega_0)(t - z/v_{gr}(\omega_0))} \right] \frac{d\omega}{\omega - \omega_0} \right\}
$$

This equation represents a signal beginning progressively at $t = 0$ and arriving at z at a time $t = z/v_{gr}$, ending at $t = T$ and $t = T + z/v_{gr}$, respectively. The velocity of the wave front is now equal to the group velocity $v_{gr}(\omega_0)$.

Assuming an evanescent medium having a purely imaginary wave number K, which is independent of frequency in the range $\omega_0 \pm \Delta\omega$ it follows

$$
f(t, z) = \frac{1}{2\pi} e^{-Kz} \Re \left\{ e^{i\omega_0 t} \int_{\omega_0 - \Delta\omega}^{\omega_0 + \Delta\omega} \left[e^{i(\omega - \omega_0)(t - T)} - e^{i(\omega - \omega_0)t} \right] \frac{d\omega}{\omega - \omega_0} \right\} \tag{1}
$$

The signal is exponentially damped traversing a distance z, however, without spending time. The wave-front still propagates with the group velocity, how fast it may become in traversing the evanescent region.

5.3.2 The measured tunneling velocity.

The center of the wave packets was measured to propagate superluminal through an evanescent region. The same holds for the energy velocity, since the impedance of the wave-guide at the measuring location behind the tunneling region has a real value. In general, the energy velocity is defined by the energy flow divided by the energy density. It represents for instance the rate at which energy flows along a wave-guide. In vacuum and in wave-guides having real characteristic impedances $v_{gr} = v_E$ holds. Accordingly, if a wave packet has arrived and been detected beyond the evanescent region, the energy has to be there at that same time. ($v_{gr} = v_E$ is not valid in the case of standing wave patterns in the medium as has been mentioned above, however, this condition was taken into consideration in the experiments discussed here.)

The measured small delay time, which was found to be independent of barrier length in the opaque regime (Fig. 11), is caused by the interference of the incident and reflected waves at the boundaries of an evanescent region. Inside this region there is no phase shift and thus no further delay time [11, 14].

According to Eq. (1) an amplitude-modulated signal restricted to a narrow frequency range tunnels instantaneously, because the wave number is purely imaginary (neglecting the phase change at the boundaries). Thus, the effective velocity, derived from the delay can be made as large as desired [10, 11, 13, 18].

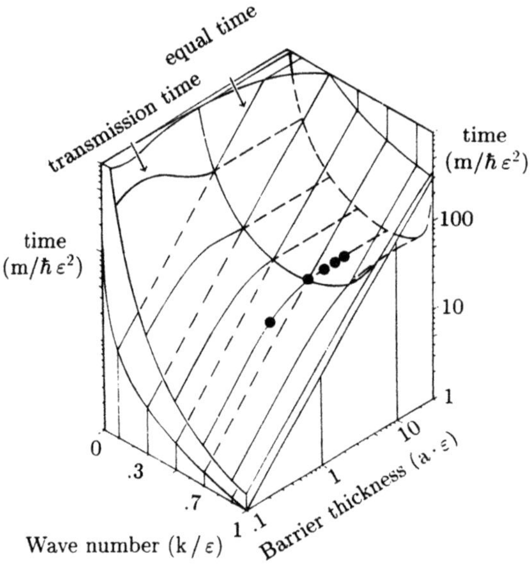

Figure 13. Graphs of the calculated particle transmission time as a function of barrier thickness. Where m is the particle mass, $\hbar$ the Planck constant, and k/ϵ is the incident wave number normalized to ϵ, the wave number equivalent to the potential barrier height. The dots represent the appropriately scaled experimental data of transmission time of evanescent electromagnetic waves.

This becomes evident from inspecting Fig. 13, where the calculated phase time data for particle tunneling of Hartman [18] together with the measured tunneling data (from Fig. 11) are presented. In the regime of opaque barrier thickness the transmission time of the wave packets becomes constant, i.e. independent of barrier thickness. This means in the case of electromagnetic wave packets they propagate superluminal.

Using an evanescent stopping band region (Fig. 2) or the frequency range between two resonant lines (Fig. 10) and Ref. [21] the dispersion is small and symmetrical with respect to the center frequency. Thus the deformation of a signal with a narrow frequency band only becomes negligible. A corresponding frequency limited wave packet was sent through the evanescent region presented in Fig. 2 at a speed of $4.7 \cdot c$ [2].

These special evanescent regions do not act like an effective accelerator by accentuating the high-frequency components of the signal as in the case of a normal tunnel barrier. This very signal is only attenuated, its shape and its start is not depending on the barrier's length. A point which was not considered by Steinberg et al. [16] in interpreting qualitatively their experimental results of photon tunneling by a puls reshaping due to the dispersion of a tunnel barrier. They did not use such a barrier in their experiment, but a flat and symmetrical dispersion of a stopping band.

6 SUMMING UP

Recently photonic crystals have been constructed and investigated in the range of microwave frequencies. They represent dielectric structures with a high potential for new devices in optoelectronics and for a significant improvement of laser diodes, since spontaneous emission can be suppressed.

Photonic tunneling experiments allowed for the first time to measure the delay time in a tunneling process. The experiments have shown, that the phase time approach of Wigner and Hartman decribes the dynamics of a tunneling process. In consequence of the formal analogy of the Helmholtz and the Schrödinger equations the results are valid for tunneling in general.

The evanescent region has shown to be a mysterious medium, where group, energy and, according to the Sommerfeld and Brillouin definition signal velocities can exceed the vacuum velocity of light. However, the question is still open, does the superluminal tunneling of wave packets restricted to a narrow frequency range violate causality? Martin and Landauer [10] have stated recently, that such a signal has very little information, however, enough to transmit "Mozart Forty" superluminal for a short, but finite distance [21].

REFERENCES

[1] E.Yablonovitch, J.Phys.Condens.Matter 5(1993)2443

[2] G.Nimtz, A.Enders and H.Spieker, J.Phys.France I 4(1994)565

[3] R. de L.Kronig and W.G.Penney, Proc.Roy.Soc.(London) 130(1931)499

[4] S.John, Phys.Rev.Letters 58(1987)2486

[5] E.Yablonovitch, T.J.Gmiter, R.D.Meade, A.M.Rappe and K.D.Brommer
and J.D.Joannopoulos, Phys.Rev.Letters 67(1991)3380

[6] E.M.Purcell, Phys.Rev. 69(1946)681

[7] R.G.Hulet et al., Phys.Rev.Letters 55(1985)2137

[8] J.P.Dowling, Found.Phys. 23(1993)895

[9] F.B.Seeley et al., Am.J.Phys. 61(1993)545

[10] Th.Martin and R.Landauer, Phys.Rev.A 45(1992)2611

[11] A.Enders and G.Nimtz, J.Phys.France I 2(1992)1693

[12] E.H.Hauge and J.A.Stovneng, Rev.Mod.Phys. 61(1989)917,
V.S.Olkhovsky and E.Recami, Phys.Rep. 214(1992)339
and R.Landauer and Th.Martin, Rev.Mod.Phys. 66(1994)217

[13] A.Enders and G.Nimtz, J.Phys.France I 3(1993)1089

[14] A.Enders and G.Nimtz, Phys.Rev.B 47(1993)9605
and Phys.Rev.E 48(1993)632

[15] A.Ranfagni et al., Phys.Rev.E 48(1993)1453

[16] A.Steinberg et al., Phys.Rev.Lett. 71(1993)708

[17] E.P.Wigner, Phys.Rev. 98(1955)145

[18] T.Hartman, J.Appl.Phys. 33(1962)3427

[19] C.R.Leavens and G.C.Aers, Phys.Rev.B 39(1989)1202

[20] F.J.Harris, Proc.IEEE 66(1978)51

[21] G.Nimtz, in Proc. Symposium on the Foundations of Modern Physics, Cologne, Eds. P.Busch, P.Lahti, P.Mittelstaedt, World Scientific (1993)309

[22] L.Brillouin, Wave Propagation and Group Velocity, Academic Press, New York and London (1960)

EXPERIMENTAL TESTS OF BELL'S INEQUALITIES

A. Aspect and P. Grangier

Institut d'Optique Théorique et Appliquée
Bâtiment 503, B.P. 147, F91403 Orsay Cedex - France

INTRODUCTION

Bell's Inequalities provide a quantitative criterion to test some reasonable Supplementary Parameters Theories versus Quantum Mechanics. Thanks to Bell [1], the debate about the possibility of completing Quantum Mechanics by an underlying substructure has been brought into the experimental domain.

The motivations for considering supplementary parameters will be found in the analysis of the famous Einstein-Podolsky-Rosen Gedankenexperiment [2]. Introducing a reasonable Locality Condition, one can derive Bell's theorem, which states :

(i) that Local Supplementary Parameters Theories are constrained by Bell's Inequalities;

(ii) that certain predictions of Quantum Mechanics sometimes violate Bell's Inequalities.

We will point out that a fundamental assumption for the conflict is the Locality assumption. We will show that in a more sophisticated version of the E.P.R. thought experiment ("timing experiment"), the Locality Condition may be considered a consequence of Einstein's Causality, preventing faster-than-light interactions.

The purpose of this discussion is to convince the reader that the formalism leading to Bell's Inequalities is very general and reasonable. What is surprising is that it conflicts with Quantum Mechanics.

As a matter of fact, situations exhibiting such a conflict are very rare, and it was necessary to design specific experiments for getting a sensitive test. Experiments that follow closely

Advances in Quantum Phenomena, Edited by E.G. Beltrametti
and J.-M. Lévy-Leblond, Plenum Press, New York, 1995

the ideal scheme of the Gedankenexperiment have been carried out using atomic cascades, and more recently using parametric pair production. We will review these experiments, and their results.

WHY SUPPLEMENTARY PARAMETERS ?
THE EINSTEIN-PODOLSKY-ROSEN-BOHM GEDANKENEXPERIMENT

A. Experimental scheme

Let us consider the optical variant of the E.P.R. Gedankenexperiment modified by Bohm [3]. A source S emits a pair of photons with different energies, ν_1 and ν_2 counterpropagating along $\pm O\vec{z}$ (Fig. 1). Suppose that the polarization part of their state vector is :

$$|\Psi(\nu_1, \nu_2) >= (1/\sqrt{2})\,[|x, x > +|y, y >] \tag{1}$$

where $|x, x >$ and $|y, y >$ are linear polarizations states.

We perform on these photons linear polarization measurements. The analyzer I in orientation $\vec{a}$, followed by two detectors, gives + or - result, corresponding to a linear polarization found parallel or perpendicular to $\vec{a}$. Similarly acts analyzer II, in orientation $\vec{b}$ [1].

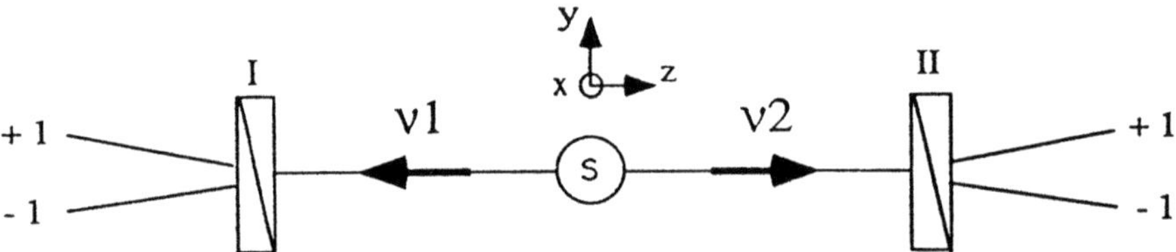

Fig. 1. Einstein-Podolsky-Rosen-Bohm Gedankenexperiment with photons. The two photons ν_1 and ν_2, emitted in the state (1), are analyzed by linear polarizers in orientations $\vec{a}$ and $\vec{b}$. One can measure the probabilities of single or joint detections after the polarizers.

It is easy to derive the Quantum Mechanical predictions for these measurements, single or in coincidence.

Let $P_\pm(\vec{a})$ be the probability of getting the result $\pm$ for ν_1 ; similarly $P_\pm(\vec{b})$ is related to ν_2. Quantum Mechanics predicts :

$$P_+(\vec{a}) = P_-(\vec{a}) = 1/2$$
$$P_+(\vec{b}) = P_-(\vec{b}) = 1/2 \tag{2}$$

Let $P_{\pm\pm}(\vec{a}, \vec{b})$ be the probability of joint detection of ν_1 in channel $\pm$ of I (in orientation $\vec{a}$), and of ν_2 in channel $\pm$ of II (in orientation $\vec{b}$). Quantum Mechanics predicts :

----

[1]There is a one-to-one correspondance with the Gedankenexperiment dealing with a pair of spin 1/2 particles, in a singlet state, and analyzed by two Stern-Gerlach filters [3]

$$P_{++}(\vec{a},\vec{b}) = P_{--}(\vec{a},\vec{b}) = \frac{1}{2}cos^2(\vec{a},\vec{b})$$

$$P_{+-}(\vec{a},\vec{b}) = P_{-+}(\vec{a},\vec{b}) = \frac{1}{2}sin^2(\vec{a},\vec{b}) \tag{3}$$

B. Correlations

In the special situation $(\vec{a},\vec{b}) = 0$ one finds

$$P_{++}(\vec{a},\vec{b}) = P_{--}(\vec{a},\vec{b}) = 1/2 \tag{4}$$

while :

$$P_{+-}(\vec{a},\vec{b}) = P_{-+}(\vec{a},\vec{b}) = 0 \tag{5}$$

So, if ν_1 is found in + the channel of I (the probability of which is 0.5), then we are sure to find ν_2 in the + channel of II (and similarly for the - channels) : there is a strong correlation between the results of measurements on ν_1 and ν_2. A convenient way of displaying these correlations is the polarization correlation coefficient :

$$E(\vec{a},\vec{b}) = P_{++}(\vec{a},\vec{b}) + P_{--}(\vec{a},\vec{b}) - P_{+-}(\vec{a},\vec{b}) - P_{-+}(\vec{a},\vec{b}) \tag{6}$$

The prediction of Quantum Mechanics is:

$$E_{MQ}(\vec{a},\vec{b}) = cos2(\vec{a},\vec{b}) \tag{7}$$

For $(\vec{a},\vec{b}) = 0$, we find $E_{MQ}(0) = 1$, i.e. a complete correlation.

C. Supplementary parameters

Correlations between distant measurements on two systems that have separated may be easily understood in terms of some common properties of the two systems. Let us consider again the correlations of polarization measurements in the case $(\vec{a},\vec{b}) = 0$. When we find + for ν_1, we are sure to find + for ν_2. We are thus led to admit that there is some property (Einstein said "an element of physical reality") pertaining to this particular pair, and determining the result ++. For another pair, the results will be −− ; the invoked property is different.

Such properties, differing from one pair to another one, are not taken into account by the Quantum Mechanical state vector $|\Psi(1,2) >$ which is the same for all pairs. This is why Einstein concluded that Quantum Mechanics is not complete. And this is why such properties are referred to as "supplementary parameters" (sometimes called "hidden-variables"). As a conclusion, one can hope to "understand" the E.P.R. correlations by such a classical- looking picture, involving supplementary parameters differing from one pair to another one. It can be hoped to recover the Quantum Mechanical predictions when averaging over the supplementary parameters. It seems that so was Einstein's position [4]. At this stage, a commitment to this view point is just a matter of taste.

Remark - Since Einstein spoke of "an element of the physical reality", some authors call "Realistic Theories" these theories involving supplementary parameters [5].

BELL'S INEQUALITIES

A. Formalism

Bell tried to translate into mathematics the preceding discussion, by introducing explicit supplementary parameters, denoted λ. Their distribution on an ensemble of emitted pairs is specified by a probability distribution $\rho(\lambda)$, such that:

$$\rho(\lambda) \geq 0 \quad \text{and} \quad \int d\lambda \rho(\lambda) = 1 \tag{8}$$

For a given pair, charactarized by a given λ, the results of measurement will be:

$$A(\lambda, \vec{a}) = \pm 1 \quad \text{at analyzer I (orientation } \vec{a})$$
$$B(\lambda, \vec{b}) = \pm 1 \quad \text{at analyzer II (orientation } \vec{b}) \tag{9}$$

A particular theory must be able to afford explicit the functions $\rho(\lambda), A(\lambda, \vec{a})$ and $B(\lambda, \vec{b})$. It is then easy to express the probabilities of various results, for instance $P_+(\vec{a}) = \frac{1}{2} \int d\lambda \rho(\lambda) [A(\lambda, \vec{a} + 1] \, etc \ldots$ In particular, we will use the correlation function:

$$E(a, b) = \int d\lambda \rho(\lambda) A(\lambda, \vec{a}) B(\lambda, \vec{b}) \tag{10}$$

An essential hypothesis in Bell's formalism is that in eq.(10) $\rho(\lambda)$ does not depend on the orientations $\vec{a}, \vec{b}$, of the remote polarizers and similarly, $A(\lambda, \vec{a})$ does not depend on $\vec{b}$, and $B(\lambda, \vec{b})$ does not depend on $\vec{a}$. This "Locality Condition" will be discussed again in section D below.

B. Bell's Inequalities

Let us consider the quantity :

$$s = A(\lambda, \vec{a}).B(\lambda, \vec{b}) - A(\lambda, \vec{b'}).B(\lambda, \vec{b'}) + A(\lambda, \vec{a'}).B(\lambda, \vec{b}) + A(\lambda, \vec{a'}).B(\lambda, \vec{b'})$$
$$= A(\lambda, \vec{a})[B(\lambda, \vec{b}) - B(\lambda, \vec{b'})] + A(\lambda, \vec{a'})[B(\lambda, \vec{b}) + B(\lambda, \vec{b'})] \tag{11}$$

Remembering that the four numbers A, B take only the values ± 1, we find that :

$$s(\lambda, \vec{a}, \vec{a'}, \vec{b}, \vec{b'}) = \pm 2 \tag{12}$$

The average over λ is therefore included between $+ 2$ and $- 2$, i.e.

$$-2 \leq \int d\lambda \rho(\lambda).s(\lambda, \vec{a}, \vec{a'}, \vec{b}, \vec{b'}) \leq 2 \tag{13}$$

According to (10) and (11), we rewrite this :

$$-2 \leq S(\vec{a}, \vec{a'}, \vec{b}, \vec{b'}) \leq 2 \tag{14}$$

with :

$$S = E(\vec{a}, \vec{b}) - E(\vec{a}, \vec{b'}) + E(\vec{a'}, \vec{b}) + E(\vec{a'}, \vec{b'}) \tag{15}$$

These are B.C.H.S.H. inequalities, i.e. Bell's inequalities generalized by Clauser, Horne, Shimony, Holt [6]. They bear upon a combination of four polarization correlation coefficients, measured in four orientations of the polarizers. S is thus a measurable quantity.

C. Conflict with Quantum Mechanics

Let us take the particular set of orientations of Fig.2a. Replacing the $E's$ by their Quantum Mechanical values (7) for pairs in state (1), we obtain :

$$S_{MQ} = 2\sqrt{2} \tag{16}$$

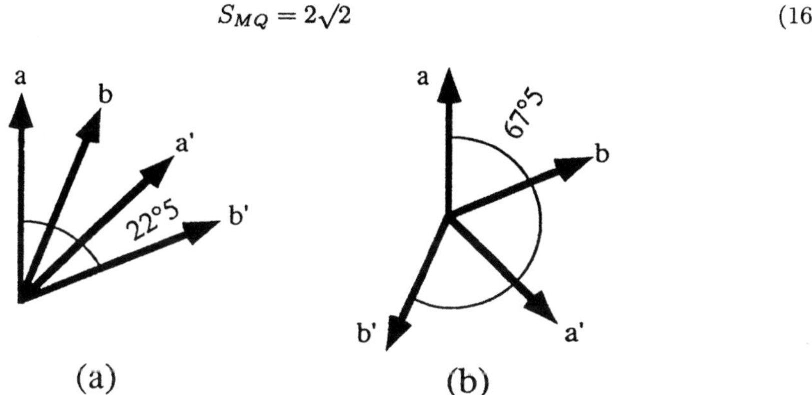

Fig. 2. Orientations yielding the largest conflict between Bell's Inequalities and Quantum Mechanics.

This Quantum Mechanical prediction strongly violates the upper limit of inequalities (14). We thus find it impossible to reconcile the formalism defined in eq.(8)-(10) with the predictions of Quantum Mechanics for the particular (E.P.R. type) state (1).

D. Gedankenexperiment with variable analyzers : the locality condition as a consequence of Einstein's causality

In static experiments, in which the polarizers are held fixed for the whole duration of a run, the Locality Condition must be stated as an assumption. Although highly reasonable, it is not prescribed by any fundamental physical law. To quote J. Bell "the settings of the instruments are made sufficiently in advance to allow them to reach some mutual rapport by exchange of signals with velocity less than or equal to that of light". If such interactions existed, the Locality Condition would no longer hold for static experiments, nor would Bell's Inequalities.

Bell thus insisted upon the importance of "experiments of the type proposed by Bohm and Aharonov [4], in which the settings are changed during the flight of the particles" [2]. In such a timing-experiment, the locality condition would become a consequence of Einstein's Causality that prevents any faster- than-light influence.

As shown in our 1975 proposal [9], it is sufficient to switch each polarizer's orientation between two particular settings ($\vec{a}$ and $\vec{a'}$ for I, $\vec{b}$ and $\vec{b'}$ for II). It then becomes possible to test experimentally a larger class of Supplementary Parameters Theories, those obeying

---

[2]The idea was already expressed in Bohm's book [3]

Einstein's Causality. In such theories, the response of polarizer I at time t is allowed to depend on the orientation $\vec{b}$ (or $\vec{b'}$) of II at time $t - L/c$ (L being the distance between the polarizers). A similar retarded dependance is considered for the way in which pairs are emitted at the source (characterized by the supplementary parameters distribution). For random switching times, with both sides uncorrelated, the predictions of these more general theories are constrained by generalized Bell's Inequalities [9].

On the other hand, it is easy to show that the polarization correlations predicted by Quantum Mechanics depend only on the orientations $\vec{a}$ or $\vec{a'}$ and $\vec{b}$ or $\vec{b'}$ at the very time of the measurements, and do not involve any retardations terms such as L/c. For a suitable choice of the set of orientations (sets displayed in Fig. 2), the Quantum Mechanical predictions still conflict with generalized Bell's Inequalities. Such a timing−experiment with variable analyzers would thus provide a test of Supplementary- Parameters−Theories obeying Einstein's Causality, versus Quantum Mechanics.

E. First experiments using atomic cascades

The first static experiment was realized in Berkeley by Clauser and Freedman [12]. The $4p^{21}S_0 - 4s4p^1P_1 - 4S^{21}S_0$ cascade of Calcium was excited by ultraviolet absorption towards a $^1P_1$ upper state. Since the signal was weak, and spurious cascades occurred, it took more than 200 hours of measurement for a significant result. The experiment upheld Quantum Mechanics, and violated Bell's inequalitites by several standard deviations.

At the same time, in Harvard, Holt and Pipkin [11] found a result in disagreement with Quantum Mechanical predictions, and in agreement with Bell's Inequalities. They excited the $9^1P_1 \rightarrow 7^3P_1 \rightarrow 6^3P_0$ cascade in Mercury 200 by an electron beam. The data accumulation lasted 150 hours.

Clauser [13] repeated their experiment in Mercury 202. He found an agreement with Quantum Mechanics, and a violation of Bell's Inequalities.

In 1976, in Houston, Fry and Thompson [14] used the $7^3S_1 \rightarrow 6^3P_1 \rightarrow 6^3S_0$ cascade in Mercury 200. Their selective excitation involved a C.W. single-line-laser. The signal was several order of magnitude larger than in previous experiments, allowing them to collect the data in a period of 80 minutes. Their result was in excellent agreement with Quantum Mechanics and violated generalized Bell's inequalities by 4 standard deviations.

ORSAY EXPERIMENTS (1980-1982)

A. The source

Since our aim was to use more sophisticated experimental schemes, we had first to build a high-efficiency and very stable and well controlled source. This was carried out by a two-photon-excitation of the $4p^{21}S_0 - 4s4p^1P_1 - 4s^{21}S_0$ cascade of calcium [15]. This cascade is very well suited to coincidence counting experiments since the lifetime τ_r of the intermediate level is rather short (5 ns). If one can reach an excitation rate about $1/\tau_r$, then an optimum signal-to-noise ratio for this cascade is attained.

We have achieved this optimum rate with the use of a Krypton laser ($\lambda_K = 406.7nm$) and a dye laser ($\lambda_D = 581nm$) tuned to resonance for the two-photon process. Both lasers ar single-mode operated. They have parallel polarizations, and they are focused onto a Calcium atomic beam (laser beam waists about $50\mu m$).

Two feedback loops provide the required stability of the source (better than 0.5% for several hours) : the first loop controls the wavelength of the tunable laser to ensure the maximum fluorescence signal ; a second loop controls the power of one laser and compensates all the fluctuations. With a few tens of milliwatts from each laser, the cascade rate is about $N = 4 \times 10^7 s^{-1}$. An increase beyond this rate would not significantly improve the signal-to-noise ratio for coincidence counting, since the accidental coincidence rate increases as N^2, while the true coincidence rate increases as N.

B. Detection - Coincidence counting

The fluorescent light is collected by large-aperture aspherical lenses, followed by a set of lenses and the polarizers. The overall detection efficiency, including collection and photo-multipliers efficiencies, is about 2.10^{-3}.

The photomultipliers feed the coincidence-counting electronics, that includes a time-to- amplitude converter and a multichannel analyzer, yielding the time-delay spectrum of the two-photon detections. This spectrum involves a flat background due to accidental coincidences (i.e. between photons emitted by different atoms). True coincidences yield a peak around the null-delay, with an exponential decrease (time constant τ_r).

Additionnally, a standard coincidence circuit with a 19 ns coincidence window monitors the rate of coincidences around null delay, while a delayed-coincidence channel monitors the accidental rate. It is then possible to check that the true coincidence rate obtained by substraction is equal to the signal in the peak of the time-delay spectrum.

In the second and third experiments, we have used a fourfold coincidence system, involving a fourfold multichannel analyzer and four double-coincidence circuits. The data were automatically gathered and processed by a computer.

C. Experimental with one-channel polarizers [16]

Our first experiment was carried out using one-channel pile-of-plates polarizers, made of ten glass-plates at Brewster angle. Thanks to our high-efficiency source, the statistical accuracy was better than 2% in a $100s$ run (with polarizers removed). This allowed us to perform various checks.The test of Bell's inequalities has yielded a violation by 9 standard deviations, in good agreement with the Quantum Mechanical predictions, which have been checked in a full 360° range of orientations.

We have repeated these measurements with the polarizers at 6.5 m from the source. This distance is larger than the coherence lengths of the wave packet associated with the lifetimes of the upper and intermediate states of the cascade (respectively 4.5 m and 1.5 m). No modification of the experimental results was observed.

D. Experiment with two–channel analyzers [17]

With single–channel polarizers, the measurements of polarization are inherently incomplete. When a pair has been emitted, if no count is obtained at one of the photomultipliers, there is no way to know whether the photon "has been missed" by the detector or whether it has been blocked by the polarizer (only the later case corresponds to a result — for the measurement). This is why one had to resort to auxilliary experiments, and indirect reasoning, in order to test Bell's inequalities.

With the use of two—channel polarizers, we have performed an experiment following much more closely the ideal scheme of Fig. 1. Our polarizers were polarizing cubes transmitting one polarization (parallel to $\vec{a}$, or respectively to $\vec{b}$) and reflecting the orthogonal one. Such a polarization splitter, and the two corresponding photomultipliers, are mounted in a rotatable mechanism. This device (polarimeter) yields + and — results for linear polarization measurements along $\vec{a}$ (respectively $\vec{b}$). It is an optical analog of a Stern—Gerlach filter for spin 1/2 particles.

With polarimeters I and II in orientations $\vec{a}$ and $\vec{b}$, and the fourfold coincidence counting system, we are able to measure in a single run the four coincidence rates $R_{\pm\pm}(\vec{a},\vec{b})$. We then get directly the correlation coefficient for the measurement along $\vec{a}$ and $\vec{b}$:

$$E(\vec{a},\vec{b}) = \frac{R_{++}(\vec{a},\vec{b}) + R_{--}(\vec{a},\vec{b}) - R_{+-}(\vec{a},\vec{b}) - R_{-+}(\vec{a},\vec{b})}{R_{++}(\vec{a},\vec{b}) + R_{--}(\vec{a},\vec{b}) + R_{+-}(\vec{a},\vec{b}) + R_{-+}(\vec{a},\vec{b})} \tag{17}$$

It is then sufficient to repeat the same measurement for three other orientations, and the B.C.H.S.H. inequality (14) can directly be tested.

This procedure is sound if the measured values (17) of the correlation coefficients can be taken equal to the definition (6), i.e. if we assume that the ensemble of actually detected pairs is a faithfull sample of all emitted pairs. This assumption is very reasonable with our very symetrical scheme, where the two measurements + 1 and — 1 are treated in the same way (the detection efficiencies in both channels of a polarimeter are equal). Moreover, we have checked that the sum of the four coincidence rates $R_{\pm\pm}(\vec{a},\vec{b})$ is constant when changing the orientations, although each rate strongly varies. The size of the selected sample of pairs is thus found constant.

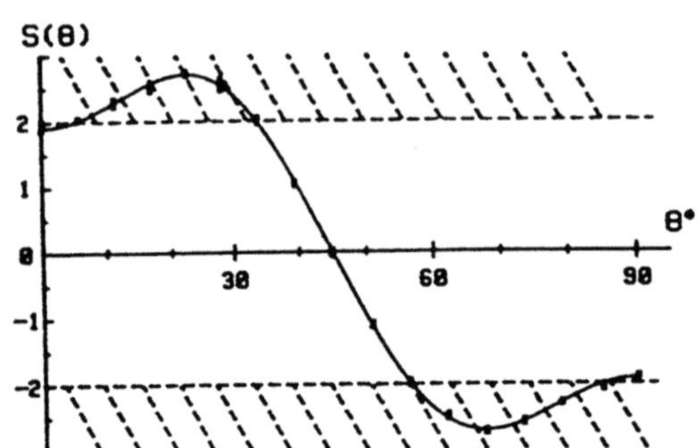

Fig. 3. Experiment with two—channels polarizers : Bell's S parameter (eq. 15) for $\Theta = (\vec{a},\vec{b}) = (\vec{b},\vec{a'}) = (\vec{a'},\vec{b'})$. Eq. 18 corresponds to $\Theta = 22°5$ (see Fig. 2). The indicated errors are ± 2 standard deviations. The solid curve is not a fit to the data, but Quantum Mechanical predictions for the actual experiment. All points in the dashed area violate *eq.14.*

The experiment has been done at the set of orientations of Fig. 4, for which the greatest conflict is predicted. We have found :

$$S_{exp} = 2.697 \pm 0.015 \tag{18}$$

violating the inequalities (14) ($\mid S \mid \leq 2$) by more than 40 standard deviations. This result is in excellent agreement with the predictions by Quantum Mechanics (for our polarizers and solid angles) :

$$S_{QM} = 2.70 \pm 0.05 \tag{19}$$

The uncertainty on S_{QM} accounts for a slight lack of symetry of both channels of a polarizer ($\pm 1\%$). The effect of these dissymetries has be computed and cannot create a variation of S_{QM} greater than 2%. We have also performed measurements of $E(\vec{a}, \vec{b})$ in various orientations, for a direct comparison with the predictions of Quantum Mechanics (Fig. 3). The agreement is clearly excellent.

E. Timing experiment [18]

As stressed in paragraph II, an ideal E.P.R. type experiment would involve the possibility of switching a random times the orientation of each polarizer. We have done a step towards such an ideal experiment by using the modified scheme displayed in Fig. 4.

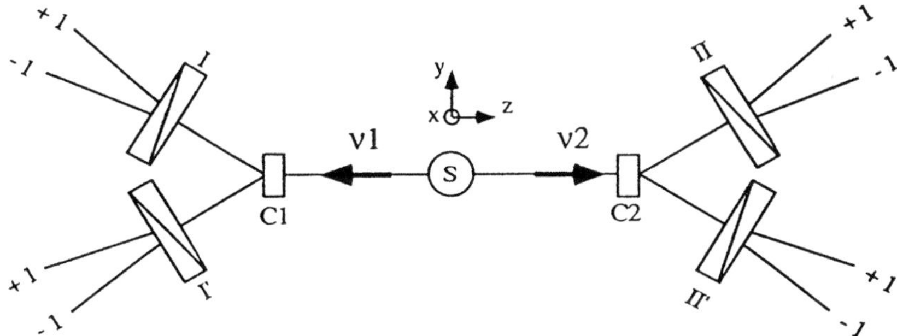

Fig. 4. Timing–experiment with optical switches (C_1 and C_2). A switching occurs each 10 ns. The two switches are independently driven.

Each (single–channel) polarizer is replaced by a setup involving a switching device followed by two polarizers in two different orientations : $\vec{a}$ and $\vec{a'}$ on side I and $\vec{b}$ and $\vec{b'}$ on side II. The optical switch is able to rapidly redirect the incident light from one polarizer to the other one. Each setup is thus equivalent to a variable polarizer switched between two orientations. The distance L between switches is 12 m.

The switching of the light is effected by acousto–optical interaction of the light with an ultrasonic standing wave in water. The incidence angle (Bragg angle) and the acoustic power are adjusted for a complete switching between the 0th and 1^{rst} order of diffraction. At an acoustical frequency of 25 MHz, the switching frequency is 50 MHz. A change of orientation of the equivalent variable polarizer then occurs each 10 ns. Since this period (10 ns) as well as lifetime τ_r (5 ns) are small compared to L/c (40 ns), a detection event on one side and the corresponding change of orientation on the other side are separated by a space–like interval.

With the large beams used in the experiment, the commutation was not complete, since the incidence angle was not exactly the Bragg angle. Instead of being 0, the minimum of transmitted light in each channel was 20%.

Since we had to reduce the divergence of the beams, the detected coincidence rates were weaker by an order of magnitude than in our previous experiments. Accordingly, the duration of data accumulation was longer.

The test of Bell's Inequalities involves a total of 8 000 s of data accumulation with the 4 polarizers in the orientations of Fig. 2. A total of 16 000 s was devoted to auxilliary calibration measurements with half or all polarizers removed.

In order to compensate the effects of systematic drifts, the data accumulation was alternated between the various configurations each 400 s. The average yields a violation of Bell's inequalities by 5 standard deviations, and a good agreement with the Quantum Mechanics predictions.

According to these results, Supplementary—Parameters Theories obeying Einstein's Causality seem to be untenable. To escape this conclusion, one might argue that the switching was not complete. However, a large fraction of the pairs undergoes forced switching. If Bell's Inequalities were obeyed by these pairs, it is hard to believe that we would not have observed a significant discrepancy between our results and the Quantum Mechanical predictions.

Our experiment differs from the ideal scheme in another respect : the switching are not truly at random, since the acoustico—optical switches are driven by quasi—periodic generators. Nevertheless, the two generators on the two sides function in a completely uncorrelated way, specially with respect to their frequency drifts.

EXPERIMENTAL TESTS OF BELL'S INEQUALITIES USING PARAMETRICALLY GENERATED PHOTON PAIRS

As we said previously, the overall efficiency of the atomic cascade experiments is quite weak, i.e., many emitted pairs of photons are missed by the detectors. Then, in order to get a valid test, one must assume that the ensemble of actually detected pairs is a faithfull sample of all emitted pairs. This "supplementary assumption" can be given different forms, but the only way to avoid it would be to detect all emitted pairs, i.e., each photon detected on one side of the apparatus should give rise to a coincidence. This condition can be achieved if the angular correlation between both photons of a pair is high. This is the case in pair production using parametric down-conversion, due to the phase-matching condition $\vec{k}_s + \vec{k}_i = \vec{k}_p$ in the splitting of one pump photon "p" into two down-converted photons "s" and "i". Another motivation for looking at parametric down-conversion is to investigate possibilities of violation of Bell's inequalities, which would not use polarization correlations. Nevertheless, all known Bell's inequalities violation in optics can be traced back to a (non-local) interference effect between two two-photon probability amplitudes, the so-called "fourth-order interference effects".

Therefore, during recent years, several tests of Bell's inequalities have been carried out using pairs of photons emitted in parametric fluorescence rather than in atomic cascades. The first experiments were using polarization correlations [19] [20], but some of the pairs could not be detected by definition due to the state preparation procedure. Improved experiments were then realized using "time-energy" correlations [21] [22], but again some of the coincidences have to be rejected by a time-selection procedure. The best experiment so far, which allows one to detect in principle all emitted pairs, uses "phase-momentum" correlations and gave a violation of Bell's inqualities by 10 standard deviations [23]. However, due to technical difficulties, the overall efficiency of these experiments is still quite weak, and atomic cascade experiments still gave the best signal-to-noise ratios and largest violation so far. Nevertheless, improved experiments allowing "loophole-free" tests Bell's inequalities are maybe not out of reach of most recent experimental techniques [24].

An implementation of the EPR paradox (using continuous rather than discrete variables) was also achieved using quantum optical techniques, close to the ones used in squeezed light production [25]. It should be noticed that here there is no violation of Bell's inequalities, because no inequalities are available for continuous variables, though this experiment may be the closest implementation of the original EPR paradox.

CONCLUSION

Some loopholes thus remain open for the advocates of Supplementary−Parameters Theories obeying Einstein's Causality. Improved experiments will probably become feasible in the future [25] [26], but we already have an impressive agreement with Quantum Mechanics. Supplementary Parameters Theories obeying Einstein's Causality and compatible with our results appear somewhat artificial, since the experimental results would have to change dramatically (disagreement with Quantum Mechanics) with certain technical improvements (such as an increase of the efficiencies of the photomultipliers).

According to Bell [27], we are thus forced to admit :

(i) either that there are, at the level of the supplementary parameters, faster−than−light influences. We must emphasize that in a timing experiment−even ideal−these hypothetical faster−than−light influences cannot be controlled for practical telegraphy [14] ;

(ii) or to renounce an explanation in terms of supplementary parameters.

The second position seems a priori more confortable. But, to quote Mermin [29], "I challenge the reader to suggest any ... other way to account for what happens" (i.e. the observed strong correlations).

I hope that even those who are not committed to such discussions will be convinced that Einstein has pointed out one of the most extraordinary property of Quantum Mechanics. We must thank J. Bell to have provided us with the possibility of experimentally evidencing this property.

REFERENCES

[1] J.S. Bell, On the Einstein$-$Podolsky$-$Rosen Paradox, Physics 1 : 195 (1964). J.S. Bell, Introduction to the Hidden$-$Variable Question, in: Foundations of Quantum Mechanics, B. d'Espagnat ed., Academic, N.Y. (1972).

[2] A. Einstein, B. Podolsky and N. Rosen, Can Quantum$-$Mechanical description of physical reality be considered complete ?, Phys. Rev. 47 : 777 (1935). See also Bohr's answer : N. Bohr, Can Quantum$-$Mechanical description of physical reality be considered complete?, Phys. Rev. 48 : 696 (1935).

[3] D. Bohm, Quantum Theory, Prentice$-$Hall, Englewoods Cliffs, N.J. (1951)

[4] D. Bohm and Y. Aharonov Discussion of Experimental Proof for the paradox of Einstein, Rosen and Podolsky, Phys. Rev. 108 : 1070 (1957)

[5] J.F. Clauser and A. Shimony, Bell's Theorem : Experimental Tests and Implications, Rep. Progr. Phys. 41 : 1881 (1978)

[6] J.F. Clauser M.A. Horne, A. Shimony and R.A. Holt, Proposed experiment to test local hidden$-$variable theories, Phys. Rev. Lett., 23 : 880 (1969).

[7] J.F. Clauser M.A. Horne,Experimental consequences of objective local theories, Phys. Rev. D, 10 : 526 (1974).

[8] A. Fine, Hidden Variables, Joint Probability, and the Bell Inequalities, Phys. Rev. Lett., 48 : 291 (1982).

[9] A. Aspect, Proposed experiment to Test Separable Hidden$-$Variable Theories, Phys. Lett., 54A : 117 (1975). A. Aspect, Proposed Experiment to test the nonseparability of Quantum Mechanics, Phys. Rev. D14 : 1944 (1976).

[10] E.S. Fry, Two$-$Photon Correlations in Atomic Transitions, Phys.Rev. A8:1219 (1973).

[11] F.M. Pipkin, Atomic Physics Tests of the Basic Concepts in Quantum Mechanics, in : Advances in Atomic and Molecular Physics, D.R. Bates and B. Bederson, ed., Academic (1978).

[12] S.J. Freedman and J.F. Clauser, Experimental test of local hidden$-$variable theories, Phys. Rev. Lett. 28 : 938 (1972).

[13] J.F. Clauser, Experimental Investigation of a Polarization Correlation Anomaly, Phys. Rev. Lett. 36 : 1223 (1976).

[14] E.S. Fry and R.C. Thompson, Experimental Test of Local Hidden$-$Variable Theories, Phys. Rev. Lett. 37 : 465 (1976).

[15] A. Aspect, C. Imbert and G. Roger, Absolute Measurement of an Atomic Cascade Rate Using a Two Photon Coincidence Technique. Application to the $4p^{2A}S_0 - 4s^{21}S_0$ Cascade of Calcium excited by a Two Photon Absorption, Opt. Comm. 34 : 46 (1980).

[16] A. Aspect, P. Grangier and G. Roger, Experimental Realization of Einstein$-$Podolsky$-$Rosen$-$Bohm Gedankenexperiment : A New Violation of Bell's Theorem, Phys. Rev. Lett. 47 : 460 (1981).

[17] A. Aspect, P. Grangier and G. Roger, Experimental Realization of Einstein$-$Podolsky$-$Rosen$-$Bohm Gedankenexperiment : A New Violation of Bell's Inequalities, Phys. Rev. Lett. 49 : 91, (1982).

[18] A. Aspect, J. Dalibard and G. Roger, Experimental Test of Bell's Inequalities Using Variable Analyzers, Phys. Rev. Lett. 49, 1804, (1982).

[19] Y.H. Shih and C.O. Alley, Phys. Rev. Lett. 61, 2921 (1988)

[20] Z.Y. Ou and L. Mandel, Phys. Rev. Lett. 61, 50 (1988)

[21] J. Brendel, E. Mohler and W. Martienssen, Europhys. Lett. 20, 575 (1993)

[22] P.G. Kwiat, A.M. Steinberg and R.Y. Chiao, Phys. Rev. A 47, R272 (1993)

[23] J.G. Rarity and P.R. Tapster, Phys. Rev. Lett. 64, 2495 (1990)

[24] Z.Y. Ou, S.F. Pereira, H.J. Kimble, and K.C. Peng, Phys. Rev. Lett. 68, 3663 (1992)

[25] P. Eberhardt, P. G. Kwiat and R. Chiao, private communication

[26] T.K. Lo and A. Shimony, Proposed Molecular Test of Local Hidden−Variables Theories, Phys. Rev. A 23 : 3003 (1981).

[27] J.S. Bell, Atomic−cascade Photons and Quantum−Mechanical Nonlocality, Comments on Atom. Mol. Phys. 9 : 121 (1980).

[28] A. Aspect, Expériences basées sur les Inégalités de Bell, J. Physique Colloque C2 : 63 (1981).

[29] N.D. Mermin, Bringing home the atomic work : Quantum mysteries for anybody, Am. J. Phys. 49 : 940 (1981).

QUANTUM CORRELATIONS BEYOND BELL'S INEQUALITIES

Anton Zeilinger

Institut für Experimentalphysik
Universität Innsbruck
Technikerstraße 25
A-6020 Innsbruck, Austria

INTRODUCTION

In 1964, John Bell[1] achieved a remarkable feat. His theorem states that a very general class of local realistic theories is incompatible with quantum mechanics. This accomplishment is of an exceptional nature, because John Bell did not have to know explicitly the details of the individual theories which were excluded by his theorem. The theorem is usually stated in terms of the spin correlations either between two fermions or between two photons, where in both cases both particles enjoy a two-dimensional spin space. (See the early review by Clauser and Shimony.[2]) The most often used state to discuss these correlations is the two-fermion singlet state

$$|\psi\rangle = \frac{1}{\sqrt{2}}(|\uparrow\rangle_1|\downarrow\rangle_2 - |\downarrow\rangle_1|\uparrow\rangle_2). \tag{1}$$

Whenever we write such a state, as a product of kets we imply the tensor product because each particle is defined in its own Hilbert space. Nearly all the experiments with spin correlations exploited the correlations between two photons in atomic cascades. The first such experiment was performed by Clauser and Freedman[3] with the development culminating in the experiments of Aspect et al.[4] In the last one of these experiments, time-varied measurement was used in such a way that the photons were switched to different polarizers while in-flight. This was intended to rule out any conspiratory type of local theory where, through some yet unknown communication channels between detectors and source, the quantum correlations are established. Yet, due to an unfortunate numerical coincidence between switching frequency and photon flight time,[5] that experiment cannot be called conclusive. Furthermore, this experiment used periodic switching, and it is easy to see that the definitive experiment would have to use a purely random switch. Another rather well-known problem of all existing experiments is the fact that only a small fraction of the number of both particles was really detected. It is, thus, so far always possible to assume that the whole set of pairs emitted by the source obeys quantum mechanics while the subset of particles detected violates Bell's inequalities. In such a way, one could retain local realism. (For a specific example of such a theory see F. Selleri and A. Zeilinger.[6]) While it is very unlikely that such an explanation will prevail, it is nevertheless important to rule it out experimentally. This goal has provided a major motivation for inventing new kinds of correlation schemes. Some of these will be discussed below.

NON-SPIN CORRELATIONS

Already back in 1935 Einstein, Podolsky and Rosen[7] used a momentum and position correlated two-particle source in their argument, intending to demonstrate that quantum mechanics is incomplete. In their argument, both the position variable and the momentum variable were considered to be continuous. On the other hand, it was pointed out by Bell that for an experimental test of his inequality, one has to employ dichotomic variables. Operationally, this means that the result of the measurement should have two possible outcomes only. It has subsequently been shown by Horne and Zeilinger[8] that it is actually possible to invent a situation where one gets dichotomic observables even in a state where the originally correlated quantity is a continuous one. This is done by projecting that state onto a two-dimensional subspace. Fig. 1 shows the generic setup of such a device. The state emitted by the source is

$$|\psi\rangle = \frac{1}{\sqrt{2}}(|\vec{K}\rangle_1|-\vec{K}\rangle_2\rangle + |\vec{K'}\rangle_1|-\vec{K'}\rangle\rangle_2) \tag{2}$$

where e.g. $|\vec{K}\rangle$ describes particle 1 having momentum $\vec{K}$. It is straightforward to see that the quantum mechanical predictions of that state violate Bell's inequality with the four free phases of Fig. 1 assuming the role of the polarizer orientations in spin-correlated states.

The advent of down-conversion as a source of correlated photons[10,11] has actually made such an experiment possible.[12,13]

ENTANGLEMENT

It has first been explicitly formulated by Schrödinger[14] in his reaction to the Einstein-Podolsky-Rosen paper that correlated states of the type of Eq.(1) or Eq.(2) imply significant novel properties of quantum systems not shared by classical systems. For these novel properties Schrödinger coined the term "Entanglement" (in the German original "Verschränkung"). This term, to Schrödinger, means that in quantum mechanics it can very well be that a system containing more than one particle has a well-defined quantum state while the individual particle's state is not well defined. This is a direct consequence of the uncertainty principle in the sense that exact definition of a joint fact like total spin or total momentum is compatible with maximal uncertainty of the related properties of the individual particles. To Schrödinger, this means that the individual particles don't enjoy properties on their own. They are very intimately correlated to the other particle, much stronger than any classical correlation can be.

Take, for example, the state of Eq.(1). This state is rotationally invariant and therefore the spin of either particle is maximally uncertain. In other words, a measurement could find the spin with equal probability along any direction. Yet, it is immediately clear that upon measurement of the spin of one of the two particles, the other one is always found along the opposite direction. It is a remarkable feature of the quantum

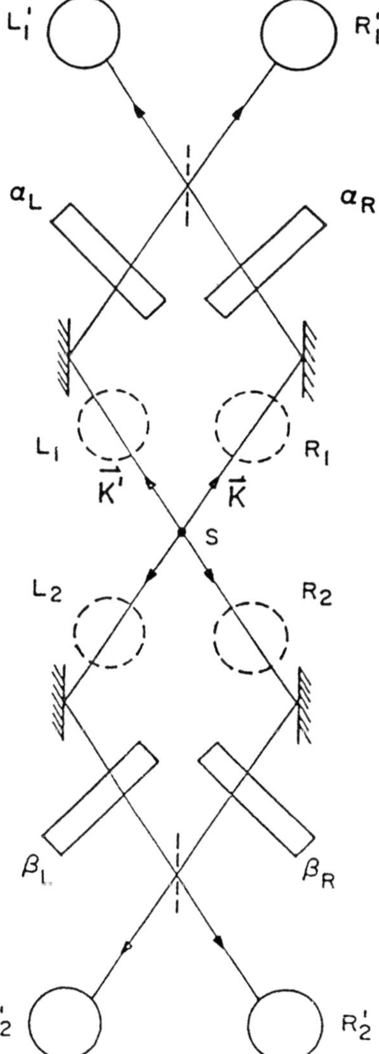

Figure 1: Schematic sketch of an EPR experiment using Mach-Zehnder interferometers. Correlations between the counts registered by the detector $L'_1(R'_1)$ and $L'_2(R'_2)$ reveal a conditional interference pattern upon variation of the phase shifts α_L, α_R and β_L, β_R. The unprimed detector inserted at the positions $L_1(R_1)$ and $L_2(R_2)$ may serve to establish the momentum anticorrelation of the particle pairs emitted by the source (from [9]).

formalism that it can describe such situations where only joint properties of particles (subsystems) are well defined, while the properties of the individuals are maximally undefined. From the conceptual point of view, this implies the rather strange property that while a particle might not "know" in a spin-measurement which of the two possible results to trigger, the particle very well "knows" that, upon measurement, it has to give the exactly opposite result of the other particle. The reader might permit me to use this rather simplistic description, but it contains in its heart the essential mystery of entangled states. Actually, Schrödinger thought that entangled states carry *the* essence of quantum mechanics. In fact, one can understand any experiment with unentangled states in such a way that it is possible to still adopt a rather realistic understanding of the experiments. Formally, this is demonstrated explicitly by Bohm's interpretation of quantum mechanics[15] where unentangled states lead to a picture which is local and realistic. In the naive picture, one could assume that a wave of some sort propagates from the source to the detector with the new feature being that the wave cannot be detected in a continuous way, but only through quantum particles. So, in a sense, quantization only enters at the moment of detection. In fact, as done by Bohm's interpretation, the quantum wave defines the trajectories of individual particles in a local way. For entangled states, such a picture has to be nonlocal, and thus its intuitive goal is certainly much less appealing.

In the experiment, the most obvious way to produce an entangled state is to have a source which emits correlated particles such that the entanglement exists from the source on. Two explicit examples of such a way to produce entanglement would be the decay of a spin-zero particle to produce the state of Eq.(1) or the emission of two particles with total momentum zero to produce the state of Eq.(2).

A totally different way to produce an entangled state is to project an initially unentangled state, say, $|\psi\rangle$ onto an entangled state, say, $|\psi_{ent}\rangle$

$$|\psi_{ent}\rangle\langle\psi_{ent}|\psi\rangle \Rightarrow \alpha|\psi_{ent}\rangle. \tag{3}$$

Evidently, this procedure always works when the entangled state is not orthogonal to the initial state $|\psi\rangle$. A specific example for spin would be to take $|\psi\rangle = |\uparrow\rangle_1|\downarrow\rangle_2$ and for $|\psi_{ent}\rangle$ the state of Eq.(1). Then, projection of $|\psi\rangle$ onto the entangled state results in

$$\frac{1}{2}(|\uparrow\rangle_1|\downarrow\rangle_2 - |\downarrow\rangle_1|\uparrow\rangle_2)((\langle\uparrow|_1\langle\downarrow|_2 - \langle\downarrow|_1\langle\uparrow|_2)|\uparrow\rangle_1|\downarrow\rangle_2$$
$$= \frac{1}{2}(|\uparrow\rangle_1|\downarrow\rangle_2 - |\downarrow\rangle_1|\uparrow\rangle_2) \tag{4}$$

which is clearly entangled. This procedure always results in an entangled state unless the entangled state to be produced is orthogonal to the original unentangled one.

ENTANGLEMENT SWAPPING

A specifically interesting case is entanglement swapping, where one can entangle originally uncorrelated particles which don't even share any common past. The basic idea is shown in Fig. 2 for the specific case of photons created by parametric down-conversion.

In Fig. 2, it is assumed that we have two parametric down-conversion sources, PDC-I and PDC-II.[16] The resulting four-photon state is

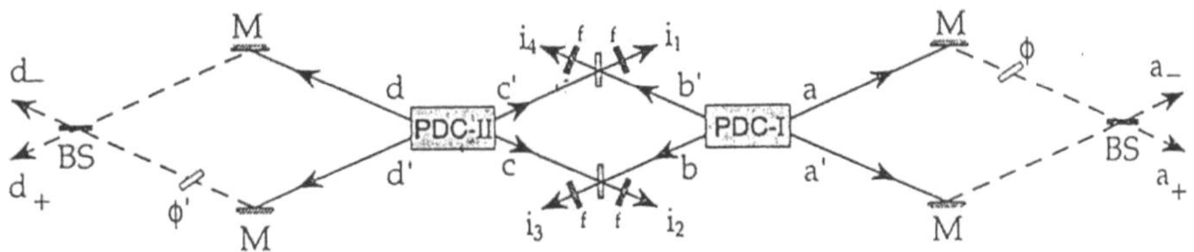

Figure 2: Principle of entanglement swapping. Two down-conversion sources PDC-I and PDC-II emit a photon pair each. The specific geometry of each PDC source is obtainable by a suitable arrangement of mirrors and apertures. The initially independent signal photons in beams a, a', d, d' get entangled by coincident registration of the idlers (M mirrors; BS beam splitters; f filters; ϕ, ϕ' local phase shifters), (from [16]).

$$\frac{1}{2}(|a\rangle|b\rangle + |a'\rangle|b'\rangle)(|c\rangle|d\rangle + |c'\rangle|d'\rangle). \tag{5}$$

Clearly, in that state, the photons emitted from source I are entangled with each other as are the photons emitted from source II while the photons emitted from different sources are not entangled with each other. Now we consider measurement of two photons in coincidence at detectors in the beams i_1 and i_2. Such a measurement can be interpreted in the following way: Either the photon registered at i_1 came via beam b', then the photon registered at i_2 came via c, or the photon registered at i_1 came via c', then the photon registered at i_2 came via b. Thus, the state of these two photons is the entangled state

$$\frac{1}{\sqrt{2}}(|b\rangle|c'\rangle + |b'\rangle|c\rangle). \tag{6}$$

We should briefly remark here that the two photons registered could also have come from the same source, meaning that both photon pairs were emitted from the same crystal, but then signal photons are detected at the same end of the apparatus and hence, tautologically, the question of an entangled state between a and d does not arise.

The collapse of the idlers into the state of Eq.(6) implies that the signals — which are the two photons emerging from the beams a, a' and d, d' — are also collapsed into the entangled state

$$\frac{1}{\sqrt{2}}(|a\rangle|d'\rangle + |a'\rangle|d\rangle). \tag{7}$$

This entangled state will then exhibit nonlocal correlations if an experiment of the type described above is performed.

It is not at all obvious which kind of coincidence gates has to be used in such an experiment. Detailed calculation[16] shows that one has to collect the two idlers within a time window significantly shorter than their coherence time. Thus, the use of narrowband filters in the experiment is essential. Otherwise the time of registration still carries path information and, hence, destroys the entangled state of Eq.(7). This is because the four photons in general are not registered exactly simultaneously, and the idler photon registered earlier can be associated with the signal photon registered earlier and vice versa. This is a clear example of how Welcher-Weg information can help in analyzing a quantum optical experiment.

MULTIPORTS AND CORRELATIONS IN HIGHER-DIMENSIONAL HILBERT SPACES

It is a well-known fact that any arrangement built out of non-absorbing beam splitters and phase shifters can be described by a unitary operator, and unitarity already requires that such a device has the same number N of input ports and output ports. The reverse is not at all obvious, namely, whether or not an explicit experimental realization is possible for any $N \times N$ unitary operator. That this is indeed the case has recently been demonstrated in a constructive proof.[17] The basic idea is that one can rewrite any unitary operator $U(N)$ as a sequence of operators which act only on a two-dimensional subspace of the whole space associated with the unitary operator. Any $N \times N$ unitary operator can then be realized in the experiment using just beam splitters and phase shifters.

From this constructive proof it immediately follows that it is possible to implement specifically rotation operators in higher dimensional Hilbert spaces in the laboratory and to do experiments corresponding to a spin larger than 1/2. All one has to do is to build the appropriate beam splitter device with N input ports and N output ports, where the incident photons are in the appropriate superposition of the N inputs. It is obvious that any such N-mode state is equivalent to a system with spin $(N-1)\hbar/2$. Fig. 3 shows a possible realization of a multiport with dimension N.

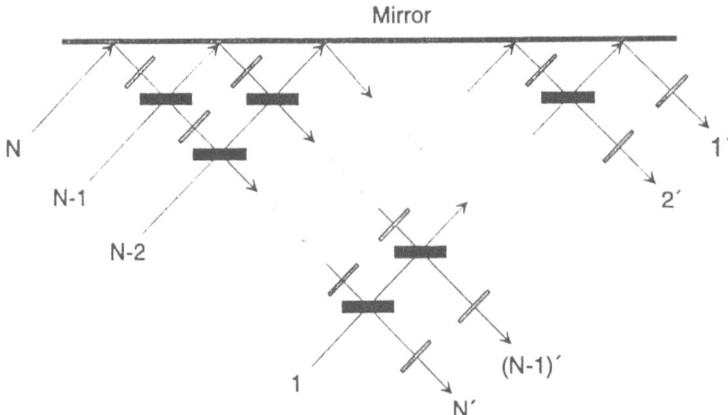

Figure 3: A triangular array of beam splitters implements any $N \times N$ unitary matrix as an optical multiport. The beams are solid lines. A suitable beam splitter is at each crossing point of the beams. Phase shifters are at one input of each beam splitter and at the outputs $(1', ..., N')$ of the multiport. Each diagonal row of beam splitters performs a transformation reducing the effective dimension of the Hilbert space by one (from [17]).

It is specifically interesting if such multiports are fed by correlated two-photon states. For the case of an ordinary four-port, i.e. a beam splitter, the so-called Hong-Ou-Mandel[18] dip results, which can easily be calculated as follows. Consider (Fig. 4) a 50/50 beam splitter where we have two photons incident simultaneously via each of the two input ports a_1 and a_2.

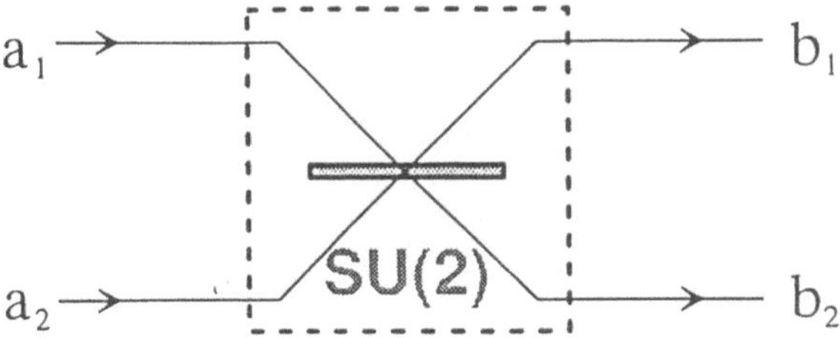

Figure 4: The beam-splitter with the two input ports a_1 and a_2 and the two output ports b_1 and b_2 can be represented by a unitary operator (from [23]).

Then, the input state is

$$|\psi\rangle = \frac{1}{\sqrt{2}}(|a_1\rangle_1|a_2\rangle_2 + |a_2\rangle_1|a_1\rangle_2).$$ (8)

This is an entangled state because we don't know which one of the two photons came via which input or, in other words, the input state has to be symmetrized such that it shows bosonic symmetry. Now consider the action of the beam splitter on the input state. It is important to realize that any beam splitter has an unavoidable phase action which is a direct consequence of unitarity.[19] In its simplest form, one assumes that both reflected beams achieve a phase shift of $\pi/2$ with respect to the transmitted beams. Therefore, the action of the beam splitter can be described via the transition rules

$$|a_1\rangle \rightarrow \frac{1}{\sqrt{2}}(i|b_1\rangle + |b_2\rangle)$$

$$|a_2\rangle \rightarrow \frac{1}{\sqrt{2}}(|b_1\rangle + i|b_2\rangle)$$ (9)

and, thus, the emerging state becomes

$$\frac{1}{\sqrt{2}}(|b_1\rangle_1|b_1\rangle_2 + |b_2\rangle_1|b_2\rangle_2).$$ (10)

This is also an entangled state, but now such that both photons are always found in the same output beam. The reader will easily be convinced that this fact is a direct consequence both of the bosonic symmetry of the incident state and of the phase factor i by inserting the relations of Eq.(9) into the incident state (Eq.(8)). In the case of fermionic symmetry, the two particles would never emerge together at the same output port.[20–22] For a multiport with incident two-particle radiation, novel quantum correlation effects are expected some of which have recently been realized.[23]

A particularly interesting experimental situation results if the two-photon source is used such that each photon feeds one multiport. Then, Einstein-Podolsky-Rosen experiments in higher-dimensional Hilbert space become immediately possible.[24] Fig. 5 shows a generic experiment of that kind. A parametric down-conversion source produces two photons. Photon 1 is emitted into the modes a, c, e and photon 2 is emitted into the modes b, d, f. The modes for each photon are then coherently superposed with each other, using the two six-port devices which, according to our proof, can represent any $U(3)$ operator. Finally, detectors at both ends register the photon correlations. Such an experiment is totally equivalent to an Einstein-Podolsky-Rosen experiment

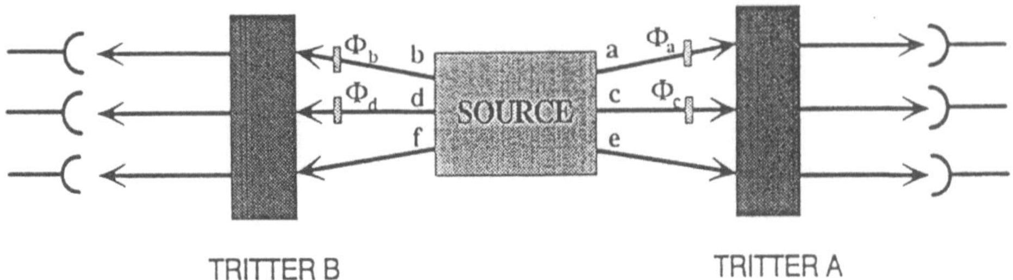

Figure 5: A two-photon three-state interferometer. The source emits one photon to the left into three beams and one photon to the right into three beams. The tritters on each side provide a mixing of all three amplitudes. Three different types of EPR correlations can occur (from [24]).

measuring the correlations between two spin-one particles. An experiment of that kind is presently in progress in our Innsbruck laboratory.

FRUSTRATED DOWN-CONVERSION: THE INFLUENCE OF EXTERNAL BOUNDARY CONDITIONS

A specific feature of the two photons created in the down-conversion process is that they are emitted in an entangled state, i.e. a superposition of product states. Even if only one of the products effectively is present in the experiment, interesting possibilities arise because then any phase shift applied to one photon is actually a phase shift experienced by the product state, that is, by both photons together. A case where this is most clearly seen in an experiment is what we call frustrated down-conversion.[25] Consider the series of setups shown in Fig. 6. Initially (Fig. 6a), we have a nonlinear crystal pumped from one side. For the down-conversion pair only the mode s_1 for the signal photon and the mode i_1 for the idler photon are selected.

We then (Fig. 6b) reflect the pump back onto itself, therefore the crystal is now pumped by two laser beams with exactly opposite momentum. This means that we have another possibility to create the photon pair. The signal might now also be emitted into beam s_2 and the idler might be emitted into beam i_2. The total two-photon state therefore is a superposition of two product states. Note from Fig. 6 that the angles of the idler and the signal photons with respect to the pump are not equal. This implies that the photons have slightly different wavelengths, a feature which we later chose to identify some wavelength-dependent effects.

One might now naively assume that inserting one mirror into only one of the beams of the down-conversion photons, for example the idler, would result in interference fringes registered by the counter in the idler beam i. Yet, one can immediately see by Welcher-Weg consideration that such interference fringes would violate complementarity. This is because the photons are always emitted in pairs. So it would suffice to insert detectors into s_1 and s_2. Then, depending on which of the two signals detectors fires, we have information about the path into which the idler photon was emitted. Thus no interference can arise. Therefore, we conclude, interference can only arise if we also destroy the path information in the signal beams. This can be done by external mirrors for idler and signal in the way as shown in Fig. 6c. Then, neither registration

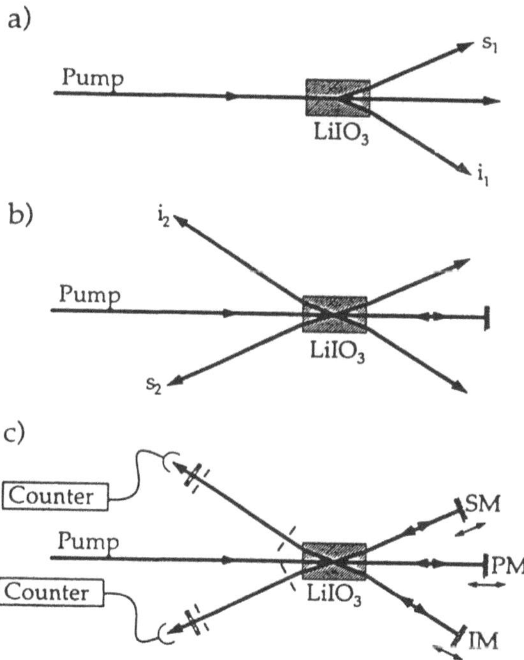

Figure 6: Principle of the experimental setup: (a) In the standard setup of parametric down-conversion a correlated photon pair may be emitted into the two outgoing modes shown. (b) The pump is reflected back onto itself such that another possibility for creating the photon pair arises. (c) The first modes of the down-converted photons are also reflected back. Thus the two possible ways of creating the photon pair may now interfere. The diaphragms serve to define one single mode and filters select out energy matched pairs (from [25]).

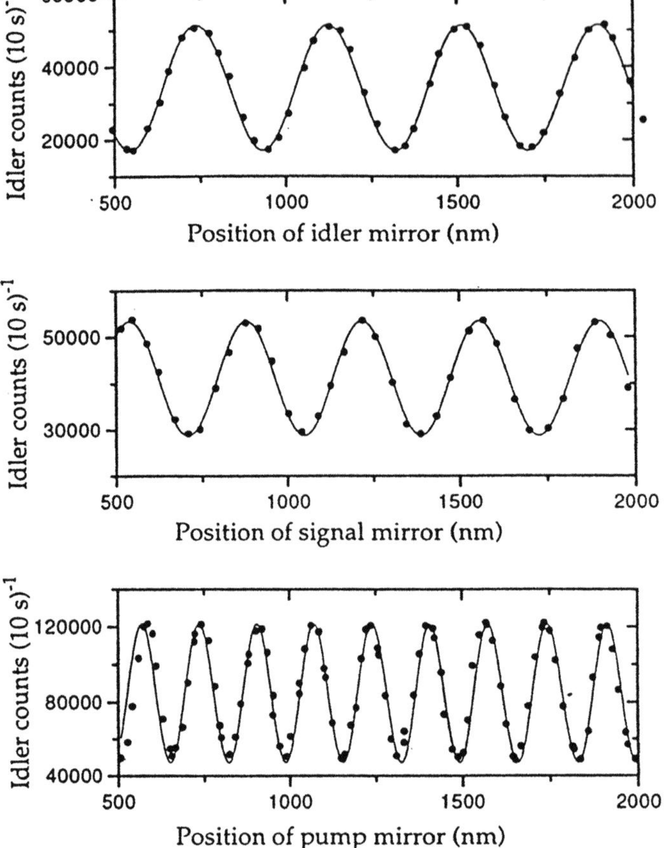

Figure 7: The idler count rate is shown in dependence on the displacement of signal, idler, and pump mirrors. The oscillation period of these idler fringes in all three cases is given by half of the wavelength of the photons in the beam whose mirror is translated (from [25]).

of the idler photon in beam i nor registration of the signal photon in beam s provides any path information, and interference fringes arise. The resulting two-photon state is then[25]

$$|\psi\rangle = \alpha(e^{i\phi_p} + e^{i(\phi_s+\phi_i)})|1\rangle_s|1\rangle_i \tag{11}$$

where ϕ_p, ϕ_s and ϕ_i are the phases accumulated by the pump, the signal, and the idler respectively on their way from the crystal to the mirror and back.

Note that the phaseshifts of both the signal and the idler beam show up as a sum only, and the phase factor actually changes the phase of the product state of signal and idler. As mentioned above it is therefore not possible to assign a phase clearly to either beam in the product state. A characteristic set of results of that experiment is shown in Fig. 7. In all three cases shown, one clearly sees the variation of the idler count rate as a function of the position of either of the three mirrors. Note that, most interestingly, the difference in the oscillation length when the idler mirror or the signal mirror are varied. While, as is usually expected, in the top figure the variation of the

224

idler count rate is periodic with the idler wavelength, the center figure shows that, if the signal mirror is varied, the idler count rate varies with the signal wavelength, again, a consequence of the state of Eq.(11).

THREE-PARTICLE CORRELATIONS

It is an important feature of the reasoning around Bell's inequalities that one takes the existence of perfect correlations — which, for example, arise in the spin-correlated case when the two polarizers are oriented parallel — as the justification to introduce local properties of the two individual particles. Such local properties are assumed to pre-exist prior to and independent of any observation. It is then deduced that these properties are in contradiction to quantum mechanics if one tries to describe statistical correlations which arise when the two polarizers are oriented at an angle with respect to each other. It has therefore been assumed implicitly by many physicists over a long time that the contradiction between quantum mechanics and a local realistic world view is a statistical one, and that classical mechanics and quantum mechanics agree whenever quantum mechanics makes non-probabilistic, definitive predictions. That such a viewpoint is untenable, is a consequence of the three-particle correlations commonly called GHZ correlations.[26–28]

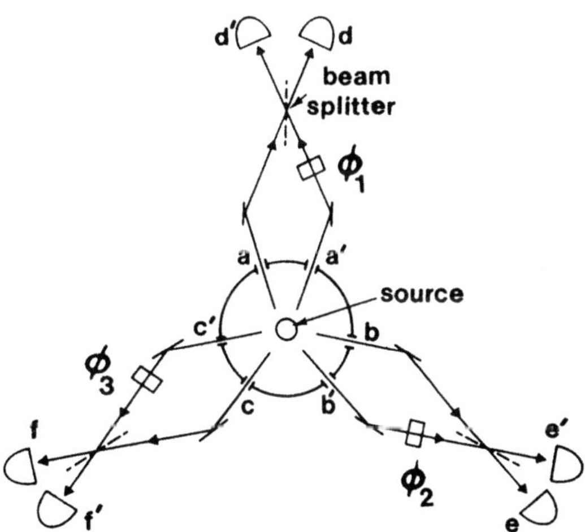

Figure 8: A gedanken three-particle interferometer. The source emits a triple of particles, 1, 2, and 3, into six beams. A phase shift ϕ_1 is imparted on beam a' of particle 1, and beams a and a' are brought together on a beam splitter before illuminating detectors d and d'. Likewise for particles 2 and 3 (from [27]).

Consider, for example, the three-particle source shown in Fig. 8 which emits the three particles into the state

$$|\psi\rangle = \frac{1}{\sqrt{2}}(|a\rangle_1|b\rangle_2|c\rangle_3 + |a'\rangle_1|b'\rangle_2|c'\rangle_3). \tag{12}$$

This state implies that each of the three particles, which very well could be photons, is emitted into two possible beams. For particle 1 for example, these beams are a and a'. While it is thus completely uncertain into which of its two possible beams a given particle is emitted, there is perfect correlation in the sense that whenever particle 1 is emitted into, say, beam a, it is certain that particle 2 is emitted into beam b and particle 3 is emitted into beam c and so on. We then have three phase shifters which are used to change the relative phases between the two beams of each particle and beam splitters to superpose these amplitudes. Finally, triple coincidences are registered in the detectors behind these beam splitters. A detailed analysis then shows that a continuous set of conditions for the phases ϕ arises, for which perfect correlations between three detectors are predicted. The interesting feature now is that these perfect correlations are mutually inconsistent if a local realistic picture is adopted. For a more detailed analysis and explanation, we refer the reader to the references Greenberger, Horne, Shimony and Zeilinger[27] and Mermin.[28]

CONCLUDING COMMENTS

The present overview of some quantum experiments with correlated photons would not be complete without mentioning some of the basic motivation behind such experiments. It is certainly correct to say that quantum mechanics is the most fantastically confirmed theory ever invented in the sciences. Its range of validity extends over many orders of magnitude and it reaches unbelievable precision. The problem with quantum mechanics is that it still lacks a firm epistemological or philosophical foundation. We might compare the situation with, for example, the theory of special relativity where such a foundation is laid by the relativity principle, which requests that all laws of physics have to be the same in all inertial reference frames. It is also our observation that too many physicists still have too "classical" pictures, like waves, particles and the like, in their minds. It is our hope that experiments of the kind described above will contribute to a better intuitive understanding of quantum mechanics, and therefore might guide us towards the missing foundational principle.

This work was supported by the Austrian Science Foundation (FWF), grant number S06502 and by the US National Science Foundation, Grant # PHY92 - 13964.

References

1. J.S. Bell, On the Einstein-Podolsky-Rosen Paradox. Physics 1, 195-200 (1964), reprinted in J.S. Bell, "Speakable and Unspeakable in Quantum Mechanics" (Cambridge U.P., Cambridge, 1987).

2. J.S. Clauser and A. Shimony, Bell's Theorem: Experimental Tests and Implications. Rep. Prog. Phys. 41 (1978) 1881.

3. S.J. Freedman and J.S. Clauser, Experimental Test of Local Hidden-Variable Theories. Phys. Rev. Lett. 28 (1972) 938- 941.

4. A. Aspect, J. Dalibard and G. Roger, Experimental Tests of Bell's Inequalities Using Time-Varying Analyzers. Phys. Rev. Lett. 49 (1982) 1804-1807.

5. A. Zeilinger, Testing Bell's Inequalities with Periodic Switching. Phys. Lett. A 118 (1986) 1.

6. F. Selleri, A. Zeilinger, Local Deterministic Description of Einstein-Podolski-Rosen Experiments. Found. Phys. 18 (1988) 1141.

7. A. Einstein, B. Podolsky and N. Rosen, Can the Quantum-Mechanical Description of Physical Reality be Considered Complete? Phys. Rev. 47 (1935) 777-780.

8. M.A. Horne, A. Zeilinger, A Bell-Type Experiment Using Linear Momenta. Symposium on the Foundations of Modern Physics, Joensuu 1985, P. Lahti and P. Mittelstaedt (eds.), World Scientific Publ. (Singapore), p. 435.

9. M. Horne, A. Zeilinger, A Possible Spin-Less Experimental Test of Bell's Inequality. In "Microphysical Reality and Quantum Formalism", A. van der Merwe, F. Selleri, G. Tarozzi (eds.), Kluwer (Dordrecht), 1988, p. 401.

10. D.C. Burnham, D.L. Weinberg, Observation of Simultaneity in Parametric Production of Optical Photon Pairs. Phys. Rev. Lett. 25 (1970) 84.

11. C.O. Alley, Y.H. Shih, in "Proc. Symp. on Foundations of Modern Physics", P. Lahti, P. Mittelstaedt, eds., World Scientific, Singapore (1985), p. 435; Y.H. Shih, C.O. Alley, Phys. Rev. Lett. 62, (1988) 2921.

12. M.A. Horne, A. Shimony, A. Zeilinger, Two-Particle Interferometry. Phys. Rev. Lett. 62 (1989) 2209; M.A. Horne, A. Shimony, A. Zeilinger, Two-Particle Interferometry, Nature 347 (1990) 429.

13. J.G. Rarity, P.R. Tapster, Experimental Realization of Bell's Inequalities Based on Phase and Momentum. Phys. Rev. Lett. 64 (1990) 2495.

14. E. Schrödinger, Die gegenwärtige Situation in der Quantenmechanik, Naturwissenschaften 23 (1935) 807-812; 823- 828; 844-849, English translation in Proceedings of the American Philosophical Society, 124 (1980) 323-338.

15. D. Bohm, "Quantum Theory", Prentice-Hall, Englewood Cliffs, NJ (1951) 614-623. Phys. Rev. 85 (1952) 166.

16. M. Żukowski, A. Zeilinger, M.A. Horne and A.K. Ekert, Event-Ready-Detectors Bell Experiment via Entanglement Swapping, Phys. Rev. Lett. 71 (1993) 4287.

17. M. Reck, A. Zeilinger, H.J. Bernstein and P. Bertani, Experimental Realization of Any Discrete Unitary Operator. Phys. Rev. Lett. 73 (1994) 58.

18. C.K. Hong, Z.Y. Ou and L. Mandel, Phys. Rev. Lett. 59 (1987) 2044.

19. A. Zeilinger, General Properties of Lossless Beam Splitters in Interferometry. Am. J. Phys. 49 (1981) 882.

20. R. Loudon, in "Coherence and Quantum Optics VI", J.H. Eberly et al. (eds.), Plenum Press, New York (1990) pp. 703-708; R. Loudon, in "Disorder and Condensed Matter Physics", J.A. Blackman and J. Taguena (eds.), Clarendon Press, Oxford (1991) pp. 441-454.

21. A. Zeilinger, H.J. Bernstein and M.A. Horne, Information Transfer with Two-State, Two-Particle Quantum Systems. J. Modern Optics (in press).

22. A. Zeilinger, Probing Higher Dimensions of Hilbert Space in Experiment. Acta Phys. Polonica A85 (1994) 717.

23. K. Mattle, M. Michler, H. Weinfurter, A. Zeilinger and M. Żukowski, Nonclassical Statistics at Multiport Beamsplitters. Appl. Phys. B (in press).

24. A. Zeilinger, H.J. Bernstein, D.M. Greenberger, M.A. Horne and M. Żukowski, Controlling Entanglement in Quantum Optics, in "Quantum Control and Measurement", H. Ezawa, Y. Murayama (eds.), Elsevier Science Publishers (1993) p. 9.

25. T.J. Herzog, J.G. Rarity, H. Weinfurter and A. Zeilinger, Frustrated Two-Photon Creation via Interference, Phys. Rev. Lett. 72 (1994) 629.

26. D. Greenberger, M.A. Horne, A. Zeilinger, Going Beyond Bell's Theorem, in "Bell's

Theorem, Quantum Theory, and Conceptions of the Universe", M. Kafatos (Ed.), Kluwer, Dordrecht (1989) p.69; D.M. Greenberger, M.A. Horne and A. Zeilinger, Multiparticle Interferometry and the Superposition Principles, Physics Today 46, 8 (August 1993) 22.

27. D.M. Greenberger, M.A. Horne, A. Shimony, A. Zeilinger, Bell's Theorem Without Inequalities, Am. J. Phys. 58 (1990) 1131.

28. N.D. Mermin, Quantum Mysteries Revisited. Am. J. Phys. 58 (1990) 731.

QUANTUM NON-DEMOLITION MEASUREMENTS IN THE OPTICAL DOMAIN

Philippe Grangier, Jean-Philippe Poizat and Jean-François Roch

Institut d'Optique Théorique et Appliquée
Bâtiment 503, B.P. 147, F91403 Orsay Cedex - France

INTRODUCTION

A quantum measurement usually perturbs the quantity which is measured, adding "back-action noise" to the system under study. However, "quantum non-demolition" (QND) measurements can be performed, leaving the observed quantity unperturbed, while adding the back-action noise into another complementary observable. Quantum non- demolition measurements were first proposed to monitor the motion of mechanical oscillators [1]- [3], but it was then realized that they were easier to implement in the optical domain [4]- [6], [29], [30]. In this case, the measurements are made on a mode of the electromagnetic field considered as a quantum harmonic oscillator, and the coupling responsible for the measurement is created using either $\chi^{(2)}$ or $\chi^{(3)}$ optical non-linearities [7]- [9]. In this paper, we will focus on the QND measurements of "travelling wave" light beams, rather than fields stored in high Q cavities [10]. Moreover, we will assume that the quantum fluctuations of the beams are very small compared to the mean intensities, so that a linear treatment of these quantum fluctuations is possible [11]. A "travelling wave" optical QND scheme usually involves the coupling of two field modes : one "signal" mode, which is to be measured, and one "meter" mode, which is directly detected and yields some informations about the signal mode. A perfect QND measurement device should leave the measured quadrature component of the signal mode unchanged, and turn the detected meter quadrature component into some kind of "macroscopic copy" of the signal. However, such a device does not exist at present time, and it appears necessary to be able to characterize the "non-ideality" of practical QND

Advances in Quantum Phenomena, Edited by E.G. Beltrametti
and J.-M. Lévy-Leblond, Plenum Press, New York, 1995

schemes. We will briefly recall the criteria [12]- [15] which have been proposed for this characterization, and we will define quantitative limits with respect to these criteria, delimiting "classical" and "quantum" domains of operation. Finally, two experiments, meeting these criteria for QND measurements, will be described in the last sections of the paper.

LINEAR MODEL FOR THE QUANTUM FLUCTUATIONS

For usual laser beams, the quantum intensity fluctuations are very small compared to the mean intensities, so that a linear treatment of these fluctuations is possible [11]. For example, the ratio between the quantum noise and the mean intensity of a laser beam with 1 mW power is about 10^{-6} for a 0.1 ms measurement time, or equivalently a 10 kHz measurement bandwidth. A standard way used in quantum optics for describing these small intensity or phase fluctuations is to use two non-commuting operators, that we will denote X and Y, describing respectively the quantum fluctuations of the sine and cosine parts of the electric field. Since these "quadrature" operators do not commute, the product $\Delta X \Delta Y$ is restricted by Heisenberg inequalities. In the case where the phase origin is chosen such that the quadrature operators describe respectively the quantum fluctuations in-phase and 90°-out-of-phase with the mean value of the field, X and Y will be interpreted as describing respectively intensity and phase fluctuations.

Since several fields will be involved, the operators for the various quadrature components of these fields will be written in column matrices in the frequency domain, denoted $F(\omega)$. All over the paper, the subscripts s and m will denote respectively the signal and meter modes :

$$F(\omega) = \begin{pmatrix} X_s(\omega) \\ Y_s(\omega) \\ X_m(\omega) \\ Y_m(\omega) \end{pmatrix} \tag{1}$$

In a linearized model (relatively to the quantum fluctuations), the relation between the input fluctuations $\delta F^{in}(\omega)$ and the output fluctuations $\delta F^{out}(\omega)$ can be written [11]- [15]:

$$\delta F^{out}(\omega) = \lambda(\omega)\delta F^{in}(\omega) + \delta F^{ad}(\omega) \tag{2}$$

where $\lambda(\omega)$ is a four by four "gain" matrix, and $\delta F^{ad}(\omega)$ is an extra noise added by other degrees of freedom involved in the coupling.

The gain $\lambda(\omega)$ and the covariance matrix of $\delta F^{ad}(\omega)$ are not independent, because the transformation (2) preserves the commutations relations between the input and output modes. In some cases, $\delta F^{ad}(\omega)$ is zero : it is the case for instance for pure "parametric" processes, involving transparent $\chi^{(2)}$ or $\chi^{(3)}$ media. On the other hand, $\delta F^{ad}(\omega)$ is non-zero if either losses or excess noise in the non linear media are involved, due for instance to resonant atomic non-linearities, or to Brillouin or Raman processes in solids.

In all cases, we will assume that the input beams and the QND device are independent, so that $\delta F^{in}(\omega)$ and $\delta F^{ad}(\omega)$ are not correlated. The fluctuations of the beams will be characterized using symetrically ordered covariances, which fit the Wigner distribution. The output covariance W^{out} is then given as a function of the input covariance W^{in} by [11] :

$$W^{out} = \lambda(\omega) W^{in} \lambda(\omega)^\dagger + W^{ad} \tag{3}$$

The expressions of $\lambda(\omega)$ and W^{ad} are obtained from a quantum treatment for each specific device. This description must ensure the relations between $\lambda(\omega)$ and W^{ad} needed to preserve the commutation relations between the input and output fields [11]. If the input light is in a coherent state (which is usually used to modelize laser light), the input covariance matrix W^{in} is diagonal, and its elements are conventionnally taken equal to one (shot-noise level).

CHARACTERIZATION OF A QND DEVICE AS A QUANTUM STATE PREPARATION DEVICE

In QND experiments, the performance of the device has usually been assessed from a measurement of some components of the output covariance W^{out}, which contains the information about the noises and the quantum correlations of the two output beams. This procedure allows one to characterize the "quantum state-preparation" (QSP) properties of the device. A convenient quantity for that purpose is the conditional variance [13] W_{QSP} of the output signal (taken here as X_s^{out}), given the output meter (taken here as Y_m^{out}). As it will be shown below, the conditional variance is closely related to a degree of squeezing, and it can be defined from a suitable combination of the various matrix elements of the output covariance W^{out} :

$$W_{QSP} = W_{XsXs}^{out} - \mid W_{XsYm}^{out} \mid^2 / W_{YmYm}^{out} \tag{4}$$

An important advantage of W_{QSP} is that it can be measured directly in an experiment, as a noise level on a spectrum analyser. For that purpose, the signal and meter photocurrents must be recombined, introducing an electronic gain on the meter photocurrent. This gain (or attenuation) must be optimized, in order to obtain the maximum reduction below the shot-noise level of the signal beam. The optimum value of the attenuation depends of the amount of correlation which has been established between the two beams, and it is given by :

$$g(\omega) = W_{XsYm}^{out} / W_{YmYm}^{out} \tag{5}$$

The criterion $W_{QSP} < 1$ means that the device is able to establish sub-shot-noise correlations between the two output beams. It is instructive to give the value of W_{QSP} for a very simple model, corresponding to the following "parametric" coupling between the quadrature components X and Y of the signal and meter modes :

$$\begin{aligned}
X_s^{out} &= X_s^{in} & Y_s^{out} &= Y_s^{in} + G X_m^{in} \\
X_m^{out} &= X_m^{in} & Y_m^{out} &= Y_m^{in} + G X_s^{in}
\end{aligned} \tag{6}$$

This model can be implemented for instance using crossed Kerr effect between the two modes, obtained using two-photon non-linearities [13], [16]- [18]. In this case, X and Y describe respectively the intensity and phase fluctuations, defined relatively to the mean values of the fields. Eq.(6) means clearly that the intensity fluctuations $X_{s,m}$ of each beam are not changed (the Kerr medium is assumed to be perfectly transparent), while the phase fluctuations $Y_{s,m}$ of one beam have "read-out" the intensity fluctuations of the other one. For a shot-noise limited meter beam, one obtains using eqs. (4) and (6) :

$$W_{QSP} = 1 - G^2/(1 + G^2) = 1/(1 + G^2) \tag{7}$$

A perfect QSP device is obtained for an infinite value of G. Another useful model is a simple beamsplitter ("linear coupler"), in which the input meter beam can be in a squeezed state [8]. Denoting by r, t the amplitude reflection and transmission coefficients of the beamsplitter, the input-output transformation is simply :

$$\begin{aligned} X_s^{out} &= t X_s^{in} + r X_m^{in} & Y_s^{out} &= t Y_s^{in} + r Y_m^{in} \\ X_m^{out} &= -r X_s^{in} + t X_m^{in} & Y_m^{out} &= -r Y_s^{in} + t Y_m^{in} \end{aligned} \tag{8}$$

In that case, the information on the signal amplitude X_s^{in} is obtained on the meter output amplitude X_m^{out}. Assuming that the input signal beam is shot-noise limited ($W_{XsXs}^{in} = 1$), and that the corresponding quadrature of the input meter beam has a noise power squeezing $W_{XmXm}^{in} = S^{in}$, the QSP properties of this scheme is given by :

$$W_{QSP} = (t^2 + r^2 S^{in}) - \frac{r^2 t^2 (S^{in} - 1)^2}{(r^2 + t^2 S^{in})} = \frac{S^{in}}{(r^2 + t^2 S^{in})} \tag{9}$$

If the input meter beam is in the vacuum state $S^{in} = 1$, one obtains immediately $W_{QSP} = 1$, for any value of r and t . On the other hand, the inequality $W_{QSP} < 1$ is obtained as soon as the meter beam is squeezed ($S^{in} < 1$), and perfect QSP is obtained for $S^{in} = 0$. This evidences the close relationship between the conditional variance and a degree of squeezing. We will take therefore the value $W_{QSP} = 1$ as a limit between "classical" and "quantum" regime for an optical coupler, as far as quantum state preparation is concerned. However, this criterion does not characterize completely the QND properties of the device, because one must also prove that the output signal and meter beams really yield an information about the input signal beam [13], [15]. This point will be discussed in the following section.

CHARACTERIZATION OF A QND DEVICE USING EQUIVALENT INPUT NOISES

The basic idea that we will use [15] is that two types of quantities can actually be measured experimentally. The first one is the output covariance matrix W^{out}, as stated previously. The second quantity is the gain matrix $\lambda(\omega)$, which represents the transfer function of the system, and can be determined using weak classical probe beams, as shown by Eq.(2). Knowing $\lambda(\omega)$ and W^{out}, one may write linear "estimators" for the measured quadrature component of the signal. For instance, in the parametric case, defined by Eq.(6), one has:

$$G^{-1} Y_m^{out} = X_s^{in} + G^{-1} Y_m^{in} \tag{10}$$

$$G^{-1} W_{YmYm}^{out} (G^{-1})^* = W_{XsXs}^{in} + N_m^{eq} \tag{11}$$

$$N_m^{eq} = |G|^{-2} W_{YmYm}^{out} - W_{XsXs}^{in} \tag{12}$$

The quantity N_m^{eq} appears as the noise added to input signal variance, and is therefore an "equivalent input noise". Here, N_m^{eq} is due to the shot-noise of the meter beam, which appears as a noise even in the parametric case. For real devices, a second noise source is due to the excess noise added by the non-linear system, and a third noise source appears to be the quadrature of the signal conjugated from the observed one. This "unwanted" quadrature can be coupled in the output signal or meter by the non-ideal device. These three degradations of the measurement efficiency can then be included in N_m^{eq}, equivalent input noise on the meter channel [15]- [19]. Similar definition can also be used for relating the output signal to the input signal, corresponding to N_s^{eq}, equivalent input noise on the signal channel.

Another possible characterization of the same properties is obtained by looking at the faithfulness of the transfer of an input signal-to-quantum-noise ratio (SNR or simply R in the equations), toward the signal and meter output channels [7] [20] [29] [30]. The input SNR is defined as the ratio of the power of a classical modulation at a given frequency by the quantum noise power at the same frequency for a given observable X,

$$R = \frac{\langle X \rangle^2}{\langle \Delta X^2 \rangle} , \tag{13}$$

where all quantities are defined in the frequency domain [15]. The signal non-degradation property is therefore evaluated by (the subscript s refers to the signal, and m to the meter)

$$T_s = \frac{R_s^{out}}{R_s^{in}} = \frac{W_s^{in}}{W_s^{in} + N_s^{eq}} , \tag{14}$$

where W_s^{in} is the variance of the input light beam normalized to the shotnoise ($W_s^{in} = 1$ for a coherent state), and N_s^{eq} is the equivalent input noise for the signal [15]. The measurement capability is given by

$$T_m = \frac{R_m^{out}}{R_s^{in}} = \frac{W_s^{in}}{W_s^{in} + N_m^{eq}} , \tag{15}$$

where N_m^{eq} is the equivalent input noise for the meter channel. It is clear from these equations that for a perfect measurement the transfer coefficients $T_{s,m}$ approach one, while the equivalent input noises $N_{s,m}^{eq}$ approach zero.

Let us finally give the values of N_m^{eq} and N_s^{eq} for our model devices. In the parametric crossed-Kerr effect model, one obtains :

$$N_s^{eq} = 0 \qquad N_m^{eq} = |G|^{-2} \tag{16}$$

In the linear coupler with squeezed meter beam, one has :

$$N_m^{eq} = (t/r)^2 S^{in} \qquad N_s^{eq} = (r/t)^2 S^{in} \tag{17}$$

$$N_s^{eq} N_m^{eq} = (S^{in})^2 \tag{18}$$

If the input meter beam is in the vacuum state $S^{in} = 1$, the product of the equivalent input noises is equal to 1, for any value of r and t .

CLASSICAL LIMITS FOR QND MEASUREMENTS

In order to discriminate various modes of operation of a "quantum coupler", it is interesting to know where the "quantum limits" are. As shown above, a useful reference is given by a simple beamsplitter with a meter beam in the vacuum state, for which one has $W_{QSP} = 1$ and $N_s^{eq}.N_m^{eq} = 1$ .We have shown more generally that the domain of operation of a "classical coupler", described as a phase-insensitive device, can be defined by the two following inequalities [19]- [20] :

$$W_{QSP} \geq 1 \tag{19}$$

$$N_s^{eq}.N_m^{eq} \geq 1 \tag{20}$$

where 1 is by convention the standard shot-noise level. The inequality (19) means that "conditional squeezing" is forbidden in the classical domain, just like ordinary squeezing. The inequality (20) means that the *product* of the signal and measurement equivalent noises is restricted : in the classical domain, small signal degradation implies poor measurement. We note that the definition of "classical" used in inequality (20) is not the usual one in quantum optics (based on the properties of the P function), but is rather the one connected to phase-sensitive vs phase-insensitive amplifiers in the theory of linear amplifiers [20].

Another form of the same inequality was introduced in Ref. [21], for the particular case of a beamsplitter, using the signal to noise ratios (SNR) on the signal and meter channels :

$$(SNR)_s^{out} + (SNR)_m^{out} \leq (SNR)_s^{in} \tag{21}$$

or equivalently

$$T_s + T_m \leq 1 \tag{22}$$

This alternative form is equivalent to the previous one, under some conditions which are fulfilled in general. A device violating this "classical" limit will be called a "quantum optical tap". On the other hand, obtaining $W_{QSP} < 1$ is "quantum state preparation". A true QND device will be defined as a quantum optical tap yielding also quantum state preparation. Finally, let us emphasize that the two conditions (19) and (20) are independant, and cannot be linked by any "triangular" inequality. This can be shown by realizing [19] that a "twin-beam" generator will yield W_{QSP} arbitrarily small, but $N_s^{eq}.N_m^{eq} > 1$. On the other hand, it is easy to check [19] that a phase sensitive amplifier followed by a beamsplitter will yield $N_s^{eq}.N_m^{eq}$ arbitrarily small, but $W_{QSP} > 1$. Therefore, both inequalities must be considered to ensure "quantum operation" of a QND device.

QND EXPERIMENTS WITH THREE-LEVEL SYSTEMS

The basic idea of this experiment is to use the optical Kerr effect to couple the intensity fluctuations of the signal beam to the phase fluctuations of the meter beam ("crossed-phase modulation effect") [9], [22] . The information is then read out using some kind of phase-sensitive (interferometric) detection scheme of the meter beam.

The first attempt to use the optical Kerr effect for QND measurements has been made using non-resonant $\chi^{(3)}$ in optical fibers [4], [23], [24]. Other experiments were done using resonant $\chi^{(3)}$ in solid-state systems [13], [25]. Despite the simplicity of the scheme, it has not been possible to obtain noiseless $\chi^{(3)}$ media: either guided acoustic wave Brillouin scattering (GAWBS) in optical fibers, or spontaneous emission noise in resonant systems, contribute to W^{ad}.

In nearly resonant three-level systems, it is possible to obtain a pure crossed-Kerr coupling, without self-phase modulation, by using two-photon non-linearities [6], [16]- [18]. For a proper choice of the detunings between the two laser beams and the two transitions, the two-photon dispersion can be made dominant, and the ideal transformation given above can be recovered in principle. We will now describe an experiment, using a cascade three-level system in a sodium atomic beam [29]. This atomic system has the advantage that the resonant enhancement of the non-linear effect is much larger than in solid- state non-linear systems, and the phase damping of the transitions much smaller.

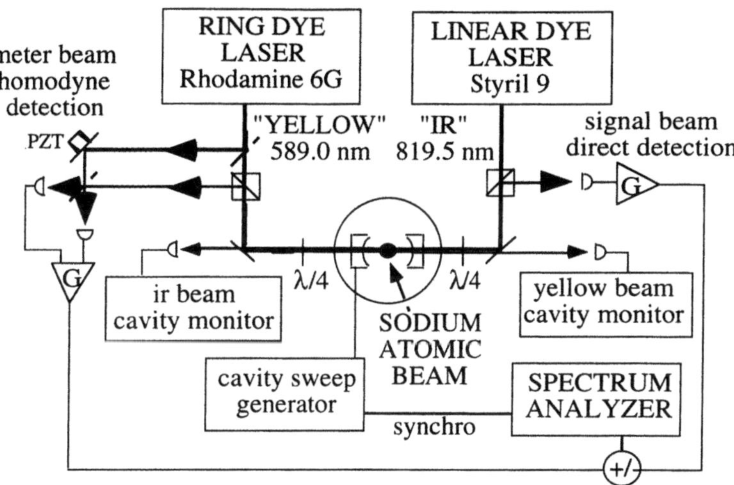

FIG. 1. Experimental set-up. Two lasers at 589.0 nm ("meter" beam) and 819.5 nm ("signal" beam) are tuned close to the resonances of the 3s1/2-3p3/2-3d5/2 cascade in a sodium atomic beam. The quarter-wave plates are used to polarize circularly the two beams in the interaction region, and also to separate the input and output beams.

The experimental set-up used for this experiment (fig.1) is an improvement of those used in Refs [6], [20]. The non-linear coupling of the signal and meter fields is obtained in a beam of three-level atoms in a ladder configuration ($3s_{1/2} - 3p_{3/2} - 3d_{5/2}$ levels of sodium atoms). The atomic beam provides a density of 5.10^{11}atoms/cm^3, over an interaction length of 1 cm, with a Doppler width of about 200 MHz (FWHM). Two cw electronically stabilized dye lasers are tuned around 589 nm (meter beam) for the lower transition and 819.5 nm (signal beam) for the upper one. The non-linearity is enhanced by placing the non-linear medium in a doubly resonant optical cavity (5.2 cm long, 5 cm mirror radius) with a good finesse (100 for each wavelength) and low losses. One of the mirrors is a high reflector at 589 nm ($t^2 = 7.10^{-4}$) and is the input-output mirror for the 819.5 nm beam ($t^2 = 7.10^{-2}$) while the other one is high reflector at 819.5 nm ($t^2 = 5.10^{-4}$) and is the input-output mirror for the 589 nm beam ($t^2 = 7.10^{-2}$), so that the cavity is single ended at each wavelength. The total losses are characterized by the on-resonance cavity reflexion coefficient which is 0.95 for the 819.5 nm beam and 0.90 for the 589 nm laser beam. One of the mirror is piezoelectrically mounted, allowing the length of the cavity to be adjusted. In order to have a phase-sensitive low-intensity detection, a homodyne detection with local oscillator power of $800\mu W$ and fringe visibility of 0.92 is used on the meter beam, while the signal output is directly detected. The detectors are $p - i - n$ silicon photodiodes with quantum efficiency of 0.78 at 589 nm and 0.93 at 819.5 nm followed by low-noise preamplifiers.

We use the *ghost transition* configuration of ref [28] (see also ref [31]). The meter beam is tuned around the lower atomic transition and has a very small power ($5\mu W$) compared to the signal beam (about $300\mu W$) which is tuned around the upper transition. In this configuration, the absorption of the signal beam is very small, and two-photon bistability effects [27] are avoided due to the weakness of the signal beam. Therefore, this configuration guarantees a theoretically nearly perfect non-demolition interaction, with quite large coupling between the amplitude of the signal and the phase of the meter beam [28]. The sign of the detunings are chosen in order to be in the most favorable conditions taking into account the hyperfine structure of the sodium atom. Typical Rabi frequencies normalized to the transverse linewidth of the intermediate level are 8 for the meter beam, and 80 for the signal beam, while the detunings are respectively -300 and $+100$ in the same units (resp. -1.5 GHz and +0.5 GHz in frequency units).

The cavity and the lasers are adjusted so that both fields are simultaneously resonant in the cavity. A classical amplitude modulation at 10 MHz is applied by an electro optical modulator on the signal beam and the measurement is read out on the phase of the 589 nm laser beam. The levels of the modulations transferred onto the output signal and meter beams are then compared with the signal input modulation level using a spectrum analyser. The level of the signal modulation reflected off the offresonant cavity is taken as the reference input modulation level.

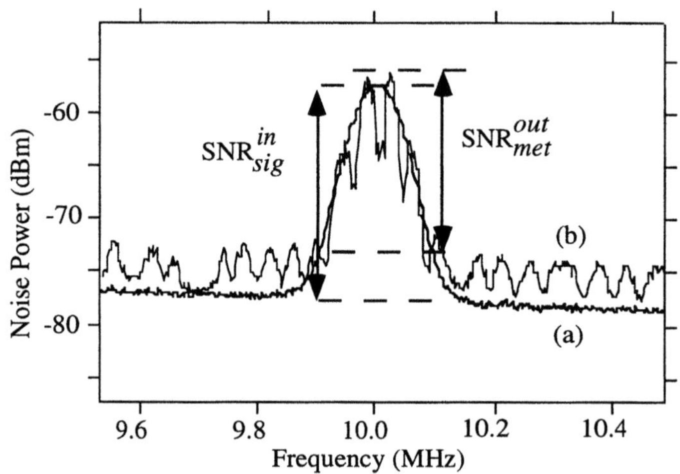

FIG. 2. Transfer of the Signal-to-Noise Ratios from the signal input (smooth curve) to the meter output (wigglering curve, due to the scan of the homodyne detection phase). The meter output SNR is obtained as the superior enveloppe of this curve.

Typical results are presented in fig.2. The level difference between the peak and the background noise corresponding to the signal beam (recorded off-resonance) gives the input SNR_{sig}^{in} (trace (a)). The read out of the meter beam is achieved via the homodyne detection, whose phase is continually scanned, giving the fringes in trace (b) of fig.2. The transferred SNR_{met}^{out} is then measured as the level difference between the peak and background noise of the phase quadrature, which corresponds to the superior envelope of the fringes. From fig.2, the raw transfer from the signal onto the meter is of $-3dB$ which gives after corrections due to amplifiers noise a measurement transfer coefficient $T_{met} = 0.45$. On the other hand, no significant degradation of the signal could be observed. Therefore the signal degradation can be estimated according to overall losses due to the imperfection of optical elements along the propagation of the beam. This gives a raw SNR transfer for the signal $T_{sig} = 0.9$. As a consequence, we obtain a sum of the signal and measurement SNR transfers $T_{sig} + T_{met} =$

1.35, which can be seen as a 1.3dB improvement of the overall information transfer of any classical optical tap (a beam splitter for example). A perfect quantum optical tap would have $T_{sig} = T_{met} = 1$ which would give a perfect 3dB improvement, and could also be regarded as a quantum duplicator.

The experimental results concerning the quantum state preparation ability of this device can be characterized by the conditional variance W_{QSP} introduced in section 3. This quantity is measured by the noise reduction of the signal when corrected by the adequately attenuated photocurrent coming from the meter beam [19], [20]. In the experiment, the noise has been reduced by $0.6dB$ below the shot-noise limit, leading to a conditional variance $W_{QSP} = 0.85$.

QUANTUM OPTICAL REPEATERS USING SEMICONDUCTOR EMITTERS AND RECEIVERS

It has been known for several years that the conversion from light to electrical current and from electrical current to light can be realized below the optical standard quantum limit, using high efficiency semiconductor electro-optical devices. Indeed, the ability of photodiodes to detect sub-shot-noise light by turning it into sub-shot-noise current is a fundamental property of quantum optics [32], and has been experimentally demonstrated by the observation of squeezing [33].

The quantum properties of the inverse transformation are due to the work of Yamamoto and coworkers on semiconductor lasers [34], [43](see also Refs. [35], [36]). More recently, the appearance of high efficiency Light Emitting Diodes (LED) made possible to use these simple devices for generating non-classical light [37], [38], [39]. A crucial idea in these experiments is that the noise of an electrical current at RF frequencies is usually dominated by thermal (Johnson-Nyquist) noise [41], whose power can be made much smaller than the shot-noise level corresponding to the same average current intensity [36] [37] [40]. In this regime, the electrical current behaves classically, and can therefore be measured, duplicated, or amplified without the quantum constraints attached to a light field. If one is concerned by the non-destructive measurement of the intensity of a light beam, one may imagine to convert the light via a photodiode into an electrical current, to measure this current, eventually to amplify it, and to convert it back to a light field using a semiconductor light emitter [42]. This would be a perfect quantum-nondemolition (QND) device [3] if both the conversion rates were unity.

A check of these ideas can be carried out using the experimental set-up [45] schematically depicted in Fig.3. The measurement apparatus itself consists in a large area p-i-n silicon photodiode PD_1 of quantum efficiency $\eta = 0.90 \pm 0.05$, followed by a low noise electronic amplifier, and a light emitting diode LED_2. The current to current conversion rate of the LED - PD system is $\epsilon = 0.16$ at room temperature and $\epsilon = 0.28$ at 100K . The detectors are placed as close as possible to the LEDs to maximize light collection, and introduced in a liquid nitrogen cryostat. The bandwidth of operation is from 150kHz to 350kHz, the lower limit being due to electronic cut-off, and the upper one to the finite response time of the LEDs. From these numbers, it appears that if the LED is series connected with the photodiode, without any amplifier, the transfer coefficients for a shot-noise-limited input will be respectively $T_s = \eta\epsilon$ and $T_m = \eta$, leading to a rather small effect ($T_s + T_m = 1.15 \pm 0.06$ at 100K) mainly due to the low conversion rate of the LED. In order to overcome this limitation, the current coming from the photodiode is amplified prior to being sent to the LED, which makes the conversion back to light less sensitive to vacuum fluctuations [44], and therefore greatly improves the non-degradation property of the system without changing the others.

For a shot-noise-limited input, and assuming a negligible amplifier noise, the transfer coefficients read

$$T_m = \eta \tag{23}$$

$$T_s = \eta \, \frac{1}{1 + \frac{1}{g_{\text{eff}}} \frac{(1-\epsilon)}{\epsilon}} \cdot \tag{24}$$

where $g_{\text{eff}} = g^2/G$ with g being the AC gain of the amplifier and G the DC gain. The distinction between AC and DC gain, allowed in our experimental set-up, is an extra degree of freedom given by electronics compared to optics. Note that for practical purposes in information transmission, one has to choose $G = g$, which leads to $g_{\text{eff}} = g$. In the high gain limit ($g_{\text{eff}} \to +\infty$), the sum of these two coefficient goes to $T_s + T_m = 2\eta$, and does not depend on LED_2's conversion rate ϵ anymore.On the other hand, the conditional variance is limited by the conversion rate of LED_2, and is given by

$$W_{QSP} = 1 - \epsilon. \tag{25}$$

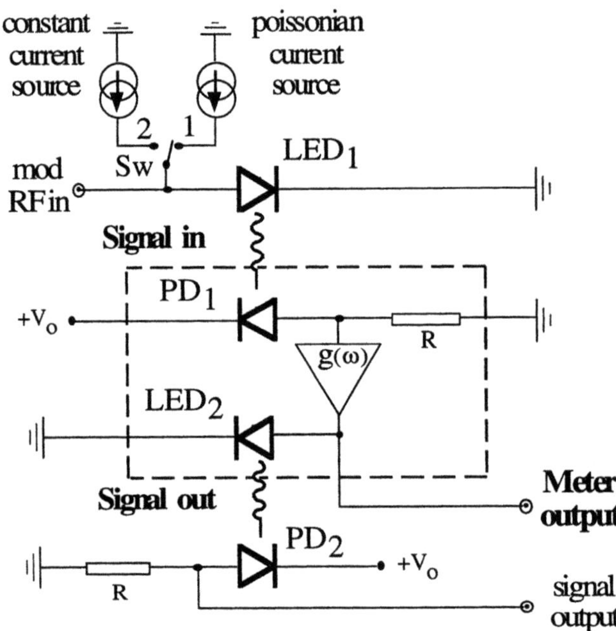

FIG. 3.   Schematic experimental set-up. The measurement device is enclosed in the dashed box. $LED_{1,2}$ and $PD_{1,2}$ are cooled at 100K. The poissonian current source consists in three parallel connected photodiodes, illuminated by white incandescent lamps. When switch Sw is in position 1, LED_1 is driven by a shotnoise limited current; when in position 2, LED_1 is driven by a constant (filtered) current, and shines therefore squeezed light on PD_1. LED_2 can be also driven by the poissonian current source, in order to get a shotnoise reference on PD_2. Note that, for clarity, all the electrical filtering is not represented in the figure.

In our experiment, the input light beam is the light coming from LED_1, and can therefore be either squeezed (switch Sw in position 2), or shot-noise limited (Sw in position 1). The evaluation of the transfer coefficient is performed by modulating the input signal light (at 250 kHz) by addition of a small RF modulation to LED_1 poissonian driving current. The input signal-to-noise-ratio R_s^{in} is deduced from the direct measurement of R_m^{out} on a spectrum analyser, taking into account the quantum efficiency of PD_1, and the SNR of the signal output R_s^{out} is visualised on the output of PD_2. For high gain ($g_{\text{eff}} = 100$), no distinction between R_m^{out} and R_s^{out} can be made within our experimental precision (about 3%), and therefore $T_s = T_m = \eta = 0.90 \pm 0.05$, i.e. $T_s + T_m = 1.8 \pm 0.1$. Moreover, the experimental evolution of T_s/η as a function of the gain g_{eff} was found in very good agreement with the theoretical expression given by Eq. 24.

The conditional variance was measured by subtracting the meter output from the signal output, using a $0 - 180^\circ$ RF power combiner, in order to get the signal fluctuations down to below the shotnoise. This requires a careful adjustment of the relative RF phase, since the fluctuations to be subtracted are both about 20dB above the shotnoise. We obtain thus a noise reduction down to 1.2 ± 0.1dB below the shot-noise of the signal beam. This yields $W_{QSP} = 0.76 \pm 0.02$, while the theoretical value is $W_{QSP} = 0.72$.

Owing to the possibility to have an amplitude squeezed signal input light, when LED_1 is driven by a constant current, we are also able to perform QND measurement of a squeezed light. In the high gain regime, the signal output beam is far above shot noise, but a squeezed output beam can be recovered by decreasing the gain g_{eff} to unity, where our system still operates in the quantum domain, with $T_s + T_m = 1.15$. The input variance W_s^{in} can be inferred from the measured degree of squeezing on the meter channel ($W_m^{out} = 0.80$), as it is usually done for a non-ideal photodetector. The directly measured signal output squeezing is then $W_s^{out} = 0.93$, while the experimental value of the conditional variance is $W_{QSP} = 0.79 \pm 0.02$ (-1.0 ± 0.1dB). All these results clearly show that our scheme does operate in the QND regime, while having a squeezed signal beam, both at the input and the output.

It is worth pointing out that having a signal gain close to one is an extra requirement with respect to the QND criteria, which appears when one is concerned by preserving the squeezing of the input signal. On the other hand, the first QND criteria would rather lead to strong amplification of the intermediate electrical current, because this improves the value of the SNR transfer, by making the system less sensitive to downstream losses. We propose to use the term "quantum repeater" for a device which fulfills the QND criteria with large signal gain, and therefore is not a QND scheme in the original sense. Finally, the value of the conditional variance does not depend on the gain, but only on the quantum efficiency of the reemitting LED and detection system. The perspective of emergence of very high efficiency single-mode LED [46], or alternatively the technological realization of low threshold laser diodes, makes this type of system very promising for a realistic implementation in ultra-low noise optical telecommunication networks.

CONCLUSION

On figure 4, the results obtained in some optical QND experiments realized to present date are plotted as a function of the values of the conditional variance and of $T_s + T_m$. Some other experiments [24] [25] [47], which are quite interesting but did not reach the quantum domain, are not represented here.

The best QND result in the usual sense (signal gain close to unity) is the three-level experiment described above (point f), while the LED experiment just described (point g) appears to be the best "quantum repeater" experiment (large signal gain). However, a fair comparison should also take into account also the frequency bandwidth over which the QND

effects are achieved. On that respect, pulsed experiments (e.g. point e) have been so far the
most successful.

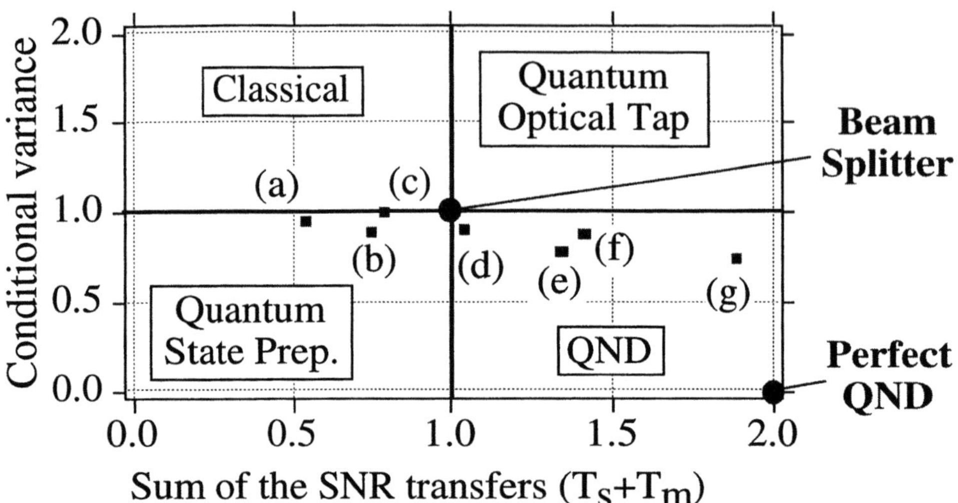

FIG. 4.   Experimental results plotted in the plane $x = T_s + T_m, y = W_{QSP}$. The point ($x$=1, y=1)
is simply a beamsplitter (lossless classical coupler, see Section 3). Points (a), (b), (c), (d), (e) correspond
respectively to the experiments described in Refs. (4), (5), (6), (20) and (30). Points (f) and (g) are the
results described in the present paper (see also Refs. (29) and (45)), which fulfill all criteria for QND
measurements.

ACKNOWLEDGEMENTS

This work was completed in the framework of the EC Contract ESPRIT III BRA 6934,
with partial support from the Centre National d'Etude des Télécommunications (France
Telecom). We acknowledge efficient technical assistance of G. Roger, F. Contet, J. C. Rodier
and A. Villing. Institut d'Optique Théorique et Appliquée is URA 14 of the Centre National
de la Recherche Scientifique.

REFERENCES

[1] C.M. Caves, K.S. Thorne, R.W.P. Drever, V.D.Sandberg and M.Zimmermann, Rev.Mod.Phys. **52**, 341 (1980).

[2] "Quantum Optics, Experimental Gravitation and Measurement Theory", eds P. Meystre and M.O. Scully (Plenum, New York, 1983).

[3] V.B. Braginskii, Soviet Physics Uspekhi **31**, 836 (1988); V.B. Braginsky, Y.I. Vorontsov and K.S. Thorne, Science **209**, 547 (1980).

[4] M.D. Levenson, R.M. Shelby, M. Reid and D.F. Walls, Phys.Rev.Lett. **57**, 2473 (1986).

[5] A.LaPorta, R.E.Slusher and B.Yurke, Phys.Rev.Lett. **62**, 28 (1989).

[6] P Grangier, J.F. Roch and G. Roger, Phys. Rev. Lett. **66**,1418 (1991).

[7] B. Yurke, J.Opt.Soc.Am B**2**, 732 (1985).

[8] J.H. Shapiro, Optics Letters **5**, 351 (1980).

[9] G.J. Milburn and D.F. Walls, Physical Review A28, 2065 (1983).

[10] M. Brune, S. Haroche, V. Lefèvre, J.M. Raimond and N. Zagury, Phys. Rev. Lett. **65**, 976 (1990).

[11] J.M. Courty, P.Grangier, L. Hilico, S. Reynaud, Optics Comm. **83** , 251 (1991).

[12] N. Imoto and S. Saito, Physical Review A **39**, 675 (1989).

[13] M.J. Holland, M. Collett, D.F Walls, M.D. Levenson, Phys. Rev A **42**, 2995 (1990).

[14] C.A. Blockley and D.F. Walls, Optics Comm. **79**, 241 (1990).

[15] P. Grangier, J.M. Courty and S. Reynaud, Optics Comm. **89**, 99 (1992).

[16] P.Grangier, J.F. Roch and S. Reynaud, Optics Comm. **72**, 387 (1989).

[17] P.Grangier and J.F. Roch, Quantum Optics **1**, 17 (1989).

[18] P.Grangier and J.F. Roch , Optics Comm. **83**, 269 (1991).

[19] J. F. Roch, Thèse de l'Université Paris XI, (unpublished), (1992).

[20] J.F. Roch, G. Roger, P. Grangier, J.M. Courty and S. Reynaud, Appl. Phys. B **55**, 291 (1992).

[21] Y. Yamamoto, Trans. Inst. Electron. Inf. Commun. Eng., Sect. E **73**, 1598 (1990).

[22] N. Imoto, H.A. Haus and Y. Yamamoto, Phys. Rev., A **32**, 2287 (1985).

[23] N. Imoto, S. Watkins and Y. Sasaki, Optics Comm. **61**, 159 (1987).

[24] H.A. Bachor, M.D. Levenson, D.F. Walls, S.H. Perlmutter and R.M. Shelby, Phys. Rev. A **38**, 180 (1988).

[25] M.D. Levenson, M.J. Holland, D.F. Walls, P.J. Manson, P.T.H. Fisk, and H.A. Bachor, Phys. Rev. A **44**, 2023 (1991).

[26] J.P. Poizat, M.J. Collett and D.F. Walls, Phys. Rev A **45**, 5171 (1992).

[27] P. Grangier, J. F. Roch, G. Roger, L. A. Lugiato, E. M. Pessina, G. Scandroglio, and P. Galatola, Phys. Rev. A. **46**, 2735 (1992)

[28] K.M. Gheri, P. Grangier, J. Ph. Poizat, and D. F. Walls, Phys. Rev. A. **46**, 4276 (1992).

[29] J.Ph. Poizat and P. Grangier, Phys. Rev. Lett. **70**, 271 (1993)

[30] J.A. Levenson, I. Abram, T. Rivera, P. Fayolle, J.C. Garreau and P. Grangier, Phys. Rev. Lett. **70**, 267 (1993)

[31] H. A. Bachor and P. T. H. Fisk, Applied Phys. B **49**, 291 (1989).

[32] R.J. Glauber, in Ecole d'Eté des Houches 1964, C. De Witt, A. Blandin and C. Cohen-Tannoudji Eds., (Gordon and Breach, 1965).

[33] Special issues on squeezed light, J. Opt. Soc. Am. B **4**, 1450 (1987), J. Mod. Optics **34** 709 (1987), Applied Physics B **55** 189-303 (1992).

[34] S. Machida and Y. Yamamoto, Optics Comm. **57**, 290 (1986); S. Machida,Y.Yamamoto, and Y. Itaya, Phys. Rev. Lett. **58**, 1000 (1987).

[35] M. C. Teich and B. E. A. Saleh, J. Opt. Soc. Am. B **2**, 275 (1985).

[36] P. R. Tapster, J. G. Rarity, and J. S. Satchell, Europhys. Lett. **4**, 293 (1987).

[37] P. J. Edwards, Int. J. Optoelectronics **6**, 23 (1991); P. J. Edwards and G. H. Pollard, Phys. Rev. Lett. **69**, 1757 (1992).

[38] H.-A. Bachor, P. Rottengatter, and C. M. Savage, Appl. Phys. B **55**, 258 (1992).

[39] E. Goobar, A. Karlsson, G. Björk, and P-J. Rigole, Phys. Rev. Lett. **70**, 437 (1993).

[40] R. H. Koch, D. J. Van Harlingen, and J. Clarke, Phys. Rev. B **26**, 74 (1982).

[41] J. B. Johnson, Phys. Rev. **32**, 97 (1928); H. Nyquist, Phys. Rev. **32**, 110 (1928).

[42] F. Capasso and M. C. Teich, Phys. Rev. Lett. **57**, 1417 (1986).

[43] W. H. Richardson, S. Machida, and Y. Yamamoto, Phys. Rev. Lett. **66**, 2867 (1991).

[44] H. P. Yuen, Phys. Rev. Lett. **56**, 2176 (1986).

[45] J.-F. Roch, J.-Ph. Poizat, and P. Grangier, Phys. Rev. Lett. **71**, 2006 (1993).

[46] E. Yablonovitch, T. J. Gmitter, R. D. Meade, A. M. Rappe, K. D. Brommer, and J. D. Joanopoulos, Phys. Rev. Lett. **67**, 3380 (1991).

[47] S.R. Friberg, S. Machida and Y. Yamamoto, Phys. Rev. Lett. **69**, 3165 (1992).

QUANTUM CRYPTOGRAPHY AND COMPUTATION

Artur K. Ekert

Clarendon Laboratory, University of Oxford
Parks Rd. Oxford OX1 3PU, U.K.

WHAT IS WRONG WITH CLASSICAL CRYPTOLOGY?

Human desire to communicate secretly is at least as old as writing itself and goes back to the beginnings of our civilisation. Methods of secret communication were developed by many ancient societies, including those of Mesopotamia, Egypt, India, and China, but details regarding the origins of cryptology[1] remain unknown [1].

We know that it was the Spartans, the most warlike of the Greeks, who pioneered military cryptography in Europe. About 400 BC they employed a device known as a *scytale*. The device, used for communication between military commanders, consisted of a tapered baton around which was wrapped a spiral strip of parchment or leather containing the message. Words were then written lengthwise along the baton, one letter on each revolution of the strip. When unwrapped, the letters of the message appeared scrambled and the parchment was sent on its way. The receiver wrapped the parchment around another baton of the same shape and the original message reappeared.

Julius Caesar allegedly used, in his correspondence, a simple letter substitution method. Each letter of Caesar's message was replaced by the letter that followed it alphabetically by three places. The letter A was replaced by D, the letter B by E, and so on. For example, the English word COLD after the Caesar substitution appears as FROG. This method is still called the Caesar cipher, regardless of how many letters shift is used for the substitution.

These two simple examples already contain the two basic methods of encryption which are still employed by cryptographers today namely *transposition* and *substitution*. In transposition (*e.g. scytale*) the letters of the *plaintext*, the technical term for the message to be transmitted, are rearranged by a special permutation. In substitution (*e.g.* Caesar's cipher) the letters of the plaintext are replaced by other letters, numbers or arbitrary symbols. In general the two techniques can be combined [2].

---

[1]The science of secure communication is called cryptology from Greek *kryptos* hidden and *logos* word. Cryptology embodies cryptography, the art of code-making, and cryptoanalysis, the art of code-breaking.

Advances in Quantum Phenomena, Edited by E.G. Beltrametti
and J.-M. Lévy-Leblond, Plenum Press, New York, 1995

Originally the security of a cryptotext depended on the secrecy of the entire encrypting and decrypting procedures; however, today we use ciphers for which the algorithm for encrypting and decrypting could be revealed to anybody without compromising the security of a particular cryptogram. In such ciphers a set of specific parameters, called a *key*, is supplied together with the plaintext as an input to the encrypting algorithm, and together with the cryptogram as an input to the decrypting algorithm. This can be written as

$$\hat{E}_K(P) = C, \text{ and conversly, } \hat{D}_K(C) = P, \tag{1}$$

where P stands for plaintext, C for cryptotext or cryptogram, K for cryptographic key, and $\hat{E}$ and $\hat{D}$ denote an encryption and a decryption operation respectively.

The encrypting and decrypting algorithms are publicly announced; the security of the cryptogram depends entirely on the secrecy of the key, and this key must consist of any *randomly chosen*, sufficiently long string of bits. Probably the best way to explain this procedure is to have a quick look at the Vernam cipher, also known as the one-time pad.

If we choose a very simple digital alphabet in which we use only capital letters and some punctuation marks such as

A	B	C	D	E	...	...	X	Y	Z	?	,	.
01	02	03	04	05	...	...	24	25	26	27	28	29	30

we can illustrate the secret-key encrypting procedure by the following simple example:

O	X	F	O	R	D		U	N	I	V	E	R	S	I	T	Y
15	24	06	15	18	04	27	21	14	09	22	05	18	19	09	20	25
19	14	26	25	29	17	28	12	01	18	27	03	23	05	10	21	24
04	08	02	10	17	21	25	03	15	27	19	08	01	24	19	11	19

In order to obtain the cryptogram (sequence of digits in the bottom row) we add the plaintext numbers (the top row of digits) to the key numbers (the middle row) , which are randomly selected from between 0 and 30, and take the remainder after division of the sum by 30. This operation is called addition modulo 30. For example, the first letter of the message "O "becomes a number "15 "in the plaintext, then we add $15 + 19 = 34$; $34 = 1 \times 30 + 4$, therefore we get 04 in the cryptogram. The encryption and decrytion can be written as $P + K \pmod{30} = C$ and $C - K \pmod{30} = P$ respectively.

The cipher was invented by Major Joseph Mauborgne and AT&T's Gilbert Vernam in 1917 [1] and we know that if the key is secure, as long as the message, truly random and never reused this cipher is really unbreakable! So what is wrong with classical cryptography?

There is a snag. It is called the key distribution. Once the key is established, subsequent communication involves sending cryptograms over a public channel which is vulnerable to total passive eavesdropping (e.g. public announcement in mass-media). However in order to establish the key, two users, who share no secret information initially, must at a certain stage of communication use a reliable and a very secure channel. Since the interception is a set of measurements performed by the eavesdropper on this channel, however difficult this might be from a technological point of view, *in principle* any classical key distribution can always be passively monitored, without the legitimate users being aware that any eavesdropping has taken place.

Mathematicians have tried hard to solve the key distribution problem. The 1970s, for example, brought a clever mathematical discovery in the shape of "public key" systems [3, 4]. In these systems users do not need to agree on a secret key before they send the message. They work on the principle of a safe with two keys, one public key to lock it, and another private one to open it. Everyone has a key to lock the safe but only one person has a key that will open it again, so anyone can put a message in the safe but only one person can take it out. These systems exploit the fact that certain mathematical operations are easier to do in one direction than the other. Mathematicians use *trap-door one-way functions* to design such public key cryptosystems. The systems avoid the key distribution problem but unfortunately their security depends on unproved mathematical assumptions, such as the difficulty of factoring large integers.[2] This means that if and when mathematicians or computer scientists come up with fast and clever procedures for factoring large integers—making it obvious to anyone that, for example, "29083" is equivalent to "127 times 229"—the whole privacy and discretion of public-key cryptosystems could vanish overnight. Indeed, recent work in quantum computation shows that quantum computers can factorize much faster than classical computers! [5]

In the following I will describe quantum channels which distribute the key with perfect security. In the second part I will introduce quantum computers and comment on their future role in breaking public key cryptosystems. All together it is supposed to be a brief story about quantum code-making and quantum code-breaking.

QUANTUM CODE-MAKING

Quantum cryptography cannot prevent eavesdropping, but any eavesdropping attempt can be detected by the legitimate users of the communication channel. This is because eavesdropping affects the quantum state of the information carriers and results in an abnormal error rate. Let me now describe how quantum mechanics solves the key distribution problem.

There are at least three main types of quantum cryptosystems for the key distribution, these are:

- Cryptosystems with encoding based on two non-commuting observables proposed by S. Wiesner (1970), and by C.H. Bennett and G. Brassard (1984) [6, 7].

- Cryptosystems with encoding built upon quantum entanglement and the Bell Theorem proposed by A.K. Ekert (1990) [8, 9].

- Cryptosystems with encoding based on two non-orthogonal state vectors proposed by C.H. Bennett (1992) [10].

Although I definitely have my favorite quantum cryptosystem (guess which one), however, for the purpose of this pedagogical presentation I will describe conceptually the simplest one, which relies on partial indistinguishability of any two non-orthogonal states [10].

---

[2]RSA - the most popular public key cryptosystem named after the three inventors, Ron Rivest, Adi Shamir, and Leonard Adleman [4] - gets its security from the difficulty of factoring large numbers.

Here is the scenario with the usual three *dramatis personae*: Alice, Bob (legitimate users), and Eve (eavesdropper). Alice and Bob want to establish a cryptographic key. Alice starts the key distribution with a quantum transmission, sending to Bob a random sequence of quanta in two non-orthogonal states u and v, which represent bits 0 and 1 respectively [10, 11, 12].

By a suitable choice of phases and of the basis, the two quantum states u and v, whose overlap is $|\langle u|v \rangle|^2 = \sin^2 2\alpha$, can always be written as

$$ u = \begin{pmatrix} \cos\alpha \\ \sin\alpha \end{pmatrix} \qquad \text{and} \qquad v = \begin{pmatrix} \sin\alpha \\ \cos\alpha \end{pmatrix}. \tag{2} $$

On the receiving side, Bob has to distinguish between these two non-orthogonal states for each incoming carrier. Of course he cannot do that with certainty. He can, however, perform a test which sometimes fails to provide an answer, but once it does give an answer it is always the correct one. To achieve this result, the simplest (but not the most efficient) approach for Bob is to measure, say with the same probability, one of the projection operators

$$ P_{\neg u} = 1 - uu^{\dagger} \qquad \text{or} \qquad P_{\neg v} = 1 - vv^{\dagger}. \tag{3} $$

Think about operators $P_{\neg u}$ and $P_{\neg v}$ as two filters that select states orthogonal to u and v respectively. Filter $P_{\neg u}$ will definitely suppress u *i.e.* bit value 0 but may with certain probability transmit v *i.e.* bit value 1; filter $P_{\neg v}$ will operate likewise. For each incoming carrier Bob decides randomly and independently of Alice whether to use filter $P_{\neg u}$ or $P_{\neg v}$.

A positive result for $P_{\neg u}$ indicates with certainty that the information carrier was in the v state, and vice versa. A null result is of no use if only unambiguous conclusions are acceptable. (It only gives the *a posteriori* probabilities for u and v.) The probability of getting this inconclusive result is $(1 + \langle u|v \rangle^2)/2$. That probability can be reduced somewhat by a more sophisticated measurement process [13, 14] which uses an auxiliary quantum system prepared in a known initial state. After a suitable unitary evolution of the combined system, the latter is left in an entangled state for which the probability of an inconclusive answer is only $\langle u|v \rangle$ (recall that $1 + x^2 > 2x$ for any $x \neq 1$). In the language of modern measurement theory [15, 16], Bob uses a *positive operator valued measure* (POVM) consisting of the operators

$$ A_u = P_{\neg v}/(1+S), \qquad A_v = P_{\neg u}/(1+S), \qquad \text{and} \qquad A_? = 1 - A_u - A_v, \tag{4} $$

where $S = \langle u|v \rangle = \sin 2\alpha$, and the index "?" refers to an inconclusive test. The probability of obtaining the answers u, v, or ?, following the preparation of a carrier in any state ρ, is $\mathrm{Tr}(\rho A_u)$, $\mathrm{Tr}(\rho A_v)$, and $\mathrm{Tr}(\rho A_?)$, respectively. (Note that the various A_i do not commute, contrary to von Neumann's projection operators in elementary measurement theory. Moreover, the final state of the carrier is *not* in general an eigenstate of A_i, so that the result of such a generalised measurement is not, in general, repeatable.)

In the following, I assume that Bob uses this more efficient detection method, so that the probability of an inconclusive result is

$$ R_0 = \sin 2\alpha \quad \text{with} \quad 0 \leq \alpha \leq \pi/4. \tag{5} $$

After completion of the quantum transmission, Alice and Bob communicate in public and discard all inconclusive results, so that, in the absence of external disturbances, the remaining instances are perfectly correlated: they consist entirely of cases in which Alice sent 0 and Bob detected 0, or Alice sent 1 and Bob detected 1.

This correlation is perfect only if there was no eavesdropping on the line. The eavesdropper does not know in advance about Alice's and Bob's choices, so she is bound to introduce some errors as she cannot build an apparatus that would definitely distinguish between Alice's u and v and any substitution of a faked signal may cause to give a positive result when it should not. Alice and Bob check for eavesdropping revealing to each other in public some random subsequence of bits which they subsequently discard. If the test is negative, the distribution must be set up again; if the test is positive the remaining unrevealed bits form the key.

An analysis of eavesdropping in quantum cryptosystems is by no means trivial. In principle, any non-zero error rate Q may indicate eavesdropping; however, in practice, the transmission is subject to noise, and there will be some discrepancies even in the absence of eavesdropping. The problem facing Alice and Bob is now to decide on the maximum tolerable discrepancy allowed Q_{max}, which would still give them a secure key. This border line error rate has been estimated at [12]

$$Q_{\mathrm{max}} = \sin^2 \alpha. \tag{6}$$

From Alice and Bob's point of view, the existence of a strategy which effectively discloses Bob's key to Eve, while creating an error rate Q_{max}, means that all quantum transmissions with an error rate above Q_{max} should be abandoned. The converse is also true: for all error rates Q below Q_{max}, there exist error-correcting codes allowing Alice and Bob to establish a perfectly correlated, secret string of bits, if they wish to do so. Therefore, Q_{max} provides us with a well defined border line: whenever $Q \geq Q_{\mathrm{max}}$ the key distribution failed (there exists a strategy which discloses all the information to Eve); whenever $Q < Q_{\mathrm{max}}$ it is in principle possible for Alice and Bob to communicate in perfect secrecy.

We see that when $\langle u|v \rangle \ll 1$, a small error rate is enough to invalidate the key distribution. For example when $\alpha = \pi/16$, corresponding to an overlap $\langle u|v \rangle = 0.38$, the maximum error rate is only 4%. In order to obtain a robust scheme, Alice and Bob need to use a large overlap: an overlap of about 95%, corresponding to $\alpha = \pi/5$, gives a maximum tolerable error rate of 35%. The disadvantage of using such a large overlap is that it increases the fraction of inconclusive results, to about 95% for $\alpha = \pi/5$, so that most of the bits sent by Alice are useless. This would significantly reduce the speed of the key distribution.

QUANTUM CRYPTOGRAPHY - PRACTICALITIES

Two non-orthogonal states can be realised, for example, by sending a single photon into two different arms of an interferometer. Each "path taken" can be viewed as a quantum state and two different paths denote two orthogonal states. One can then prepare a coherent superposition of the two different paths using an appropriate beam-splitter. The resulting state of the photon in the interferometer can be written as

$$u = \sqrt{t}\,\big| 1^{\mathrm{st}}\mathsf{PATH} \big\rangle + i\sqrt{r}\,\big| 2^{\mathrm{nd}}\mathsf{PATH} \big\rangle, \tag{7}$$

where t and r denote respectively the transmission and the reflection coefficients of the

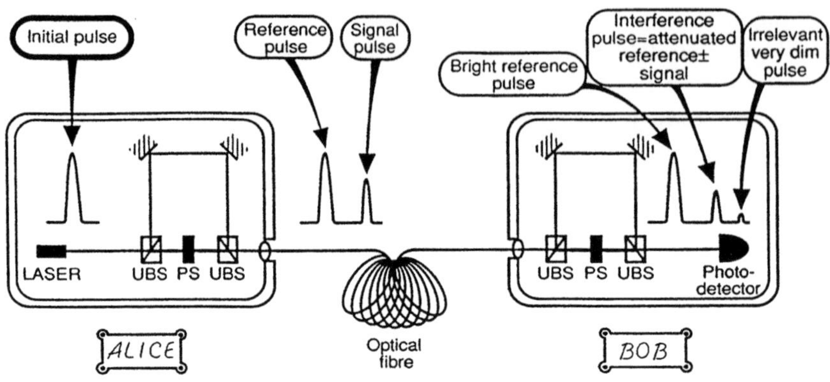

Figure 1: Interferometric implementation of the quantum key distribution.

beam-splitter. For generating the second of the two non-orthogonal signal states all we need is a phase-shifter in one of the arms of the interferometer so that

$$\mathbf{v} = e^{i\phi}\sqrt{t}\left|1^{\text{st}}\text{PATH}\right\rangle + i\sqrt{r}\left|2^{\text{nd}}\text{PATH}\right\rangle. \tag{8}$$

Switching the phase between 0 and π generates respectively either state $\mathbf{u}$ or state $\mathbf{v}$ with the overlap $|\langle \mathbf{u}|\mathbf{v}\rangle|^2 = (r - t)^2$. Of course, for symmetric beam-splitters ($r = t$) one has to switch the phase, say between 0 and $\frac{\pi}{2}$ (why ?).

The two non-orthogonal states $\mathbf{u}$ and $\mathbf{v}$ can be both produced and detected in Mach-Zehnder interferometers. A possible experimental set up is shown in Figure (1). Alice and Bob are in possession of identical Mach-Zehnder interferometers with phase shifters (PS) in the short arms and unsymmetric beam splitters (UBS) such that incoming pulses are split into dim pulses which travel in the short arms and bright pulses which are delayed in the long arms. Phase shifters can be switched in each interferometer between two different positions to permit encoding on Alice's side and filtering on Bob's side. If Alice's and Bob's phase shifts are equal, constructive interference between the signal pulse and the attenuated reference pulse will occur in Bob's interferometer, which can result in a registered photocount in Bob's detector. If Alice's and Bob's phase shifts differ, the destructive interference will never produce a photocount in the detector. The delayed bright pulse at Bob's detector serves to confirm that the reference pulse has actually arrived and excludes the possibilities of various attacks by substitutions and suppressions.

To build such cryptosystem one needs a single photon source, a low loss telecommunication fibre, and a good single-photon detector (and some know-how of course). Let us look what is available.

- A pretty good and simple single photon source can be obtained by attenuating light pulses from a semiconductor laser. The laser light exhibits the well-known Poisson photon distribution. When attenuated so that the *average* number of photons per pulse, $\bar{n}$, is of the order 0.1 then the probability of finding two or more photons per pulse is of the order $\bar{n}^2/2 = 0.005$. The remaining probability ≈ 0.9 describes the situation with zero photons in the pulse. Using this method on average 100 pulses out of 1000 will contain exactly one photon, 895 will be empty, and the remaining 5 pulses will contain more than one photon. There are some other methods of producing single photons [17] and it is only a matter of convenience and money which of them is finally used.

- There are good optical fibres on the market. What we want is a low loss telecommunications fibre. Losses in fibres have two main sources: Rayleigh scattering and absorbtion. The exact chemical composition of the fibre varies from place to place causing a fluctuation in refractive index through the material. If the scale of this fluctuation is of the order $\lambda/10$ or less, each irregularity acts as a point source scattering centre. This scattering, knowns as Rayleigh scattering, gives rise to an effective absorbtion coefficient that varies as $1/\lambda^4$. Absorption losses in the visible and near-infrared regions are due to impurities, particularly traces of metal ions and hydroxyl (OH^-) ions. The near-infrared region is of special interest here because many high capacity telecommunication links now operate at 1.3μm or 1.5μm.

- Assuming that the optical crypto-system will operate at 1.3μm or 1.5μm we have to look for appropriate photodetectors operating at these wavelengths. Fortunately suitable detectors - avalanche photodiodes (APD) - have become increasingly available over the last few years. They are semiconductor devices made of indium gallium arsenide or germanium. A photon striking the APD can create an electron-hole pair, and an applied electric field accelerates the electron and the hole in opposite directions. The electron moves towards the avalanche region (multiple quantum wells) where it can knock out more electrons into the conduction band. The freed electrons knock out more electrons and the resulting avalanche is detected at the bottom of the chip. The quantum efficiency of such devices, and consequently the data transmission rate, is quite low ($\approx 5\%$).

  Apart from the eavesdropping, the error rate is mainly due to dark counts in photodetectors i.e. avalanches triggered by thermally generated electrons and holes. Cooling the detector (to about 77 K) reduces the dark counts but still one has to rely on time synchronisation between the laser pulses and the photocounts to discriminate between the signal counts and the dark counts. Electrons and holes can be also trapped at defect sites in semicondutor and subsequently released by thermal excitation, leading to a new avalanche. This effect, known as "afterpulsing" is obviously undesirable and must be taken into account in the final data analysis.

Assuming that you managed to persuade your sponsors to buy you the quantum cryptographer DIY kit you can now set out to build a quantum transmission line for a secure key distribution. By no means it is an easy job but rapid experimental progress in this field during the last two years is quite encouraging. The first experimental device at IBM distributed the key over a modest distance of 30 cm; at the time the primary motivation was to show that quantum cryptography is experimentally feasible rather than to use it for a long distance crypto-communication [18]. Experimental quantum cryptography was later revitalised, with much more advanced technology, by a joint research effort of the British Defence Research Agency and the University of Oxford [9]. Finally the British Telecom became the first company to take quantum cryptography seriousely by establishing (in collaboration with DRA, Malvern) a single photon interference channel which can transmit the key data at the rate of 20 kbits/s over a distance of 10 km [19]. Other research institutions are joining in [20].

Although the fully operational quantum cryptosystem (with error correcting codes, privacy amplification[3] and various eavesdropping detection techniques) still remains to be built, the message is clear - perfect security of communication can be achieved in practice! Quantum cryptography is not any more a theoretical curiosity and the exclusive work of a few eccentric individuals (including yours truly).

WHAT IS QUANTUM COMPUTATION?

A computation is a process that produces *outputs* that depend in some desired way on given *inputs*. For the purpose of this presentation I define quantum computers as physical devices whose unitary dynamics can be regarded as the performance of a computation. Consider, for example, a function

$$f : \{0, 1, \ldots m - 1\} \longrightarrow \{0, 1, \ldots n - 1\}, \tag{9}$$

where m and n are natural numbers.

A classical computer computes f by evolving each labelled input, $0, 1, \ldots m - 1$ into a respective labelled output, $f(0), f(1), \ldots f(m - 1)$. Quantum computers, due to the unitary (and therefore reversible) nature of their evolution, compute functions in a slightly different way. In order to compute functions which are not one-to-one and to preserve the reversibility of computation, quantum computers have to keep the record of the input. Here is how it is done.

Imagine a computational device with two quantum registers; the first register to store the input data, the second one for the output data. Each possible input x is represented by $|x\rangle$ - the quantum state of the first register. Analogously, each possible output $y = f(x)$ is represented by $|y\rangle$ - the quantum state of the second register. Vectors $|x\rangle$ belong to the m-dimensional Hilbert space $\mathcal{H}_1$, and vectors $|y\rangle$ belong to the n-dimensional Hilbert space $\mathcal{H}_2$. States corresponding to different inputs and different outputs are orthogonal, $\langle x|x'\rangle = \delta_{xx'}$, $\langle y|y'\rangle = \delta_{yy'}$. The function evaluation is then determined by the evolution of the two registers,

$$\mathcal{H}_1 \otimes \mathcal{H}_2 \xrightarrow{U_f} \mathcal{H}_1 \otimes \mathcal{H}_2, \tag{10}$$

of the form

$$|x\rangle |0\rangle \xrightarrow{U_f} |x\rangle |f(x)\rangle. \tag{11}$$

Now comes the marvel of quantum computation; we can prepare a superposition of all input values as a single state and by running the computation U_f only *once*, we compute *all* of the m values $f(0), \ldots, f(m - 1)$,

$$\frac{1}{\sqrt{m}} \left[\sum_{k=0}^{m-1} |k\rangle \right] |0\rangle \xrightarrow{U_f} |0\rangle |f(0)\rangle + |1\rangle |f(1)\rangle + \ldots + |m - 1\rangle |f(m - 1)\rangle. \tag{12}$$

---

[3]As a matter of fact, with proper error correcting codes the privacy amplification is obsolete.

It looks too good to be true so where is the catch? How much information about f does the state

$$| f \rangle = | 0 \rangle | f(0) \rangle + | 1 \rangle | f(1) \rangle + \ldots + | m-1 \rangle | f(m-1) \rangle \tag{13}$$

contain?

Unfortunately no quantum measurement can extract the m values $f(0), \ldots, f(m-1)$ from $| f \rangle$. However, there are measurements that provide us with information about joint properties of all the output values $f(x)$. To illustrate this let us consider a simple example.

Suppose that for some $f : \{0,1\} \to \{0,1\}$, i.e. x and $y = f(x)$ can attain only two values 0 and 1, we wish to know whether the two values $f(0)$ and $f(1)$ are the same or different. Function f can be given to us as a physical device, a black box which can compute this function. The black box could be either classical, say a reversible computing machine which does not accept a superposition of inputs, or quantum, a reversible machine which accepts a superposition of inputs. We know that there are only four different functions f which map set $\{0,1\}$ into itself but we do not know which of the four functions the black box computes. Our task is to use the machine only *once* and to determine whether $f(0) = f(1)$ or $f(0) \neq f(1)$. We will show that the accomplishment of this task depends on the type of the black box i.e. whether it is classical or quantum.

Let us emphasise that we are not interested in the specific values $f(0)$ and $f(1)$, we are interested in the joint property: are the two values same or not. Nevertheless, with the classical black box we have to compute $f(0)$ and $f(1)$ separately and compare the two values. This implies performing computation twice, once for each argument of f. Surprisingly, if we are given the quantum black box instead we can solve the problem, albeit with probability 50%, in one run. We can prepare an equally weighted superposition of the two possible inputs and after a single computation we obtain,

$$\left[\frac{1}{\sqrt{2}} (| 0 \rangle + | 1 \rangle) \right] | 0 \rangle \xrightarrow{U_f} \frac{1}{\sqrt{2}} (| 0 \rangle | f(0) \rangle + | 1 \rangle | f(1) \rangle). \tag{14}$$

The device computes one of the four functions f; let us label them f_1 (constant at zero, $f(x) = 0$), f_2 (constant at one $f(x) = 1$), f_3 ($f(x) = x$), and f_4 ($f(x) = 1 - x$). Depending of which function was computed the two registers after the evolution end up in one of the four possible states,

$$| f_1 \rangle \;=\; \frac{1}{\sqrt{2}} (| 0 \rangle | 0 \rangle + | 1 \rangle | 0 \rangle), \tag{15}$$

$$| f_2 \rangle \;=\; \frac{1}{\sqrt{2}} (| 0 \rangle | 1 \rangle + | 1 \rangle | 1 \rangle), \tag{16}$$

$$| f_3 \rangle \;=\; \frac{1}{\sqrt{2}} (| 0 \rangle | 0 \rangle + | 1 \rangle | 1 \rangle), \tag{17}$$

$$| f_4 \rangle \;=\; \frac{1}{\sqrt{2}} (| 0 \rangle | 1 \rangle + | 1 \rangle | 0 \rangle). \tag{18}$$

To find out whether $f(0)$ and $f(1)$ are the same or different we measure observable $\hat{G}_1$ on the first register and observable $\hat{G}_2$ on the second register, where $\hat{G}_1$ and $\hat{G}_2$ are defined by their eigenvectors $1/\sqrt{2}(| 0 \rangle \pm | 1 \rangle)$ in $\mathcal{H}_1$ and $\mathcal{H}_2$ respectively for $\hat{G}_1$ and $\hat{G}_2$. Alternatively, we can specify one observable $\hat{G} = \hat{G}_1 \otimes \hat{G}_2$ with the eigenvectors

$$|\text{THE SAME}\rangle = \tfrac{1}{2}(|0\rangle + |1\rangle)(|0\rangle - |1\rangle), \tag{19}$$

$$|\text{DIFFERENT}\rangle = \tfrac{1}{2}(|0\rangle - |1\rangle)(|0\rangle - |1\rangle), \tag{20}$$

$$|\text{FAIL}\rangle = \tfrac{1}{2}(|0\rangle + |1\rangle)(|0\rangle + |1\rangle), \tag{21}$$

$$|\text{ERROR}\rangle = \tfrac{1}{2}(|0\rangle - |1\rangle)(|0\rangle + |1\rangle). \tag{22}$$

The measurement performed on our quantum black box after the unitary evolution (computation) projects the state of the device onto one of the four states above. If the outcome of the measurement is "THE SAME" (*i.e.* the eigenvalue corresponding to the state $|\text{THE SAME}\rangle$) then the two values $f(0)$ and $f(1)$ are indeed the same. If the outcome is "DIFFERENT" then $f(0)$ and $f(1)$ are different. If the outcome is "FAIL" then nothing can be inferred about the two values and the outcome "ERROR" denotes the wrong answer. In order to see this let us express the four possible (but mutually exclusive) final states of the two registers in terms of the eigenstates of the observable $\hat{G}$,

$$|f_1\rangle = \frac{1}{\sqrt{2}}(|\text{THE SAME}\rangle + |\text{FAIL}\rangle), \tag{23}$$

$$|f_2\rangle = \frac{1}{\sqrt{2}}(|\text{THE SAME}\rangle - |\text{FAIL}\rangle), \tag{24}$$

$$|f_3\rangle = \frac{1}{\sqrt{2}}(|\text{DIFFERENT}\rangle + |\text{FAIL}\rangle), \tag{25}$$

$$|f_4\rangle = \frac{1}{\sqrt{2}}(|\text{DIFFERENT}\rangle - |\text{FAIL}\rangle). \tag{26}$$

Thus after running the computing device once and performing the measurement $\hat{G}$ we obtain either the correct answer (with probability 50%) or the inconclusive answer (also with probability 50%). We never observe the wrong answer (no projection on $|\text{ERROR}\rangle$).

One may wonder whether we gained anything with our quantum device which, although never lies, provides inconclusive outcomes *on average* every second run. The *mean* running time of the device equals the time needed to solve the problem using any classical device, which has to compute function f twice. True, however, we did gain something. Let us imagine that the two outcomes "THE SAME" and "DIFFERENT" represent a crucial decision which must be made within 24 hours, such as buying or selling on the stock market. Assume that computing a single value $f(k)$ takes almost one day. Within this scenario a classical computer is of no use; the decision made by the time both $f(0)$ and $f(1)$ have been computed is obsolete. In contrast, the quantum device allows to make the correct decision on average every second day. Clearly you are much better off buying a quantum computer.

The power of quantum computation lies in quantum superpositions. A single quantum computer can follow many distinct computational paths all at the same time and produce a final result depending on the interference of all of them. After the first seminal paper on the Universal Quantum Turing Machine by Deutsch [21] there has been much recent progress in understanding how the possibility of such interference makes the theory of quantum computation differ from its classical counterpart. The most spectacular result, due to Peter W. Shor from AT&T Bell Lab, employs quantum computation to perform an efficient factorisation of big composite numbers, the code-breakers dream for the last two decades [5].

Here is the problem on which security of many classical public key cryptosystems is based: given natural number N find its prime factors. It is easy to factorise small numbers such as 10 or 60, but when the numbers are getting bigger and bigger the time needed for factorisation grows exponentially. How long would it take you to figure out that $252601 = 41 \times 61 \times 101$?

Finding an efficient way to resolve a given number into prime factors is not an easy task. The most direct way, of course, is trial division; divide by each number in turn up to its square root and see if the remainder is zero. This is not efficient, however. Imagine a (classical) computer capable of performing a million trials per second. It will take about a day to factorise a 30-digit number; a million years for a 40-digit number; and more than the age of the Universe (10^{10} years) for a 50-digit number. To factorise big numbers we must forget about trial division and look for more clever techniques. Currently, the best factoring algorithms are: the Multiple Polynomial Quadratic Sieve [24], for numbers less than 120 decimal digit long, and the Number Field Sieve [25], particularly good for numbers more than 110 decimal digits long. Still, even the fastest algorithms would need a couple of billions years to factorise a 200-digit number. If N is the number being factored (and N has more than 120 decimal digits) the Number Sieve algorithm takes about

$$\exp[(\ln N)^{1/3}(\ln(\ln N))^{2/3}] \tag{27}$$

elementary operations. In contrast we are going to describe an algorithm for quantum computers which takes only about $(\log_2 N)^2$ elementary operations. Skipping the rudiments of the computational complexity we only mention that computer scientists have a rigorous way of defining what makes an algorithm fast (and usable) or slow (and unusable). For an algorithm to be fast, the time it takes to execute the algorithm must increase no faster than a polynomial function of the size of the input. Informally think about the input size as the total number of bits needed to specify the input to the problem, for example, the number of bits needed to encode the number we want to factorise. The Number Sieve algorithm, although the best one we have at the moment, is regarded as slow and therefore not usable for cracking public key ciphers such as RSA. The quantum factorisation algorithm, if implemented in practice, can factorise large composite numbers very fast. Before going into details let us refresh our memory about arithmetic we learned at school.

We have already mentioned modular arithmetic in the first section of this paper (addition modulo 30), here we will need the power of a number x modulo another number m,

$$x^a \bmod m. \tag{28}$$

The result of this operation is the remainder from the division of x^a by m. In practice, because the operation is distributive, it is faster to do the exponentiation as a stream of successive multiplications, taking the modulus every time. The exponentiation modulo m is an easy operation and can be performed very fast [26]. Now, take any numbers x and m and consider a sequence $x^n \bmod m$ where n goes from 0 to infinity, i.e.

$$1, \ x \bmod m, \ x^2 \bmod m, \ x^3 \bmod m, \dots \tag{29}$$

This sequence is periodic and the period after which the sequence repeats for the first time is called the order of x modulo m. For example, the increasing powers of 2 modulo 7 go like $1, 2, 4, 1, 2, 4, 1, \dots$ and so on. Clearly the order of 2 modulo 7 is 3. In general, the order of x modulo m is defined as the smallest $r \neq 0$ for which

$$x^r \bmod m = 1. \tag{30}$$

It turns out that if you know how to calculate the order of a number modulo N, then you can easily find factors of N. If x is chosen at random then with probability greater than half x is coprime with N and r is even and $(x^{r/2} \pm 1) \bmod N \neq 0$. On the other hand Eq.(30) implies that

$$(x^{r/2} + 1)(x^{r/2} - 1) \bmod N = 0, \tag{31}$$

so N divides (without any reminder) the expression above, but it does not divide each of the two factors separately. It is then enough to compute the greatest common divisor of N and $x^{r/2} \pm 1$ to find a non-trivial factor of N. Fortunately an easy and very efficient algorithm to compute the greatest common divisor has been known since 300 BC. The algorithm, known as the Euclidean algorithm, is described in Euclid's *Elements*, the oldest Greek treatise in mathematics to reach us in its entirety. Instead of consulting the seventh book of the *Elements* try your school textbooks first.

After this crash-course in modular arithmetic and number theory (for more see [27]) let us analyse the quantum algorithm for factorisation. The algorithm finds the order of a randomly chosen number x modulo N. Before we go into details one more useful definition; the Fourier unitary transformation, $\mathcal{F}$, is defined as

$$\mathcal{F} : |\, a \bmod m \rangle \longrightarrow \frac{1}{\sqrt{m}} \sum_{b=0}^{m-1} e^{\frac{2\pi i a b}{m}} |\, b \bmod m \rangle . \tag{32}$$

This operation will be performed to recover the period of the function $x^a \bmod N$ where $a = 0, 1, 2, 3 \dots, m - 1$. When m is *smooth*, i.e. m is a product of small prime factors, the Fourier transform is very fast. In the following m is the dimension of the Hilbert space associated with one of the registers, so we can fix m to be smooth and greater than N^2 (we will need the second condition later) and skip the analysis of the computational complexity of the Fourier transform.

We are now prepared to discuss the main steps of the algorithm. As before, our machine has two quantum registers.

- The input register is prepared in the equally weighted superposition of all possible inputs,

$$(\mathcal{F}\,|\,0\rangle)\,|\,0\rangle = \frac{1}{\sqrt{m}} \sum_{a=0}^{m-1} |\, a \bmod m \rangle\,|\,0\rangle . \tag{33}$$

- A random x is chosen and function $x^a \bmod N$ is computed,

$$\frac{1}{\sqrt{m}} \sum_{a=0}^{m-1} |\, a \bmod m \rangle\,|\,0\rangle \longrightarrow \frac{1}{\sqrt{m}} \sum_{a=0}^{m-1} |\, a \bmod m \rangle\,|\, x^a \bmod N \rangle . \tag{34}$$

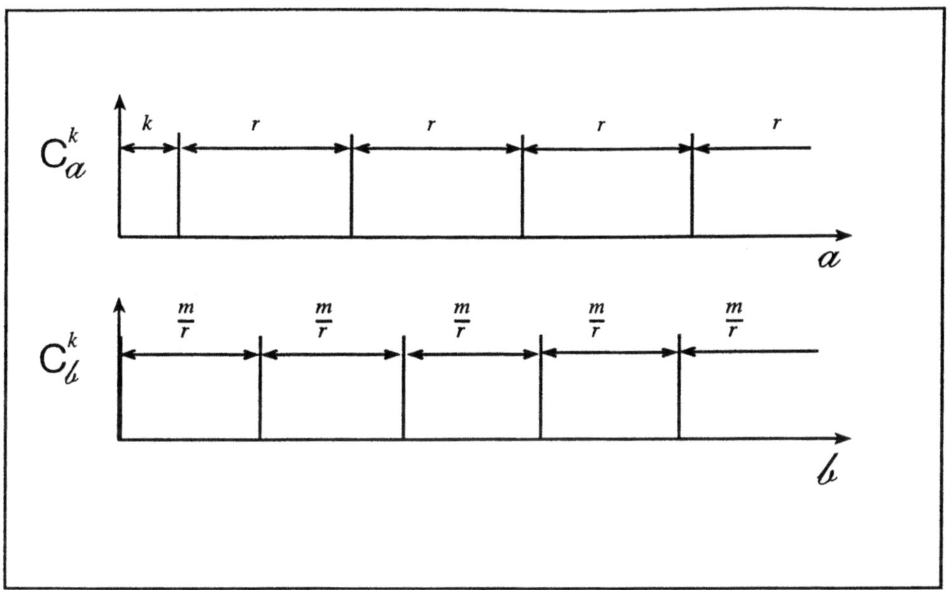

Figure 2: Functions c_a^k and c_b^k.

- The measurement is performed on the second register (projection on $\left| x^k \right\rangle$ for some k) so that the state of the first register is reduced to the respective relative state labelled by k. It can be written as,

$$\sqrt{\frac{r}{m}} \sum_{a=0}^{m-1} c_a^k \left| a \bmod m \right\rangle , \tag{35}$$

where the coefficients c_a^k, for a fixed k are given by

$$c_a^k = \sum_l \delta_{a,lr+k}, \tag{36}$$

with $l = 0, 1, 2 ... m/r$ (see Fig.(2)).

The counter-intuitive nature of quantum computation unfolds now; having so laboriously prepared the second register, the algorithm never consults it again! It can be discarded, reset, left alone, it does not really matter; it is the first register that matters. The reason we have computed function $x^a \bmod N$ is that it is periodic with period r and consequently there are only r different values of $x^a \bmod N$, each being repeated roughly m/r times when a goes from 0 to $m-1$. For any k the coefficients c_a^k are periodic with the same period r, so the outcome of the measurement on the second register is irrelevant (it does not even provide the value of k because solving $x^k \bmod N = y$ is as difficult as factoring N). In order to retrieve r from the first register we need the Fourier transformation.

- The Fourier unitary transformation is performed on the first register,

$$\mathcal{F} : \sqrt{\frac{r}{m}} \sum_{a=0}^{m-1} c_a^k \left| a \bmod m \right\rangle \longrightarrow \frac{\sqrt{r}}{m} \sum_{b=0}^{m-1} \sum_{a=0}^{m-1} e^{\frac{2\pi i a b}{m}} c_a^k \left| b \bmod m \right\rangle \tag{37}$$

. The state of the register after the transformation can be written as

$$\sqrt{\frac{r}{m}} \sum_{b}^{m-1} c_b^k \, | \, b \bmod m \rangle , \tag{38}$$

where,

$$c_b^k = \frac{1}{\sqrt{m}} \sum_{a=0}^{m-1} c_a^k e^{\frac{2\pi i a b}{m}} . \tag{39}$$

- The Fourier transform is followed by the measurement of $| \, b \bmod m \rangle$. The probability that the outcome of the measurement is a particular value b is given by

$$|c_b^k|^2 = \frac{r}{m^2} \left| \sum_{a=0}^{m-1} c_a^k e^{\frac{2\pi i a b}{m}} \right|^2 . \tag{40}$$

The rest of the algorithm is purely classical. Values of b for which

$$-\frac{r}{2} \leq rb \bmod m \leq \frac{r}{2} \tag{41}$$

will have especially high probabilities of occurance because for those values all the terms in the sum in the Fourier transform (40) will lie in similar directions in the complex plane. The condition (41) can be rewritten as

$$-\frac{1}{2m} \leq \frac{b}{m} - \frac{b'}{r} \leq \frac{1}{2m}, \tag{42}$$

where b' is an integer. For $m > 10N^2$ and $0 \leq r \leq N$ there is only one fraction $\frac{b'}{r}$ which is such a good approximation of $\frac{b}{m}$. Values b' and, more importantly, r can be found by approximating $\frac{b}{m}$ with fractions whose denominator is less than N (this can be done efficiently using continued-fraction expansions). From r we obtain factors of N completing the computation in not more than $(\log_2 N)^2$ elementary steps!

QUANTUM COMPUTERS - PRACTICALITIES

The appropriate question to ask now is whether it would ever be practical to build physical devices to perform such computations, or whether they would forever remain theoretical curiosities. Quantum computers require a coherent, controlled evolution for a period of time which is necessary to complete the computation. Many view this requirement as an insurmountable experimemntal problem. I believe that progress in the techniques of nano-construction and in manipulation of atomic systems by electromagnetic fields will soon make such devices feasible.

Mathematical models of quantum computation such as Quantum Turing Machines, although very convenient for analysing quantum complexity, provide no clue how to implement quantum computation in practice. Fortunately an alternative model has been proposed. Deutsch described quantum circuits composed of elementary logic gates connected together by wires [28]. Yao showed that any Quantum Turing Machine can simulate and can be simulated by uniform families of polynomial size quantum circuits, with at most polynomial slowdown [29]. It was also shown that there exists a universal quantum gate operating on three bits [28] such that any logical operation can be composed of this basic primitive. The gate performs a unitary operation on one input bit provided the logical AND of the remaining two control bits has value 1, and does nothing otherwise. From the experimental point of view all we need to build quantum circuits is a conditional dynamics of physical bits, i.e. we want to perform a unitary transformation on one physical subsystem conditioned upon the quantum state of the other subsystem,

$$U = |0\rangle \langle 0| \otimes U_0 + |1\rangle \langle 1| \otimes U_1 + \ldots + |k\rangle \langle k| \otimes U_k, \tag{43}$$

where the projectors refer to quantum states of the control subsystem and the unitary operations U_i are performed on the controlled subsystem.

Here is one idea how to induce this conditional dynamics [30]. What I am going to describe is one representative technology - the best current candidate, I believe - among several. But I expect that the techniques and considerations that I describe are generic and would also apply to various other technologies including the technology that is finally used. The model is based on inducing conditional dynamics (logic) in an array of quantum dots via the selective driving of (infrared) resonances and making use of the confined Stark effect [31] and the dipole-dipole interaction between the adjacent dots. It is a modified version of the original idea due to Mahler, Obermayer, and Teich [32] subsequently analysed in the contex of quantum coherent computation by Lloyd [33]. Another promising models may include an array of nuclear spins with spin-spin interaction and selective driving of resonances by a radio-frequency field or, to implement quantum logic gates, Rydberg atoms interacting with quantised electromagnetic field in high Q cavities [34].

A quantum dot is a semiconductor nanostructure in which the charge carriers are confined in all three dimensions of space. Consequently they have discrete energy eigenstates. The energy levels can be adjusted over a broad range by varying parameters such as the geometry of the dot, the material used, etc. I shall consider quantum dots each containing a single electron. In order to separate the two lowest-energy states, to be used for holding information, from other states, the potential well can be made cigar-shaped, i.e. longer in one direction, say x, than in the perpendicular directions y and z. Fig.(3) shows a putative array of such quantum dots.

At low energies the electron behaves like a particle in a one-dimensional square-well potential. Its charge distributions in the ground state $|0\rangle$ and first excited state $|1\rangle$ (named to indicate the binary digits that they will represent) are shown in inset A of Fig.(3). In both these states, the charge distributions are symmetrical about the centre of the well. However, in the presence of an external static electric field in the x-direction (inset B), the charge distribution of the ground state is shifted in the direction of the field whilst the charge distribution of the first excited state is shifted in the opposite direction (the *quantum-confined Stark effect* [31]). When a dot has been given a charge asymmetry in this way, we shall say that it is active. We choose coordinates in which the dipole moments of an active dot in states $|0\rangle$ and $|1\rangle$ are $\pm d$, where d may vary from one dot to another.

The electric field from the electron in one active quantum dot may shift the energy levels of an adjacent one, but to a good approximation it does not cause transitions. That is because the total Hamiltonian

$$\hat{H} = \hat{H}_1 + \hat{H}_2 + \hat{V}_{12} \tag{44}$$

is dominated by a dipole-dipole interaction term V_{12} that is diagonal in the four-dimensional state space spanned by eigenstates $\{|\epsilon_1\rangle, |\epsilon_2\rangle\}$ of the free Hamiltonian

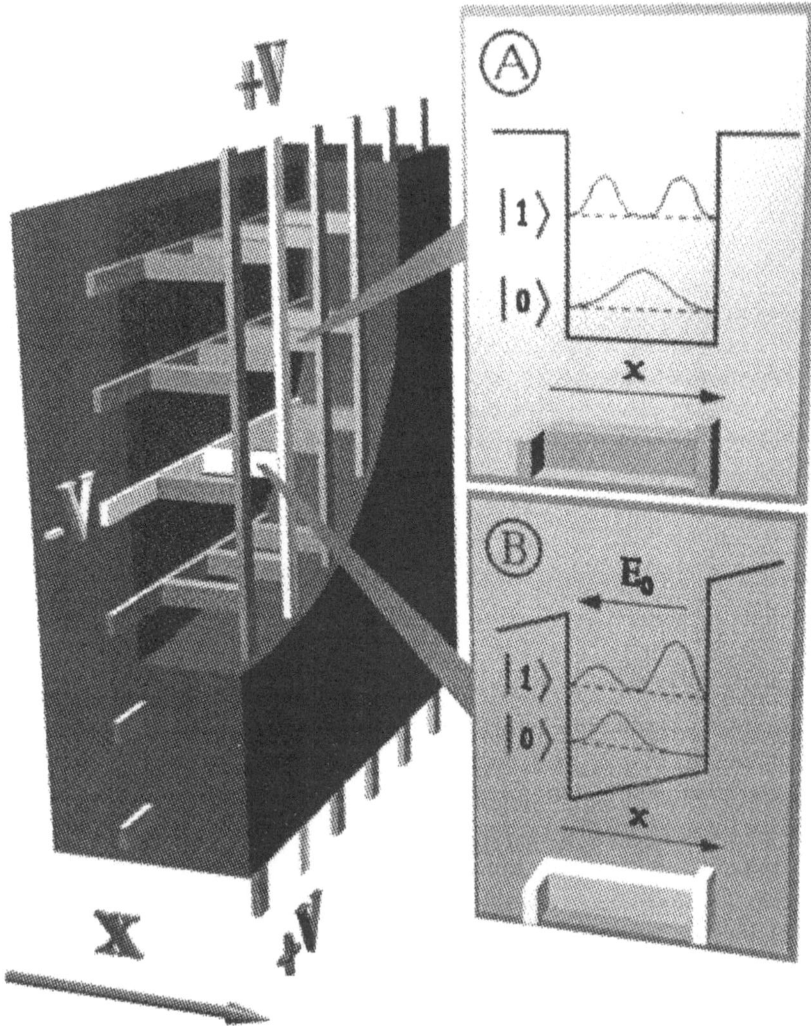

Figure 3: How an array of quantum dots might be used to perform quantum computation. A dot is "activated" by applying suitable voltage to the two metal wires that cross at that dot. Similarly, several dots may be activated at the same time. Because the states of an active dot are asymmetrically charged, two adjacent active dots are each exposed to an additional electric field that depends on the other's state. The resonant frequency for transitions in one dot therefore depends on the state of the other. Thus light at certain frequencies will cause transitions only in (say) adjacent, activated dots that are in certain states. The transitions are coherent quantum processes, and are elementary quantum computations.

$\hat{H}_1 + \hat{H}_2$, where ϵ_1 and ϵ_2 range over 0 and 1. Specifically,

$$(\hat{H}_1 + \hat{H}_2)\,|\,\epsilon_1\rangle\,|\,\epsilon_2\rangle = \hbar(\epsilon_1\omega_1 + \epsilon_2\omega_2)\,|\,\epsilon_1\rangle\,|\,\epsilon_2\rangle\,, \tag{45}$$

and

$$\hat{V}_{12}\,|\,\epsilon_1\rangle\,|\,\epsilon_2\rangle = (-1)^{\epsilon_1+\epsilon_2}\hbar\bar{\omega}\,|\,\epsilon_1\rangle\,|\,\epsilon_2\rangle\,, \tag{46}$$

where ω_1 and ω_2 are the resonant frequencies for transitions between states $|\,0\rangle$ and $|\,1\rangle$ and of the respective dots when their neighbours are inactive, and

$$\bar{\omega} = -\frac{d_1 d_2}{4\pi\epsilon_0 R^3}\,, \tag{47}$$

where d_1 and d_2 are the dipole moments of the first and the second dot respecively and R is the separation between the dots. It follows that the resonant frequency for transitions between the states $|\,0\rangle$ and $|\,1\rangle$ of an active dot whose neighbour is also active *depends on the neighbour's state*. The resonant frequency for the first dot becomes $\omega_1 \pm \bar{\omega}$ according as the second dot is in state $|\,0\rangle$ or $|\,1\rangle$ respectively. Similarly the second dot's resonant frequency becomes $\omega_2 \pm \bar{\omega}$, depending on the state of the first dot.

This effect could be used to perform coherent quantum computations, as follows. External light pulses tuned to one of the four frequencies $\omega_1 \pm \bar{\omega}$ or $\omega_2 \pm \bar{\omega}$, would be applied for suitable durations. A pulse whose intensity and duration are precisely such as to interchange the states $|\,0\rangle$ or $|\,1\rangle$ for some dot is known as a π-pulse for that dot. Thus a π-pulse at frequency $\omega_2 + \bar{\omega}$ causes two adjacent active dots to undergo a coherent evolution corresponding to the logical controlled-not operation. This conditional dynamics can be extended to transfer quantum states from one dot to another so that a chain of quantum dots could play the role of connecting wires in a computational network. Such a network would not be predefined by its hardware, but would be defined on the fly by sequences of voltages applied to the metallic wires and by light pulses, all controlled by an external classical computer. Pulses of the same frequencies, but shorter or less intense than π-pulses, would cause pairs of neighbouring active dots to perform elementary quantum computational operations with no classical analogues. To make short pulses we need to manipulate the laser action either by shutter techniques called Q-switching or by correlating laser modes of different frequencies in a method called modelocking. There are several techniques to produce properly shaped and tunable pulses of this sort [35]. By composing operations, three dots could be made to perform the operation of a universal quantum gate, and it follows that an array of dots could perform completely general quantum computations, subject only to limitations of stability and memory capacity.

However, these limitations would, at least in the medium-term future of this technology, be quite severe. The two critical parameters which determine the character of the dynamics are the decoherence time of the computer T_d and the typical time of an elementary coherent computational step - the clock cycle T_c. For a non-trivial quantum computation we require $T_d \gg T_c$, however, even a coherent evolution may create errors e.g. the pulses which are not properly tuned may not be selective enough and may affect different transitions and different quantum dots. There are several other parameters that must be considered in conncetion with the proposed device.

- The minimal clock cycle T_c is of the order of the typical pulse length T_p. The pulses must be approximately monochromatic and selective which requires $T_p \gg 1/\omega$ and $T_p \gg 1/\bar{\omega}$, where ω is the carrier frequency of the pulse. Clearly higher carrier frequencies allow shorter pulses and consequently the shorter clock cycle.

- The decoherence time T_d depends on the interaction with the enviroment. Electrons in quantum dots are coupled to the quantised electromagnetic vacuum which leads to the spontaneous emission. Typical lifetime of the excited state is of the order

$$4\pi\epsilon_0 \frac{3\hbar c^3}{4D^2\omega^3},\tag{48}$$

  where D is the dipole moment between states $|0\rangle$ and $|1\rangle$, the carrier frequency ω is tuned to the transition frequency between the two states (the off-diagonal dipole moment D is different from the diagonal one d). The interaction between the neighboring quantum dots will also have some off-diagonal terms, which were not included in V_{12}, and which induce the propagation of excitons in the array. This process reduces the lifetime of the excited states of the quantum dots and shortens the decoherence time.

For quantum dots of nanometer size, separated by $\approx 10^{-8}$m, and the resonant wavelength of the order of $\approx 10^{-6}$m ($\approx 10^{14}$Hz) realistic estimates could be $T_c \approx 10^{-6}$s and $\bar{\omega} \approx 10^{12}$Hz. Tunable vibronic solid-state laser are available at this wavelengths [36]; for example, the titanium-sapphire lasers with modelocking can produce adequate short pulses $T_p \approx 10^{-13} - 10^{-10}$s with nanosecond intervals. The applied voltage which activates quantum dots and determines the computational network can be changed on much slower time scale without affecting the speed of computation. Assuming that the clock cycle is of the order 10^{-9}s, we should be able to squeeze about 10^4 coherent steps. Nota bene, one coherent step of the whole device implies many elementary transitions performed in parallel on many activated dots.

Fortunately, the problem of factorising integers, and Shor's quantum algorithm for solving it, seem almost ideally suited for minimising the impact of these limitations while still harnessing quantum parallelism of a tremendous degree. Perhaps in this particular case the experimental realisations may come sooner than we anticipate!

ACKNOWLEDGMENTS

Many thanks to A. Barenco, K. Burnett, G. Castagnoli, D. Deutsch, B. Huttner, R. Jozsa, G. Mahler, G. Musso, G.M. Palma, A. Peres, U. Vazirani, K-A. Suominen for discussions, comments, help, and partial collaboration on this paper and to P.W. Shor for sending me his preprint on quantum factorisation.
This research was supported by the Royal Society, London.

References

[1] D. Kahn, *The Codebreakers: The Story of Secret Writing*, Macmillan, New York (1967).

[2] There are several good textbooks on cryptology where the modern techniques are explained, my favourite titles include: D. Welsh, *Codes and Cryptography*, Clarendon Press, Oxford (1988); B. Schneier, *Applied cryptography: protocols, algorithms, and source code in C.*, John Wiley & Sons, New York (1994); G. Brassard, *Modern Cryptology: A Tutorial*, Springer-Verlag, Berlin (1988); D.E. Denning, *Cryptography and Data Security*, Addison-Wesley (1982).

[3] W. Diffie and M.E. Hellman, *IEEE Trans. Inf. Theory* **IT-22**, 644 (1977).

[4] R. Rivest, A. Shamir, and L. Adleman, *On Digital Signatures and Public-Key Cryptosystems*, MIT Laboratory for Computer Science, Technical Report, MIT/LCS/TR-212 (January 1979).

[5] P.W. Shor, *Algorithms for quantum computation* draft of April 18, 1994 manuscript, AT&T Bell Laboratories.

[6] S. Wiesner, *SIGACT News*, **15**, 78 (1983); original manuscript written *circa* 1970.

[7] C. H. Bennett and G. Brassard, in "Proc. IEEE Int. Conference on Computers, Systems and Signal Processing", IEEE, New York, (1984).

[8] A. K. Ekert, *Phys. Rev. Lett.* **67**, 661 (1991).

[9] A. K. Ekert, J. G. Rarity, P. R. Tapster, and G. M. Palma, *Phys. Rev. Lett.* **69**, 1293 (1992).

[10] C. H. Bennett, *Phys. Rev. Lett.* **68**, 3121 (1992).

[11] A.K. Ekert, *Nature* **358**, 14 (1992).

[12] A.K. Ekert, B. Huttner, G.M. Palma, and A.Peres, *Phys. Rev. A* (in press).

[13] I. D. Ivanovic, *Phys. Lett. A* **123**, 257 (1987).

[14] A. Peres, *Phys. Lett. A* **128**, 19 (1988).

[15] P. Busch, P. J. Lahti, and P. Mittelstaedt, *The Quantum Theory of Measurement*, Springer, Berlin (1991).

[16] A. Peres, *Quantum Theory: Concepts and Methods*, Kluwer, Dordrecht (1993), Chapt. 9.

[17] P. Grangier and A. Aspect, *Phys. Rev. Lett.* **54**, 418 (1985).

[18] C.H. Bennett, F. Bessette, G. Brassard, L. Salvail, and J. Smolin, *J. Crypt.* **5**, 3 (1992).

[19] P. D. Townsend, J. G. Rarity, and P. R. Tapster, *Electron. Lett.* **29**, 634 *ibid.* 1291 (1993).

[20] A. Muller, J. Breguet, and N. Gisin, *Europhys. Lett.*, **23**, 383 (1993); R. Hughes (Los Alamos, Progress Report).

[21] D. Deutsch, *Proc. R. Soc. London A* **400**, 97 (1985).

[22] C.H. Bennett, *IBM J. Res. Dev.* **17**, 525 (1973).

[23] R. Feynman, *Int. J. Theor. Phys.* **21**, 467 (1982); D. Deutsch and R. Jozsa, *Proc. R. Soc. London A* **439**, 553 (1992); A. Berthiaume and G. Brassard, in *Proc. 7th IEEE Conference on Structure in Complexity Theory.* (1992); E. Bernstein and U. Vazirani, in *Proc. 25th ACM Symp. on Theory of Computation.* (1993); D.R. Simon (private communication).

[24] R.D. Silverman, *Math. Comp.* **48**, 329 (1987).

[25] A.K. Lenstra, H.W. Lenstra Jr., M.S. Manasse, and J.M. Pollard, in *Proc. 22nd ACM Symposium on the Theory of Computing*, pp.564-572 (1990).

[26] D. Knuth, *The Art of Computer Programming: Volume 2, Seminumerical Algorithms.* Addison-Wesley (1981).

[27] M.R. Schroeder, *NumberTheory in Science and Communication.* Springer-Verlag (1984).

[28] D. Deutsch, *Proc. R. Soc. London A* **425**, 73 (1989).

[29] A.Yao,*Proc. 34th IEEE Symp. on Foundation of Computer Science* pp.352-360, IEEE Computer Society Press, Los Alamitos, (1993).

[30] D. Deutsch, A.K. Ekert, and A. Barenco (in preparation)

[31] D.S. Chemla and D.A.B. Miller in *Heterojuction band discountinuities, physics and device applications*, eds. F. Capasso and G. Margaritondo, North-Holland, Elsevier Science Publishing Co., New York, (1987).

[32] K. Obermayer, W.G. Teich, and G. Mahler, *Phys.Rev. B* **37**, 8096 (1988); W.G. Teich, K. Obermayer, and G. Mahler, *ibid.* p.8111; W.G. Teich and G. Mahler, *Phys.Rev. A* **45**, 3300 (1992).

[33] S. Lloyd, *Science* **261**, 1569 (1993).

[34] S. Haroche and J.M. Raimond, *Advances in Atomic, Molecular and Optical Physics; Supplement* **2** 123 (1994).

[35] W.S. Warren, *Science* **242**, 878 (1988).

[36] J. Hecht, *The Laser Guidebook.* McGraw-Hill, Inc. (1992).

QUANTUM MECHANICS WITH SPONTANEOUS LOCALIZATION AND EXPERIMENTS

F. Benatti,[1] G. C. Ghirardi,[1,2] and R. Grassi[3]

[1] Dipartimento di Fisica Teorica, Università di Trieste

Strada Costiera 11, I-34100 Trieste, Italy

[2] International Centre for Theoretical Physics, Trieste, Italy

[3] Dipartimento di Fisica, Università di Udine

Via delle Scienze 208, I-33100 Udine, Italy

1. INTRODUCTION

One of the most intriguing features of Quantum Mechanics (QM) is that it provides too good a description of microphenomena to be considered a mere recipe to predict probabilities of prospective measurement outcomes from given initial conditions and, yet, as a fundamental theory of reality, it gives rise to serious conceptual problems, among which the impossibility, in its orthodox interpretation, to think of all physical observables of the system under consideration as possessing in all instances definite values.

This fact can be (and actually is) accepted as far as microsystems are involved, far less so if we consider that individual macrosystems, as aggregates of microsystems, should also be thoroughly described by QM, whence they too ought to be associated with vectors $|\psi>$ in some Hilbert space.

In fact, if Q is a macroobservable with two eigenstates $|\pm>$ such that:

$$Q|\pm> = \pm|\pm>, \quad Q = P_+ - P_-, \quad P_\pm = |\pm><\pm|, \tag{1}$$

then, according to QM, Q does not possess a definite value unless the macrosystem is in one of the eigenstates $|\pm>$, with the consequence that linear superpositions of macrostates like

$$|\psi_\pm> = \frac{1}{\sqrt{2}}\{|+> \pm |->\}, \tag{2}$$

do conflict with macrorealism, namely with the sensible prejudice about reality that macroobservables possess definite values in all macrostates independently of any obser-

Advances in Quantum Phenomena, Edited by E.G. Beltrametti
and J.-M. Lévy-Leblond, Plenum Press, New York, 1995

vation that can be carried out on the corresponding macrosystems. This motivates the search for a theory allowing a macroobjective description of the world.

In standard QM the objectification of the considered observable requires that, at the ensemble level (see, however footnote 1), the description given by the actual statistical operator ρ associated with the ensemble be equivalent to the one given by the transformed statistical operator:

$$T[\rho] = P_+\rho P_+ + P_-\rho P_- .\tag{3}$$

Just to give an example, in the case of $|\psi_\pm>$, the statistical operators are $P_\pm^\psi$, namely pure states (projection operators) corresponding to homogeneous statistical ensembles. If we represent them with respect to the basis $|\pm>$, the interference terms due to the coherent superposition of the macrostates $|\pm>$ give rise to nonzero off-diagonal terms:

$$P_\pm^\psi = \frac{1}{2}\begin{pmatrix} 1 & \pm 1 \\ \pm 1 & 1 \end{pmatrix}.\tag{4}$$

According to the above discussion, it should, for one reason or another, be legitimate to replace such statistical operators by

$$T[P_\pm^\psi] = P_+|\psi_\pm><\psi_\pm|P_+ + P_-|\psi_\pm><\psi_\pm|P_- .\tag{5}$$

$T[P_\pm^\psi]$ are not pure states (not projection operators) and are interpreted as corresponding to the inhomogeneous statistical ensemble made up by the equal weights mixture of homogeneous statistical ensembles associated with the systems in the state $|+>$, respectively $|->$. $T[P_\pm^\psi]$ is an incoherent mixture of the macrostates $P_\pm$ without interference between them:

$$T[P_\pm^\psi] = \frac{1}{2}P_+ + \frac{1}{2}P_- = \begin{pmatrix} 1 & 0 \\ 0 & 1 \end{pmatrix}.\tag{6}$$

Coming back to our general argument we mention that the macroobjectification problem, which is as old as QM itself, has been attacked in a number of different ways.

1. QM has not to be taken as a theory of physical reality, but, rather, as a means to calculate probabilities of outcomes of prospective measurements.

This is an extreme instrumentalistic position which does not question the quantum mechanical scheme, but simply refuses asking what kind of reality brings about the regular patterns in nature.

2. QM is an incomplete theory of physical reality. Hilbert space vectors such as $|\psi_\pm>$ encompass only a partial knowledge of physical states which would be fully characterized by specifying some set of unknown parameters (hidden variables) λ.

It is suggested that QM should be completed (e.g. as in the de Broglie-Bohm Pilot Wave Theory) so that all results of QM are reproduced, but, at the same time, it is conceptually correct to think that a macroobservable like Q possesses a definite value in each of the systems (specified by the hidden variable λ) that make up the inhomogeneous statistical ensemble associated with the linear superpositions $|\psi_\pm>$.

3. QM is a complete theory, but, within QM, macrosystems are singled out as special objects completely defined by sets of commuting observables.

It is suggested that on macroscopic objects only classical observables can be measured. The restriction to sets of commuting macroobservables means that, when dealing with macrosystems, pure states like $P_\pm^\psi$ cannot be operationally distinguished from non-pure states like $T[P_\pm^\psi]$. Indeed, the interference terms (off-diagonal elements in (4)) can be seen only by measuring observables like $R = |+><-|+|-><+|$ which do not commute with $P_\pm$. However, the assumption of the impossibility of measuring noncommuting macroobservables is certainly questionable in principle; among other implications it requires to assume that the Hamiltonian describing the system-apparatus interaction in a measurement process is not observable[1].

4.   QM is a complete theory, within which one should however treat all systems, especially macrosystems, as not isolated from their environment.

It is suggested that macrosystems (S) are unavoidably and uncontrollably coupled with their environment (E) and entangled with it by interactions. We must then consider the compound systems

SYSTEM+ENVIRONMENT (S+E).

In order to extract the state (density matrix) ρ_S of the system of interest out of the state ρ_{S+E} of the whole system, we have to perform a partial trace Tr_E with respect to any orthonormal basis in the Hilbert space $\mathbb{H}_E$ of the environment:

$$\rho_S = \mathrm{Tr}_E\, \rho_{S+E}\,. \tag{7}$$

By getting rid of the environment degrees of freedom, interference effects among (macro)states of S may be eliminated due to the fact that they are very likely to become entangled with orthogonal states (vectors in $\mathbb{H}_E$) of the environment. The mechanism is quite simple: let $|\pm>$ and $|i>$, $i = 1,2$ be orthogonal vectors in the system, $\mathbb{H}_S$, and environment Hilbert space $\mathbb{H}_E$, respectively. If, initially, the state of the compound system S+E is

$$|\Psi_{S+E}> = \Big(c_+|+> +c_-|->\Big) \otimes |0>, \tag{8}$$

and the common dynamics is such that, after some time,

$$|\Psi_{S+E}> \longrightarrow c_+|+> \otimes|1> +c_-|-> \otimes|2>, \tag{9}$$

then, taking the partial trace to single out the state of the system yields

$$\rho_S = \mathrm{Tr}_E\, \rho_{S+E} = |c_+|^2\,|+><+| + |c_-|^2\,|-><-|\,. \tag{10}$$

Obviously this cannot be considered as solving, in principle, the objectification problem but simply showing that the above is not a problem in practice, because practice is not accurate enough and may be never will be.

5.   QM is a complete theory, but its evolution equation should be changed in such a way that, on a suitable time-scale, macroproperties result objectified as a consequence of the dynamics and not by virtue of an external assumption like the wave packet reduction postulate, while the behaviour of microsystems is disturbed as little as possible.

In this approach it is required that the unitary evolution for statistical operators

$$\partial_t \rho_t = -\frac{i}{\hbar}\Big[H\,,\rho_t\Big] \tag{11}$$

be changed so that microsystems keep on behaving according to (11) on any reasonable interval of time, whereas a reduction mechanism like in (3), on a time-scale suited to a macrorealist position, is consistently supplied by the unique dynamics itself.

A simple example: let the physical system be described in a two-dimensional Hilbert space $\mathbb{H} = \mathbb{C}^2$, $\sigma_x, \sigma_y, \sigma_z$ be the Pauli matrices and choose a Hamiltonian H

$$H = \hbar\frac{\omega}{2}\sigma_y = \hbar\frac{\omega}{2}\begin{pmatrix} 0 & -i \\ i & 0 \end{pmatrix}\,. \tag{12}$$

If the initial state is P_+, then at time t, $P_+(t)$ will in general have nonvanishing off-diagonal terms, since

$$|+> \longrightarrow \exp\left(-it\frac{\omega}{2}\sigma_y\right)|+> = \cos\frac{\omega}{2}t|+> + \sin\frac{\omega}{2}t|->\,. \tag{13}$$

Adding a term $\lambda T[\rho_t]$ to (11) (compare (3)), where λ^{-1} is some characteristic time, and a counterterm $-\lambda\rho_t$ to assure that $\operatorname{Tr}\rho_t = 1$ for all $t \geq 0$, the new evolution equation

$$\partial_t\rho_t = -\frac{i}{\hbar}\big[H\,,\rho_t\big] - \lambda\rho_t + \lambda T[\rho_t]\,, \tag{14}$$

is such that, for times much larger than λ^{-1}, the off-diagonal elements of ρ_t with respect to the basis $|\pm>$ will be negligible. Indeed, if we can neglect the commutator (generator of the Schrödinger evolution) in (14), the solution ρ_t with initial condition $\rho(0) = \rho$ is:

$$\rho_t = e^{-\lambda t}\rho + \big[1 - e^{-\lambda t}\big]T[\rho] \overset{t\to+\infty}{\Longrightarrow} \rho_\infty = T[\rho]\,. \tag{15}$$

A modified QM incorporating a dynamical reduction (DR) mechanism can be called a Collapse Theory to be contrasted with a Non-Collapse Theory like the one sketched in point 4 above.

In the following, after a sketchy presentation of dynamical reduction models[2,3,4,5], we shall examine explicit models of this type i.e. Quantum Mechanics with Spontaneous Localization (QMSL) and Continuous Spontaneous Localization (CSL), respectively, in connection with some experimental situations:

Dynamical Reduction and Non-Collapse Theories

Dynamical Reduction and the Quantum Telegraph

Dynamical Reduction and Proton Decay.

2. THE DYNAMICAL REDUCTION PROGRAM

2.1 Quantum Mechanics with Spontaneous Localization

We start considering one-particle systems described in $\mathrm{IH} = L^2(\mathrm{IR}^3)$.

QMSL1. At random times, with mean frequency λ, any system is affected by localization processes in position around points $\mathbf{x} \in \mathrm{IR}^3$ occurring with probability $P_{Loc}(\mathbf{x})$ and transforming their states $\Psi(\mathbf{q}) \in \mathrm{IH}$ according to

$$\Psi(\mathbf{q}) \longrightarrow \frac{\Phi_\mathbf{x}(\mathbf{q})}{\|\Phi_\mathbf{x}\|} \tag{16}$$

$$\Phi_\mathbf{x}(\mathbf{q}) = \left(\frac{\alpha}{\pi}\right)^{\frac{3}{4}} \exp\left(-\frac{\alpha}{2}(\mathbf{q}-\mathbf{x})^2\right)\Psi(\mathbf{q}) \tag{17}$$

$$P_{Loc}(\mathbf{x}) = \|\Phi_\mathbf{x}\|^2\,, \tag{18}$$

the characteristic length $\alpha^{-\frac{1}{2}}$ and the frequency of localization being chosen to be

$$\frac{1}{\sqrt{\alpha}} = 10^{-5}\,cm\,, \qquad \lambda = 10^{-16}\,sec^{-1}\,. \tag{19}$$

The projection operator $P^\Psi = |\Psi><\Psi|$ undergoes a process similar to the one in (5):

$$P^\Psi \longrightarrow T_{GRW}[P^\Psi] = (\frac{\alpha}{\pi})^{\frac{3}{2}} \int_{\mathbb{R}^3} d\mathbf{x}\, e^{-\frac{\alpha}{2}(\mathbf{q}-\mathbf{x})^2} P^\Psi\, e^{-\frac{\alpha}{2}(\mathbf{q}-\mathbf{x})^2}\,, \tag{20}$$

q being the position operator of the system.

In spite of the fact that the conceptual relevance of QMSL, embodied in (16)-(18) resides in its yielding reduction at the individual level [1], its main physical consequences can also be studied by considering what the dynamics looks like at the ensemble level. The extension of (20) to the density matrices ρ corresponding to the considered physical states is, in fact, easily obtained by linearity with the result that the localization processes depress on the scale of $\dfrac{1}{\sqrt{\alpha}}$ the off-diagonal matrix elements in position representation. In fact, $|\mathbf{x} - \mathbf{y}|^2\alpha \gg 1$ implies:

$$| < \mathbf{x}|T_{GRW}[\rho]|\mathbf{y} > | = \exp\left(-\frac{\alpha}{4}(\mathbf{x}-\mathbf{y})^2\right)| < \mathbf{x}|\rho|\mathbf{y} > | \ll 1\,. \tag{21}$$

With the above assumptions one gets the evolution equation for the statistical operator

$$\partial_t\rho_t = -\frac{i}{\hbar}\Big[H\,,\,\rho_t\Big] - \lambda\rho_t + \lambda T_{GRW}[\rho_t]\,. \tag{22}$$

With the choice (19) the dynamics of microsystems determined by (22) is almost never distinguishable from (11), the usual one.

If we consider an interval of time $[t\,,\,t + \delta t]$ where the effects of the Hamiltonian evolution are negligible compared with the localization processes ($\lambda\delta t \gg 1$), we get:

$$< \mathbf{x}|\rho_{t+\delta t}|\mathbf{y} > \;\simeq\; \exp\left(-\lambda\delta t(1 - e^{-\frac{\alpha}{4}(\mathbf{x}-\mathbf{y})^2})\right) < \mathbf{x}|\rho_t|\mathbf{y} >$$

$$\simeq\; < \mathbf{x}|\rho_t|\mathbf{y} > \begin{cases} e^{-\lambda\delta t} & |\mathbf{x} - \mathbf{y}|^2\alpha \gg 1 \\[2mm] e^{-\frac{\alpha\lambda}{4}\delta t |\mathbf{x}-\mathbf{y}|^2} & |\mathbf{x} - \mathbf{y}|^2\alpha \ll 1 \end{cases}\,. \tag{23}$$

Then, we are led to single out two basic parameters of QMSL:

$$\textbf{Coherence Time:}\qquad \tau = \lambda^{-1} = 10^{16}\, sec \tag{24}$$

$$\textbf{Decoherence Rate:}\quad \Delta = \frac{\alpha\lambda}{2} = 10^{-6}\, cm^{-2}\, sec^{-1}\,. \tag{25}$$

In going from microsystems to macrosystems we first consider the case of aggregates of N distinguishable particles described in $\mathbb{H} = L^2(\mathbb{R}^{3N})$.

QMSL2. Each i-th particle is assumed to undergo an independent localization process as in (16)-(18). There follows that the evolution equation for the density matrix of an N-particle system becomes:

$$\partial_t\rho_t \;=\; -\frac{i}{\hbar}\Big[H\,,\,\rho_t\Big] - N\lambda\rho_t + \lambda T_{QMSL}[\rho_t] \tag{26}$$

$$T_{QMSL}[\rho_t] \;=\; \sum_{i=1}^{N}\int_{\mathbb{R}} d\mathbf{x}\, e^{-\frac{\alpha}{2}(\mathbf{q}_i-\mathbf{x})^2} \rho_t\, e^{-\frac{\alpha}{2}(\mathbf{q}_i-\mathbf{x})^2}\,. \tag{27}$$

---

[1] There is an unavoidable ambiguity with density matrices. A state like in (6) can be read as an incoherent mixture of the states $|\pm><\pm|$ as well as of $|\psi_\pm><\psi_\pm|$ with equal weights. On the contrary, the Hilbert space processes (16-18) allow to conclude that states which are linear (coherent) superpositions of states widely separated on the scale of $\alpha^{-1/2}$ lose coherence in positions (see also the detailed discussion of CSL in Subsection 2.2).

$\mathbf{q}_i$ being the position operator relative to the i-th particle.

As a main consequence, the centre of mass motion is such that the corresponding density matrix, where the internal motion has been traced out, obeys an evolution equation similar to (22) with a frequency λ_N of localization which is N times larger than the microscopic λ. This is reasonable for macroscopic rigid bodies since each localization process, one in λ^{-1} seconds, suffered by any of the N particles is felt as a localization process by the center of mass of their aggregate. Hence, for macroscopic bodies made up of N constituents:

$$\textbf{Coherence Time:} \qquad \tau_N = \frac{\tau}{N} \tag{28}$$

$$\textbf{Decoherence Rate:} \quad \Delta_N = N\Delta. \tag{29}$$

The effects of the QMSL dynamics (26) are such that the interference between states of almost rigid macroscopic bodies in which the centre of mass is differently located with respect to the length $\alpha^{-\frac{1}{2}}$ is suppressed after times of the order of τ_N or less:

Table 1.

System	Radius (cm)	N	$\Delta_N[cm^{-2}sec^{-1}]$	$\tau_N[sec]$
Dust Particle	10^{-3}	10^{15}	10^{9}	10
Bowling Ball	10	10^{27}	10^{21}	10^{-11}

2.2 Quantum Mechanics With Continuous Spontaneous Localization

Macroscopic bodies composed by one or more species of indistinguishable particles are dealt with by a generalization of QMSL called CSL. In CSL a random potential generates a continuous, stochastic evolution of Hilbert space vectors. The probability of occurrence of the random potential depends on the corresponding evolved state vector through a "cooking procedure" which makes the dynamics nonlinear.

After averaging out the stochasticity, density matrices change in time according to the following evolution equation with localization:

$$\partial_t \rho_t \;=\; -\frac{i}{\hbar}\Big[H\,,\,\rho_t\Big] + \gamma T_{CSL}[\rho_t] \tag{30}$$

$$T_{CSL}[\rho_t] \;=\; \sum_k \int_{\mathbb{R}^3} d\mathbf{x}\left\{ N_k(\mathbf{x})\rho_t\, N_k(\mathbf{x}) - \frac{1}{2}\Big(N_k^2(\mathbf{x})\rho_t + \rho_t\, N_k^2(\mathbf{x})\Big)\right\} \tag{31}$$

$$N_k(\mathbf{x}) \;=\; \Big(\frac{\alpha}{2\pi}\Big)^{\frac{3}{2}} \int_{\mathbb{R}^3} d\mathbf{y}\, e^{-\frac{\alpha}{2}(\mathbf{y}-\mathbf{x})^2}\, a_k^\dagger(\mathbf{y})a_k(\mathbf{y})\,, \tag{32}$$

where the index k numbers the types of particles and $a_k^\dagger(\mathbf{y})$, $a_k(\mathbf{y})$ are creation and annihilation operators of particles of type k at points $\mathbf{y} \in \mathbb{R}^3$.

Let us consider, for the moment, a physical system made up of just one species of particles ($k = 1$). With the choice $\gamma = \lambda(\frac{4\pi}{\alpha})^{\frac{3}{2}}$, (31) reduces to (27) on the one-particle sector. If $N > 1$, denoting with $|\mathbf{q}>= |\mathbf{q}_1,\ldots,\mathbf{q}_N>$ the improper eigenstates of the position operators of the N particles, then

$$\gamma <\mathbf{q}|T_{CSL}[\rho_t]|\mathbf{q}'> = \;-\;\frac{\lambda}{2}\sum_{j=1}^{N}\sum_{i=1}^{N}\Big\{ e^{-\frac{\alpha}{4}(\mathbf{q}_j-\mathbf{q}_i)^2} + e^{-\frac{\alpha}{4}(\mathbf{q}'_j-\mathbf{q}'_i)^2}$$
$$-\; 2e^{-\frac{\alpha}{4}(\mathbf{q}_j-\mathbf{q}'_i)^2}\Big\}\, <\mathbf{q}|\rho_t|\mathbf{q}'>\,, \tag{33}$$

whereas QMSL, which does not take into account particle indistinguishability, gives:

$$< \mathbf{q}| - N\lambda\rho_t + \lambda T_{QMSL}[\rho_t]|\mathbf{q}' > = -\lambda \sum_{j=1}^{N} \left\{ 1 - e^{-\frac{\alpha}{4}(\mathbf{q}_j - \mathbf{q}'_j)^2} \right\} < \mathbf{q}|\rho_t|\mathbf{q}' > . \qquad (34)$$

While (22) (see footnote 1) diagonalizes with respect to a preferred basis identifiable with the improper eigenstates of the position operators, (30) induces decoherence of macrostates with macroscopically different densities over volumes $\alpha^{-\frac{3}{2}}$.

For normal densities the decoupling rate for a superposition of two macrostates completely spatially separated is much faster than for QMSL; compare

$$\gamma < \mathbf{q}|T_{CSL}[\rho_t]|\mathbf{q}' > \simeq -\lambda N 10^8 < \mathbf{q}|\rho_t|\mathbf{q}' > , \qquad (35)$$

against

$$< \mathbf{q}| - N\lambda\rho_t + \lambda T_{QMSL}[\rho_t]|\mathbf{q}' > \simeq -\lambda N < \mathbf{q}|\rho_t|\mathbf{q}' > . \qquad (36)$$

In fact, of the three terms in (33), the third one is negligible ($|\mathbf{q}_i - \mathbf{q}'_j|^2\alpha \gg 1$) for all $i, j = 1, \ldots, N$, while the other two, for given $j = 1, \ldots, N$, are significant only for those i's that label particles within a volume $\alpha^{-\frac{3}{2}}$ around the j-th one and there are about 10^8 such particles in normal density conditions. Accordingly the values of τ_N of table 1 decrease by a factor 10^8.

3. ENVIRONMENT INDUCED REDUCTION PROCESSES

We begin by recalling that points of view 4 and 5 concerning the macroobjectification problem in QM are conceptually quite different.

One of the most interesting attempts to justify an environment-induced macroobjectification has been made by Joos and Zeh[6] who considered reasonable interactions between a macroobject like a pointer and a background of photons and/or neutrinos, air molecules and so on. The specific modification of the standard dynamics has not a universal character, but depends on the details of the process, in particular on which kind of environment exhibits relevant interactions with the system.

A collapse theory points, instead, to a modified QM whereby it is a fundamental new dynamics that eliminates the interferences on a macroscopic scale and is not merely devised to account for some unavoidable ignorance of all the details intervening in the Schrödinger evolution for S+E.

In particular, contrary to QMSL which, as the true dynamics, introduces two new fundamental constants of nature, α and λ, experimentally testable in principle, the reduction mechanisms of non-collapse theories depend crucially on the specific environmental situation. It might then be useful to compare the "reduction rates" and the physical consequences of specific examples of decoherence mechanisms of non-collapse theories with those of QMSL and CSL. Such a comparison has been presented in an interesting recent paper[7] we are going to discuss.

In it the environment is felt by the physical system of interest as a background noise due to the (instantaneous) scattering of photons, neutrinos or air molecules (various environments) off a system, the effect on the compound initial state ρ_{S+E}^i being determined by a transition matrix T

$$\rho_{S+E}^i \longrightarrow \rho_{S+E}^f = T\rho_{S+E}^i T^\dagger . \qquad (37)$$

Let $\mathbf{p}$, $\mathbf{k}$ be the momenta of the system and of a background incident particle, respectively, and $a_{\mathbf{pk}}(\mathbf{q})$ the probability amplitude that the momentum transferred to the system is $\mathbf{q}$. Reasonable assumptions on T and approximations about the nature of the scattering processes are the following.

EIRM1.   Conservation of energy and momentum:

$$< \mathbf{p}', \mathbf{k}'|T|\mathbf{p}, \mathbf{k} >= \delta(\mathbf{p}' + \mathbf{k}' - \mathbf{p} - \mathbf{k})a_{\mathbf{pk}}(\mathbf{p}' - \mathbf{p}) \,. \tag{38}$$

EIRM2.   Independence of $a_{\mathbf{pk}}(\mathbf{q})$ from the motion of the system due to the higher velocity of the incident particle:

$$a_{\mathbf{pk}}(\mathbf{q}) = a_{\mathbf{k}}(\mathbf{q}) \,. \tag{39}$$

EIRM3.   The system is supposed to be exposed to a constant particle flux Φ per unit area and unit time scattered off with a total scattering cross section σ and a temporal distribution modelled by a Poisson process with intensity $\Lambda = \sigma\Phi$. Furthermore, the background incident particles are supposed to be in momentum eigenstates or incoherent mixtures of them with probability momentum distribution $\mu(\mathbf{k})$.

If $\rho_S^{i,f}$ are the initial (before) and final (after a scattering process) states of the system obtained as in (7), and

$$P_{\mathbf{k}}(\mathbf{q}) \;\; = \;\; |a_{\mathbf{k}}(\mathbf{q})|^2 \tag{40}$$

$$\hat{P}_{\mathbf{k}}(\mathbf{x}) \;\; = \;\; \frac{1}{(2\pi)^3} \int_{\mathbb{R}^3} \mathbf{dq}\, e^{-i\mathbf{q}\mathbf{x}} P_{\mathbf{k}}(\mathbf{q}) \tag{41}$$

are the probability distribution of momentum transfers, respectively its Fourier transform, it follows that, in coordinate representation,

$$< \mathbf{x}|\rho_S^f|\mathbf{y} > \;\; = \;\; \hat{P}(\mathbf{x} - \mathbf{y}) < \mathbf{x}|\rho_S^i|\mathbf{y} > \tag{42}$$

$$\hat{P}(\mathbf{x} - \mathbf{y}) \;\; = \;\; \int_{\mathbb{R}^3} \mathbf{dk}\, \mu(\mathbf{k})\, \hat{P}_{\mathbf{k}}(\mathbf{x} - \mathbf{y}) \,. \tag{43}$$

Taking into account the third assumption above and denoting by $T_{Sca}[\rho]$ the ρ_S^f in (42), in the time interval $[t, t + dt]$ a scattering induced process occurs with probability Λdt. Consequently the unitary Schrödinger evolution turns into a master equation very much similar to (22):

$$\rho_{t+\delta t} = -\frac{i}{\hbar}\Big[H\,,\rho_t\Big]\delta t + (1 - \Lambda\delta t)\,\rho_t + \Lambda\delta t T_{Sca}[\rho_t] \,. \tag{44}$$

A comparison of $T_{Sca}[\rho]$ and $T_{GRW}[\rho]$ is possible by looking at the expansions of the Gaussian damping factor that appears in (21) and the factor $\hat{P}(\mathbf{x})$, which plays an analogous role in (42) ($|\hat{P}(\mathbf{x})| \leq 1$), in powers of $\mathbf{x} = (x_1, x_2, x_3)$:

$$< \mathbf{x}|T_{GRW}[\rho]|\mathbf{y} > \simeq \Big(1 - \frac{\alpha}{4}|\mathbf{x} - \mathbf{y}|^2 + \dots\Big) < \mathbf{x}|\rho|\mathbf{y} > \tag{45}$$

$$< \mathbf{x}|T_{Sca}[\rho]|\mathbf{y} > \simeq \Big(1 - i\sum_{j=1}^{3}(x_j - y_j)\int_{\mathbb{R}^3} \mathbf{dq}\, q_j P(\mathbf{q}) -$$

$$-\frac{1}{2}\sum_{j=1}^{3}\sum_{k=1}^{3}(x_j - y_j)(x_k - y_k)\int_{\mathbb{R}^3} \mathbf{dq}\, q_j q_k P(\mathbf{q}) + \dots\Big) < \mathbf{x}|\rho|\mathbf{y} > \,. \tag{46}$$

EIRM4. The incident particles are considered isotropically distributed with

$$\mu(\mathbf{k}) = \frac{1}{4\pi \mathbf{k}^2}\lambda_0 \nu(\lambda_0|\mathbf{k}|)\,, \tag{47}$$

where λ_0 is a typical wavelength of the hitting particle and $\nu(x)$ is a probability distribution on the positive real axis.

Consequently, the linear term in (46) vanishes and the quadratic one (covariance matrix) is completely determined by

$$l_{eff}^{-2} = \int_{\mathbb{R}^3} d\mathbf{q}\, q_i^2 P(\mathbf{q}) = \frac{\partial^2}{\partial x_i^2}\hat{P}(\mathbf{x})|_{\mathbf{x}=0}\,, \quad i = 1,2,3\,, \tag{48}$$

which has dimension cm^{-2} and introduces a characteristic length l_{eff} of the reduction processes to be compared with $(\frac{\alpha}{2})^{-\frac{1}{2}}$ of QMSL. Moreover, a natural time scale is given by $\tau = \Lambda^{-1}$. Solving (44) for short times $\Lambda \delta t \gg 1$, yields as in (23)

$$< \mathbf{x}|\rho_{t+\delta t}|\mathbf{y} > \simeq \exp\left(-\Lambda \delta t(1 - \hat{P}(\mathbf{x}-\mathbf{y}))\right) < \mathbf{x}|\rho_t|\mathbf{y} > \,. \tag{49}$$

Off the diagonal ($|\mathbf{x}-\mathbf{y}|l_{eff}^{-1} \gg 1$) the damping is dominated by $\exp(-\Lambda \delta t)$, whereas near the diagonal ($|\mathbf{x}-\mathbf{y}|l_{eff}^{-1} \ll 1$) the damping goes as $\exp(-\frac{1}{2}\frac{\Lambda}{l_{eff}^2}\delta t|\mathbf{x}-\mathbf{y}|^2)$. We can then introduce:

$$\textbf{Coherence Time:} \quad \tau = \Lambda^{-1} \tag{50}$$

$$\textbf{Decoherence Rate:} \quad \Delta = \frac{\Lambda}{l_{eff}^2}\,. \tag{51}$$

The decoherence time τ is fixed by the total scattering cross section σ and the flux of incident particles per unit area and unit time, while the decoherence rate Δ by the differential cross section that enters the expression of $P_\mathbf{k}(\mathbf{q})$ in (40). The author calculates the values of l_{eff} and τ for a microsystem (electron) in different physical backgrounds.

Table 2.: l_{eff} for various scattering processes and $\tau_{electron}$.

Cause of collapse	$l_{eff}[cm]$	$\Phi[cm^{-2}sec^{-1}]$	$\tau_{electron}[sec]$
300K air at 1 atm	10^{-9}	10^{24}	10^{-13}
300K air in lab vacuum	10^{-9}	10^{11}	1
Sunlight on earth	9×10^{-5}	10^{17}	10^7
300K photons	2×10^{-3}	10^{19}	10^5
Background radioactivity	10^{-12}	10^{-4}	10^{18}
Quantum gravity	$10^5 - 10^{12}$	10^{109}	30
GRW effect	10^{-5}		10^{16}
Cosmic microwave background	2×10^{-1}	10^{13}	10^{11}
Solar neutrinos	10^{-9}	10^{11}	10^{33}
Cosmic background neutrinos	3×10^{-1}	10^{13}	10^{51}

The absence of a second figure in the GRW effect row emphasizes the deep conceptual difference between collapse and noncollapse theories: there is no particle flux scattered off the system inducing the localization mechanism on the length scale $l_{eff} = (\frac{\alpha}{2})^{-\frac{1}{2}}$, the effect being due to a completely new dynamics and not to the environment. Analogously the author gives estimates of Δ for various objects in different backgrounds.

Table 3.: Δ in $cm^{-2}sec^{-1}$ for various scattering processes.

Cause of apparent wave function collapse	Free electron	Dust particle	Bowling ball
300K air at 1 atm	10^{31}	10^{37}	10^{45}
300K air in lab vacuum	10^{18}	10^{23}	10^{31}
Sunlight on earth	10	10^{20}	10^{28}
300K photons	1	10^{19}	10^{27}
Background radioactivity	10^{-4}	10^{15}	10^{23}
Quantum gravity	10^{-25}	10^{10}	10^{22}
GRW effect	10^{-6}	10^{9}	10^{21}
Cosmic microwave background	10^{-10}	10^{6}	10^{17}
Solar neutrinos	10^{-15}	10	10^{13}

As it is evident from Table 3., the GRW effect, e.g. for a free electron, is weaker than those of air molecules in lab vacuum or photons on earth by a factor in between 10^{26} and 10^{6}, respectively, hence masked by them.

Therefore, to put into evidence effects due to spontaneous decoherence mechanisms one should isolate the physical system of interest from the environment to a presently hardly attainable degree of accuracy.

However, the figures in Table 3. might lead to erroneous conclusions, if not correctly understood. In fact, e.g. for bound electrons, QMSL can, as we shall discuss later, induce, as a result of the localization mechanism, transitions (which are not considered in the preceding analysis of environment induced decoherence) leading to excitations or dissociations of the composed systems to which the electrons belong.

As an example, in a recent paper[8] it is argued that the Lyman-α ultraviolet radiation emitted by hydrogen atoms (about $1 - 10$ photons per second per mole) as a consequence of the spontaneous localizations suffered by electrons could be detected by an appropriate experimental setup. On the other hand, on the basis of energy balance considerations, it can be shown that an analogous effect due to collisions of an atom with 300 Kelvin air molecules at 1 atmosphere is much smaller in comparison.

4. DYNAMICAL REDUCTION AND THE QUANTUM TELEGRAPH

In this section we examine not a suggested experimental test for the DR-program, but, rather, some recent claims that there exists some already available empirical evidence that spontaneous collapses are necessary, but of a nature different from either QMSL or CSL. We quote directly from A. Shimony[9]:

"A great weakness of the investigations carried out so far in search of modifications of quantum dynamics is the absence of empirical heuristic. To be sure there is a grand body of empirical fact which motivates all the advocates of nonlinear modifications: that is, the occurrence of definite events, and, in particular, the achievement of definite outcomes of measurement. But this body of fact is singularly unsuggestive of the details of a reasonable modification of Quantum Mechanics. What is needed are phenomena which are suggesting and revelatory...

No more promising phenomena for this purpose have been found than the intermittency of resonant fluorescence of a three-level atom."

4.1 The Phenomenology of the Quantum Telegraph

The physical system consists of two laser beams of intensities I_1, I_2 scattered off a single trapped atomic system which can be dealt with as a three-level system with a ground state $|0>$ and two excitable states: a higher level $|1>$ and a metastable lower level $|2>$, with mean lives

$$\beta_1^{-1} \simeq 10^{-8}\,sec \ll \beta_2^{-1} \simeq 1\,sec. \tag{52}$$

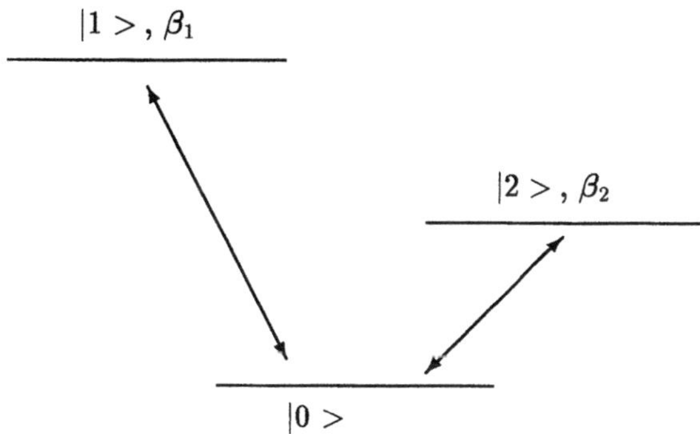

Fig. 1: atomic system in the quantum telegraph phenomenon.

The laser beams are tuned so that the one of intensity I_1 excites the atom from the ground state $|0>$ to $|1>$, that of intensity I_2 provokes the transition $|0> \rightarrow |2>$, followed by emissions of blue, respectively red photons and return to $|0>$.

The emission pattern of an experiment conducted with $I_1 \gg I_2$, nearly 10^8 vs. 10 photons per second, reveals an intermittent blue fluorescence randomly interrupted by periods of darkness.

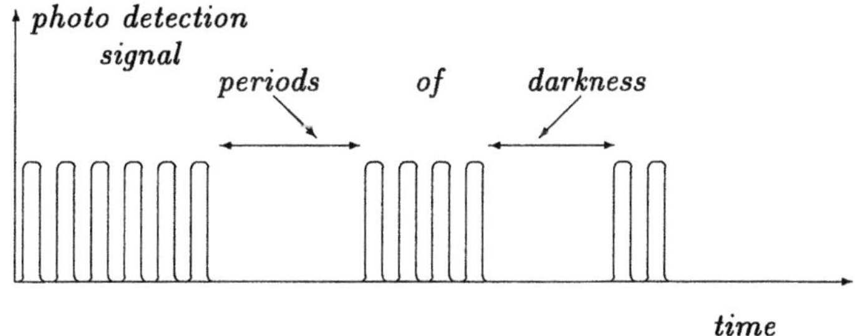

Fig. 2: emission pattern in the quantum telegraph phenomenon.

4.2 Quantum Mechanical Interpretations

At a first glance, the experimental evidence seems to be explainable by a naive argument based on the concepts of photons and of transitions among energy levels.

A.   Because of the higher intensity of the beam I_1, the atom is most of the time excited to the short-lived level $|1>$ from which it jumps down to the ground state in approximately 10^{-8} seconds with the emission a blue photon. But, every now and then, a red photon from the beam I_2 sneaks in and the atom is excited to the metastable state

$|2 >$, where it gets shelved for approximately 1 second before emitting a red photon and starting again a period of blue fluorescence.

However, these conclusions are far too classical (a la Bohr) and underestimate a relevant quantum effect, namely the interference between blue and red photons in the laser beams which is propagated to the atom by the linearity of the quantum evolution and results in the emergence of linear superpositions of the atomic levels.

B. After interacting with the laser beams, the atom, initially in its ground state $|0 >$, evolves in t seconds into a new state $|\phi(t) >$ which is the coherent superposition of the three levels:

$$|0 > \longrightarrow |\phi(t) >= c_0(t)|0 > +c_1(t)|1 > +c_2(t)|2 >, \qquad (53)$$

the probability P_i of a spontaneous emission corresponding to the jump $|i > \rightarrow |0 >$ being $P_i = \beta_i |c_i(t)|^2$.

Being the amplitude $|c_2(t)|^2 \ll |c_1(t)|^2$ for almost all t, one expects an emission pattern consisting in continuous fluorescence and, sometimes, the emission of a red photon, periods of darkness resulting extremely unlikely.

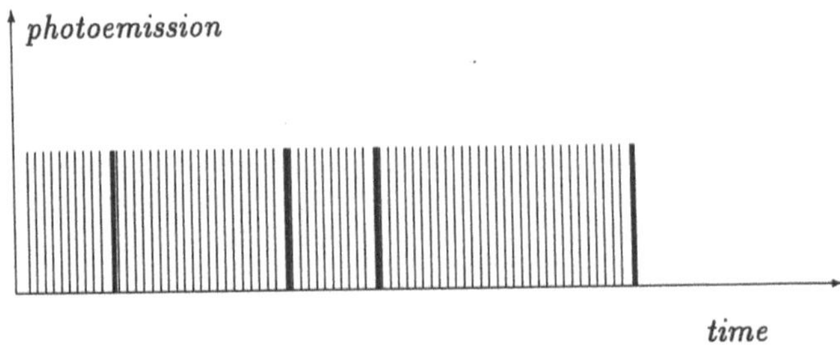

Fig. 3: emission pattern according to argument B, thick lines correspond to *red* photons.

The above attempt to embody linearity does not explain the occurrence of intermittency in the emission pattern and in ref.[10] (compare also corresponding references in [9]) it is suggested that an explanation is only possible if a reduction mechanism corresponding to null measurements (seeing no photons) is introduced into the game.

In view of these facts A. Shimony concludes[9]:

"Two propositions seem to me to suggest themselves quite strongly. The first is that a stochastic modification of quantum dynamics is a natural way to accommodate the jumps from a period of darkness to a period of fluorescence. The second is that the natural locus of the jumps is the interaction of a physical system with the electromagnetic vacuum. Whether stochasticity is exhibited when the system in question is simple and microscopic like a single atom, or only when it is macroscopic and complex like the phosphor of a photo-detector, is not suggested preferentially by the quantum telegraph, for the simple reason that the single trapped atom and the photo-detector are both essential ingredients in the phenomenon...

But, whichever choice is made points to a stochastic modification of quantum dynamics which has little to do with spontaneous localization."

4.3 The Correct Quantum Argument

Concerning the alleged impossibility of explaining the intermittent fluorescence of the quantum telegraph by resorting to a dynamical reduction model with localization, it must be stressed that

C. The presence of periods of darkness in the emission pattern can be deduced within a purely unitary quantum mechanical scheme[11], by taking into account the whole system

$$ATOM + RADIATION\ FIELD$$

without any need of invoking reduction processes induced by detecting no photons.

The analysis to be correct must consider states $|\psi(t)>$ of the form

$$|\psi(t)>= \sum_{i=0}^{2}\sum_{\{n\}} c_{i,\{n\}}(t)\,|i> \otimes |\{n\}>\,, \tag{54}$$

where $|\{n\}>$ is a state with n scattered photons. Then, with

$$P = \Big(\sum_{i=0}^{2} |i><i|\Big) \bigotimes |\{0\}><\{0\}| \tag{55}$$

the orthogonal projection onto the Fock sector with no scattered photons, the probability $P(t)$ of periods of darkness extending in the interval of time $[0\,,\,t]$ when, initially, $|\psi(0)>= |0> \otimes |\{0\}>$, is:

$$P(t) = \|P|\psi(t)>\|^2 = \sum_{i=0}^{2} |c_{i,\{0\}}|^2\,. \tag{56}$$

The study of $P(t)$ leads to the correct prediction of periods of darkness in a purely quantum dynamical context. Moreover, during a period of darkness the state of the system ATOM+RADIATION FIELD is

$$\Big(c_0(t)|0> +c_1(t)|1> +c_2(t)|2> \Big) \otimes |\{0\}>\,, \tag{57}$$

so that periods of darkness can end with the emission of both red and blue photons with an emission pattern like in the following picture:

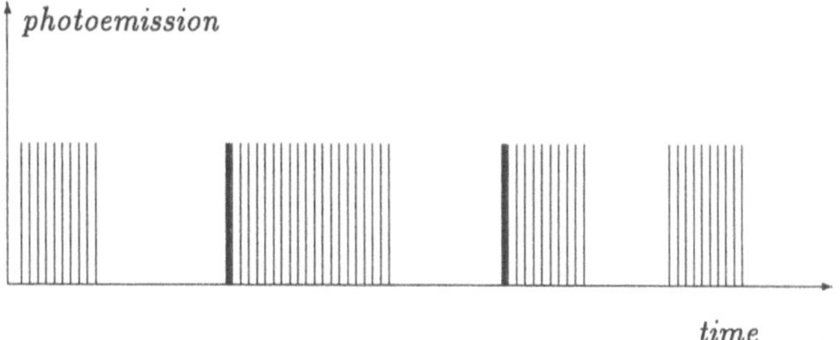

Fig. 4: emission pattern according to argument C.

As a further remark, it must be noted that a complete account of the quantum telegraph experiment ought to include the macroscopic detectors that are involved in the measurement of the emission pattern. Consequently, the physical system to be dealt with is

$$ATOM+RADIATION\ FIELD+MACROSCOPIC\ DETECTORS$$

Within the DR-program the actualization of the different macrostates of the detecting apparatuses is accounted by the new dynamics and the corresponding objectification of macroproperties is thus obtained. One could raise the question whether this can have any appreciable influence on the quantum telegraph phenomenon.

It is sufficient to observe that, according to the analysis of the previous Section, the effects of the reduction mechanism are comparable with those of environment-induced reductions that occur at the detectors level.

Indeed, on the basis of the agreement of the correct quantum mechanical computations of the probability of occurrence of periods of darkness and of their duration with the experimental results, one can safely conclude that

    i. QMSL and CSL dynamics play, for the process under consideration, exactly the same role as for any macroscopic detection process, namely they objectify macroproperties.

    ii. There is no need to require that new nonlinear and stochastic modifications of standard QM become effective at the microscopic level to account for the quantum telegraph phenomenology.

    iii. In particular, nothing in the considered experiments suggests that reductions take place with respect to an "energy" rather than to a "position" preferred basis.

5. DYNAMICAL REDUCTION AND NUCLEON DECAY

The presence in nature of a mechanism that localizes particles would be accompanied by a corresponding spreading in their momenta. It is thus interesting to study its effect on the stability of atoms and nuclei. It is possible to get a rough idea of the consequences of QMSL by modelling atoms and nuclei as one dimensional systems moving in a harmonic potential so that their ground states can be approximated by Gaussian wave functions $\Psi_G(q)$ of appropriate width γ^{-1}:

$$\Psi_G(q) = (\frac{2\gamma^2}{\pi})^{\frac{1}{4}} e^{-\gamma^2 q^2}, \qquad \begin{array}{ll} \gamma \simeq 10^8 \, cm^{-1} & \text{for an atom} \\ \gamma \simeq 10^{12} \, cm^{-1} & \text{for a nucleus} \end{array} \tag{58}$$

If the particle undergoes a localization around x as in (16), $\Psi_G(q)$ changes into:

$$\frac{\Phi_x(q)}{\|\Phi_x\|}, \quad \Phi_x(q) = (\frac{\alpha}{\pi})^{\frac{1}{4}} e^{-\frac{\alpha}{2}(q-x)^2} \Psi_G(q). \tag{59}$$

From (18), the probability density that a localization takes place at x is given by $\|\Phi_x\|^2$. Accordingly, the probability that, if a hitting process occurs, the state of the system is still $\Psi_G(q)$ is given by:

$$P_{ND} = \int \mathbf{dx} \, | < \Psi_G | \Phi_\mathbf{x} > |^2 = \frac{1}{\sqrt{1 + \dfrac{\alpha}{4\gamma^2}}} \simeq \begin{cases} 1 - 10^{-7} & \text{for an atom} \\ 1 - 10^{-15} & \text{for a nucleus} \end{cases} \tag{60}$$

Since microsystems are supposed to undergo one localization every 10^{16} seconds, the transition rate Q_{E+D} to an excited or dissociated state is

$$Q_{E+D} = \lambda(1 - P_{ND}) \simeq \frac{\lambda\alpha}{8\gamma^2} = \begin{cases} 10^{-23} & \text{for an atom} \\ 10^{-31} & \text{for a nucleus} \end{cases} \tag{61}$$

In[8] Pearle examines the case of the hydrogen atom and compares the flux of Lyman-α ultraviolet photons emitted by intergalactic hydrogen with the one expected if a GRW mechanism were at work[5,12,13], the latter turning out to be much weaker than the one observed.

Applying the above argument to the quark model of proton we would get a decaying time of the same order of magnitude of the one of a nucleus ($10^{31}\,sec$), whereas the proton lifetime is estimated longer than 10^{31} years, that is $10^{38}\,sec$.

This fact seems to indicate that the decoherence rate Δ should be decreased by a factor 10^7.

However, the consequences of $\Delta \simeq 10^{-13}\,cm^{-2}sec^{-1}$ would be unacceptable. In fact, being $\alpha^{-\frac{1}{2}} = 10^{-5}\,cm$ a reasonable value for the localization length (l_{eff} in Table 2.), it would yield the value $\tau_N \simeq N10^{-23}\,sec$ for the macroscopic coherence time in (28). Thus, linear superpositions of spatially separated states of any reasonable macroscopic "pointer" ($N \simeq 10^{23}$) would typically take times of the order of the second to be suppressed.

5.1 Reconsidering the Argument within the CSL Approach

In[14] the authors let an initial bound state $\rho_B = |\Psi_B ><\Psi_B|$ evolve into ρ_t according to the dynamics (30) of CSL and study the transition probability $P(t)$ to an excited orthogonal state $|\Psi_E >$:

$$\frac{\mathrm{d}}{\mathrm{dt}}P(t)|_{t=0} = \partial_t <\Psi_E|\rho_t|\Psi_E > |_{t=0} \,. \tag{62}$$

The only contribution is that from the reducing term $\gamma T_{CSL}[\rho_t]$ in (31). By developing up to the first order in α one gets:

$$\frac{\mathrm{d}}{\mathrm{dt}}P(t)|_{t=0} \simeq \sum_j \frac{\alpha\lambda N_j^2}{2}| <\Psi_E|\mathbf{Q}_j|\Psi_B > |^2, \quad \mathbf{Q}_j = \frac{1}{N_j}\sum_{i=1}^{N_j}\mathbf{q}_{ij}, \tag{63}$$

j numbering the species of identical particles making up the system, $\mathbf{q}_{ij}$ being the position operators of the j-th type of particles and $\mathbf{Q}_j$ their centre of mass.

For just one nucleon the result agrees with that of QMSL, while, for macrosystems, due to the more efficient decoupling rate, it appears that a correction of $\Delta = \alpha\lambda/2$ which would lead to no conflict with the proton lifetime is possible.

However, such a change of the value of Δ would lead us to the limit of acceptability for small objects: the dynamics will not reduce within the perception time an object containing 10^{15} particles like a carbon particle of radius $10^{-3}\,cm$. Similarly, the argument of[15] about the perception mechanism would no longer be correct.

5.2 The Pearle and Squires Argument

Considering a macroscopic body of total mass M made up of different types of identical particles of mass m_j, one may think of relating the reduction process to the mass density operator

$$M(\mathbf{x}) = \sum_k m_k(\frac{\alpha}{2\pi})^{\frac{3}{2}} \int_{\mathbb{R}^3} \mathrm{d}\mathbf{y}\, e^{-\frac{\alpha}{2}(\mathbf{y}-\mathbf{x})^2} a_k^\dagger(\mathbf{y})a_k(\mathbf{y})\,, \tag{64}$$

rather than considering independent stochastic processes for the various classes of particles [2].

Specifically, one should replace in equation (30) $\gamma T_{CSL}[\rho_t]$ with $\gamma T_M[\rho_t]$

$$T_M[\rho_t] = \frac{1}{m_0^2} \int_{\mathbb{R}^3} d\mathbf{x}\Big\{ M(\mathbf{x})\rho_t M(\mathbf{x}) - \frac{1}{2}\big(M^2(\mathbf{x})\rho_t + \rho_t M^2(\mathbf{x})\big)\Big\}, \tag{65}$$

---

[2] The advantages of taking such a position have been discussed also by A. Rimini in these proceedings.

m_0 being some typical reference mass.

The first order analysis that has led to (63) can be similarly carried out in this case, only, one has to consider the total centre of mass $\mathbf{Q}$ of the system and not only the various species centres of mass $\mathbf{Q}_j$

$$\mathbf{Q} = \frac{1}{M} \sum_j m_j \sum_{i=1}^{N_j} \mathbf{q}_{ji} \,,$$

whence the rate of internal excitation and/or dissociation appears as a second order effect. Indeed, the total centre of mass $\mathbf{Q}$ cannot excite any internal degree of freedom:

$$\frac{\mathrm{d}}{\mathrm{d}t} P(t)|_{t=0} \simeq \frac{M^2 \alpha \lambda}{2m_0^2} | < \Psi_E | \mathbf{Q} | \Psi_B > |^2 = 0 \,. \tag{66}$$

If one takes the QMSL value of λ for nucleons, that is the reference mass m_0 is identified with the nucleon mass, one has a remarkable decrease of the QMSL rate Q_{D+E} in (61) for atoms: from one atom per mole being either excited or dissociated every second to one of these possible events per mole every 10^{12} seconds (10^5 years). Nevertheless, all the important features of QMSL are preserved:

The decoupling of macroscopically distinguishable states is taken care of by the nucleons of the macroscopic bodies.

The energy increase is almost identical to the acceptable figure which is a consequence of the QMSL localization.

The collapse induced decay probability of a proton is depressed by a factor 10^{-16} making the CSL predicted lifetime well compatible with the experimental data.

Some concluding comments are in order at this point: the above analysis has appropriately pointed out the nice features deriving from relating reduction to the mass of the elementary particles. However,

i. The DR program has made clear that one can try to follow a new line to solve the conceptual problems that QM meets with macroobjects and measurement processes, namely, modifying the dynamics so that the physics of microsystems remains unaltered, while macrosystems exhibit an acceptable behaviour. For the time being, the program still requires many improvements, in particular the crucial problem to be faced is getting reasonable relativistic generalizations. Being the quark dynamics fundamentally relativistic, and due to the great difficulties haunting the so called "nonrelativistic quark models", simply applying the specific models of QMSL and CSl to nucleons seems a little bit too hazardous.

ii. Other collapse theories are still under consideration. A model presented in[16] tries to relate the decoherence mechanism to gravity and to reduce the number of new fundamental constants characterizing CSL. In particular, the reduction mechanism is linked to the mass of the systems involved, already meeting the basic request of[14].

iii. The difficulties connected with nucleon decay might also be avoided by slight modifications of the standard CSL, for instance, by using a higher power $N^{\frac{4}{3}}(\mathbf{x})$ of the smeared number operator appearing in (32).

16

REFERENCES

1- P. Busch, Macroscopic quantum systems and the objectification problem, *in* "Symposium on the Foundations of Modern Physics," P. Lahti and P. Mittelstaedt. eds., World Scientific, Singapore (1991).

2- G.C. Ghirardi, A. Rimini and T. Weber, Unified dynamics for microscopic and macroscopic systems, *Phys. Rev. D* 34:470 (1986).

3- P. Pearle, Combining stochastic dynamical statevector reduction with spontaneous localization, *Phys. Rev. A* 39:2277 (1989).

4- G.C. Ghirardi, P. Pearle and A. Rimini, Markov processes in hilbert space and continuous spontaneous localization of identical particles, *Phys. Rev. A* 42:78 (1990).

5- G.C. Ghirardi, A. Rimini, Old and new ideas in quantum theory of measurement, *in* " Sixty-Two Years of Uncertainty: Historical, Philosophical and Physical Inquiries into the Foundations of Quantum Physics," A. I. Miller, ed., Plenum Press, New York (1990).

6- E. Joos and H. D. Zeh, The emergence of classical properties through interaction with the environment, *Z. Phys. B*59:223 (1985).

7- M. Tegmark, Apparent wave function collapse caused by scattering, *Found. Phys. Lett.* 6:571 (1993).

8- P. Pearle, Reality checkpoint, to be published, *in: Acta Enciclopedia (Istituto della Enciclopedia Italiana).*

9- A. Shimony, Desiderata for a modified quantum dynamics, *in:* "PSA 1990," Vol.2, A. Fine, M. Forbes and L. Wessels, eds., Philosophy of Science Association, East Leasing, Michigan (1991).

10- M. Porrati and S. Putterman, Coherent intermittency in the resonant fluorescence of a multilevel atom, *Phys. Rev. A*39:3010 (1989).

11- C. Cohen-Tannoudji and J. Dalibard, Single-atom laser spectroscopy. Looking for dark periods in fluorescence light, *Europhys. Lett.*1:441 (1986).

12- F. Benatti, G.C. Ghirardi, A. Rimini and T. Weber, Operations involving momentum variables in nonhamiltonian evolution equations, *Il N. Cim. B*101:333 (1988).

13- E. Squires, Wave function collapse and ultraviolet photons, *Phys. Lett. A*158:431 (1991).

14- P. Pearle and E. Squires, Consequences of nucleon decay experiments for models of wavefunction collapse, to be published in *Phys. Rev. Lett..*

15- F. Aicardi, A. Borsellino, G.C. Ghirardi and R. Grassi, Dynamical models for statevector reduction: do they ensure that measurements have outcomes?, *Found. Phys. Lett.*4:109 (1991).

16- G.C. Ghirardi, R. Grassi and A. Rimini, A continuous spontaneous reduction model involving gravity, *Phys. Rev. A*42:1057 (1990).

QUANTUM PHENOMENA AT LARGE

Jean-Marc Lévy-Leblond

Physique Théorique, Université de Nice Sophia-Antipolis
Parc Valrose, 06108 Nice Cedex, France

QUANTUM THEORY AND THE MACROSCOPIC WORLD

The advent of quantum theory was closely related to investigations of the microscopic world, and motivated by the necessity of understanding the stability of individual atoms, the nature of their spectra, the photoelectric effect, etc. As a result, quantum phenomena are commonly thought to be essentially microscopic ones, in the realms of atomic and molecular, or nuclear and subnuclear physics. And, inded, we are currently witnessing, as in this Colloquium, wonderful advances in our mastering ever more subtle quantum effects on the microscopic scale — one should even speak nowadays of the 'nanoscopic' and 'picoscopic' scales. It is my purpose here to stress that on our macroscopic scale as well (and even on the 'megascopic' one), essential phenomena require a genuine quantum understanding; after all, let us not forget that quantum theory found one of its roots in the explanation of the blackbody radiation, a phenomenon which certainly belongs to the macroscopic world. In a sense, of course, the statement that most natural phenomena, as we commonly experience them, are genuine evidence for quantum theory is rather trivial, since no chemical reactions and no biological transformations can be explained without recourse to quantum notions. But I have in mind here phenomena much less subtle and much more basic, that is, the very existence of matter in bulk, with its gross properties of cohesion and density.

The treatment will be heuristic, based on order-of-magnitude estimates and rough approximations, in the spirit of 'qualitative physics'; it may be viewed as a tribute to the modern masters of this craft, Fermi, Weisskopf[1], Wheeler, Feynman among them. As a matter of fact, quantum theory will be invoked here under the sole guise of the Heisenberg inequalities. Namely, it will be sufficient to keep in mind that any physical magnitude A of a quantum system is in general characterized not by a single numerical value (except for the special case of the eigenstates of this quantity), but by a spectrum with some width or dispersion ΔA; position and momentum then have their respective dispersions correlated by the standard Heisenberg inequality:

$$\Delta r \Delta p \succ \hbar, \tag{1}$$

Advances in Quantum Phenomena, Edited by E.G. Beltrametti
and J.-M. Lévy-Leblond, Plenum Press, New York, 1995

where $\hbar$ is the quantum constant.[*] After decades of mistaken formulations, it may not be totally preposterous to stress in passing that the Heisenberg inequalities, far from being 'uncertainty' relations limiting our knowledge, are, quite on the contrary, most useful tools offering us a trustworthy simulation of the results to be achieved through rigorous quantum calculations.[2]

Our results will be given in terms of a few fundamental constants, which we now list, in order to fix the notation:

— $\hbar$ is the quantum constant,

— m is the electron mass,

— M is the proton mass,

— e^2 is the electromagnetic coupling constant (that is, in the SI system, $e^2 = q_e^2/4\pi\varepsilon_0$ where q_e is the elementary charge),

— G is the Newton gravitational coupling constant,

Besides these main ingredients of our considerations, we will occasionally mention :

— k , the Boltzmann constant,

— c , the invariant velocity,which we will not use much, as most of our considerations will be valid within the Galilean-Newtonian (so-called 'nonrelativistic') picture of spacetime.

COLLECTIVE STABILITY OF MATTER

In order to achieve some understanding of the structure and behaviour of matter, a necessary condition obviously is to check that the stability of atomic systems is ensured; such systems should possess a ground state, that is, their energy must exhibit a finite lower bound. It is not generally realized, however, that this condition by andlarge is not sufficient, insofar as we are interested by the energy spectrum of large systems. A crucial requirement concerns the behaviour of the ground state energy for an N-particle system, $E_0(N)$, as a function of the particle number N; it should be linearly bounded from below:

$$E_0(N) \geq -N\mathcal{E} \tag{2}$$

where $\mathcal{E}$ is some energy standard. Indeed, such a condition is necessary for the energy of a large system to be an extensive quantity, and for a thermodynamical treatment to be possible at all. This condition can be rewritten:

$$\left|E_0(N)\right|/N \leq \mathcal{E}, \tag{3}$$

expressing the reqirement that the binding energy per particle be limited independently of the number of particles. In other words, no more than a constant amount of energy ($\approx \mathcal{E}$) is to be recovered by incorporating an additional particle to the system, however large it is; everything must happen as if any particle would only interact effectively with a limited amount of other ones. For this reason, conditions (2) or (3) are often referred to as the "saturation of forces" within the system.

One may also interpret (3) as stating that the removal of a particle (or of a subsystem with fixed number n of particles) has a constant energy cost $\mathcal{E}$ (respectively $n\mathcal{E}$) which does not depend on the size N of the system. Suppose instead we were to deal with

---

[*] We use the symbol $\succ$ to denote an order-of-magnitude bound (that is, an inequality within a numerical constant not *too* different from unity), and the symbol $\approx$, similarly, to denote order-of-magnitude equality, up to such a constant, or in some asymptotic limit. The symbol $\cong$ corresponds to approximate numerical equality, and $\equiv$ to a definition.

nonsaturated forces, so that the ground state energy would exhibit the following behaviour as a function of the particle number (for large values of this number at least):

$$E_0(N) \approx N^\alpha \mathcal{E}, \text{ with } \alpha > 1. \tag{4}$$

Then,the removal of a given bit of matter with n particles ($n \ll N$) would require an expenditure of energy:

$$E_0(N - n) + E_0(n) - E_0(N) \approx \alpha N^{\alpha-1} n \mathcal{E}, \tag{5}$$

increasing with the size of the initial system. In Nature as we know it on our scale however, it is as easy to pick a cherry from a small or a big cherry tree, or to tear off a speck of stone from a gravel or a large rock. The effective cohesion forces of matter may be considered as surfaces forces, which means that when a piece of matter is broken, the energy cost depends only on the amount of locally disrupted bonds. This is what allows for the existence of matter in bits and pieces, making the world such a varied and interesting place. If the forces responsible for the cohesion of matter were nonsaturated, the most stable state of an N-particle system would consist of a single condensed lump (fortunately, we would not be there anyway to live in such a dull place).

COULOMB FORCES AND ORDINARY MATTER

Let us now consider ordinary matter, that is, large systems of electrons and nuclei bound by their electrostatic mutual interactions. As we are interested in the main aspects of such systems, we lightheartedly disregard all subtleties, that is magnetic (spin) interactions, einsteinian corrections and so on. The question is: do the sheer consideration of the coulombic interactions suffice to prove, under the quantum spell, of course, the collective stability of matter? Are Coulomb forces saturated? The question, strangely enough, was asked rather late in the history of quantum physics, by Dyson and Lenard,[3] who were able to give it a positive answer through a rigorous and most sophisticated analysis,which has since been improved, but remains quite a feat of mathematical physics.[4] We will be content here to give a very rough idea of the situation, and to put the emphasis on its main ingredients. For the sake of simplicity, we will only consider the case of hydrogenic matter, consisting of an equal number N of electrons and protons, leaving to the reader the generalization to the case of a system consisting of N (A, Z)-nuclei and NZ electrons. The total energy of our system, its hamiltonian, then writes:

$$H = \sum_{i=1}^{N} \frac{p_i^2}{2m} + \sum_{j=1}^{N} \frac{P_j^2}{2M} + \sum_{i=1}^{N} \sum_{i'<i} \frac{e^2}{\left|r_i - r_{i'}\right|} + \sum_{j=1}^{N} \sum_{j'<j} \frac{e^2}{\left|R_j - R_{j'}\right|} - \sum_{i=1}^{N} \sum_{j=1}^{N} \frac{e^2}{\left|r_i - R_j\right|}, \tag{6}$$

with obvious notations for the positions and momenta of the particles.

We would like to check whether or not a bound such as (2) holds true for this hamiltonian. Observe that we are interested in establishing a *lower* bound, which is notoriously difficult to achieve, unlike upper bounds, for which the variational principle is most serviceable. Unwilling to tackle the formidable task of confronting this huge operator, and afraid to become lost in the vast Hilbert space of the system, we estimate its ground state energy by making use of a simple parametrization of the system state. Let us denote by D some average linear size of the system and by p the mean (square) momenta of the electrons and nucleons. The energy eigenvalue of the hamiltonian (6) then may be approximated by :

$$E_0(N) \approx N\frac{p^2}{2m} + N\frac{p^2}{2M} + \frac{N^2}{2}\frac{e^2}{D} + \frac{N^2}{2}\frac{e^2}{D} - N^2\frac{e^2}{D}, \tag{7}$$

since $\frac{1}{2}N^2 \approx \frac{1}{2}N(N-1)$ is the number of electron as well as of proton pairs obeying mutual repulsion, and N^2 the number of electron-proton pairs with attractive forces. For this estimate to be valid, the stringent assumption has to be made that the distribution of particles is roughly uniform within the global volume of the system (Figure 1); as a consequence, the average distances of electron pairs, nucleon pairs and electron-nucleon pairs respectively, are of the order of the size D, and, similarly, the average momenta of electrons and nucleons respectively are comparable. This spatial uniformity is only natural, taking into account that for the energy of such a Coulomb system to be as low as possible, the positive energy of repulsion between like charges has to be minimized by spreading and mixing as uniformly as possible the positive and negative charges; of course, that is

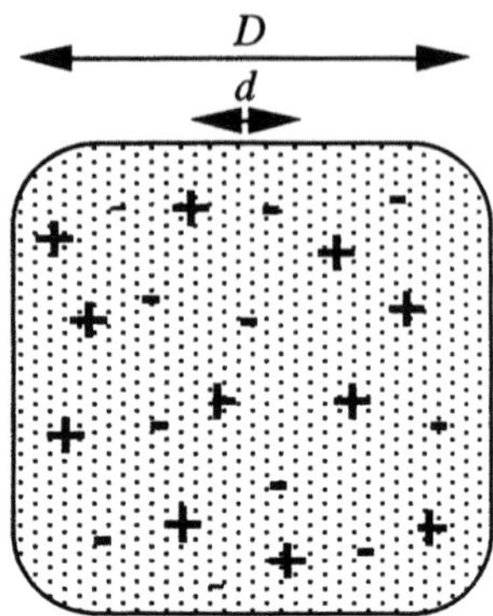

Figure 1. A Coulomb system

precisely one of the difficult statements to prove in a formal approach.[3,4] Anyway, the estimate (7) then shows that the positive (repulsion) and negative (attraction) contributions to the potential energy cancel to order N^2. This is but the well known screening effect of Coulomb forces; the attractions and repulsions exerted on a given charge by the distant like and unlike ones respectively cancel beyond its immediate neighbourhood. The net effect is that 'everything happens as if' any particle would only interact with its nearest neighbour, of the opposite sign. Finally, noting the very large mass of the protons compared to that of the electrons, and neglecting their kinetic energy accordingly, the total energy can be re-estimated as:

$$E_0(N) \approx N\frac{p^2}{2m} - N\frac{e^2}{d}, \tag{8}$$

where d now is the average distance *between nearest neighbours*. It could seem as if we had reached our goal and obtained an estimate of the groundstate energy exhibiting the expected linear dependence on the number of particles. But that is a mere illusion, as neither p nor d are constrained, since we have not yet invoked the quantum rule.

As announced, we proceed to emulate quantum theory by simply imposing the Heisenberg inequality (1), which takes here the form:

$$pD \succ \hbar, \tag{9}$$

since p and D give good estimates of the dispersion in momentum and position of the particles. We have yet to relate the average distance D, that is, the size of the system, with the average distance between nearest neighbours d. This last quantity also may be thought of as the average size of the individual volume occupied by a single particle. Taking into account the rough uniformity of the distribution, the total volume can be written as:

$$D^3 \approx Nd^3, \text{ whence } D \approx N^{1/3}d. \tag{10}$$

Combining (9) and (10), the total energy (8) obeys the inequality:

$$E_0(N) \succ N\frac{\hbar^2}{2mD^2} - N^{4/3}\frac{e^2}{D}, \tag{11}$$

as a function of the size D. Minimizing with respect to this parameter, we finally obtain the following absolute estimate:

$$E_0(N) \approx -N^{5/3}\mathcal{E}_0, \text{ with } \mathcal{E}_0 \equiv \frac{me^4}{\hbar^2} \tag{12}$$

for a system size:

$$D_0(N) \approx N^{-1/3}a_0, \text{ with } a_0 \equiv \frac{\hbar^2}{me^2} \tag{13}$$

That the standards of energy and size respectively are given by the Rydberg constant and the Bohr radius is but a trivial consequence of dimensional consistency; it is the strange power dependence on N exhibited by (12) and (13) which we are in fact interested in. Not only are these results far from the theoretical behaviour expected, but they are numerically outlandish. If they were true, the energy needed to remove one electron from 1 g of hydrogen, that is, from a system with a value $N = 6 \times 10^{23}$, would be of the order of 10^{17} electron-volts (instead of a few eV), and the system would occupy a volume with a size of a mere 10^{-16} m (instead of a few cm). Some crucial ingredient clearly is missing...

THE PAULI PRINCIPLE AND SATURATION OF COULOMB FORCES

As we now proceed to show, it is the fermionic nature of the electrons which plays a crucial role for ensuring the saturation of Coulomb forces, that is, the collective stability of ordinary matter. The physical reason is plain: the Pauli principle keeps the electrons much more apart that would be the case otherwise. The overall neutrality imposed by the Coulomb forces thus transmits, so to speak, this spreading of the distribution to the nuclei (irrespective of their own nature as fermions or bosons). The spatial scale of the system then is much larger, and the effective Coulomb forces much less effective, thus keeping the system from condensing ever more as it grows larger. A nice way of expressing these ideas in an explicit calculation is to use the Heisenberg inequality as fit to the case of a fermionic system. The simplest derivation of this too little known result, at the heuristic

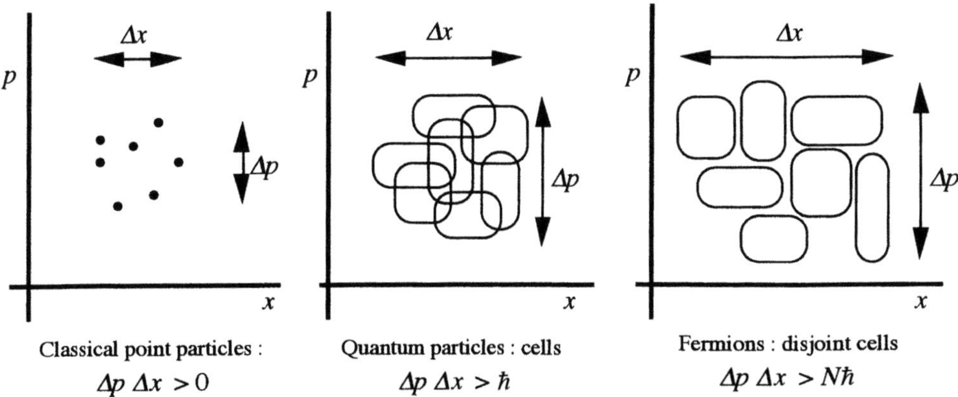

Figure 2. One-dimensional phase space and the exclusion principle

level we contemplate, is to proceed through a phase-space description of the system (Figure 2). Consider first, for simplicity, a one-dimensional N-particles system. In the corresponding 2-dimensional phase-space (position x and momentum p), the state of a classical system would be described by N points. The state of a quantum particle, however, is represented, not by a point, but, because of the standard Heisenberg inequality, by some 'cell' with dimensions δx and δp, that is, with total volume $\delta x \delta p \succ \hbar$. To the state of a quantum N-particles system, their corresponds a set of N such cells. As they are independent, however, and may overlap without restraint, the total volume in phase space of the system, expressed through its collective widths in position Δx and Δp, may be no larger than the volume of an individual cell, so that we only recover the usual Heisenberg inequality $\Delta x \, \Delta p \succ \hbar$. The situation is very different if the system consists of N identical fermions; for now, the exclusion principle prevents any overlap of the cells corresponding to the individual states out of which the collective (antisymmetrical) state is built. As a result, the total volume of phase space occupied by the system must be at least N times that of the individual cell. In other words, we now have:

$$\Delta x \, \Delta p \succ N\hbar, \quad \text{1-dimensional Heisenberg-Pauli inequality.} \tag{14}$$

The extension to the 3-dimensional case is immediate. Letting Δr and Δp be some estimates of the size of the system in space and momentum respectively, the same argument as above requires the total volume in phase space, that is, $(\Delta r)^3 (\Delta p)^3$ to be larger than the sum of the volumes of N individual cells, each one being at least $\hbar^3$. We obtain finally:

$$\Delta r \, \Delta p \succ N^{1/3}\hbar, \quad \text{3-dimensional Heisenberg-Pauli inequality.} \tag{15}$$

Let us hasten to point out that such an inequality can be proved with full rigour (and without too much difficulty).[5] The full impact of that important result is best appreciated by interpreting it as showing a drastic amplification of quantum effects by the Pauli principle: indeed, one may think of (15) as a modified Heisenberg inequality where the usual quantum constant $\hbar$ is replaced by an "effective fermionic quantum constant":

$$\hbar_{\text{eff}}^{\text{fermi}} \equiv N^{1/3}\hbar. \tag{16}$$

Consequently, in order to impose the Pauli principle on the result of some previous calculation for an N-particles quantum system, it will be sufficient to replace the true quantum constant by the effective fermionic one.

Coming back to the ground state energy of matter under Coulomb forces, we see how easy it is now to take into account the fermionic nature of the electrons; we have but to make the replacement in (12) and (13) of $\hbar$ by $\hbar_{\text{eff}}^{\text{fermi}}$, as given by (16). We immediately obtain:

$$E_0(N) \approx -N\mathcal{E}_0 \tag{17}$$

and

$$D_0(N) \approx N^{1/3}a_0. \tag{18}$$

In other words, the energy per particle is:

$$\left|E_0(N)\right|/N \approx \mathcal{E}_0 \equiv \frac{me^4}{\hbar^2}, \tag{19}$$

that is, the atomic standard of energy, a few electron-volts, and the nearest-neighbour distance is (see (10)):

$$d_0 \approx a_0 \equiv \frac{\hbar^2}{me^2}, \tag{20}$$

that is, the atomic scale, of the order of the angström, as befits real matter. Note that this result only depends on the fermionic nature of the electrons, so that it is their mass which alone determines the energy and space scale. Indeed, neither the fermionic or bosonic nature of the nuclei, nor their mass, play any role in the gross properties of matter.

One cannot stress too strongly how crucial is this result in explaining the nature of matter as we know it; we are not dealing here with minute and subtle effects. If it is the quantum nature of the universe which prevents it from collapsing down an infinite energy well, it requires the Pauli principle to keep us safely above its finite, all right, but very, very deep bottom. Suppose that the Lord, once more irritated (and rightly so) with the behaviour of humankind, decides upon some drastic punishment. Rather than provoking floods or plagues, he could, taking advice from Pauli of course, turn off the exclusion principle dial, switching so to speak, from fermi-on to fermi-off. In an immediate and unprecedented catastrophe, any body (that is, ours) with a size around the meter and a mass of 50 to 80 kilograms, containing something like 10^{28} electrons, would collapse to a size of 10^{-9} Å, liberating an energy around 10^{48} eV, or something like the equivalent of 10^{11} gigatons of TNT (Figure 3). We are all Pauli bombs!*

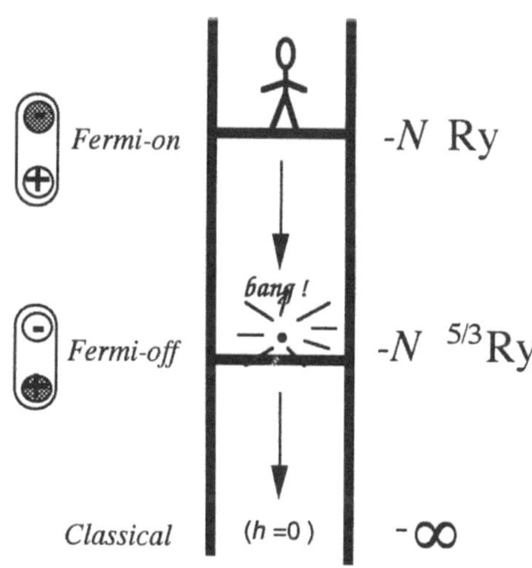

Figure 3. The energy scale of matter

NEWTON FORCES

Coulomb forces are not alone to control the behaviour of macroscopic matter; we know well that for bodies large enough, Newton forces come into play. Since they are universally attractive, we cannot expect them to obey saturation, as it was reached in the Coulomb case only through a delicate balance of attractions and repulsions. Before examining the realistic case of matter with its electrons and nuclei, let us deal with the

---

* This crucial importance of the Pauli principle may be interpreted as the most convincing proof of quantum nonseparability. Indeed, it is only because of the complete entanglement of the collective electronic wave function imposed by its antisymmetrization that matter behaves as it does. In a sense, then, one might wonder why so many decades and subtle arguments (Bell) and experiments (Aspect) have been necessary in order to convince us of such an obvious and archimportant property of matter.

simple case of a system consisting of N identical particles with mass μ. The corresponding hamiltonian reads:

$$H = \sum_{i=1}^{N} \frac{p_i^2}{2\mu} - \sum_{i=1}^{N} \sum_{i<i'} \frac{G\mu^2}{|r_i - r_{i'}|}. \tag{21}$$

Following a similar procedure to that used earlier, we may approximate the ground state energy by starting from the following evaluation of (21):

$$E(N) \approx N \frac{p^2}{2\mu} - \frac{N^2}{2} \frac{G\mu^2}{D}, \tag{22}$$

with the same notations as above. Observe that no screening effect is to be invoked here. From the Heisenberg inequality in the form (9), the minimization of (22) leads to the ground state energy:

$$E_0(N) \approx -N^3 \frac{G^2\mu^5}{\hbar^2}, \tag{23}$$

for a size:

$$D_0(N) \approx N^{-1} \frac{\hbar^2}{G\mu^3}. \tag{24}$$

We are a far cry indeed from saturation, even when allowing for the spreading effect of the Pauli principle; in the case where we would deal with fermions, we only obtain (through the use in (23) and (24) of the effective fermionic quantum constant (16)):

$$E_0^f(N) \approx -N^{7/3} \frac{G^2\mu^5}{\hbar^2}, \tag{25}$$

and

$$D_0^f(N) \approx N^{-4/3} \frac{\hbar^2}{G\mu^3}. \tag{26}$$

The situation for real matter is easily deduced from the above expressions. It is sufficient to note that, because of the disproportion in the masses of electrons and nuclei, the kinetic energy is essentially that of the electrons, and that conversely the gravitational potential energy is essentially that of the nuclei. Now, the Coulomb forces still constrain the system to local neutrality, so that electrons and protons (we limit ourselves tohydrogenic matter) have the same spatial distribution. In other words, the gravitational energy of the system is approximately the same as that of a fictitious system of neutral particles with inertial mass m and gravitational mass M:

$$E(N) \approx N \frac{p^2}{2m} - \frac{N^2}{2} \frac{GM^2}{D}, \tag{27}$$

to be compared to (22). Its ground state energy, taking into account the fermionic nature of the electrons, thus is given by:

$$E_0(N) \approx -N^{7/3} \frac{G^2 M^4 m}{\hbar^2}, \tag{28}$$

for a size:

$$D_0(N) \approx N^{-1/3} \frac{\hbar^2}{GM^2 m}. \tag{29}$$

COLD MATTER: FROM STONES TO STARS

It becomes quite clear now that matter in bulk, for systems small enough, is governed by Coulomb forces; its energy and size obey (17) and (18). For large enough systems, Newton forces take over, and relations (28) and (29) are valid. The first case is that of the objects on our own scale (or smaller), from dust specks to mountains. The second case is that of large astronomical objects, such as planets. Indeed, the transition between the Coulomb and the Newton regime takes place when their respective contributions to the energy are of the same order of magnitude. Equating (17) and (28), one obtains for the critical number of particles separating the two regimes:

$$N \approx N_c \equiv \left(\frac{e^2}{GM^2} \right)^{3/2}, \tag{30}$$

which corresponds to a critical size:

$$D \approx D_c \equiv \left(\frac{e^2}{GM^2} \right)^{1/2} \frac{\hbar^2}{me^2}. \tag{31}$$

It is interesting to note the occurrence in these formulas of the ratio e^2/GM^2 expressing the relative strength of electric and gravitational forces.

We may sum up the preceding results by Figure 4 which show the relationship between the size of a material object D and its number of particles N (which also gives its mass $\mathcal{M} = NM$). In the Coulomb regime, the size grows as the power 1/3, so that the density of matter is constant; that is certainly our experience with matter, since a stone and a mountain certainly have the same density. Bodies on that scale are held together by effective surfaces forces, which explains the other observed fact that their shape is rather arbitrary and does not affect their cohesion. As the critical number of particles is approached, gravitational forces, that is, volume forces, come into play and constrain the body to assume a spherical shape, which certainly is the case for planets, but not for small enough asteroids (with dimensions in the tens of kilometers), as many photographs now witness. It is time to put in some numbers; starting from the known result (ratio of electric to gravitational forces):

$$\frac{e^2}{GM^2} \cong 10^{36}, \tag{32}$$

we obtain:

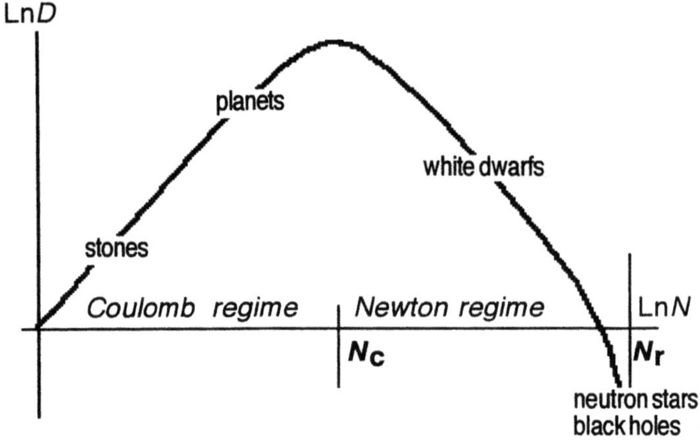

Figure 4. The behaviour of cold matter in bulk :
size against number of particles

$$N_c \cong 10^{54}, \text{ and } D_c \cong 10^8 \text{m} \tag{33}$$

It is gratifying that the critical number of particles and critical size correspond roughly to the mass and size of large planets (something in between the Earth and Jupiter). Beyond the critical number of particles, we deal with bodies held together mainly by gravitation, which accounts for the fact that the heavier they are, the smaller. Such bodies, are usually considered as dwarf stars: stars cold enough for the kinetic pressure not to play a role in their equilibrium. While their masses may be as high as that of the Sun, their sizes are comparable to that of the Earth[*]. It must be stressed that the graph of Figure 4, and our comments, deal only with such cold matter, as we have been concerned only with the ground state energy of material bodies.

We cannot leave the matter here, as a new (last but not least) phenomenon still occurs if we pursue the analysis in term of an ever increasing number of particles. For, in the Newton regime, to the average size (29) of the body, there corresponds an average momentum of the electrons given by the Heisenberg-Pauli inequality (15), to wit:

$$p_0(N) \approx N^{2/3} \frac{GM^2m}{\hbar}. \tag{34}$$

The value of the momentum thus increases with the number of particles, so that it may reach the order of magnitude of the typical einsteinian relativistic momentum mc. This transition to a 'relativistic' regime occurs for a number of particles:

$$N \approx N_r \equiv \left(\frac{\hbar c}{GM^2}\right)^{3/2} = \left(\frac{\hbar c}{e^2}\right)^{3/2} N_c \cong 10^{57}. \tag{35}$$

For such bodies, the above estimates fail, since they are based on the Galilean (low velocity) formula for the kinetic energy of the electrons, as shown by expression (27) for instance. The Einsteinian formula now must be used, leading to the estimate:

---

[*] Many white dwarfs are known, old stars which have cooled down and shrunk, but still shine enough to be seen; their observed masses and sizes are quite in accordance with the numbers derived here. Brown dwarfs, cooled to the point of darkness, and large planetoids ('Jupiters') are believed to exist and some recent evidence has been obtained for them.

$$E(N) \approx N\sqrt{p^2 + m^2c^4} - \frac{N^2}{2}\frac{GM^2}{D} \succ N\sqrt{p^2 + m^2c^4} - \frac{N^{5/3}}{2}\frac{GM^2}{\hbar}p, \tag{36}$$

which, for convenience, we have parametrized with the help of p instead of D as before. Due to the fact that the first term in the right-hand-side of (36) is, for large p, linear instead of quadratic, yielding the asymptotic expression:

$$E(N) \succ Npc - \frac{N^{5/3}}{2}\frac{GM^2}{\hbar}p \approx Npc\left[1 - \left(\frac{N}{N_r}\right)^{2/3}\right], \tag{37}$$

it is quite clear that the expression (36) for the energy does not have a finite minimum (as a function of the parameter p) when the number of particles goes beyond the relativistic value (35). In other words, the energy is no longer bounded from below at all, and the system becomes absolutely unstable against gravitational collapse. Cold bodies, for example white dwarfs, cannot be larger than defined by N_r, which is well known in astrophysics as the 'Chandrasekhar limit'.[*] Hot bodies, that is, nuclear fueled stars, escape this limit because the gravitational pull is balanced by the thermodynamical pressure. As they cool off, however, if their mass is higher than the Chandrasekhar limit, they encounter a fatal doom, due to the insufficient degeneracy pressure of the electrons in the 'relativistic' regime, and they cannot avoid the collapse to the state of neutron stars (which again, and for similar reasons, have a mass limit of the same order of magnitude) or to black holes — unless they manage, through some weightwatching program, to get rid of their dangerous mass excess.

The theoretical estimate (35) of the Chandrasekhar limit gives a mass comparable to that of the Sun, in good agreement with observations. This coincidence between the upper mass of cold stars and the typical lower mass of the hot ones is not a mere coincidence; for a mass of matter condensing under its own gravitation to become a star, that is, for nuclear reactions to start, it is necessary for the energy per particle acquired through condensation, to match the threshold of nuclear reactions, that is a few MeV, which turns out to be comparable with the electronic mass energy mc^2 (that, perhaps, is the coincidence!). Since the energy given out by the condensation precisely is given by our earlier estimate (28), the condition for nuclear ignition is given by:

$$mc^2 \prec \frac{|E_0(N)|}{N} \approx N^{4/3}\frac{G^2M^4m}{\hbar^2}. \tag{38}$$

If this condition, which is precisely equivalent to $N \succ N_r$ as expected, is fulfilled, *a star is born*. That its mass is given by the beautiful formula (35) in terms of fundamental constants, among which the quantum constant, perhaps is a sufficient proof that the spell of the quantum extends in the macroscopic and megascopic domains — up to the stars, indeed.

BACK TO EARTH; THE SCALE OF LIFE

We will now show that the preceding considerations may be extended to a somewhat more subtle problem, and that the scale of such a delicate phenomenon as life itself is

---

[*] Note that the Chandrasekhar limit may be written in the amusing form:
$$N_r = (M_{Pl}/M)^3, \text{ where } M_{Pl} \equiv (\hbar c/G)^{1/2} \text{ is the 'Planck mass'.}$$

governed by simple combinations of fundamental constants. Under what conditions, shall we ask, can a planet host living beings? We stick to the rather irrealistic fiction of purely hydrogenic matter, which, of course, could not properly generate life. But the necessary and possible extension of the reasoning will only introduce numerical rescaling by some powers of the average mass and charge numbers A and Z of the main atomic species, thereby not changing the dependence of our results on the dimensional fundamental constants, nor their rough orders of magnitudes. We will characterize the planet by its number of atoms N and radius R, and living beings by their number of atoms n and their size l (Figure 5).

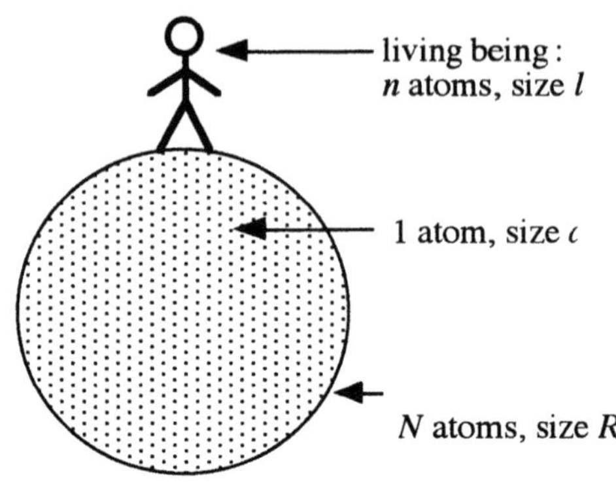

Figure 5.

Consider first the conditions that the planet must fulfill in order for life to be possible at all; of course, we stick to a conservative view of life as we know it, that is, ground based and chemically activated (Hoyle's "black cloud" form of life and other exotic science-fiction monsters notwithstanding). Two conditions then may be stated:

A. Possibility of matter exchange: The planet must be such as to allow for matter to exist in bits and pieces, so that separate living individuals can exist at all, and may move around, ingest other ones or multiply. This requires, as we have seen, the saturation of the cohesion forces of matter, namely the dominance of Coulomb forces over Newton forces. In other words, the planet must obey the Coulomb regime, and its mass be lower than the critical one (30):

$$N \prec N_c = \left(\frac{e^2}{GM^2} \right)^{3/2} \tag{39}$$

B. Surface chemistry: For numerous and complex chemical reactions to proceed at the surface of the planets so as to produce sophisticated molecules, and the living organisms built from them, the reagents have to stay there, and not to become lost in space. This requires that, during these chemical processes, the kinetic energy imparted to molecules in exothermic reactions must be much smaller than the depth of the gravitational potential well preventing them to escape the planet attraction. Since the energy exchanges during chemical reactions are of the order of the standard atomic energy $\mathcal{E}_0$, and since the potential energy is roughly that of an atom (mass M) at the surface of the planet (mass NM, radius R) the condition reads:

$$GN\frac{M^2}{R} \succ \mathcal{E}_0 . \tag{40}$$

Condition A enables us to write the radius R of the planet in terms of the number of particles N and the atomic scale a_0, as in (18):

$$R \approx N^{1/3} a_0 . \tag{41}$$

Equation (40), where use is made of (41) gives:

292

$$N^{4/3} \succ \frac{a_0 \mathcal{E}_0}{GM^2} = \frac{e^2}{GM^2}, \quad \text{or} \quad N \succ \left(\frac{e^2}{GM^2}\right)^{3/2} = N_c. \tag{42}$$

Inequalities (39) and (42) taken together show that for a celestial body to host life, it must be a large one, a planet indeed, close to the critical size[*] :

$$N \approx N_c. \tag{43}$$

We now consider the living beings themselves, and, more specifically, 'animals', that is, mobile forms of life (plants, that is, vegetative forms, would not obey so stringent restrictions), and formulate a last condition:

C. Possibility of motion: If the living organisms are to move around, their internal forces must be able to overcome the pull of gravity at the surface of the planet, so that they can jump or walk. Any mechanism acting within their body, such as the contraction of a 'muscle', operates through biochemical reactions which typically concern a number of atoms of the order of $n^{2/3}$, out of the total number n in the organism; in effect, $n^{2/3}$ is roughly the number of atoms within the area of a typical cross-section of the body. Each one will deliver, in the process, an energy of the order of $\mathcal{E}_0$. The total energy furnished must be large enough to lift the mass nM of the animal by a height comparable to his own typical dimension l against the gravity of the planet:

$$n^{2/3}\mathcal{E}_0 \succ nMgl. \tag{44}$$

There is yet another way to interpret this condition; namely, it may be understood as the requirement that a living body does not break into pieces when falling down, since the r.h.s. of (44) is the potential energy liberated by a fall from a height comparable to the size of the organism, and the l.h.s. is an estimate of the energy it takes to break it into two pieces.

Now, the gravity g at the surface of the planet is given by:

$$g \approx G\frac{NM}{R^2} = GN^{1/3}\frac{M}{a_0^2}, \tag{45}$$

where use has been made of (41). A relation similar to (41) holds for any piece of matter on the planet, such as the size of the animal itself:

---

[*] Condition (40) may be obtained also in a more roundabout, but somewhat more realistic way, by requiring:

1) that an active but nondestructive chemistry takes place on the planet; biochemical reactions must be possible, so that the temperature T should not be too low, but they should not lead to thermal equilibrium (that is, death) too rapidly. The average kinetic energy kT thus has to be of the order of the atomic excitation energy $\mathcal{E}_0$: $kT \approx \mathcal{E}_0$

2) the planet to retain an atmosphere, so that active gaseous exchanges (respiration) can take place, not to mention the necessity of sheer mechanical protection offered by the atmosphere against the radiations coming from space, or its role in balancing the inner pressure of the being. Observe that this satement commits us to the consideration of rather elaborate forms of life, well beyond the unicellular stage, since many bacterial species are anaerobic and live in non-atmospheric environments. The temperature of the planet thus must be low enough for thermal escape of the atoms to be impossible; the gravitational potential of an atom at the surface of the planet must overcome its thermal kinetic energy: $GNM^2/R \succ kT$.

Combining these two conditions, we recover condition (40).

$$l \approx n^{1/3} a_0. \tag{46}$$

Together, (44), (45) and (46) yield:

$$n^{2/3} \prec \frac{a_0 E_0}{GM^2} N^{-1/3} = N_c^{2/3} N^{-1/3}. \tag{47}$$

But, as we have proved, the size of the planet must of the order of the critical size — see (43). We thus obtain the final and most interesting result:

$$n \prec n_c \approx N_c^{1/2} = \left(\frac{e^2}{GM^2} \right)^{3/4}. \tag{48}$$

Accordingly, the maximal mass and size scale of animals, that is, most probably, the scale of the more complex ones, us included, is *absolutely fixed* in terms of the fundamental constants:

$$m_c \approx n_c M \approx \left(\frac{e^2}{GM^2} \right)^{3/4} M \tag{49}$$

and

$$l_c \approx n_c^{1/3} a_0 \approx \left(\frac{e^2}{GM^2} \right)^{1/4} \frac{\hbar^2}{me^2}. \tag{50}$$

Putting in the numbers (see (32)), we obtain:

$$n \cong 10^{27}, \quad m \cong 1 \text{ kg}, \quad l \cong 10 \text{ cm}. \tag{51}$$

Not bad, for an estimate of the scale of life, that is, *our* scale! The result is perhaps even a little *too* good, considering the roughness of many of our assumptions, first among which is the limitation to hydrogen atoms (which certainly would not suffice to allow for the appearance of life); then, a number of "fudge factors" should be introduced in most of our estimates, for instance (this is one of the most obvious cases) in (40) and (45), where the typical energies for biochemistry are certainly overestimated by a factor of a hundred or so. Nevertheless, as already said, such corrections would only introduce pure number factors which, probably, due to several cancellations or reductions (through small powers), would not affect the results (51) by many orders of magnitude. But, obviously, the expressions (51) cannot be modified as far as concerns their dimensional dependence on the fundamental constants. Anyway, we know that, concerning the size l of animals, the result in (51) is right in the middle of the biological realm, which extends roughly by a factor of one hundred both ways (from the millimeter to the decameter).[*]

---

[*] There *is* a way for life to beat the scale limits imposed by (51), and that is to deal away with the gravitational constraints, by trying to move without needing to struggle against gravity. Crawling, that is moving as horizontally as possible, is a first solution, but not a very good one, at least on a planet like Earth where the ground is so rough and bumpy that one has to move up and down anyway (do not forget, though, that snakes may reach quite an appreciable length, around 10 m). The real solution, of course, is to play Archimedes against newton, by staying in, or going back to water, as whales have understood. This also works for downscaling, where life in fluids offers many opportunities to microscopic life.

CONCLUSIONS

It is interesting to point out that the number n_c expressed by (48) is but the number of atoms characterizing pieces of matter on our scale, that is, after all, something like *the Avogadro-Loschmidt number*! In other words, we have shown that, far from being purely conventional, this number, which makes the connection between the microscopic and the macroscopic world, is fixed by the fundamental constants themselves, and more precisely, depends only on the ratio between electric and gravitational forces.[*] Of course, the precise definition of the Avogadro number does contain an element of convention; instead of choosing the number of atoms in 1 g of hydrogen, one could as well have chosen 1 kg as the reference mass, bringing the value of the Avogadro number in exact coincidence with (50).

That *we* are quantum phenomena, now appears quite explicitly on the expression (50) giving our human size scale in terms of the quantum constant. A nice way of stating the results in (50) is the following:

$$\text{macroscopic} = \sqrt{\text{microscopic} \times \text{megascopic}} \, ,$$

or, more pictorially:

$$\text{Human scale} = \sqrt{\text{Atomic scale} \times \text{Planet scale}} \, .$$

That we are half-way between the 'infinitely small' and the 'infinitely large' of course has been known to many for quite some time, and put in particularly eloquent terms by Pascal. The preceding results may be viewed as a definite answer to his query: « Why have limits been set upon (...) my height (...)? For what reason did nature make it so, and choose this rather than that number from the whole of infinity, when there is no more reason to choose one rather than another, as none is more attractive than another? ».[6] We have simply given substance to a known answer, and old as well, stating that our world is neither too large, nor too small, « but just right ».[7]

REFERENCES

1. See,e.g., V. Weisskopf, *Science*, **187**, 605 (1975), for considerations related to the present ones.
2. See J.-M. Lévy-Leblond & F. Balibar, *Quantics (Rudiments)*,North-Holland, 1990, and J.-M. Lévy-Leblond, *Bull.Soc.Fr.Phys., Enc. Péd.***1** (1973)
3. F. J. Dyson & Lenard, J. Math. Phys. **8**, 423 (1967) and **9**, 698 (1968).
4. E. H. Lieb& H. Thirring, *Phys. Rev. Lett.* **35**, 687-689 (1975) ; E. H. Lieb, *Bull. Amer. Math. Soc.* **22**,1-49 (1990).
5. J.-M. Lévy-Leblond, *J. Math. Phys.* **10**, 806 (1969) and *J. de Phys.* **30**, C3-43 (1969).
6. Blaise Pascal, *Pensées* ; English translation by A. J. Krailshamer, Penguin Classics, 1966.
7. *Goldilocks and the Three Bears*. See also the opinion of W. Leibniz.

<u>Note added in proof</u>. After the completion of this paper, it has been brought to my attention that results very close to those of the last Section and more detailed discussions of other macroscopic quantities, have been published by B.J. Carr & M.J. Rees, *Nature* **278**, 605 (1979) and W.H. Press & A.P. Lightman, *Phil. Trans. R. Soc. Lond.* **A 310**, 323 (1983).

---

[*] This is not to say that nuclear forces, strong and weak, play no crucial role in the fabric of the universe; without them, there would be no complex nuclei, that is, no atoms nor molecules of the many types needed to generate the variety and complexity of the world. It may be safely asserted, though, that they play no role in fixing the scale of most macroscopic phenomena. This is, after all, only natural, since these forces are short-range interactions. It is left to the long-range ones, gravitation and electromagnetism, to rule the scale of the world at large.

SPECIFIC TOPICS

QUANTUM FLUCTUATIONS AND SUPERCONDUCTIVITY

R.Fazio[1] and A.Tagliacozzo[2]

[1] Istituto di Fisica, Facoltá di Ingegneria,
Universitá di Catania
via A. Doria 6, I-95129 Catania Italy
[2] Dipartimento di Scienze Fisiche,
Universita' di Napoli
Mostra d'Oltremare Pad.19,
I-80125 Napoli,Italy

also GNSM (CNR) and INFM,Italy

Soon after Kamerlingh Omnes succeded in liquifying helium, he discovered (1911) that resistance in samples of mercury dropped suddenly to zero below the critical temperature of 4.15 $^\circ K$. Since then, it was readily realized that superconductivity is a quite common feature of metals with some exceptions. Twentytwo years later it was found that in applying a magnetic field to a superconducting sample, below some critical value, the magnetic flux is excluded from its bulk (Meissner effect) [1, 2, 3].

A big effort was required in understanding the underlying physics of conventional low temperature superconductors and the mechanism of pairing in the recently discovered high temperature superconductors is still unknown. Quantum mechanics is the fundamental ingredient for the explanation of the phenomenon and viceversa, superconductivity contributes enormously to the developement of quantum mechanics itself. The collective behavior of electrons in a metal in the superconducting state is responsible for these peculiar properties, leading to a direct detection of quantum behavior at the macroscopic level. This is what is named *coherence* of a superconductor, although it is hard to provide a precise definition for this concept.

The breakthrough in understanding the structure of the superconducting ground state and of the nature of its excitations was BCS theory in 1957 which provided an interpretation of the thermodynamic and transport properties in terms of a condensate of Cooper pairs[4]. In the first section, we are using the BCS ground state to introduce the very general description of a superconductor in terms of an order parameter which is space and time dependent as well as of the electromagnetic field inside the superconductor. In essence, the superconducting coherence implies that, first of all, an amplitude of a superconducting

Advances in Quantum Phenomena, Edited by E.G. Beltrametti
and J.-M. Lévy-Leblond, Plenum Press, New York, 1995

order parameter is required at each point and secondary, the correlations of the order parameter are long ranged.

The thermal fluctuations in bulk superconductivity are confined to a very narrow temperature interval around T_{co} (the critical region), outside of which the order parameter and the electromagnetic field obey the classical equation of motion, but with $\hbar$ built in. Outside the critical region, fluctuations of the phase can take place while it is very costly to change the modulus of the order parameter away from the value that minimizes the free energy. Usually both thermodynamics and transport properties are explained by means of classical statistical mechanics for the order parameter and the electromagnetic field.

However, in this chapter we focus our attention not on this classical description of superconductive samples, which is still being developed with beauteful results and applications, but on quantum properties at very low temperatures. Reduced dimensionality or special geometries as small junctions can highlight the zero point motion of the order parameter and quantum fluctuations can strongly affect the classical superconducting behavior. It is important to study the fluctuations of the order parameter, in order to ascertain when they destroy the phase coherence.

The theory is far from being complete. Here we only try to motivate some of its predictions, using simple physical arguments. To stress the role of dimensionality on the long range superconducting order, we concentrate on a single junction between two superconductors (Josephson Junction [5]) as a prototype of a zero dimensional system and on a two dimensional arrays of such junctions. In this work the charging energy is always taken as a small correction to the Josephson coupling energy, although of crucial importance to classify these phenomena as quantum. Therefore, a representation of states in the phase coordinate (instead of the number of charges) is the most suitable. The opposite case is not considered here[3]

In the case of arrays, additional cooperative effects take place, as outlined in the second part, allowing for a striking observation of quantum properties in the response of the system at the macroscopic level. Arrays offer a practical realization of models of statistical mechanics whose quantum version is currently investigated and we expect they will provide a wealth of applications in the future. The parameters of the system can be tuned in the fabrication process over a wide range of values. Some arrays behave as it is expected classically, that is they become superconducting at low temperatures. Others however go insulating, as shown in Fig. 1, taken form ref.[6], what is attributed to quantum effects. To see such phenomena, arrays are required,consisting of very small Josephson Junctions of high quality (usually Al junctions of area $\leq 0.1\mu m^2$) and temperatures of the order of milliKelvin. Analogous behaviors have been found in granular thin films [7].

The last section is devoted to the predicted quantum interference of a vortex in the configuration of phases distributed over the array (Aharonov-Casher effect). The challenge is to create the conditions under which a vortex constitutes a quantum particle as a whole, in spite of the fact that it involves an huge number of microscopic variables. Were this the case, this is one more step toward the observation of Macroscopic Quantum Mechanics.

The latter assumes that a single macroscopic variable can describe the state of a whole system (e.g. the voltage at a Josephson Junction, or the flux in an *rf-SQUID* ring) and that its quantum fluctuations cannot be disregarded. The problems underlying these views are the content of one more chapter in this book by G. Ghirardi and are not touched here.

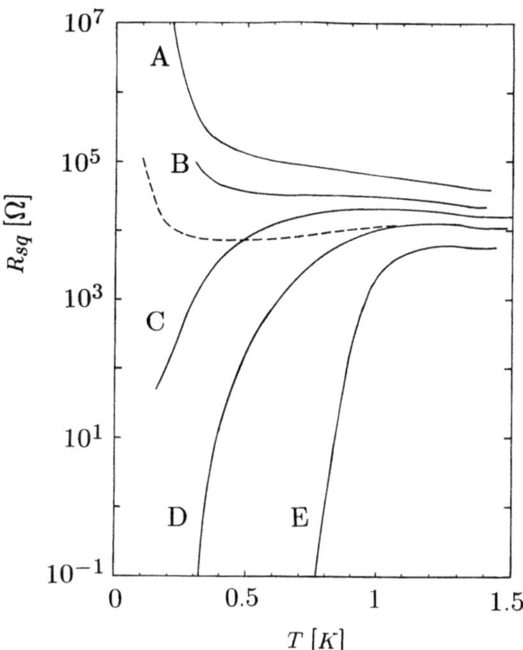

Figure 1: Resistivity vs. temperature for various arrays of small area junctions ($0.01\mu m^2$ ($E_C \approx 0.84°K$)) corresponding to different normal state resistance (after Geerligs et al.[8]). The parameter values of the different samples can be found in [9].

The discussion found here is only introductory to the field. Some references are given, where more details can be found.

SUPERCONDUCTING ORDER PARAMETER

In the grancanonical ensemble a model Hamiltonian that gives rise to superconductivity is:

$$\hat{H} - \mu\hat{N} = \sum_{\sigma} \int dr \psi_\sigma^\dagger(r)(t(r) - \mu)\psi_\sigma(r) - \frac{g}{2} \int dr \; \psi_\downarrow^\dagger(r)\psi_\uparrow^\dagger(r)\psi_\uparrow(r)\psi_\downarrow(r) \qquad (1)$$

Here μ is the chemical potential and $\hat{N}$ the number operator. The fields ψ_σ are field operators for the electrons and $t(r)$ is the single particle Hamiltonian including the kinetic energy and possibly some crystal potential. The pairing interaction of strength $-g$ is represented by the second term and is taken to be local. It is effective only in a small shell of energies around the Fermi energy.

A variational Ansatz to the ground state of the pairing Hamiltonian of eq.(1) for an electron gas was proposed in 1957 by Bardeen, Cooper and Schrieffer:

$$|\theta> = \prod_k (u_k + v_k e^{i\theta} c_{-k\downarrow}^\dagger c_{k\uparrow}^\dagger)|0> \qquad (2)$$

Here the u_k, v_k's are variational parameters and the operators $c_{k\sigma}^\dagger$ create particles in plane wave states of wavevector $\vec{k}$ and spin σ on the vacuum state $|0>$. The phase θ does not depend on k and does not affect the energy of the state. It labels the BCS state, however.

This state and the excited states, that can be constructed starting from it, can explain at least qualitatively the properties of low temperature superconductors.The excited states corresponding to low excitation energies, are found to be single particle excitations and there is a gap Δ in the energy excitation spectrum.

The state of eq.(2) has the remarkable feature that its two-particles density matrix:

$$< r_1' \uparrow r_2' \downarrow \,|\hat{\rho}|\, r_1 \uparrow r_2 \downarrow> = Tr\left\{e^{-\beta H}\psi_\downarrow^\dagger(r_2)\psi_\uparrow^\dagger(r_1)\psi_\uparrow(r_1')\psi_\downarrow(r_2')\right\} \tag{3}$$

displays the so called Off Diagonal Long Range Order (ODLRO)[8].

To understand what ODLRO implies and why it is a genuine quantum feature, let us examine the quantity defined in eq.(3), with the unprimed variables taken equal to the corresponding primed ones (i.e. the diagonal matrix elements in the space representation of the density matrix operator $\hat{\rho} = exp(-\beta H)$). It represents the space correlation between two particles, when all the other particle positions have been integrated out. This is a well defined quantity also in classical statistical mechanics. The off diagonal matrix elements do not have a classical analogue and they usually tend to vanish rapidly, when moving away from the diagonal. At zero temperature they can be easily computed on the state of eq.(2) giving the surprising result that they do not vanish in this limit:

$$\lim_{|r_1-r_1'|\to\infty} \lim_{|r_2-r_2'|\to\infty} < \theta|r_1' \uparrow r_2' \downarrow |\hat{\rho}|r_1 \uparrow r_2 \downarrow |\theta > = \frac{N}{2}\,\phi^*(r_2'-r_1')\phi(r_2-r_1) \tag{3a}$$

$$\phi(r_2-r_1) = e^{-i\theta}\frac{1}{V}\sum_k u_k v_k^* e^{ik(r_2-r_1)} \equiv < \psi(r_2)\psi(r_1) >_\theta \tag{3b}$$

Due to the translational invariance of the electron gas, the function ϕ doesn't depend on the center of mass of the pair, $R = (r_2+r_1)/2$, but just on the coordinate related to its internal structure $x = r_2-r_1$. It is localized over a typical distance $|r_2-r_1| \sim \xi_o$ which is called the 'coherence length'. In case the medium is not uniform in space over a length scale which is large compared to ξ_o, the dependence on the center of mass coordinate is the only one relevant instead. Then, eq.(3a) is suggestive of a Bose condensate of particles ("pairs"), all described by the single particle wave function $\phi(R, x \approx 0)$. The extension of the r.h.s. of eq.(3a) to the non uniform case could be reinterpreted as the corresponding single particle density matrix $\rho_{Bose}^{(1)}(R, R')$. This leads to the definition of $\phi(R)$ as the superconducting order parameter. Because the BCS wavefunction of eq.(2) does not conserve the particle number, the density matrix $\hat{\rho}$ does not commute with $\hat{N}$ in the subspace in which θ is fixed. This allows for a non zero expectation value of the order parameter:

$$\phi(R) = < \hat{\phi}(R) >_\theta \equiv < \psi(R)\psi(R) >_\theta = (n_s(R))^{1/2}\,e^{i\theta(R)} \tag{4}$$

The square modulus $n_s(R)$ is a measure of the superconducting condensate at each point R while the space gradients of its phase $\theta(R)$ are related to the superconducting currents. Let us consider the thermal averages at finite temperature within the subspace in which θ is fixed and the partition function $\mathcal{Z} = Tr^{(\theta)}\{\hat{\rho}\}$ (we denote the restricted trace with a superscript (θ)). Then we require:

$$< \hat{\phi}(R) >_\theta = -\frac{1}{2\mathcal{Z}}Tr^{(\theta)}\left\{\hat{\rho}[\hat{N},\hat{\phi}(R)]\right\} = -\frac{1}{2\mathcal{Z}}Tr^{(\theta)}\left\{[\hat{\rho},\hat{N}]\hat{\phi}(R)\right\} \neq 0 \; , \tag{5}$$

in spite of the fact that the Hamiltonian of eq.(1) commutes with the number operator. This is usually called symmetry breaking in the sense that , while the Hamiltonian conserves the number of particles, its eigenstates do not. It follows that coherence of the phase θ of the order parameter over a volume implies non conservation of the number of particles within that volume.

Nevertheless a superconducting state with a fixed number of particles $|N>$ can be constructed by superposition of states of any phase:

$$|N> = \int_0^{2\pi} \frac{d\theta}{2\pi} e^{-iN\theta} |\theta> \tag{6}$$

Viceversa the BCS state can be viewed as a superposition of states $|N>$:

$$|\theta> = \sum_{N=0}^{\infty} e^{iN\theta} |N> \tag{7}$$

It will be used in the following that the representation of the operator $\hat{N}$ onto states of the type of eq.(7) is $\hat{N} = -i\partial/\partial\theta$, while the representation of $\hat{\theta}$ onto states of the $|N>$ type is $\hat{\theta} = i\partial/\partial N$.

At finite temperatures, model expressions for the free energy $F = -\beta^{-1}\ln\mathcal{Z}$ can be conceived on phenomenological grounds (Ginzburg-Landau theory) as functionals of a space dependent order parameter ϕ and, possibly, of the vector potential $\vec{A}$ giving rise to a magnetic field (if below its critical value H_c) . The average value of the order parameter, $\overline{\phi}$, is obtained by minimization under the appropriate boundary conditions. At thermal equilibrium the order parameter fluctuates around its average value $\overline{\phi}$[1, 9]. Fluctuations of the modulus are prominent in approaching the critical temperature T_{co} from above. Below the critical temperature, fluctuations are only important in a very tiny temperature region close to T_{co} . Infact, because they extend over volumes that cannot be smaller than ξ^3, according to the Ginzburg Landau theory, they are very costly (ξ can be several hundreds of $\mathring{A}$), unless the modulus of the order parameter is quite small. One finds (Ginzburg criterion):

$$\frac{< \left(\phi - \overline{\phi}\right)^2 >}{\overline{\phi}^2} \sim \qquad TT_{co}/T_F^2 \qquad \text{for } T \sim 0$$

$$\sim (T_{co}/T_F)^2 \left[T_{co}/(T_{co} - T)\right]^{1/2} \qquad \text{for } T \leq T_{co}$$

We now concentrate on the effect of fluctuations on the correlations of the phase.

In a neutral superfluid described by an order parameter $n_s^{1/2}e^{i\theta}$, low lying collective excitations are compressional waves. The correlations of the phase at different points and equal times range all over the volume at any temperature.

The case of two-dimensions is rather peculiar and will be treated in some detail in the second part of the chapter for the special case of an array of junctions. In one dimension, instead, they decay exponentially at finite T and with a power law at zero temperature. This is an example of the well established fact that fluctuations are enhanced in lowering the dimensionality of the system.

In the case of a superconductor, long range Coulomb forces between the electrons cannot be ignored. They make the sound mode disappear and the eigenmode of the phase at $k = 0$

is found at $\omega = \omega_p$. It follows that a bulk superconductor has no low lying collective excitations and that the only excitations are the single particle ones with an energy gap Δ in the spectrum. Thus, in a bulk superconductor, thermal fluctuations of the phase from its average uniform value are not expected far from the critical temperature.

Josephson junctions: quantum versus classical behavior

Consider a tunnel junction between two superconductors. Its simplest realization is by deposition of two superconducting thin films in a cross geometry, forming the tunnel barrier with the oxidation of the bottom layer. Today's technology provides high quality, long lasting junctions made of Nb, but the fabrication process is more involved. At zero temperature, a voltage difference V applied to the junction, can produce a quasiparticle current, if it is large enough to break a pair. In a junction, in which both sides are made of the same superconducting material, this happens for $V_g = 2\Delta/e$, where Δ is the gap in the excitation spectrum. Infact single particle excitations have energy $E_k = \sqrt{\epsilon_k^2 + \Delta^2}$ where ϵ_k is the band energy. At finite temperature there can be thermally excited quasiparticles anyhow, whose distribution in energy, at a given temperature, is given by the Fermi function $f(E_k) = 1/(\exp \beta E_k + 1)$. They contribute to the current at all voltages.

In Fig. 2 the quasiparticle branch of the I/V characteristic of a $Nb/AlO_x/Nb$ junction is reported [10]. The subgap region is amplified to show small steps at $V_g/2$ and $V_g/3$, which are supposed to be due to two or three quasiparticles tunneling simultaneously.

When the insulating barrier in the junction is thin enough (about $10\mathring{A}$), electron pairs of the condensate can tunnel across it and a supercurrent can flow even in the absence of a voltage drop at the junction. This current, first predicted by Josephson [11], depends on the phase difference of the order parameters of the two superconductors.

Its simplest description assumes that the junction is connected to leads of infinite impedance at all frequencies, so that the damping would remain zero and no capacitive loading is present on the junction. Consider first the two isolated superconductors. They can be characterized jointly by states labeled by the number of pairs N_A and N_B , respectively. Because they are isolated, the states $|N_A, N_B >$ are all degenerate with energy E_o . When the weak link between the two is created, an hopping term has to be added to the Hamiltonian, possibly depending on the vector potential $\vec{A}$ in the form suggested by Peierls:

$$\mathcal{H}_t = \frac{E_J}{2} e^{-2e\,i \int_A^B \vec{A}\cdot d\vec{l}/\hbar c} \sum_{N_A,N_B} |N_A + 1, N_B - 1 >< N_A, N_B| + h.c.$$

It has been assumed implicitly that just one pair at a time can cross the junction. The gauge chosen for the vector potential is $\vec{\nabla} \cdot \vec{A} = 0$ so that it vanishes well inside the two superconductors. This restricts the line integral in the phase just to the barrier region. E_J is the Josephson coupling energy.

The Ansatz for the state to solve the stationary Schrödinger equation, $\mathcal{H}_t \Psi^{(N)} = (E - E_o)\Psi^{(N)}$ $(N = N_A + N_B)$, is

$$|\Psi^{(N)}\rangle >= \sum_p a_p \left| \frac{N}{2} + p, \frac{N}{2} - p \right\rangle \tag{8}$$

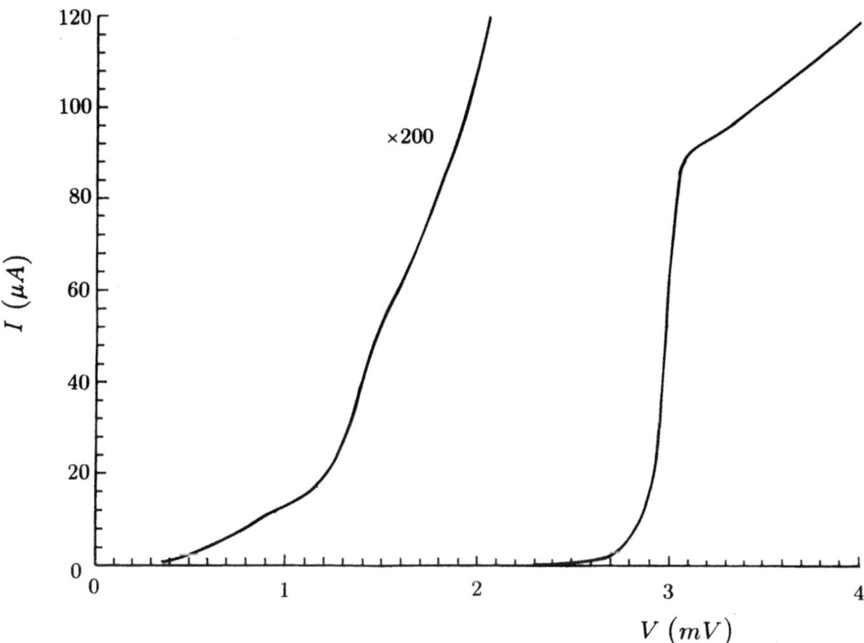

Figure 2: Quasiparicle branch of the I/V characteristic of a $Nb/AlO_x/Nb$ junction with the small voltage region magnified (after R.Cristiano et al.[16]).

with p integer. It is straightforward to show that the choice of eq.(8) indeed solves the Schrödinger equation with $a_p = \exp(ip\varphi)$ (where φ will be fixed by the boundary conditions) and

$$E(\vec{A}) = E_o - E_J \cos\left(\varphi - \frac{2e}{\hbar c}\int_A^B \vec{A}\cdot d\vec{l}\right) \qquad (9)$$

The state $|\Psi^{(N)}\rangle$ of eq.(8) is fully delocalized as a function of the charge difference between the two superconductors $Q = 2e\,p$ because $|a_p| = 1$ for any p. Using eq.(6) one can rewrite $|\Psi^{(N)}>$ as the superposition of states in the phase representation:

$$|\Psi^{(N)}> = \sum_{m,m'} \delta_{m+m',N}\; e^{i(m-m')\varphi/2}|m,m'> = \int_0^{2\pi} \frac{d\theta}{2\pi}\; e^{-iN\theta}|\theta + \varphi/2, \theta - \varphi/2 > \qquad (10)$$

where use of the representation of the Kronecker δ function has been made: $\delta_{n,n'} = (2\pi)^{-1}\int_0^{2\pi} d\theta \exp -i(n-n')\theta$. Eq.(10) provides the desired identification of φ with the phase difference between the A and B superconductors.

The variation of the energy of eq.(9) with respect to the vector potential gives the current I :

$$\frac{1}{c}\,I = \frac{\delta E(A)}{\delta A} = \frac{2e}{\hbar c}E_J\,\sin\left(\varphi - \frac{2e}{\hbar c}\int_A^B \vec{A}\cdot d\vec{l}\right)\;. \qquad (11)$$

This is the Josephson current at zero voltage.

Up to now the transfer of pairs across the weak link was assumed not to change the neutrality of both sides. In a real situation, because the geometric structure of the junction acts also as a capacitor, a charging energy should be added and a dissipative channel for the transport is present, as well. From the conservation of the current it is easy to get the following Langevin equation (RSJ model):

$$C\ddot{\varphi} + \frac{1}{R}\dot{\varphi} + \frac{\partial u}{\partial \varphi} = \frac{2e}{\hbar} I_N(t) \tag{12}$$

Here $U = (\hbar/2e)^2\, u$ is the so called 'washboard 'potential:

$$U(\varphi) = -E_J(\cos\varphi - 1) - \hbar I\varphi/2e \tag{13}$$

This has a minimum for $\varphi = 0$ if the biasing current I is less than the critical current $I_{co} = 2eE_J/\hbar$.

An intrinsic thermal noise has been added, of ohmic type, by means of a stochastic source of zero average and white correlation:

$$< I_N(t) >= 0 \qquad\qquad < I_N(t)I_N(0) >= \frac{2k_BT}{R}\,\delta(t)$$

The typical parameter to describe dissipation is $Q = \omega_J RC$ defined in terms of the plasma frequency: $\omega_J = (2eI_c/\hbar C)^{1/2}$, with $I_c = (I_{co}^2 - I^2)^{1/2}$ (ranging between 10 and 100 GHz).

The switching from the Josephson branch of the I/V characteristics to the quasiparticle branch can be hysteretic (underdamped case: $Q >> 1$) or non hysteretic (overdamped case: $Q << 1$). Thermal noise can induce activation over the potential barrier from the zero voltage state, to the resistive state in the underdamped case. In the overdamped case, phase diffusion along the washboard potential can be activated by the thermal bath, rounding the switch out of the Josephson branch. In very underdamped junctions of high normal state resistance, both features can coexist[12].

At low temperatures (below $1°K$) one expects that quantum tunneling of the phase can take place, instead of its thermal activation over the barrier between different minima of the washboard potential. This is also enhanced if the capacitance is reduced. Infact, in eq.(12) the latter acts as the mass of the particle, whose position is described by φ. Therefore, when it is smaller, quantum effects become prominent[2]. Recent experiments on underdamped junctions seem to confirm these expectations[13].

The quantum mechanics for a Josephson Junction in presence of dissipation is derived from a microscopic Hamiltonian in ref.[14]. Here we resort only to the macroscopic picture and put aside dissipation for the time being, what corresponds to the ideal situation of a junction connected to leads of infinite impedance at all frequencies. Actually this is very far from reality, because at frequencies corresponding to micron wavelength ($\approx 10^{14}Hz$), $Z(\omega)$ will be inevitably dominated by radiation phenomena and will be of the order of the impedance of the free space $Z_v = (\mu_o/\epsilon_o)^{1/2} \sim 377\Omega$. Nonetheless, it is useful to understand what happens and we discuss briefly the effect of the dissipation afterwards.

It is easy to construct an Hamiltonian for the isolated junction, depending on the macroscopic variables Q and φ:

$$\mathcal{H} = \frac{Q^2}{2C} - E_J(\cos\varphi - 1) \tag{14}$$

This is the generalization of eq.(9) (with $\vec{A} = 0$), which includes the charging energy. Because $Q/2e$ is the momentum conjugated to the phase, quantization implies that they do not commute: $[\varphi, Q/2e] = i$. This leads to the expected Heisenberg equations of motion:

$$\frac{\hbar}{2e}\dot{\varphi} = 2ei[\mathcal{H}, \varphi] = \frac{Q}{C} = V$$
$$\frac{dQ}{dt} = \frac{i}{\hbar}[\mathcal{H}, Q] = \frac{2e}{\hbar}E_J \sin\varphi$$

The first is the celebrated Josephson relation, while the second corresponds to the classical equation of motion, eq.(12), in the limit of $R \to \infty$ and $I_N(t) = 0$.

The addition of the charging energy lifts the degeneracy of the states in the sum of eq.(8). It follows that the wavefunction that solves the Schrödinger equation is no longer fully localized in the phase representation. Anderson was the first to discuss the possibility that the charging energy could spoil the phase coherence [15]. Suppose the unbiased junction has been prepared in a state with average value of the phase $< \varphi >= 0$. One can get a feeling of the spread in phase and in charge of the state due to the zero point motion in the case of $E_J >> E_C = e^2/2C$, assuming that the bottom of the potential well is approximately harmonic. The phase and charge mean square fluctuations estimated in this way are: $\sqrt{< \varphi^2 >_{osc}} = 2(E_C/E_J)^{1/4} << 2\pi$ $\sqrt{< (\Delta N)^2 >_{osc}} = \sqrt{\left\langle (Q/2e)^2 \right\rangle_{osc}} = I_{co}/e\omega_J >> 1$.

Nowadays very small Al junctions can be fabricated which endanger these inequalities. Their typical characteristics are a junction area $\sigma \approx 0.02\mu m^2$, a capacitance of $C \approx 1fF$ and a resistance in the normal state of $R_n \approx 20k\Omega$. In these junctions the energy corresponding to pair breaking is rather high $2\Delta \approx 0.4meV$ as compared to the charging energy $E_C \approx 80\mu eV$ (corresponding to a temperature of $0.9°K$)and to the Josephson coupling energy which can be estimated from the Ambegaokar-Baratoff formula [16] at zero temperature($E_J \approx h\Delta/8e^2R_n \approx 33\mu eV$). Below $1°K$ the charging effects are not washed out by thermal fluctuations.

Delocalization of the phase in a Josephson Junction

When the phase is delocalized the full cosine potential has to be accounted for in the hamiltonian of eq.(14). As previously stated, we will only consider the case $E_J >> E_C$, so that the phase representation is still the most convenient one. The solutions of the Schrödinger equation are Mathieu functions labelled by some parameter q according to the property:

$$\Psi_q(\varphi - 2\pi m) = e^{imq}\Psi_q(\varphi) \tag{15}$$

and corresponding to eigenvalues $E(q)$ which are 2π-periodic in q.

Because phases which differ by 2π are indistinguishible in an isolated junction, it is desiderable to take the corresponding states to be periodic. This can be achieved by a gauge transformation which has the effect of suppressing e^{imq} in eq.(15) by shifting the momentum conjugate to the phase, $Q \to Q - 2eq/2\pi$ in the hamiltonian of eq.(14). This allows to relate q to an external charge difference Q_x which could be present at the junction[17]: $q = \frac{2\pi Q_x}{2e}$.

Quantum interference between states localized at different minima of the potential, thought of as a sequence of phase slips between these two points, ideally requires that an external charge is feeded at the junction of all possible values, to compensate all possible voltages induced by these processes. Viceversa, one could think of fixing the external charge. Then, the energy is no longer independent of the number of pairs which move across the junction, but it acquires a dispersion with the parameter Q_x due to the delocalization of the phase on various minima of the washboard potential. An analogous result can be obtained by studying quantum phase slips in quasi one-dimensional systems, as saddle point solutions of the G-L type Lagrangian [18]. It is a subtle, matter, wether Q_x can be a continuous variable, or it can only change by multiples of $2e$, what trivially washes out the dependence of the energy on Q_x. The discussion can be found in the chapter of this book by M.Devoret.

Anyhow, the model of a junction which is isolated from any electromagnetic environment is highly unrealistic. Dissipation in the leads of the polarizing circuit tends to suppress the tunneling of the phase between different minima of the washboard potential and to stabilize the phase. We now discuss in an heuristic way the effect of dissipation, from which the following conclusion arises.

The phase can be localized and the quantum effects disappear, depending on the frequency dependence of the real part of the impedance. In particular the admittance of the circuit is expected to be a power law at low frequencies: $\Re Y(\omega) \sim \omega^\nu \exp(-\omega/\omega_c)$ what classifies dissipation as superohmic $(\nu > 0)$, ohmic $(\nu = 0)$ or subohmic $(-1 < \nu < 0)$ (ω_c is some cutoff frequency). At zero temperature, the phase is localized by the dissipation in the subhomic case, while it is not in the superohmic case. The ohmic case, which is the most interesting one because it corresponds to the classical RSJ model, is marginal. Infact, in the ohmic case, a localization-delocalization transition takes place, increasing the value of the resistance $R \approx (\Re Y)^{-1}$. The threshold between the two phases is the quantum resistance $R_Q = h/4e^2 \sim 6.5k\Omega$. One finds delocalization for $R_Q < R$ and localization otherwise.

A very qualitative argument illustrates what happens in the ohmic case. Let us assume that the circuitry on which the junction is closed has a discharge time roughly equal to RC. This has to be compared with the expected time scale for quantum fluctuations of the charge at the junction: $\hbar/4E_C$. If this last time is much longer than the time of discharging, the charge can fluctuate wildly at the junction without accumulating, what implies localization of the phase. This occurs if $h/4e^2 >> R$.

It has become very common to include dissipation in a quantum mechanical frame following the approach of Caldeira and Leggett [19]. They assume that the environment can be described as a collection of harmonic oscillators of a given spectral density, linearly coupled to the phase.

They start from a generalization of the RSJ model, given by eq.(12):

$$C\ddot{\varphi} + \int_{-\infty}^{t} dt'\, Y(t - t')\dot{\varphi}(t') + \frac{\partial}{\partial\varphi}u(\varphi) = \frac{e}{\hbar}I_N(t) \tag{16}$$

This equation can be derived from the Hamiltonian:

$$\mathcal{H} = \frac{1}{2M}\left(P_\varphi\right)^2 + u(\varphi) + \sum_\alpha \frac{1}{2m_\alpha}\left(p_\alpha + \frac{e}{c}C_\alpha\varphi\right)^2 + \frac{1}{2}\sum_\alpha m_\alpha\omega_\alpha^2 q_\alpha^2 \tag{17}$$

where α labels the collection of oscillators, whose conjugate momentum is coupled to the variable of the junction φ via appropriate constants C_α. The current noise I_N is related to the dynamics of the $\dot{q}_\alpha$'s. To obtain eq.(16) the spectral density of the oscillators has to be properly chosen. In particular the real part of the admittance as a function of the frequency is:

$$\Re\{Y(\omega + i0^+)\} = \frac{\pi}{2}\left(\frac{e^2}{\hbar c}\right)\sum_\alpha \frac{C_\alpha^2}{m_\alpha}[\delta(\omega - \omega_\alpha) + \delta(\omega + \omega_\alpha)] \tag{18}$$

Dealing with the washboard potential is rather cumbersome, but, in a perturbative scheme, the golden rule gives the probability per unit time, $\Gamma(E)$, that the system moves from one well to the other, exchanging energy E with the environment. It is a reasonable approximation if the phase is delocalized, but coherence effects between the wells can be ignored, and if the frequencies of the oscillators involved are much larger than the bare tunneling frequency, $|\mathcal{T}|/\hbar << \overline{\omega} < \omega$ (adiabatic approximation).In this range of frequencies one gets:

$$\Gamma(E) \sim \left(\frac{|\mathcal{T}|^2}{4\hbar}\exp-\int_{\overline{\omega}}^\infty \frac{d\omega}{\omega}\frac{\pi\hbar}{e^2}\Re\{Y(\omega + i0^+)\}\right)\cdot P(E) \tag{19}$$

where the prefactor can be viewed as a renormalized tunneling rate $|\mathcal{T}_1|^2$ and the function

$$P(E) = \frac{1}{\hbar}\int_{-\infty}^\infty dt e^{iEt/\hbar}\exp\left\{\int_{\overline{\omega}}^\infty \frac{d\omega}{\omega}\frac{\pi\hbar}{e^2}\Re\{Y(\omega)\}\left(\cos(\omega t)\coth(\beta\hbar\omega/2) - i\sin(\omega t)\right)\right\},$$

is the probability to emit energy E to the external circuit, which is discussed at length in [3]. Thus, within our approximations, the main effect of coupling to high frequency modes of the environment is to depress the bare tunneling rate $|\mathcal{T}|^2$. This allows to consider a new range of frequencies between $\overline{\omega}_1(<\overline{\omega})$ and $\overline{\omega}$, large compared to the renormalized tunneling frequency $\mathcal{T}_1/\hbar$ and their elimination can be repeated, within the adiabatic approximation. It is clear that if $d\ln|\mathcal{T}|/d\ln\omega$ is less than one, the lower cutoff frequency $\overline{\omega}$ is lowered faster than the tunneling rate in the renormalization process and the latter will tend to a finite value. In the opposite case the tunneling rate will flow to zero and localization of the phase sets in. Therefore the localization in one well takes place if there is a frequency ω_t such that for all $\omega < \omega_t$ the inequality holds:

$$\frac{d\ln|\mathcal{T}|}{d\ln\omega} = \Re\{Y(\omega)\}R_Q > 1$$

In the ohmic case the resistance R has to be less then the quantum resistance $R_Q = h/4e^2$.

Going back to a junction, it can be shown that, due to dissipation, the interaction between phase slips becomes long ranged, so that the dilute gas approximation breaks down [17]. However, the main result of the previous argument is not affected.

PHASE COHERENCE IN JOSEPHSON JUNCTION ARRAYS

After having discussed some of the properties of a single junction, the rest of this chapter will be entirely devoted to the physics of Josephson Junction Arrays (JJA). They are constituted by superconducting islands connected by tunnel junctions and placed at the lattice points of a particular two dimensional lattice. In the past ten years, thanks to the advanced lithographic techniques, it has been possible to realize experimentally arrays which may include several millions of Josephson junctions (e.g. 2000x2000). The junction parameters can be varied in a controlled way, so that, JJA offer an unique possibility to study the effect of classical and quantum phase fluctuations on the onset of superconducting phase coherence.

In a sense, effects which should be present in a single junction but are difficult to detect, are enhanced in an array, because arrays magnify the scale of the voltage and the current in the I/V characteristic. Furthermore, because each junction is surrounded by other high resistance junctions it is well decoupled from the environmental circuit. This makes charging effects visible in JJA. Moreover, what makes JJA appealing with respect to granular superconductors or amorphous films, is that in Josephson arrays disorder plays a minor role, unless it is explicitely desired.

It is impossible to cover the whole subject here, which is extended and still rapidly growing. Instead, we will content ourselves in pointing out the role of fluctuations in these systems. The interested reader can find additional information in ref. [2, 21].

In JJA it is important to make a distinction between local and global superconductivity. Cooling down the sample , each island of the array goes superconducting at a critical temperature T_{co}. However, dissipationless conduction requires phase coherence across the whole system. This could set in at a much lower temperature T_c which is by definition the superconducting transition temperature.

Phase correlations in classical arrays

Let us first consider a classical array, i.e. an array in which the dimensions of the islands are large enough, so that the charging energy can be disregarded. If we are interested in the global phase coherence ($T \ll T_{co}$), the modulus of the order parameter is pinned to its mean field (BCS) value. The only relevant dynamical variable is its phase. The Hamiltonian governing the dynamics of the system is:

$$H = -E_J \sum_{<i,j>} \cos \varphi_{ij} \tag{20}$$

Here $\varphi_{ij} = \theta_i - \theta_j$ is the phase difference between two neighboring sites. This model Hamiltonian can also be used to describe the properties of a granular film: in this case the index i labels a region in which the phase coherence is already established. For granular films, however, the additional effect of the disorder should be included. We will assume that E_J is a coupling constant that can be varied arbitrarily, throughout the discussion. In reality the Josephson coupling itself depends on the temperature according to the Ambegaokar-Baratoff formula [16].

The model Hamiltonian of eq.(20) is well known in statistical mechanics as the XY model[22]. In two dimension it has been shown that it does not possess long-range order. The order parameter $\langle \cos(\theta_i) \rangle$ is zero at any temperature different from zero. At high temperatures the correlations decay exponentially with the distance on the lattice. At very low temperatures, instead, the so called spin waves are responsable for the absence of long range order. Infact, one might imagine that the relevant configurations are those in which the phase differs slightly between neighbouring sites. If only these configurations are taken into account (spin waves), the Hamiltonian of eq. (20) can be approximated quadratically leading to a phase-phase correlation with a power-law decay:

$$\langle e^{i\theta_i} e^{-i\theta_j} \rangle \sim (|i - j|)^{-1/2\pi\beta E_J} \qquad (\beta \to \infty) \ . \tag{21}$$

A phase transition occurs at some temperature T_{KTB} in between named after Kosterlitz,Thouless and Berezinskii [23].

Infact, in JJA, the phase transition is driven by an other type of excitations which are called vortices. Actually, by assuming that only small angle configurations are important, we disregarded those configurations where the phases wind by 2π around a closed path. A typical phase configuration for a vortex centered at $\vec{r}$ is:

$$\theta_i = \Theta(\vec{r}_i - \vec{r}) = \arctan\left(\frac{y_i - y}{x_i - x}\right) \tag{22}$$

This is such that, on a closed path around $\vec{r}$, is $\int \vec{\nabla}\theta \cdot d\vec{l} = \pm 2\pi$. The core of the vortex is located within a lattice cell and, because the screening is very small (in an array the Josephson screening length is $\lambda_J \sim L$), its size extends all over the system. This last condition implies that when a magnetic field is applied perpendicularly to the sample, it penetrates (for all practical pourposes) uniformly.

Classically vortices are characterized by their charge $v_i = \pm 1$ and position in the lattice at site i. The energy of a neutral configuration of N vortices can be evaluated to give [24]

$$E_v = \pi E_J \sum_{(i,j)} v_i v_j \ln r_{ij}/a \ , \tag{23}$$

that is they interact logarithmically with the distance (a is the lattice spacing). We considered only neutral configurations because in the thermodynamical limit only these configurations have a finite energy while the energy of *unbalanced* vortices diverges.

Therefore, if the flux threading the sample is zero then the total vortex "charge" is zero and only vortex-antivortex pairs which are thermally excited are present in the system. The transition is driven by the unbinding of the pairs at the critical temperature $T_{KTB} = (2/\pi)E_J$. Above T_{KTB} unbounding of vortices enhance the fluctuations of the phase and the correlations decay exponentially with the distance. On the contrary, below T_{KTB}, vortices are bound in pairs and are irrelevant for the long range correlation of the system (see Fig. 3). As a result the phase-phase correlation is dominated by the spinwaves and decays as a power law. A simple free energy argument can easily explain the existence of the transition itself. The energy of an isolated vortex in an array of linear dimensions L is proportional to the logarithm of the system size. The free energy is:

$$F = U - TS = \pi E_J \log(L) - K_B T \log(L^2) \ .$$

The entropic term comes from the probability to put the vortex in any of the L^2 cells of the array. The free energy becomes negative at the critical temperature, indicating that vortices above T_{KTB} proliferate.

The KTB mechanism has important consequences also for the dynamical properties. Above T_{KTB} the vortices are free to move. By applying a current $\vec{I}$ to the system, the vortices experience a Lorentz force perpendicular to the direction of the current.

Thus, when an infinitesimal current is applied, vortices cross the sample generating a voltage drop (flux flow resistance), proportional to the number of free (unbound) vortices

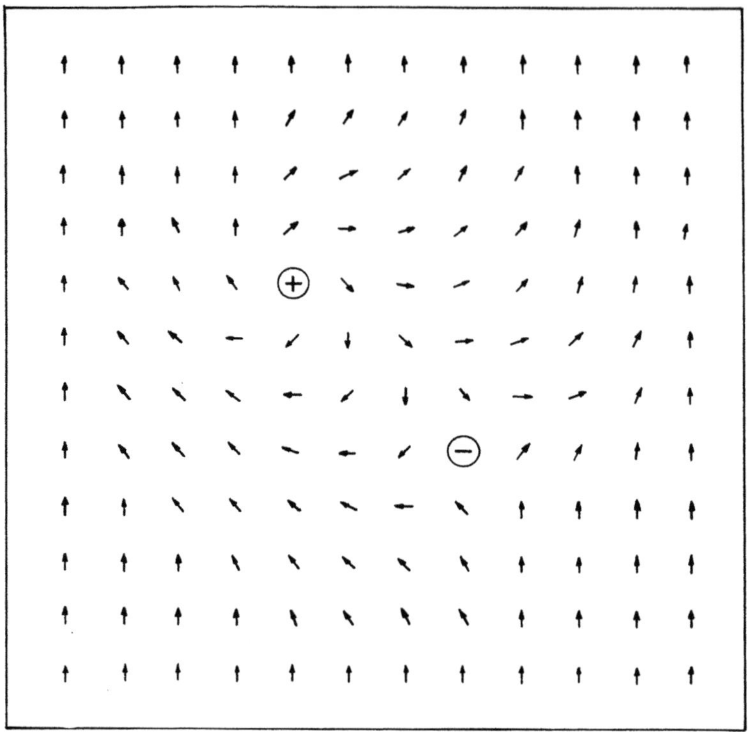

Figure 3: Phase configuration including a vortex- antivortex pair.

present in the system. The density of such vortices is of the order of the inverse correlation length and leads to a peculiar temperature dependence of the differential resistance, i.e. $R(T) = R_n \exp\left\{-1.2/\sqrt{(T/T_{KBT}) - 1}\right\}$. However, this is difficult to observe experimentally, because of the strong divergence of the correlation length, so that finite size effects are unavoidable, close to the transition temperature.

Below T_{KTB} the current should be strong enough to break up a vortex-antivortex pair, the differential resistance is zero and the I/V characteristic goes like $I \sim V^\eta$ where η is related to the superfluid density and has a jump at the transition.

If a magnetic field is applied perpendicularly to the system, the Hamiltonian of eq.(20) is modified in order to be gauge invariant, by means of the substitution $\varphi_{ij} \longrightarrow \varphi_{ij} - A_{ij}$, where A_{ij} is the line integral of the vector potential between the two neighbouring islands. Because the penetration length is much larger than typical system sizes, it is safe to assume that the field penetrates the array uniformly. The flux piercing an elementary plaquette is f in units of the flux quantum $\Phi_0 = hc/2e$. In the presence of an external magnetic field a finite number of vortices of one sign will be induced into the system. For some special values of the frustration f, they will tend to form a uniform superlattice. Because of the presence of the underlying Josephson lattice, they will be pinned in their positions and therefore no dissipation will take place. The phase transition in the presence of a magnetic field is still not fully understood.

Quantum arrays

In the presence of charging the model Hamiltonian of eq.(20) has to be supplemented with a kinetic energy:

$$H = \frac{1}{2}\sum_{ij}(Q_i - Q_{x,i})C_{ij}^{-1}(Q_j - Q_{x,j}) - E_J\sum_{<ij>}\cos(\theta_i - \theta_j),\qquad(24)$$

where Q_i is the charge on the i-th island. Charge and phase on equal sites do not commute: $[\theta_j, Q_i] = 2ei\,\delta_{ij}$. The external charges $Q_{x,i}$ are also included, which could be induced by a gate voltage on the islands, or by trapped impurities in the substrate. In the following we will ignore the tunneling of quasiparticles. They only become important at energies of the order of the superconducting gap which we are assuming to be located far up with respect to the typical energy scales of the problem. These are the Josephson coupling E_J and the charging energy $E_C = e^2/2C$ or $E_0 = e^2/2C_0$. C_0 and C are the ground capacitance and the nearest neighbor capacitance, respectively, that constitute the capacitance matrix C_{ij}. At large distances the capacitance to the ground C_0 dominates. The spatial range of the electrostatic interaction between Cooper pairs, is $\sqrt{C/C_0}$. Various features, i.e. the phase diagram and the dissipation mechanisms, depend in a crucial way on this quantity.

The partition function of this model can be cast in terms of the topological excitations of the system. These are the discrete charges q_i and vortices v_i, which form two coupled Coulomb gases [25]. Here we review the most characteristic features of the phase diagram for E_J larger than the charging energies and zero external charges, resorting to simpler approximations.

First of all it is important to stress that while in a single Josephson junction quantum fluctuations destroy the coherence at zero temperature in the case of an array the cooperative effects guarantee a range of parameters in which global superconductive coherence persists at zero temperatures. In the large E_J limit and at very low temperatures, there is phase coherence through the system and the low lying excitations can be expressed in terms of the fluctuation of the phase of wavevector k and frequency ω within the harmonic approximation (spin waves). Depending on the size of the system, the excitation spectrum can be either acustic or optical like.

In the limit of on-site interactions ($C_0 >> C$) the spinwave dispersion is acustical: $\omega_k = \bar{\omega}_p k$. Here $\bar{\omega}_p = \sqrt{8E_J E_0}$ is the plasma frequency in terms of the on site capacitance. In the case of the Coulomb interactions being long range ($C >> C_0$) the spectrum is optical, $\omega_k = \omega_p$.

At finite temperatures the phase-phase correlator goes as a power law of the distance as in the classical case, but with a renormalized Josephson energy so that the exponent appearing in eq.(21) becomes $1/[2\pi\beta E_J(1 - \beta E_0/24)]$. Thus the tendency of the system to order is weakened by the charging and the transition temperature is pushed down to lower values. The transition, however, is still of the KTB type.

On the contrary, at zero temperature, the imaginary time variable plays the role of an extra dimension, because it extends to infinity. This property is known as dimensional crossover. As a consequence, the order parameter is non vanishing, at difference with the finite temperature case.

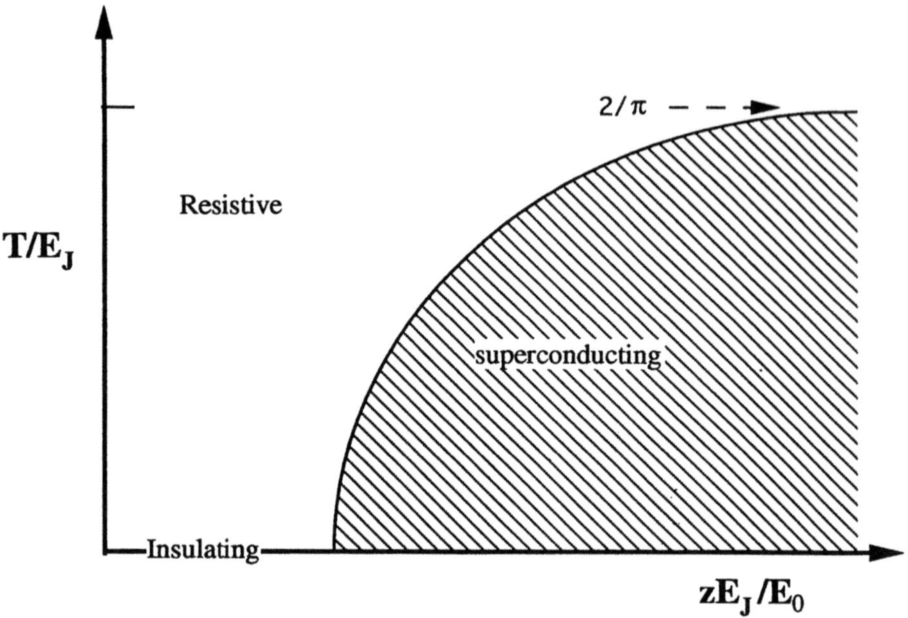

Figure 4: The phase diagram of a JJA in the case of short range Coulomb interaction (with no applied magnetic field).

At large distances the correlator in this approximation goes as:

$$\langle e^{i\theta_i(\tau)}e^{-i\theta_j(\tau)}\rangle \sim \exp\left(-K\sqrt{\frac{8E_0}{E_J}}\,\frac{1}{|i-j|}\right) \qquad (|i-j|\to\infty,\ T=0), \qquad (25)$$

where K is a numerical factor.

As the ratio E_0/E_J increases above a certain threshold, the transition temperature reaches the absolute zero and beyond that value the JJA is incoherent at any temperature. Qualitatively the phase diagram can be obtained by a mean field approach although it is only well founded at $T=0$, when long range order is established in the system. At finite T it is conceptually misleading, because thermal fluctuations always destroy the order parameter. Nevertheless its results are qualitatively reliable.

The mean field approximation consists in approximating the Hamiltonian of eq.(24) in the following way:

$$H_{MF} = \frac{1}{2}\sum_{i,j} Q_i C_{ij}^{-1} Q_j - \frac{z}{2}E_J\langle\cos(\theta)\rangle\sum_j \cos(\theta_j)$$

where z is the coordination number in the lattice and $\phi \equiv \langle\cos(\theta)\rangle$ is the order parameter which, in turn, is calculated according to $\langle\cos(\theta)\rangle = Tr\left\{\cos(\theta_i)\exp(-\beta H_{MF})\right\}$. Close to the transition point, the thermal average on the r.h.s can be approximated by expanding in powers of ϕ. To third order, a Ginzburg-Landau type equation arises:

$$\left(1 - \frac{1}{2}zE_J\int_0^\beta d\tau\langle\cos\theta_i(\tau)\cos\theta_i(0)\rangle_{ch}\right)\phi + \left(\frac{zE_J}{E_0}\right)^3 \mathcal{B}\,\phi^3 = 0 \qquad (26)$$

Here the average $< ... >_{ch}$ is performed over the eigenstates of the charging part of the Hamiltonian only and the quantity $\mathcal{B}$ entails the four point phase correlation. The changing of sign of the prefactor in the l.h.s. of eq.(26) yelds the critical temperature β_{cr}^{-1}.

Eq.(26) has a simple physical meaning. If the charging term is absent, the phase-phase correlator is one and we recover the classical result $\beta_{cr} z E_J = 2$. Due to the charging the phase starts to fluctuate and the critical temperature is depressed. At $T = 0$ the correlator is easy to evaluate (considering for simplicity the on-site case):

$$\langle \cos \theta_i(\tau) \cos \theta_i(0) \rangle_{ch} = \exp\{-E_0 \tau (1 - \tau/\beta)\} \qquad (T = 0) ,$$

thus giving the critical point at $z E_J / E_0 = 2$. For larger values of the charging energy the array is incoherent even at zero temperature. The full phase diagram is shown in Fig.4.

Similar types of mean field approaches were used to study the effect of frustration in these systems. It is worthwhile to stress that in the case of quantum systems, it is possible to introduce frustration both by means of a magnetic field (as in the classical case) and by imposing an external charge distribution. In both cases the critical point can be modulated by tuning these external fields. The effect of fluctuation beyond the mean field as well as the effect of dissipation has been studied, for instance in ref. [27].

In the disordered phase, at zero temperature, there is a gap in the excitation spectrum and no overall coherence. Therefore the Josephson array is a perfect insulator with infinite resistance! Tuning the Josephson coupling, the array is driven through a *Superconductor-Insulator* transition. This is infact what is found experimentally as shown in Fig. 1[6]. It was proposed theoretically [28] that at the critical point the system is metallic (i.e. shows a finite conductivity) and that the value of this critical resistance is universal, $R_{cr} \sim R_Q = h/4e^2$. Both theoretical and experimental [6] [7] work seems to support this result.

Quantum vortices

Let us now consider an array in the superconducting phase, at very low temperature. As mentioned already, vortices can be induced artificially, by means of a magnetic field. If the applied magnetic field is very small, very few vortices are introduced in the system and the interaction can be disregarded. When an external driving current is applied to the array, the vortices move orthogonally to it and a voltage drop is produced across the sample.

We assume that the vortex can be treated, in some sense, as a rigid body. This is strictly related to the limit in which the Josephson coupling E_J is much larger than the charging energy E_C (and, of course, larger than the temperature T).

On the other hand the charging energy provides the chance for a dynamics of the vortex. This is characterized by two quantities, inertia and viscosity, as for macroscopic bodies. In classical JJA's, because the charging energy is very small, only this second quantity is relevant. The vortex motion is diffusive and the equation of motion is simply obtained by balancing the driving force with the dissipative one, which is proportional to the velocity. In quantum arrays, where the charging effects are important, vortex dynamics can be ballistic and the possibility opens of studying macroscopic quantum properties. Infact, the electrostatic energy stored in the junction capacitances due to the vortex motion

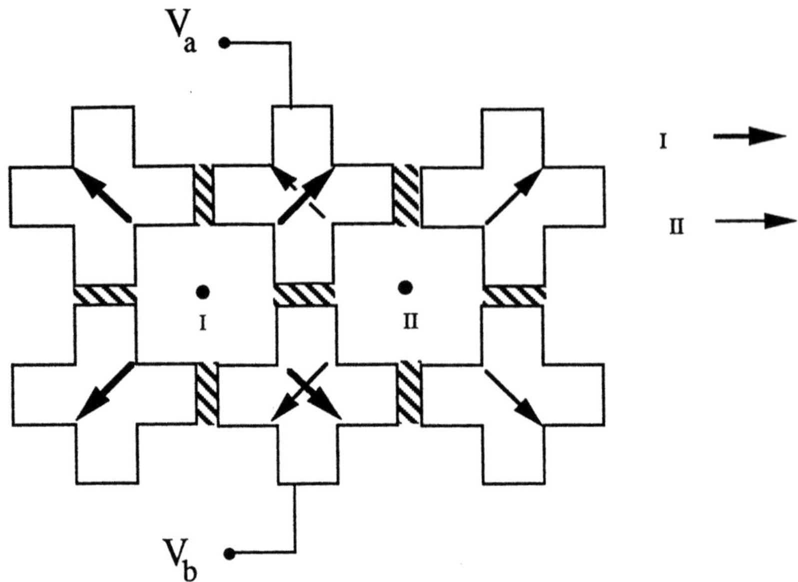

Figure 5: The distribution of the phases for a vortex in two different positions. The phase difference of the junction left behind in the vortex motion has changed by π.

can be interpreted as a kinetic energy and, as a consequence, a mass can be attributed to the vortex [29]. The drawback is that the charging energy is also responsable for mechanisms of dissipation as discussed in the next.

Let's consider a very easy argument to convince ourselves that this inertia exists indeed. A vortex motion involves the collective evolution of the phases in each island of the array. As it appears from Fig.5, if the vortex center has moved one lattice constant forward in a given time δt, the phase difference of the junction left behind has changed by π. The associated charging energy (expressed in terms of the potential drop), $E_{ch} = \frac{1}{2}\sum_{ij} V_i C_{ij} V_j$, can be evaluated using the Josephson relation and assuming that the most important contribution comes from that particular junction (what indeed happens if the junction capacitance is dominant):

$$E \sim \frac{\hbar^2 C}{4e^2 a^2}\left(\frac{\delta\varphi}{\delta t}\right)^2 \sim \frac{\pi^2\hbar^2 C}{4e^2 a^2}\dot{r}^2 = \frac{1}{2}M_{ES}\ \dot{r}^2$$

where $\dot{r}$ is the vortex velocity and a is the lattice spacing. The last equality defines $M_{ES} = \hbar^2\pi^2/4E_C a^2$ as the mass.

By applying an external current the dynamics can be studied. While moving in the JJA , vortices experience a potential due to the underlying periodicity of this artificial two dimensional space, in addition to the Lorentz force. In order for a vortex to move from one plaquette to the other it must overcome and energy barrier that has been extimated in $U_{bar} \approx 0.2E_J$ [30].

When $E_J \gg E_C$ the moving vortex can be described by a phenomenological equation of motion for the vortex center coordinate $r(t) = (x(t), 0)$ to be inserted in eq.(22)[31]. If the underlying conjectures are correct , the vortex can be treated as a macroscopic particle, in spite of the fact that it is an object which involves thousands of superconducting islands.

The strongest requirement for a vortex to be a macroscopic quantum particle, is that its interaction with the environment is non destructive. Nowadays it is possible to fabricate high quality JJA's in which ohmic dissipation is neglegibly small. If the vortex velocities are below a certain threshold (related to the superconducting gap), the quasiparticles are also frozen and do not contribute to the damping. However, the mechanism by which a classical moving vortex can loose its kinetic energy is by interacting with other vortices that are present in the system, but mainly , with quantum fluctuations of charges, that is by emitting spinwaves [32]. The dissipative term in the action cannot be expressed as a simple quadratic form as it happens in the Caldeira-Leggett model, because the damping is non ohmic and moreover it is a highly nonlinear function of the vortex velocity.

The outcome of the analysis is the following. In the limit of on-site interactions ($C_0 >>$ C), when the spinwave dispersion is acustical, the vortex experiences a Cherenkov- like effect: it emits energy quanta, if its velocity is larger than the velocity of the spinwaves.

In the case of the Coulomb interactions being long range ($C >> C_0$) the spectrum is optical, $\omega_k = \omega_p$, and the motion of the vortex is ballistic only for a velocity $v < v_{cr} = \omega_p/\pi$, which is lower than the optical threshold, due to the nonlinear coupling to the bath. However, the minimum velocity that a vortex needs in order to move over the pinning potential of the lattice is about 0.1 ω_p, which is close to v_{cr} , so that no ballistic window would be available. Fortunately, the critical velocity is even higher than the value reported above, due to the fact that the optical branch is not flat. This is a very happy chance, because it allows for a range of velocity values in which ballistic motion is possible, also in the square array case. In triangular arrays, however, a similar analysis yields a somewhat wider velocity window. In any case, the excitation modes which dissipate most of the kinetic energy of the moving vortex are the short wavelength ones, and these are made stiffer by an increase of the charging energy. This implies that the more the vortex is quantum, the more it is effectively decoupled from its environment and ballistic.

Very recently, in an important experiment, van der Zant et al. [34] have shown that such ballistic vortices indeed exist at temperatures of the order of $m°K$. In the experiment vortices were shooted in a force free region and their motion was observed. A series of voltage detectors were conceived in order to see the point where the vortex eventually comes out of the array. In the experiment, the vortex mass was found to be somewhat larger than the theoretical prediction: $M_v \sim 10^{-33} Kg$, approximatively 10^{-3} smaller than the electron mass!!

Aharonov-Casher effect

Until now we have assumed that no external charges $Q_{x,i}$, appearing in the Hamiltonian of eq. (24) are fed on the array. In particular geometries, their inclusion could lead to even more spectacular consequences which rest on the hypotetical analogy between a vortex and a macroscopic quantum particle. Here we assume that the external charges can be tuned continuosly, while the gas associated to the Cooper pairs has discrete charges Q_i. This fact has the consequence that the external charges can be screened only up to modulus $2e$ so that they cannot be washed out by charge fluctuations. This is at striking difference with classical arrays, or with arrays in which the junctions are resistively shunted.

Some time after the Aharonov-Bohm effect was predicted, Aharonov and Casher [35] (AC) suggested that an analogous interference should be observed when a neutral particle carrying a magnetic moment moves around a fixed charge. van Wees [36] proposed that an effect of AC type could be observed for an isolated vortex moving in a JJA around a charged island and interfering with itself. Although there are similarities with the *conventional* AC effect, there are important differences. In this case vortices do not carry a flux, unlike the Abrikosov vortices in a bulk superconductor. Moreover, as already stressed, JJA's form an artificial 2D space.

It can be shown rigorously [21] that, while multiples of $2e$ do not contribute to the action, the pseudo-charges $q_{x,i} = mod(Q_{x,i}, 2e)$ located on the islands interact with the moving vortex (with velocity $\dot{\vec{r}}$) giving rise to the following term in the action:

$$\mathcal{S}_{AC} = -i \sum_i \frac{q_{x,i}}{2e} \int_0^\beta d\tau \dot{\vec{r}} \cdot \vec{\nabla}\Theta\left(\vec{r}_i - \vec{r}(\tau)\right) \tag{27}$$

Eq.(27) defines a pseudo-charge gauge field $2e\vec{A}(\vec{r}) = q_{x,i}\hat{z} \times \vec{r}/r^2$, which is singular at the location of the charge. Thus the external charges act like a vector potential for the vortex.

It could be asked, what is the phase accumulated by the vortex propagator in real time, when it moves around the charge along a closed loop in a time T. The phase factor implied by eq.(27) is $\exp(i\chi)$, with:

$$\chi = \frac{1}{\hbar c} \int_0^T dt\, \dot{\vec{r}} \cdot \int d^2R\, \rho(R)\vec{A}(\vec{R} - \vec{r}) = \frac{1}{2e} \int d^2R\, \rho(R) \oint_\Gamma d\vec{r} \cdot \left(\vec{\nabla}_r \ln|\vec{R} - \vec{r}| \times \hat{z}\right)$$

$$= -\frac{1}{2e} \int d^2R\, \rho(R) \int d^2r \nabla_r^2 \ln|\vec{R} - \vec{r}| = 2\pi \sum_{i \subset \Gamma} Q_{x,i}/2e$$

where the sum extends to all islands enclosed by the trajectory Γ and the property has been exploited: $\nabla_r^2 \ln|r| = -2\pi\delta^{(2)}(r)$.

It follows that the energy levels are $2e -$ periodic functions of Q_x . Therefore a persistent voltage $V = dE_0/dQ_x$ would be associated to an array of the form of a Corbino disk, with the charge Q_x placed at the center island, in the presence of a vortex. The array is bounded by two superconducting rings. The vortex can be induced in the array also by twisting the boundary conditions for the order parameter phase by 2π around the inner ring: because it is repelled by the superconducting boundaries, it will be centered half way from the center of the disk.

One more experimental setup considers the analogue of the Young's experiment, providing the interference of a ballistic vortex set in motion by an external current and passing by a charged island along two different paths. Again a voltage should be detected at the two superconducting contacts, which is an oscillating function of the external charge. Features in the I/V characteristic of this device have been measured, in the flux flow regime. Oscillations of the dynamical resistance appear when the charge on the central island is changed [37].

ACKNOWLEDGMENTS: We would like to thank C. Bruder, G. Falci, G. Giaquinta, A. van Otterlo, G. Schön, E.Tosatti and F.Ventriglia for collaboration and acknowledge useful discussions with A.Barone, R.Cristiano, W. Elion, L.J. Geerligs, U.Geigenmüller, A.J.Leggett, J.E. Mooij and H. van der Zant.

318

References

[1] M.Tinkham, *Introduction to superconductivity*, (McGraw - Hill Inc. , New York , 1975)

[2] *Percolation, Localization and Superconductivity*, A.M. Goldman and S.A. Wolf Eds.NATO ASI Series B, vol.109 (Plenum Press, New York 1984)

[3] *Single Charge tunneling* H. Grabert and M.H. Devoret Eds., NATO ASI seires vol.294 (Plenum,NY 1992)

[4] J.Bardeen, L.N.Cooper, J.R.Schrieffer, Phys. Rev. **108**, 1175 (1957)

[5] A.Barone, G.Paterno', *Physics and applications of the Josephson effect*, (Wiley, New York, 1982)

[6] L.J. Geerligs, M. Peters, L.E.M. de Groot, A. Verbruggen and J.E. Mooij, Phys. Rev. Lett **63**, 326 (1989). In Fig.1 R_{sq} is the resistance divided by the length/width ratio 3.14. Each solid curve corresponds to an array in zero field with the following values of $R_n(k\Omega)$, $E_J/k_B(^\circ K)$, and E_J/E_C: sample $A : 36, 0.22, 3.9$; $B : 15.3, 0.51, 1.8$; $C : 14.1, 0.55, 1.5$; $D : 9.7, 0.80, 1.0$; $E : 4.8, 1.6, 0.53$. The dashed curve is for array D in magnetic field ($f = 1/2$).

[7] B.G. Orr, H.M. Jaeger, A.M. Goldman and C.G. Kuper, Phys. Rev. Lett **56**, 378 (1986); H.M.Jaeger, D.B.Haviland, B.G.Orr and A.M.Goldman, Phys.Rev.**B40**,182 (1989)

[8] C.N.Yang, Rev. Mod. Phys. **34**,694 (1962)

[9] G.Giaquinta, N.A.Mancini, Rivista del Nuovo Cimento, **1**, 1 (1978); A. Tagliacozzo and F. Ventriglia, Il Nuovo Cimento, **11D**, 141 (1989);

[10] R.Cristiano, F.Frunzio, R.Monaco, C.Nappi and S.Pagano, Phys.Rev.**B49**,429 (1994)

[11] B.D.Josephson, Phys. Lett. **1**, 251 (1962)

[12] R.Kautz and J.M.Martinis Phys. Rev. **B42**,9903 (1990)

[13] R.F.Voss and R.A.Webb, Phys.Rev.Lett.**47**,265 (1981); M.H.Devoret,J.M.Martinis and J.Clarke, Phys.Rev.Lett. **18**, 1543 (1985) and A.N.Cleland, J.M.Martinis and J.Clarke, Phys.Rev.**B37**, 5950 (1988)

[14] V.Ambegaokar, U.Eckern, G.Schön, Phys. Rev. Lett. **48**,1745, (1982) ; U.Eckern, G.Schön, V.Ambegaokar, Phys. Rev. **B30**,6419, 1984 and V.Ambegaokar, in ref.[2] p.43

[15] P.W.Anderson, in *Lectures on the many body problem*, ed. by E.R.Caianiello, (Academic Press 1964),vol.2, p.113

[16] V.Ambegaokar, A.Baratoff, Phys. Rev. Lett. **10**,486,(1963)

[17] G.Schön and A.D.Zaikin, Phys.Rep. **198**,237 (1990)

[18] A.Tagliacozzo and F.Ventriglia, in *"Microscopic Aspects of Nonlinearity in Condensed Matter "* ed. by A.R.Bishop et al. (Plenum Press, New York 1991)

[19] A.Caldeira, A.J.Leggett, Ann. Phys. (NY) **149**, 374 (1983)

[20] A.J.Leggett, S.Chakravarty, A.T.Dorsey, M.P.A.Fisher, A.Garg, W.Zwerger, Rev. Mod. Phys. **59**,1 (1987)

[21] J.E. Mooij and G. Schön in ref.[3] p.275

[22] C.Itzykson and J.M.Drouffe, *Statistical Field Theory*, Cambridge University Press (1989)

[23] V.L. Berezinskii, Sov. Phys. JETP **32**, 493, (1971); J.M. Kosterlitz and D.J. Thouless, J.Phys. **C6**, 1181 (1973)

[24] J. Villain, J. Phys. **36**, 581 (1975); J.V. José et al. Phys. Rev. **B16**, 1217 (1977) and J.E. Mooij, in ref.[2], p.325

[25] R. Fazio and G. Schön, Phys. Rev. **B43**, 5307 (1991)

[26] C. Bruder, R.Fazio, A.Kampf, A.van Otterlo and G.Schön, Phys. Scr. **T42**, 159 (1992)

[27] A. Kampf and G. Schön,Phys. Rev. **B36**, 3651 (1987);S. Chakravarty,G.L. Ingold, S.Kivelson and G.T. Zimanyi, Phys. Rev. **B37**, 3283 (1988);G. Falci, R. Fazio and G. Giaquinta, Europhys. Lett. **14**, 145 (1991)

[28] M.P.A. Fisher, G. Grinstein an S.M. Girvin Phys. Rev. Lett **64**, 587 (1990); M.C. Cha, M.P.A. Fisher, S.M. Girvin, M. Wallin and A.P. Young, Phys. Rev. **B44**, 6883 (1991)

[29] E. Simanek, Solid State Comm. **48**, 1023 (1983); U. Eckern and A. Schmid, Phys. Rev. **B10**, 6441 (1989)

[30] C.J. Lobb, D.W. Abraham and M. Tinkham, Phys. Rev. B **27**, 250 (1983)

[31] T.P. Orlando, J.E. Mooij, and H.S.J. van der Zant, Phys. Rev. **B43**, 10218 (1991)

[32] U. Geigenmüller, C.J. Lobb and C.B. Whan, Phys. Rev. B **47**, 348 (1993); U. Eckern and E. Sonin, Phys. Rev. B **47**, 505 (1993)

[33] R. Fazio, A. van Otterlo, G. Schön, Europhys.Lett. **25** 453 (1994)

[34] H.S.J. van der Zant, F.C Fritschy, T.P Orlando and J.E. Mooij, Europhys. Lett. **18**, 343 (1992)

[35] Y. Aharonov and A. Casher, Phys. Rev. Lett. **53**, 319 (1984)

[36] B.J. van Wees,Phys. Rev. Lett. **65**, 255 (1990)

[37] W.J. Elion, J.J Wachters, L.L. Sohn and J.E. Mooij, Phys. Rev. Lett. **71**, 2311 (1993)

SPONTANEOUS LOCALIZATION AND SUPERCONDUCTIVITY[♡][◇]

Alberto Rimini

Dipartimento di Fisica Nucleare e Teorica
Università degli Studi di Pavia
27100 Pavia, Italy

1. INTRODUCTION

The Schrödinger equation does not include the stochastic features which emerge in quantum mechanics during measurements and which are incorporated, in the ordinary presentation, into the reduction principle. In other words, the Schrödinger equation does not describe the particularity of the world we experience. If one is not satisfied with the ordinary presentation because of the dualism between the Schrödinger equation and the reduction principle and of the related ambiguous dualism between quantum systems and measuring apparatuses, a possible reaction is to declare that the Schrödinger wave function describes statistical ensembles rather than individual systems, and to resort to the practical impossibility of detecting interference among the different macroscopically distinguishable terms appearing in the wave function (this impossibility is often referred to as decoherence). This interpretative attitude works to some extent, but it cannot avoid certain typical inconsistencies[1,2] arising from the pretension to renounce in principle to any descriptive tool related to the result of an individual measurement. In this situation, one is induced to believe that any fully consistent theory of measurement must necessarily be based on a formulation of quantum mechanics such that the outcome of an individual measurement has a counterpart in the description of the system after the measurement (by system we mean here the quantum system plus the apparatus plus whatever else may be relevant). There are at least three types of formulations of quantum mechanics which satisfy this requirement. The first type introduces additional variables, like Bohm's pilot-wave theory.[3] That this theory allows a consistent description of measurement was first claimed by Bell.[4,5] The element of description of the system which corresponds to the outcome of a measurement is the configuration. The most refined formulation of the second type is the history approach,[6-8] in which the element of description distinguishing among the results is the indication of the term in the wave function corresponding to the specific outcome emerging in each measurement (or,

---

[♡]   Based on work done in collaboration with M. Buffa and O. Nicrosini.

[◇]   Supported in part by Istituto Nazionale di Fisica Nucleare and Ministero dell'Università e della Ricerca Scientifica e Tecnologica.

more generally, in each measurement situation). Let us call such an indication the branch label. The history approach derives in the long run from Everett's so called many–world formulation of quantum mechanics[9] supplemented by Zurek's remark[10] on the necessity of introducing a criterion to identify the branching basis. However, the use of the branch label as an element of description of the system is more explicit in the former theory than in the latter. The status of the branch label in the history approach to quantum mechanics is somewhat peculiar, because it is explicitly used as an element of description of the system i. e. as a part of the state, and yet only the wave function is introduced into the formalism from the outset, the branch label emerging later as a consequence of decoherence. Whether this is satisfactory or not can be the subject of an endless discussion.

The two formulations of quantum mechanics mentioned above have the common feature that some element of description, the configuration or the branch label, is added to the wave function. The third possibility is to insist that the wave function is the sole element of description of the system and, correspondingly, to alter the Schrödinger equation so that reduction is decribed as a real physical process affecting the wave function. The alteration must be ruled by definite, universally valid, precise assumptions. It must have practically unobservable consequences in all ordinary situations and, nevertheless, it must reduce rapidly superpositions of macroscopically distinguishable states. A theory satisfying these requirements was proposed some years ago with the name of quantum mechanics with spontaneous localization[11,12] and important improvements were introduced later.[13,14] A consequence of the theory is that the internal structure of macroscopic objects is unaffected by the modification carried out on the Schrödinger equation, an essential ingredient of the proof being that the internal wave function of the macroscopic object be sharply localized with respect to a parameter of the theory whose order of magnitude is 10^{-5} cm. This condition is well satisfied in normal states but is violated in superconducting states, so that these states need a separate consideration. A qualitative estimate of the effects of spontaneous localization on superconducting states has been given by Rae[15] in the framework of the old form of the theory. It is the purpose of this work to present the results of calculations performed within the new, more satisfactory theory.

Since the original form of spontaneous localization theory allows a much more intuitive understanding of what is going on, we shall begin in the next section by expounding the theory in this form and reporting the arguments by Rae. In section 3 we shall expose the new form of the theory and discuss some points which have to be settled. In section 4 we briefly recall a simple description of superconducting states and in section 5 we apply to it the theory of section 3.

2. LOCALIZATION OF A SYSTEM OF NONIDENTICAL PARTICLES

We assume[11,12] that each particle i of a system of N distiguishable particles undergoes at random times with mean frequency λ^i a spontaneous instantaneous process and that, in the time interval between two successive such processes, the system evolves according to the Schrödinger equation. The instantaneous process is a localization defined by

$$
\begin{aligned}
|\psi\rangle \to |\psi_x^i\rangle &= |\phi_x^i\rangle / \||\phi_x^i\| , \\
|\phi_x^i\rangle &= \hat{L}_x^i |\psi\rangle,
\end{aligned}
\tag{2.1}
$$

where

$$\hat{L}_x^i = \left(\frac{\alpha}{\pi}\right)^{\frac{3}{4}} \exp\left(-\tfrac{1}{2}\alpha\left(\hat{q}^i - x\right)^2\right). \tag{2.2}$$

We further assume that the probability density for the occurrence of x when the process affecting particle i takes place is $\|\phi_x^i\|^2$, the total probability being 1 since

$$\int d^3 x \left(\hat{L}_x^i\right)^2 = 1. \tag{2.3}$$

The length $1/\sqrt{\alpha}$ measures the sharpness of the localization and it is meant to be small on a macroscopic scale but large with respect atomic distances, say

$$1/\sqrt{\alpha} \approx 10^{-5} \text{ cm}. \tag{2.4}$$

We now study the consequences of the above assumptions. Consider first a system consisting of a single particle. Let the wave function be

$$\psi(qs) = \psi^{(1)}(qs) + \psi^{(2)}(qs), \tag{2.5}$$

where each term $\psi^{(i)}$ is localized around the point $\overline{x}_i$ within a distance smaller or much smaller than $1/\sqrt{\alpha}$ and, on the contrary, the distance $|\overline{x}_1 - \overline{x}_2|$ is much larger than $1/\sqrt{\alpha}$. Then $\hat{L}_x\psi \neq 0$ only for x around $\overline{x}_1$ or $\overline{x}_2$ and ψ is reduced by the process to $\psi^{(1)}$ or $\psi^{(2)}$ with probabilty $\|\psi^{(1)}\|^2$ or $\|\psi^{(2)}\|^2$, respectively. Of course, the idea is that this happens with a frequency λ extremely small. Consider next a system of N independent particles, e. g. a splitted beam. Then the wave function is

$$\psi(qs) = \psi_1(q_1 s_1)\ldots\psi_i(q_i s_i)\ldots\psi_N(q_N s_N), \tag{2.6}$$

where each factor ψ_i has the structure (2.5). The effect on the whole system of the localization of one particle, say i, is

$$\hat{L}_x^i\psi(qs) = \psi_1(q_1 s_1)\ldots\hat{L}_x^i\psi_i(q_i s_i)\ldots\psi_N(q_N s_N), \tag{2.7}$$

so that nothing happens to the remaining particles. If τ is the time of flight, the net result of the localization process on all particles is simply that the extremely small fraction of particles

$$\tau\frac{1}{N}\sum_i^N \lambda^i \tag{2.8}$$

is lost to interference. Consider now a system of N particles forming an object, e. g. an atom or a pointer. Then the wave function of the system can be written

$$\psi(qs) = \Psi(Q)\,\chi(rs), \tag{2.9}$$

where Ψ is the center–of–mass wave function and χ is the structural wave function.*
We further assume that the structural wave function $\chi(rs)$ be sharply (with respect to $1/\sqrt{\alpha}$) peaked around $r = r_0$. Then, putting

$$\tilde{q}_i(r) = q_i - Q, \tag{2.10}$$

---

* We assume for simplicity that the center–of–mass position be the only collective degree of freedom. Including other degrees of freedom, such as the orientation, among the collective ones would make things much more complicated without gaining physical insight.

the localization of particle i has the effect

$$
\begin{aligned}
\hat{L}_x^i \psi(q,s) &= \left(\frac{\alpha}{\pi}\right)^{\frac{3}{4}} \exp\left(-\tfrac{1}{2}\alpha\left(q_i - x\right)^2\right) \Psi(Q)\,\chi(r,s) \\
&= \left(\frac{\alpha}{\pi}\right)^{\frac{3}{4}} \exp\left(-\tfrac{1}{2}\alpha\left(Q + \tilde{q}_i(r) - x\right)^2\right) \Psi(Q)\,\chi(r,s) \\
&\simeq \left(\frac{\alpha}{\pi}\right)^{\frac{3}{4}} \exp\left(-\tfrac{1}{2}\alpha\left(Q + \tilde{q}_i(r_0) - x\right)^2\right) \Psi(Q)\,\chi(r,s) \\
&= \left(\hat{L}_{X_i}^{\text{c.m.}}\,\Psi(Q)\right)\chi(r,s),
\end{aligned}
\tag{2.11}
$$

where

$$
X_i = x - \tilde{q}_i(r_0), \qquad \hat{L}_X^{\text{c.m.}} = \left(\frac{\alpha}{\pi}\right)^{\frac{3}{4}} \exp\left(-\tfrac{1}{2}\alpha\left(\hat{Q} - X\right)^2\right).
\tag{2.12}
$$

We see from eq. (2.11) that the structural wave function is practically unaffected by the process, while, on the other hand, the localization of one single particle in our object entails the localization of the center of mass. The frequency of the localization of the center of mass is therefore

$$
\lambda^{\text{c.m.}} = \sum_{i}^{N} \lambda^i.
\tag{2.13}
$$

There follows that it is possible to choose a small value for the frequency of the process affecting the microscopic constituents and nevertheless to obtain a large value for the frequency $\lambda^{\text{c.m.}}$ for a macroscopic object. For example, in the case of a one gram object made up of protons (we disregard for the moment the fact that protons are identical), taking

$$
\lambda^i \approx 10^{-16}\,\text{s}^{-1}, \qquad 1/\lambda^i \approx (10^8 \div 10^9\,\text{years})
\tag{2.14}
$$

one gets

$$
\lambda^{(1\text{g})} \approx 10^7\,\text{s}^{-1}.
\tag{2.15}
$$

Without entering into details, we finally note that the localization process introduces some stochastic effects in the motion of macroscopic objects. It can be shown[12] that such effects are completely negligible.

As we have seen, the proof that the structural wave function is practically unaffected by the localization process is based on the assumption that such a wave function be sharply localized with respect to $1/\sqrt{\alpha}$. This assumption is certainly violated in the case of superconducting states. The problem has been first discussed by Rae,[15] who argues as follows. The wave function of the superconducting state can be written as

$$
\psi = \psi_{k_1}\psi_{k_2}\ldots\psi_{k_C}\exp[iS(r)]
\tag{2.16}
$$

where ψ_{k_j} represents the wave function of a Cooper pair with electron wave vectors $\pm k_j$ and $S(r)$ is the macroscopic phase associated with the supercurrent. Then the maximum consequence of the localization of an electron would be the break of one of the Cooper pairs. The overall rate of the process can be written $N\lambda$, where Rae uses for N the number paired electrons $\approx 10^{20}\,\text{cm}^{-3}$ and for λ the value $10^{-15}\,\text{s}^{-1}$ (somewhat larger than the value (2.14)). In this way one finds for the rate of destruction of Cooper pairs $10^5\,\text{cm}^{-3}\,\text{s}^{-1}$, which corresponds to a decay in the supercurrent well below the experimental detection limit. Moreover Rae observes

that the above argument does not take into account the possibility of recreation of
Cooper pairs which would lower the observability of the localization process even
further.

Another problem to be considered is that of the superpositions of states corresponding to two different supercurrents possibly taking place in superconducting
devices like SQUIDs. The two different supercurrents are typically separated by an
intensity of the order of 10^{-7}A and in this sense they are macroscopically different.
In this case the wave function can be written

$$\psi = \psi_{\boldsymbol{k}_1}\psi_{\boldsymbol{k}_2}\ldots\psi_{\boldsymbol{k}_C}\{\exp[iS_1(\boldsymbol{r})] + \exp[iS_2(\boldsymbol{r})]\}, \tag{2.17}$$

where the phases $S_1(\boldsymbol{r})$ and $S_2(\boldsymbol{r})$ correspond to the two different supercurrents. Rae
argues the two terms in eq. (2.17) are not located in different places, so that neither
of them is selected by the localization process. The only result is again the break up
of one of the Cooper pairs, giving rise simply to some negligible additional dissipative
effect.

The spontaneous localization theory described in this section cannot by applied
to systems of identical particles because the proper symmetry or antisymmetry of
the wave function is spoiled by the envisaged process (on the other hand, the wave
functions used by Rae, too, are not properly antisymmetrized). Another unpleasant
feature of the described theory is that the stochastic process takes place discontinu-
ously in time. Both these inconveniencies can be cured independently. In the next
section we go over directly to the continuous version for identical particles.

3. LOCALIZATION OF A SYSTEM WITH IDENTICAL PARTICLES

The spontaneous localization of a system consisting of a single kind of identical
particles can be described[13,14] through the Itô stochastic equation

$$\mathrm{d}|\psi\rangle = \left[-\frac{i}{\hbar}\hat{H}\mathrm{d}t + \sqrt{\gamma}\int\mathrm{d}^3x\,\hat{N}_\psi(\boldsymbol{x})\mathrm{d}B(\boldsymbol{x}) - \tfrac{1}{2}\gamma\int\mathrm{d}^3x\left(\hat{N}_\psi(\boldsymbol{x})\right)^2\mathrm{d}t\right]|\psi\rangle, \tag{3.1}$$

where $B(\boldsymbol{x}, t)$ is a set of Wiener processes (we drop for shortness the argument t)
such that

$$\overline{\mathrm{d}B(\boldsymbol{x})} = 0, \qquad \overline{\mathrm{d}B(\boldsymbol{x})\mathrm{d}B(\boldsymbol{x}')} = \delta^3(\boldsymbol{x} - \boldsymbol{x}')\mathrm{d}t. \tag{3.2}$$

The nonlinear operator $\hat{N}_\psi(\boldsymbol{x})$ is defined by

$$\hat{N}_\psi(\boldsymbol{x}) = \hat{N}(\boldsymbol{x}) - \langle\psi|\hat{N}(\boldsymbol{x})|\psi\rangle, \tag{3.3}$$

where $\hat{N}(\boldsymbol{x})$ is a locally averaged number density operator

$$\hat{N}(\boldsymbol{x}) = \sum_s \int \mathrm{d}^3x'\, G\left(\boldsymbol{x}' - \boldsymbol{x}\right) a^\dagger(\boldsymbol{x}', s)a(\boldsymbol{x}', s). \tag{3.4}$$

The function $G(\boldsymbol{x})$ can conveniently be taken of a Gaussian shape

$$G(\boldsymbol{x}) = \left(\frac{\alpha}{2\pi}\right)^{\frac{3}{2}}\exp\left(-\tfrac{1}{2}\alpha\,\boldsymbol{x}^2\right). \tag{3.5}$$

The stochastic process is characterized by the range $1/\sqrt{\alpha}$ of the average (3.4) and by the strength γ. The sizes of these parameters were discussed in ref. 14, and the suggested values were

$$1/\sqrt{\alpha} \approx 10^{-5}\,\text{cm}, \qquad \gamma \approx 10^{-30}\,\text{cm}^3\,\text{s}^{-1} \tag{3.6}$$

with reference to systems of protons. The equation for the statistical operator following from eq. (3.1) is

$$\frac{\mathrm{d}}{\mathrm{d}t}\rho = -\frac{i}{\hbar}\,[\hat{H}, \rho] - \tfrac{1}{2}\gamma\!\int\! \mathrm{d}^3x\,\left[\hat{N}(\boldsymbol{x}), \left[\hat{N}(\boldsymbol{x}), \rho\right]\right]. \tag{3.7}$$

If equations (3.1) or (3.7) are applied to a system consisting of one single particle the process described is essentially equivalent to a localization of the particle with accuracy $1/\sqrt{\alpha}$ taking place with mean frequency

$$\lambda = \gamma\left(\frac{\alpha}{4\pi}\right)^{\frac{3}{2}}. \tag{3.8}$$

In the case in which several kinds of identical particles are present, one can formulate a generalization of eq. (3.1) along two possible lines. In the first formulation, one independent stochastic field is assumed for each kind of particles. The stochastic equation is in this case

$$\mathrm{d}|\psi\rangle = \left[-\frac{i}{\hbar}\hat{H}\mathrm{d}t + \sum_k g_k\!\int\! \mathrm{d}^3x\,\hat{N}_\psi^k(\boldsymbol{x})\mathrm{d}B^k(\boldsymbol{x}) \right.$$
$$\left. -\tfrac{1}{2}\sum_k g_k^2\!\int\! \mathrm{d}^3x\,\left(\hat{N}_\psi^k(\boldsymbol{x})\right)^2 \mathrm{d}t\right]|\psi\rangle, \tag{3.9}$$

where the index k runs over the kinds of particles, g_k is the coupling of the averaged density $\hat{N}^k(\boldsymbol{x})$ of particles k to the corresponding stochastic field B^k, and

$$\overline{\mathrm{d}B^k(\boldsymbol{x})} = 0, \qquad \overline{\mathrm{d}B^k(\boldsymbol{x})\mathrm{d}B^{k'}(\boldsymbol{x}')} = \delta_{kk'}\delta^3(\boldsymbol{x} - \boldsymbol{x}')\mathrm{d}t. \tag{3.10}$$

In the second formulation, a single stochastic field is assumed for all kinds of particles. Then the stochastic equation is

$$\mathrm{d}|\psi\rangle = \left[-\frac{i}{\hbar}\hat{H}\mathrm{d}t + \sum_k g_k\!\int\! \mathrm{d}^3x\,\hat{N}_\psi^k(\boldsymbol{x})\mathrm{d}B(\boldsymbol{x}) \right.$$
$$\left. -\tfrac{1}{2}\sum_{kk'} g_k g_{k'}\!\int\! \mathrm{d}^3x\,\hat{N}_\psi^k(\boldsymbol{x})\hat{N}_\psi^{k'}(\boldsymbol{x})\mathrm{d}t\right]|\psi\rangle, \tag{3.11}$$

where the field B satisfies condition (3.2). An indication for choice is obtained by considering a microscopic compound system consisting, say, of n_1 particles of kind 1, n_2 particles of kind 2, ..., and requiring that the dynamical equation for the motion of the compound particle as a whole be again of the type (3.1). Let

$$Q = \frac{\sum_k \sum_i m_k q_i^k}{\sum_k n_k m_k} \tag{3.12}$$

be the center–of–mass coordinate and

$$\tilde{q}_i^k(r, s) = q_i^k - Q \tag{3.13}$$

the coordinate relative to it of particle i of kind k. If

$$\psi(q, s) = \Psi(\boldsymbol{Q})\, \chi(r, s) \tag{3.14}$$

is the complete wave function of the microscopic system, one finds

$$\hat{N}^k(\boldsymbol{x})\, \Psi(\boldsymbol{Q})\, \chi(r, s) = \sum_i^{n_k} G(\boldsymbol{Q} + \tilde{q}_i^k(r, s) - \boldsymbol{x})\, \Psi(\boldsymbol{Q})\, \chi(r, s). \tag{3.15}$$

The system being microscopic means that $\chi(r, s)$ is zero unless $|\tilde{q}_i^k(r, s)| \ll 1/\sqrt{\alpha}$, so that

$$\hat{N}^k(\boldsymbol{x})\, \Psi(\boldsymbol{Q})\, \chi(r, s) = n_k\, G(\boldsymbol{Q} - \boldsymbol{x})\, \Psi(\boldsymbol{Q})\, \chi(r, s). \tag{3.16}$$

Then, if

$$\hat{H} = \hat{H}_{\boldsymbol{Q}} + \hat{H}_{rs}, \tag{3.17}$$

one finds from both equations (3.9) and (3.11)

$$\frac{\mathrm{d}}{\mathrm{d}t}\chi(r, s) = -\frac{i}{\hbar}\,\hat{H}_{rs}\,\chi(r, s)$$

$$\mathrm{d}\Psi(\boldsymbol{Q}) = \left[-\frac{i}{\hbar}\hat{H}_{\boldsymbol{Q}}\mathrm{d}t + g\int \mathrm{d}^3\boldsymbol{x}\, G_{\Psi}(\boldsymbol{Q} - \boldsymbol{x})\mathrm{d}B(\boldsymbol{x}) \right. \tag{3.18}$$

$$\left. - \tfrac{1}{2}\,g^2\int \mathrm{d}^3\boldsymbol{x}\, (G_{\Psi}(\boldsymbol{Q} - \boldsymbol{x}))^2\mathrm{d}t \right] \Psi(\boldsymbol{Q}),$$

where

$$G_{\Psi}(\boldsymbol{Q} - \boldsymbol{x}) = G(\boldsymbol{Q} - \boldsymbol{x}) - \int \mathrm{d}^3\boldsymbol{Q}\, \Psi^*(\boldsymbol{Q})\, G(\boldsymbol{Q} - \boldsymbol{x})\, \Psi(\boldsymbol{Q}). \tag{3.19}$$

The difference between the two formulations is that eq. (3.18) is obtained from eq. (3.9) by taking, for the stochastic field pertinent to the compound system and for the coupling to it,

$$\mathrm{d}B(\boldsymbol{x}) = \frac{1}{\sqrt{\sum_k (n_k g_k)^2}}\sum_k n_k g_k \mathrm{d}B^k(\boldsymbol{x}), \qquad g = \sqrt{\sum_k (n_k g_k)^2}, \tag{3.20}$$

while, in the case of eq. (3.11), one has to take

$$g = \sum_k n_k g_k, \tag{3.21}$$

the stochastic field remaining the universal one. The relations (3.20) appear strange and unnatural, while the rule (3.21) has a simple and natural implementation by taking for any microscopic system

$$g = \sqrt{\gamma_0}\,\frac{m}{m_0}, \tag{3.22}$$

where m_0 is a reference mass (for which we take the proton mass) and γ_0 is universal. The discussion in ref. 14 can be repeated and suggests the value

$$\gamma_0 \approx 10^{-30}\ \mathrm{cm^3\, s^{-1}}. \tag{3.23}$$

Since ions are microscopic systems, on the basis of what we have just seen, the system in which we are interested can anyhow be considered to be made of ions (i) and electrons (e). Then the equation for the state vector is

$$d|\psi\rangle = \left[-\frac{i}{\hbar}\hat{H}dt + \sqrt{\gamma_0} \sum_{k}^{(=\text{i,e})} \frac{m_k}{m_0} \int d^3x\, \hat{N}_\psi^k(x)dB(x) \right.$$
$$\left. - \tfrac{1}{2}\gamma_0 \sum_{kk'}^{(=\text{i,e})} \frac{m_k m_{k'}}{m_0^2} \int d^3x\, \hat{N}_\psi^k(x)\hat{N}_\psi^{k'}(x)dt \right]|\psi\rangle, \tag{3.24}$$

and that for the statistical operator is

$$\frac{d}{dt}\rho = -\frac{i}{\hbar}\left[\hat{H},\rho\right] - \tfrac{1}{2}\gamma_0 \sum_{kk'}^{(=\text{i,e})} \frac{m_k m_{k'}}{m_0^2} \int d^3x\, \left[\hat{N}^k(x),\left[\hat{N}^{k'}(x),\rho\right]\right]. \tag{3.25}$$

We assume that the electron–ion interaction can be incorporated effectively into the electron and ion Hamiltonians, so that

$$\hat{H} = \hat{H}^{\text{i}} + \hat{H}^{\text{e}} \tag{3.26}$$

and the wave function for the whole system can be assumed of the form

$$\psi(q,Q) = \psi^{\text{e}}(q)\,\psi^{\text{i}}(Q). \tag{3.27}$$

The ion wave function has the structure

$$\psi^{\text{i}}(Q) = \prod_{1}^{n_{\text{i}}} {}_j \tilde{\delta}(Q_j - a_j), \tag{3.28}$$

where the vectors a_j indicate the lattice positions and the function $\tilde{\delta}(Q)$ is zero unless $|Q| \ll 1/\sqrt{\alpha}$. There follows that

$$\hat{N}^{\text{i}}(x)\,\psi^{\text{i}}(Q) = D_{\text{i}}\,\chi_V(x)\,\psi^{\text{i}}(Q), \tag{3.29}$$

where D_{i} is the density of ions and $\chi_V(x)$ is the characteristic function of the volume occupied by the sample. Inserting eq. (3.29) into eq. (3.25), all stochastic terms cancel except the pure electron ones, so that we get

$$\frac{d}{dt}\rho^{\text{i}} = -\frac{i}{\hbar}\left[\hat{H}^{\text{i}},\rho^{\text{i}}\right], \tag{3.30a}$$
$$\frac{d}{dt}\rho^{\text{e}} = -\frac{i}{\hbar}\left[\hat{H}^{\text{e}},\rho^{\text{e}}\right] - \tfrac{1}{2}\gamma\int d^3x\, \left[\hat{N}^{\text{e}}(x),\left[\hat{N}^{\text{e}}(x),\rho^{\text{e}}\right]\right]. \tag{3.30b}$$

In the following we shall drop the electron index e in eq. (3.30b). For γ we shall use the value

$$\gamma = 2.97\,10^{-37}\ \text{cm}^3\,\text{s}^{-1}, \tag{3.31}$$

corresponding to the value (3.23) of γ_0.

4. DESCRIPTION OF THE SUPERCONDUCTING STATE

We describe the superconducting state by the BCS quasiparticle vacuum, as an approximation to the finite–temperature state. Let $a^+_{k's'}$, a_{ks} be the creation and annihilation operators, respectively, of an electron with wave vector k and spin component s, satisfying canonical anticommutation relations. Define the quasiparticle operators

$$\alpha_k = u_k a_{k,+} - v_k a^+_{-k,-} \qquad \beta_k = u_k a_{k,-} + v_k a^+_{-k,+}, \tag{4.1}$$

where u_k, v_k are real, dependent only on $|k|$, and such that $u_k^2 + v_k^2 = 1$. The quasiparticle operators satisfy

$$\{\alpha_k, \alpha^+_{k'}\} = \{\beta_k, \beta^+_{k'}\} = \delta^3_{kk'}, \tag{4.2}$$

all other anticommutators being zero. The quasiparticle vacuum $|0\rangle$ is defined by

$$\alpha_k |0\rangle = \beta_k |0\rangle = 0. \tag{4.3}$$

The coefficients of the transformation are given by

$$u_k^2 = \tfrac{1}{2}\left(1 + \frac{\epsilon_k - \mu}{E_k}\right), \qquad v_k^2 = \tfrac{1}{2}\left(1 - \frac{\epsilon_k - \mu}{E_k}\right), \tag{4.4}$$

where ϵ_k are the single particle energies, μ is the chemical potential (such that $\langle 0|\hat{N}|0\rangle = N$), and E_k are the single quasiparticle energies given by

$$E_k = \sqrt{\Delta_k^2 + (\epsilon_k - \mu)^2} \tag{4.5}$$

in terms of the energy gap Δ_k. For this quantity we use the simple expression

$$\Delta_k = \Delta\Theta(\epsilon_D - |\epsilon_k - \mu|), \tag{4.6}$$

where Δ is the zero–temperature gap, Θ is the step function, and ϵ_D is the Debye cut–off energy $\epsilon_D = B T_D$, B being the Boltzmann constant. The zero–temperature gap is related to the critical temperature through the relation

$$\Delta = \pi \exp(-\gamma) B T_c = 1.76 B T_c, \tag{4.7}$$

following from the finite–temperature BCS theory. We identify μ with the Fermi energy and use plane–wave single–particle states and free single–particle energies $\epsilon_k = \hbar^2 k^2 / 2m$.

The numerical values given in the following section refer to the case of tin, whose relevant parameters are

$$N/V = 1.48\ 10^{23}\ \mathrm{cm}^{-3}, \qquad T_c = 3.72\ {}^\circ\mathrm{K}, \qquad T_D = 199\ {}^\circ\mathrm{K}. \tag{4.8}$$

From

$$k_F^3 = 3\pi^2 N/V, \qquad \epsilon_F = \hbar^2 k_F^2 / 2m, \tag{4.9}$$

it follows

$$k_F = 1.64\ 10^8\ \mathrm{cm}^{-1}, \qquad \epsilon_F = 1.63\ 10^{-11}\ \mathrm{erg}. \tag{4.10}$$

The gap and Debye energies are

$$\Delta = 1.76\, B\, T_c = 0.904\ 10^{-15}\ \text{erg}, \qquad \epsilon_D = B\, T_D = 2.75\ 10^{-14}\ \text{erg}, \qquad (4.11)$$

this value of Δ being very near to the measured value. Defining k_Δ and k_D by

$$\Delta = \hbar^2 k_F k_\Delta / m, \qquad \epsilon_D = \hbar^2 k_F k_D / m, \qquad (4.12)$$

one finds

$$k_\Delta = 4.51\ 10^3\ \text{cm}^{-1}, \qquad k_D = 1.37\ 10^5\ \text{cm}^{-1}. \qquad (4.13)$$

Note that one has the hierarchy of wave numbers

$$k_\Delta \underset{(30)}{\ll} \sqrt{\alpha}, k_D \underset{(1000)}{\ll} k_F, \qquad (4.14)$$

where rough values of the intervening factors are indicated.

5. EFFECTS OF THE LOCALIZATION PROCESS

Let the statistical operator be initially

$$\rho(0) = |0\rangle\langle 0|. \qquad (5.1)$$

The transition probability from the state (5.1) to an arbitrary orthogonal state is

$$P_{0\to\{l\}}(t) = \sum_l^{(\neq 0)} \langle l\,|\rho(t)|\,l\,\rangle, \qquad (5.2)$$

where $\rho(t)$ is evolved from (5.1) according to eq. (3.30b). We are interested in the initial value of the transition probability per unit time

$$\dot P = \left[\frac{\mathrm{d}}{\mathrm{d}t} P_{0\to\{l\}}(t)\right]_{t=0} = \sum_l^{(\neq 0)} \langle l\,|\dot\rho(0)|\,l\,\rangle$$
$$= \gamma \sum_l^{(\neq 0)} \int \mathrm{d}^3 x\, \langle l\,|\hat N(\boldsymbol{x})|0\rangle\langle 0|\hat N(\boldsymbol{x})|\,l\,\rangle, \qquad (5.3)$$

where we have assumed that $|0\rangle$ is an eigenstate of $\hat H$. Expressing $\hat N(\boldsymbol{x})$ in terms of quasiparticle operators in the wave–vector representation, one finds

$$\dot P = \frac{V\gamma}{(2\pi)^6} \int \mathrm{d}^3 k_1 \int \mathrm{d}^3 k_2 \exp\left[-(\boldsymbol{k}_1 + \boldsymbol{k}_2)^2/\alpha\right] (u_{k_1} v_{k_2} + v_{k_1} u_{k_2})^2$$
$$= \frac{V\gamma\alpha}{(2\pi)^4} \int_0^\infty \mathrm{d}k_1 \int_0^\infty \mathrm{d}k_2\, k_1 k_2 \left(\exp\left[-(k_1 - k_2)^2/\alpha\right]\right.$$
$$\left. - \exp\left[-(k_1 + k_2)^2/\alpha\right] \right) (u_{k_1} v_{k_2} + v_{k_1} u_{k_2})^2. \qquad (5.4)$$

An exact analitical evaluation of the double integral in the last line of eq. (5.4) is impossible, and even an approximate calculation is very cumbersome because of the different structures in the $k_1 k_2$ plane of the factors. Let us write

$$\dot P = \dot P_0 (1 + R), \qquad (5.5)$$

where $\dot{P}_0$ is $\dot{P}$ in the absence of superconducting effects ($\Delta = 0$). $\dot{P}_0$ can be calculated exactly and is given by

$$\dot{P}_0 = \frac{\gamma \alpha^2 k_{\mathrm{F}}^2}{(2\pi)^4} V \left\{ 1 - \tfrac{1}{6} \frac{\alpha}{k_{\mathrm{F}}^2} \left[1 - \exp\left(-4k_{\mathrm{F}}^2/\alpha\right) \right] \right.$$
$$\left. - \tfrac{1}{3} \exp\left(-4k_{\mathrm{F}}^2/\alpha\right) - \tfrac{2}{3}\sqrt{\pi}\frac{k_{\mathrm{F}}}{\sqrt{\alpha}} \left[\mathrm{erf}\left(2k_{\mathrm{F}}/\sqrt{\alpha}\right) - 1 \right] \right\} \qquad (5.6)$$
$$= \frac{\gamma \alpha^2 k_{\mathrm{F}}^2}{(2\pi)^4} V,$$

the correction neglected in the last line being of the order of 10^{-7}. For the correction R in eq. (5.5) we find

$$R \simeq x_1^2 \left\{ 4\ln(x_1)[\ln(x_1) - 1] - \left[\ln^2(x_2) + \ln^2(x_3) - \ln^2(x_4) \right] \Theta(1 - x_2) \right\}, \qquad (5.7)$$

where

$$x_1 = k_\Delta/\sqrt{2\alpha}, \qquad x_2 = k_D/\sqrt{\alpha}, \qquad x_3 = k_D/k_F, \qquad x_4 = \sqrt{\alpha}/k_F. \qquad (5.8)$$

The numerical values given by eq. (5.7) were checked against the results of a double numerical integration and found to be in pretty good agreement for wide ranges of values of the parameters. In the case of tin one has

$$x_1 = 3.19\ 10^{-2}, \qquad\qquad x_2 = 1.37, \qquad (5.9)$$

and

$$R \simeq 6.23\ 10^{-2}, \qquad R_{\mathrm{num}} = 6.49\ 10^{-2}, \qquad (5.10)$$

where R is calculated from eq. (5.7) and R_{num} is the "exact" value obtained numerically. As one can see, pairing does not affect the transition probability per unit time in an important way. Eq. (5.6) gives for tin

$$\dot{P}_0 = 5.13\ 10^{-4}\ V\mathrm{cm}^{-3}\,\mathrm{s}^{-1}. \qquad (5.11)$$

Eq. (5.11), apart from the correction (5.10), provides the numerical value of the rate of production of quasiparticle and particle–hole excitations due to the localization process. When compared to the number of electrons $N \approx 10^{23}V\mathrm{cm}^{-3}$ or to the number of paired electrons $N_{\mathrm{p}} \approx 10^{20}V\mathrm{cm}^{-3}$ it shows that the superconducting state is practically unaffected by the process.

It is interesting to compare our result with the previous estimate by Rae.[15] This estimate was

$$\left[\dot{P}\right]_{\mathrm{Rae}} = N\lambda = N\gamma \left(\frac{\alpha}{4\pi}\right)^{\frac{3}{2}}, \qquad (5.12)$$

where we have inserted the value (3.8) for the single–particle frequency λ. Comparison with our value (5.6), neglecting R, gives

$$\left[\dot{P}\right]_{\mathrm{ours}} = \frac{3}{2\sqrt{\pi}} \frac{\sqrt{\alpha}}{k_{\mathrm{F}}} \left[\dot{P}\right]_{\mathrm{Rae}}, \qquad (5.13)$$

which shows that the correct treatment taking into account indistinguishability of electrons gives a further reduction by a factor of the order of $\sqrt{\alpha}/k_{\mathrm{F}}$.*

We have seen that the superconducting state is not quickly destroyed by the localization process. Knowing that, we can conveniently introduce besides $\dot{P}$ another quantity measuring the size of the effect. Our sample is maintained at a fixed, nonzero temperature by the reservoir or the cooling device. This means that the excitations we have considered, which add to the excitations which are anyway there because of the finite temperature, eventually decay, filling a hole by a particle or recreating a pair, and supplying energy to the lattice. Recreation of Cooper pairs was already envisaged by Rae.[15] In this picture, the quantity which really matters is the amount of heat pumped in by the spontaneous localization process, to be pumped out by the refrigerator. In calculating this quantity, the role played by ions cannot be neglected because of their large mass, so that we come back to eq. (3.25). It has already been noted by Pearle and Squires[16] that the total rate of energy production implied by this equation can be calculated exactly (at least if the potentials contained in $\hat{H}$ are local) and the result is

$$\frac{\mathrm{d}}{\mathrm{d}t}\langle H\rangle = \mathrm{Tr}\,(\dot{\rho}H) = \frac{3\,\hbar^2\gamma_0\,\alpha^{\frac{5}{2}}}{4\,(4\pi)^{\frac{3}{2}}\,m_0^2}M \tag{5.14}$$

for any state ρ, M being the total mass. For a macroscopic body, the fraction of this energy increase due to the increase of center–of–mass energy is of the order of $a/\sqrt{\alpha}$, a representing lattice distances, so that eq. (5.14) essentially gives the increase of internal energy $\mathrm{d}E/\mathrm{d}t$. Therefore one finds

$$\frac{\mathrm{d}E}{\mathrm{d}t} \simeq 6\,10^{-14}\,\mathrm{erg\,s^{-1}}M\,\mathrm{g^{-1}}. \tag{5.15}$$

To appreciate how small is this small quantity, one can conveniently estimate the corresponding temperature increase, using as an order of magnitude of heat capacity

$$\frac{\mathrm{d}E}{\mathrm{d}T} \approx 3NB. \tag{5.16}$$

One finds

$$\frac{\mathrm{d}T}{\mathrm{d}t} \approx 2\,10^{-22}A\,^\circ\mathrm{K\ s^{-1}}, \tag{5.17}$$

A being the atomic weight.

6. CONCLUSIONS

The parameters of the spontaneous localization process we have considered were chosen in such a way that ordinary quantum mechanics remains fully valid for micro-scopic objects and for the internal structure macroscopic bodies, that the stochastic disturbance to the classical motion of macroscopic objects as a whole is also quite negligible, and that, nevertheless, superpositions of macroscopically distinguishable

---

* The value used by Rae for λ corresponds to the proton value of γ given in eq. (3.6) which is considerably larger than our value (3.31). Our analysis for systems containing different kinds of particles shows that the value (3.31) is more reliable. On the other hand he used for N the number of paired electrons N_{p}. This choice does not take into account that particle–hole excitations eventually decay, presumably affecting paired electrons.

states are rapidly suppressed. From the above analysis we conclude, conferming Rae's previous estimate, that the effects of the process are negligible also for superconducting states, in spite of the fact that the internal wave function cannot be assumed to be sharply localized in this case.

We did not discuss the problem of the possible reduction of superpositions of states corresponding to different supercurrents. Work on this point is in progress. However, we do not see any reason why the qualitative arguments by Rae should not be essentially correct also in this case.

REFERENCES

1. O. Nicrosini and A. Rimini, *in* Symposium on the Foundations of Modern Physics 1990, P. Lahti and P. Mittelstaedt, eds., World Scientific, Singapore (1991), p. 280.
2. A. Rimini, *to be published in* Proceedings of the Lanczos Centenary Conference, M. Chu, R. Plemmons, D. Brown, and D. Ellison, eds., SIAM, Philadelphia.
3. D. Bohm, *Phys. Rev.* 85: 166, 180 (1952).
4. J.S. Bell, CERN Preprint TH 1424 (1971); *Epist. Lett.* July 1978: 1 (1978).
5. J.S. Bell, *in* Quantum Gravity 2, C. Isham, R. Penrose, and D. Sciama, eds., Clarendon Press, Oxford (1981), p. 611.
6. R. Griffiths, *J. Stat. Phys.* 36: 219 (1984)
7. M. Gell-Mann and J. Hartle, *Phys. Rev. D* 47: 3345 (1993).
8. R. Omnès, *Rev. Mod Phys.* 64: 339 (1992).
9. H. Everett III, *Rev. Mod Phys.* 29: 454 (1957).
10. W. H. Zurek, *Phys. Rev. D* 24: 1516 (1981).
11. G.C. Ghirardi, A. Rimini, and T. Weber, *in* Quantum Probability and Applications II, L. Accardi and W. von Waldenfels, eds., Lecture Notes in Mathematics, vol. 1136, Springer, Berlin (1985), p. 223.
12. G.C. Ghirardi, A. Rimini, and T. Weber, *Phys. Rev. D* 34: 470 (1986); *Phys. Rev. D* 36: 3287 (1987).
13. P. Pearle, *Phys. Rev. A* 39: 2277 (1989).
14. G.C. Ghirardi, P. Pearle, and A. Rimini, *Phys. Rev. A* 42: 78 (1990).
15. A. I. M. Rae, *J. Phys, A* 23: L57 (1990).
16. P. Pearle and E. Squires, *to be published in Phys. Rev. Lett.*

PHOTON-PHOTON CORRELATIONS FROM SINGLE ATOMS*

Marlan O. Scully

Department of Physics,
Texas A&M University, College Station, TX 77843, and
Texas Laser Laboratory,
HARC, The Woodlands,TX 77831, and
Max-Planck-Institut für Quantenoptik,
D-85748 Garching, Germany

ABSTRACT

Two-photon correlation functions are calculated for simple single atom "down converters" in various configurations.

I. INTRODUCTION AND SUMMARY

Over the past decade or so several new fundamental tests of the foundations of quantum mechanics have been suggested based on photon-photon correlations[1,2]. These include tests of Young's interference of the light scattered from individual atoms[3,4,5], quantum eraser[3,4,6,7], delayed choice[8], complementarity[9,10], Bell's Theorem[11,12,13] and more[14,15]. Beautiful experiments have been carried out using trapped atoms[16] as well as nonlinear crystals, and the future promises more intriguing results along these lines.

It is therefore useful to have detailed treatments of simple two-photon cascade schemes and of the associated intensity or photon-photon correlation functions. To this end we here calculate the correlation functions for two-photon emission by a single three-level atom in a conventional cascade arrangement and compare these results with the "interrupted" two-photon cascade of the type used in the early quantum eraser studies[3].

---

* Work supported by the Office of Naval Research, the Welch Foundation, and the Texas Advanced Research Program.

The presentation will be that of a detailed technical analysis, aimed at graduate students who want to understand the theory. The results of these "simple" calculations should, it is to be hoped, provide guidance in thinking about more complicated schemes.

My interest in the problem of intensity correlations dates back to calculations in which I was seeking a potentially realizable quantum eraser scheme. I had been looking at problems associated with an observer in a physical system via a variant[17] of Wigner's Stern-Gerlach experiment. This suggested utilizing atoms in a cascade emission configuration in order to put an "observer" into the quantum measurement problem. Thus, I was naturally led to the study of the "interrupted" cascade emissions (see Fig. 5) and to the associated photon-photon correlation function.

The state of the radiation field procured in this way may be written as a product state

$$|\psi\rangle = |\gamma\rangle|\phi\rangle,$$

where $|\gamma\rangle$ and $|\phi\rangle$ are given by Eqs (18) and (19). In such a case, the photon-photon or intensity correlation function takes the factorized form given by Scully and Drühl[3]

$$G_{SD}^{(2)}(r_1,t_1;r_2,t_2) = \left|\psi_\gamma(r_1,t_1)\,\psi_\phi(r_2,t_2) + \psi_\phi(r_1,t_1)\,\psi_\gamma(r_2,t_2)\right|^2,$$

where $\psi_\gamma(r_i,t_i)$ and $\psi_\phi(r_i,t_i)$ are the amplitudes for the γ or ϕ photons exciting the ith ($i=1,2$) photoelectron at r_i at time t_i as in Eqs (21a) and (21b). These results were later derived via an operator formalism by Hillery and Scully[4]. The essential physics here is that of a rapidly decaying upper level decaying to a long lived intermediate state.

More recently, Franson[9] has been looking at photon correlation problems in which "three-level" atoms have a sharp upper level and a rapidly decaying intermediate state, see Fig.1. Note that this atomic configuration is in contrast to that of the quantum eraser problem, in which the upper state is broad and the intermediate state is sharp. The interconnection between these two problems was something that I became interested in after listening to a nice talk by Jim Franson at the 1994 Physics of Quantum Electronics Winter Conference in Snowbird, Utah. In the Franson problem, the state of the radiation field can no longer be written in the simple factorized form of the quantum eraser problem but now takes the form

$$|\psi\rangle = \sum_{\vec{k},\vec{q}} C_{\vec{k},\vec{q}} \left|1_{\vec{k}}, 1_{\vec{q}}\right\rangle,$$

where $C_{\vec{k},\vec{q}}$ is given by Eq. (9).

After carrying out the calculations on the Franson scheme, I had the good fortune to interact with Hu Huang and Girish Agarwal in Texas and Trieste and to learn of their interesting work along the same lines and I am indebted to them for giving me a preprint[2] of their paper. The correlation function for the Franson problem now takes on the more complicated form

$$G_F^{(2)}(r_1,t_1;r_2,t_2) = \left|\psi^{(2)}(r_1,t_1;r_2,t_2)\right|^2,$$

where $\psi^{(2)}$ is given by Eq. (17). It is noted, that in the limit that $\gamma_a \gg \gamma_b$, i.e., the opposite limit to that of Franson, one regains the Scully-Drühl result, that is

$$G_F^{(2)}(r_1,t_1;r_2,t_2) \xrightarrow{\quad \lim \gamma_a \gg \gamma_b \quad} G_{SD}^{(2)}(r_1,t_1;r_2,t_2)$$

as is shown at the end of these lectures.

Thus I arrived at the ERICE school on "Advances in Quantum Phenomena" full of photon-photon correlations and their application to the foundations of quantum mechanics. There I enjoyed many spirited discussions, often finding that a lack of appreciation for the technical details of the calculations was hampering the understanding of the physics.

Furthermore, upon returning to the university I was pleased to interact with several bright students, whose enthusiasm prompted me to "writeup" these lecture notes. It is, therefore, my purpose to provide detailed QED calculations for the problems at hand. I especially thank Georg M. Meyer, Ulrich W. Rathe, Chang Su, and Susanne F. Yelin for checking the calculations and for helpful comments on the present paper.

II. TWO PHOTON CASCADES

Consider the three-level cascade as in Fig.1. The state of the atom-field system is described by

$$|\psi\rangle = A(t)|a,0\rangle + \sum_{\vec{k}} B_{\vec{k}}(t)|b,1_{\vec{k}}\rangle + \sum_{\vec{k},\vec{q}} C_{\vec{k},\vec{q}}(t)|c,1_{\vec{k}},1_{\vec{q}}\rangle, \tag{1}$$

where the probability amplitudes in the interaction picture obey the equations of motion

$$\dot{A}(t) = -i\sum_{\vec{k}} g_{a,\vec{k}}(\vec{r})B_{\vec{k}} \exp\left[-i(v_k - \omega_{ab})t\right] \tag{2a}$$

$$\dot{B}_{\vec{k}}(t) = -ig_{a,\vec{k}}(\vec{r})A \exp\left[i(v_k - \omega_{ab})t\right] - i\sum_{\vec{q}} g_{b,\vec{q}}(\vec{r})C_{\vec{k},\vec{q}} \exp\left[-i(v_q - \omega_{bc})t\right] \tag{2b}$$

$$\dot{C}_{\vec{k},\vec{q}}(t) = -ig_{b,\vec{q}}(\vec{r})B_{\vec{k}} \exp\left[i(v_q - \omega_{bc})t\right] \tag{2c}$$

and the coupling constants, which depend on the atomic position $\vec{r}$, are given by

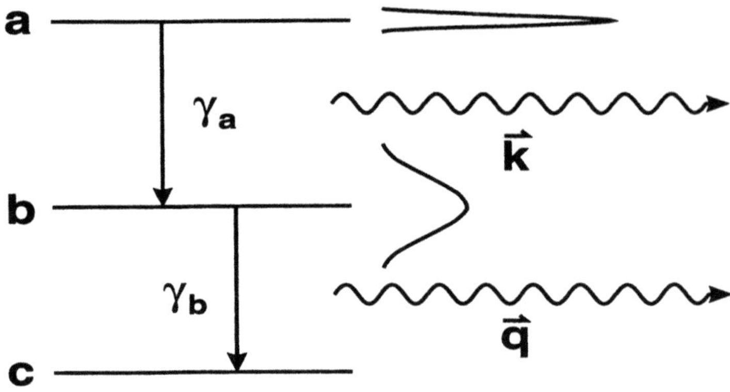

Figure 1. Sharp upper level a decays to b, which rapidly decays to ground state c.

$$g_{a,\vec{k}}(\vec{r}) = \frac{p_{ab}}{\hbar}\sqrt{\frac{\hbar v_k}{\varepsilon_o V}}e^{i\vec{k}\cdot\vec{r}} \equiv g_{a,k}e^{i\vec{k}\cdot\vec{r}} \tag{3a}$$

$$g_{b,\vec{q}}(\vec{r}) = \frac{p_{bc}}{\hbar}\sqrt{\frac{\hbar v_q}{\varepsilon_o V}}e^{i\vec{q}\cdot\vec{r}} \equiv g_{b,q}e^{i\vec{q}\cdot\vec{r}} \tag{3b}$$

We are interested in the state of the field for times $t >> \gamma_a^{-1}$ and $t >> \gamma_b^{-1}$, i.e., we want to know $C_{\vec{k},\vec{q}}(\infty)$. It is possible to "shortcut" some of the algebra by noting that $A(t) = A(0)e^{-\gamma_a t}$, so that the equations of motion become

$$\dot{A} = -\gamma_a A \tag{4a}$$

$$\dot{B}_{\vec{k}} = -ig_{a,\vec{k}}(\vec{r})e^{i(v_k-\omega_{ab})t-\gamma_a t} - i\sum_{\vec{q}} g_{b,\vec{q}}(\vec{r})C_{\vec{k},\vec{q}}e^{-i(v_q-\omega_{bc})t} \tag{4b}$$

$$\dot{C}_{\vec{k},\vec{q}} = -ig_{b,\vec{q}}(\vec{r})B_{\vec{k}}e^{i(v_q-\omega_{bc})t}. \tag{4c}$$

That is, we recognize from our previous experience with Weisskopf-Wigner theories that

$$-i\sum_{\vec{k}} g_{a,\vec{k}}(\vec{r})B_{\vec{k}}e^{-i(v_k-\omega_{ab})t} \cong -\gamma_a A. \tag{5}$$

Likewise, we may note that the second term in (4b) represents decay from b to c and write

$$-i\sum_{\vec{q}} g_{b,\vec{q}}(\vec{r})C_{\vec{k},\vec{q}}e^{-i(v_q-\omega_{bc})t} \cong -\gamma_b B_{\vec{k}} \tag{6}$$

so that we have the equations of motion in a useful final form

$$\dot{A} = -\gamma_a A \tag{7a}$$

$$\dot{B}_{\vec{k}} = -ig_{a,\vec{k}}(\vec{r})e^{i(\nu_k-\omega_{ab})t-\gamma_a t} - \gamma_b B_{\vec{k}} \tag{7b}$$

$$\dot{C}_{\vec{k},\vec{q}} = -ig_{b,\vec{q}}(\vec{r})e^{i(\nu_q-\omega_{bc})t} B_{\vec{k}}. \tag{7c}$$

Solving for $B_{\vec{k}}(t)$ from (7b) we have

$$B_{\vec{k}}(t) = -ig_{a,\vec{k}}(\vec{r})\int_0^t dt'\, e^{i(\nu_k-\omega_{ab})t'-\gamma_a t'-\gamma_b(t-t')}$$

$$= -ig_{a,\vec{k}}(\vec{r})\frac{1}{i(\nu_k-\omega_{ab})-(\gamma_a-\gamma_b)}\left[e^{i(\nu_k-\omega_{ab})t-\gamma_a t}-e^{-\gamma_b t}\right]. \tag{8}$$

Inserting (8) into (7c) we find the long-time limit of the integral to be

$$C_{\vec{k},\vec{q}}(\infty) = g_{a,k}g_{b,q}e^{-i\vec{k}\cdot\vec{r}}e^{-i\vec{q}\cdot\vec{r}}\frac{1}{i(\nu_k-\omega_{ab})-(\gamma_a-\gamma_b)}$$

$$\times\left[\frac{1}{i(\nu_k+\nu_q-\omega_{ac})-\gamma_a}-\frac{1}{i(\nu_q-\omega_{bc})-\gamma_b}\right]$$

$$= \frac{-g_{a,k}g_{b,q}e^{-i\vec{k}\cdot\vec{r}-i\vec{q}\cdot\vec{r}}}{\left[i(\nu_k+\nu_q-\omega_{ac})-\gamma_a\right]\left[i(\nu_q-\omega_{bc})-\gamma_b\right]}. \tag{9}$$

Thus the state of the radiation field is given by

$$|\psi\rangle = \sum_{\vec{k},\vec{q}}\frac{-g_{a,k}g_{b,q}e^{-i\vec{k}\cdot\vec{r}-i\vec{q}\cdot\vec{r}}}{\left[i(\nu_k+\nu_q-\omega_{ac})-\gamma_a\right]\left[i(\nu_q-\omega_{bc})-\gamma_b\right]}|1_{\vec{k}},1_{\vec{q}}\rangle. \tag{10}$$

Next we calculate the two-photon correlation function

$$G^{(2)}(\vec{r}_1,t_1;\vec{r}_2,t_2) = \langle\psi|\hat{E}^-(\vec{r}_1,t_1)\hat{E}^-(\vec{r}_2,t_2)\hat{E}^+(\vec{r}_2,t_2)\hat{E}^+(\vec{r}_1,t_1)|\psi\rangle,$$

corresponding to two detectors at points $\vec{r}_1$ and $\vec{r}_2$ and where interaction with the photon field, described by $|\psi\rangle$, is switched on at times t_1 and t_2, respectively, see Fig. 2.

We note that for the radiation from a single atom only two photons are involved so that

$$\langle\psi|\hat{E}_1^-\hat{E}_2^-\hat{E}_2^+\hat{E}_1^+|\psi\rangle = \sum_{\{n\}}\langle\psi|\hat{E}_1^-\hat{E}_2^-|\{n\}\rangle\langle\{n\}|\hat{E}_2^+\hat{E}_1^+|\psi\rangle$$

$$= \langle\psi|\hat{E}_1^-E_2^-|0\rangle\langle 0|\hat{E}_2^+\hat{E}_1^+|\psi\rangle$$

and therefore it is the two-photon "wavefunction" which is of interest

$$\psi^{(2)}(\vec{r}_1,t_1;\vec{r}_2,t_2) \equiv \langle 0|\hat{E}^+(\vec{r}_2,t_2)\hat{E}^+(\vec{r}_1,t_1)|\psi\rangle. \tag{11}$$

Using the fact that

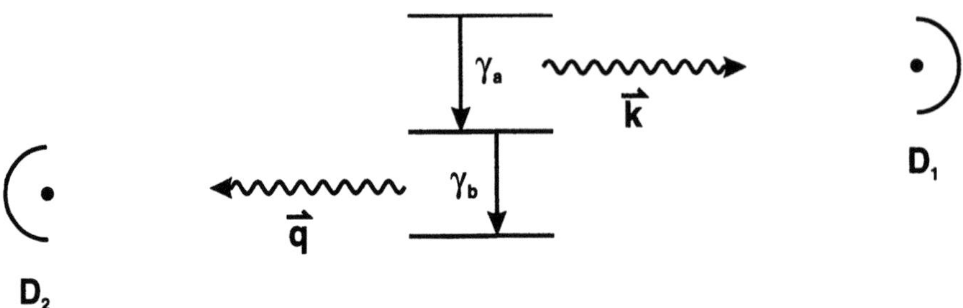

Fig. 2a. Detectors D_1 and D_2 interact with $\bar{k}$ and $\bar{q}$ photons.

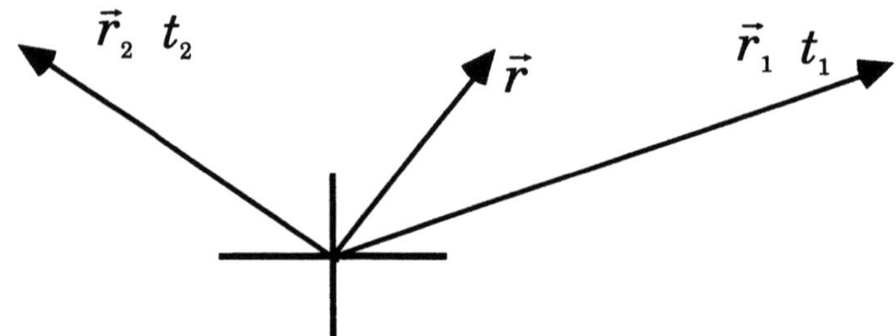

Fig. 2b. Spatial and temporal locations and turn-on times.

$$\hat{E}^{+}(\bar{r}_i,t_i) = \sum_{\bar{k}} \varepsilon_{\bar{k}}\hat{a}_{\bar{k}}e^{-iv_k t_i + \bar{k}\cdot\bar{r}_i} \qquad\qquad i = 1,2 \qquad\qquad (12)$$

and taking $|\psi\rangle$ from Eq. (10) we have

$$\psi^{(2)}(r_1,t_1;r_2,t_2) = \sum_{\bar{k},\bar{q}}\sum_{\bar{p},\bar{s}} \langle 0|\varepsilon_{\bar{p}}\varepsilon_{\bar{s}}\hat{a}_{\bar{p}}\hat{a}_{\bar{s}}e^{-iv_p t_1 + i\bar{p}\cdot\bar{r}_1}e^{-iv_s t_2 + i\bar{s}\cdot\bar{r}_2}$$

$$\times \frac{(-1)e^{-i\bar{k}\cdot\bar{r} - i\bar{q}\cdot\bar{r}}g_{a,k}g_{b,q}}{\left[i(v_k + v_q - \omega_{ac}) - \gamma_a\right]\left[i(v_q - \omega_{bc}) - \gamma_b\right]}\Big|1_{\bar{k}},1_{\bar{q}}\rangle$$

$$= \sum_{\bar{k},\bar{q}}\left[e^{-iv_k t_1 + i\bar{k}\cdot\bar{r}_1}e^{-iv_q t_2 + i\bar{q}\cdot\bar{r}_2} + e^{-iv_k t_2 + i\bar{k}\cdot\bar{r}_2}e^{-iv_q t_1 + i\bar{q}\cdot\bar{r}_1}\right]\varepsilon_{\bar{k}}\varepsilon_{\bar{q}}$$

$$\times \frac{(-1)e^{-i\bar{k}\cdot\bar{r} - i\bar{q}\cdot\bar{r}}g_{a,k}g_{b,q}}{\left[i(v_k + v_q - \omega_{ac}) - \gamma_a\right]\left[i(v_q - \omega_{bc}) - \gamma_b\right]} \,. \qquad\qquad (13)$$

In order to evaluate $\psi^{(2)}$ we:

1) Change the sums on $\bar{k}$ and $\bar{q}$ into integrals.

2) Evaluate all coupling constants and density of state factors at the atomic resonances $ck_0 = \omega_{ab}$ and $cq_0 = \omega_{bc}$.

3) Choose the "z axis" for the $\bar{k}$ and $\bar{q}$ integrations as $\Delta\vec{r}_1 = \vec{r}_1 - \vec{r}$ and $\Delta\vec{r}_2 = \vec{r}_2 - \vec{r}$ for the first term in the square brackets and $\vec{r}_2 - \vec{r}$ and $\vec{r}_1 - \vec{r}$ for the second.

In light of the above we may write Eq. (13) as

$$\psi^{(2)}(\vec{r}_1,t_1;\vec{r}_2,t_2) = -\int_0^\infty k^2 dk \int_0^\pi \sin\theta_k d\theta_k \int_0^{2\pi} d\phi_k \int_0^\infty q^2 dq \int_0^\pi \sin\theta_q d\theta_q \int_0^{2\pi} d\varphi_q \varepsilon_{\bar{k}}\varepsilon_{\bar{q}}$$

$$\times\ g_{a,k_0}g_{b,q_0}\,\sigma(k_0)\sigma(q_0)\left[e^{-ickt_1+ik\Delta r_1\cos\theta_k}e^{-icqt_2+iq\Delta r_2\cos\theta_q} + (1 \leftrightarrow 2)\right]$$

$$\times\ \frac{1}{\left[i(ck+cq-\omega_{ac})-\gamma_a\right]\left[i(cq-\omega_{bc})-\gamma_b\right]} \tag{14}$$

carrying out the angular integrations yields

$$\psi^{(2)}(r_1,t_1;r_2,t_2) = -\frac{(2\pi)^2}{(i)^2\Delta r_1\Delta r_2}\cdot k_0 g_{a,k_0}\sigma(k_0)\varepsilon_{k_0}q_0 g_{b,q_0}\sigma(q_0)\varepsilon_{q_0}$$

$$\times\ \int_0^\infty dkdq\,\frac{e^{-ickt_1}\left[e^{-ik\Delta r_1}-e^{ik\Delta r_1}\right]e^{-icqt_2}\left[e^{-iq\Delta r_2}-e^{iq\Delta r_2}\right]}{\left[i(ck+cq-\omega_{ac})-\gamma_a\right]\left[i(cq-\omega_{bc})-\gamma_b\right]}+(1 \leftrightarrow 2). \tag{15}$$

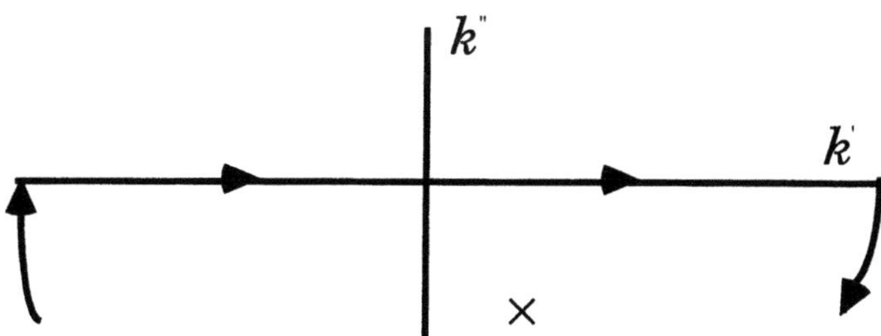

Fig. 3. Figure illustrating pole at $k = \dfrac{\omega_{ac}}{c} - q - i\dfrac{\gamma_a}{c}$ and contour for $ct > \Delta r_1$. Integral vanishes when $ct < \Delta r_1$.

We proceed to do the k integration. The limits of integration on k and q may be extended to $-\infty$ because of the sharply peaked Lorentzians. The k contour is closed in the lower half plane, as in Fig 3, when $ct > \Delta r_1$ and in the upper half plane when $ct < \Delta r_1$. Hence, neglecting the unphysical incoming waves which go as $e^{-ik\Delta r_1}$, we find

$$\psi^{(2)}(r_1,t_1;r_2,t_2) =$$

$$\times i \frac{(2\pi)^3 k_0 g_{a,k_0}\sigma(k_0)\varepsilon_{k_0} q_0 g_{b,q_0}\sigma(q_0)\varepsilon_{q_0}}{\Delta r_1 \Delta r_2} e^{-(i\omega_{ac}+\gamma_a)\left(t_1-\frac{\Delta r_1}{c}\right)}\Theta\left(t_1-\frac{\Delta r_1}{c}\right)$$

$$\times \int_{-\infty}^{\infty} dq \frac{\exp\left\{-icq\left[\left(t_2-\frac{\Delta r_2}{c}\right)-\left(t_1-\frac{\Delta r_1}{c}\right)\right]\right\}}{i(cq-\omega_{bc})-\gamma_b} + (1\leftrightarrow 2) \ . \tag{16}$$

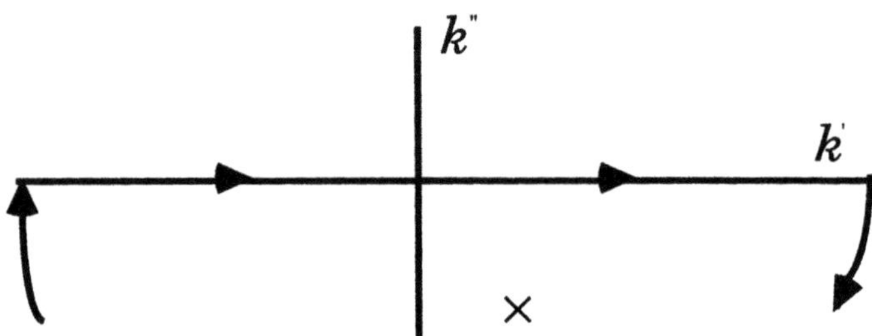

Fig. 4. Figure illustrating pole at $q_0 = \dfrac{\omega_{bc}}{c} - i\dfrac{\gamma_b}{c}$ and contour for $t_2 - \dfrac{\Delta r_2}{c} < t_1 - \dfrac{\Delta r_2}{c}$.

In like manner we carry out the q integration, via the contour of Fig. 4, to find

$$\psi^{(2)}(r_1,t_1;r_2,t_2) = \frac{-\kappa}{\Delta r_1 \Delta r_2}\exp\left[-(i\omega_{ac}+\gamma_a)\left(t_1-\frac{\Delta r_1}{c}\right)\right]\Theta\left(t_1-\frac{\Delta r_1}{c}\right)$$

$$\times \exp\left\{-(i\omega_{bc}+\gamma_b)\left[\left(t_2-\frac{\Delta r_2}{c}\right)-\left(t_1-\frac{\Delta r_1}{c}\right)\right]\right\}$$

$$\times \Theta\left[\left(t_2-\frac{\Delta r_2}{c}\right)-\left(t_1-\frac{\Delta r_1}{c}\right)\right] + (1\leftrightarrow 2) \ , \tag{17}$$

where $\kappa = (2\pi)^4 k_0 g_{a,k_0}\sigma(k_0)\varepsilon_{k_0} q_0 g_{b,q_0}\sigma(q_0)\varepsilon_{q_0}$.

III. INTERRUPTED TWO-PHOTON CASCADE

Next we consider the problem of "interrupted" emission, as in Fig. 5. The first photon, associated with the $a \leftrightarrow b$ transition, is described in the long time limit by our "old friend"

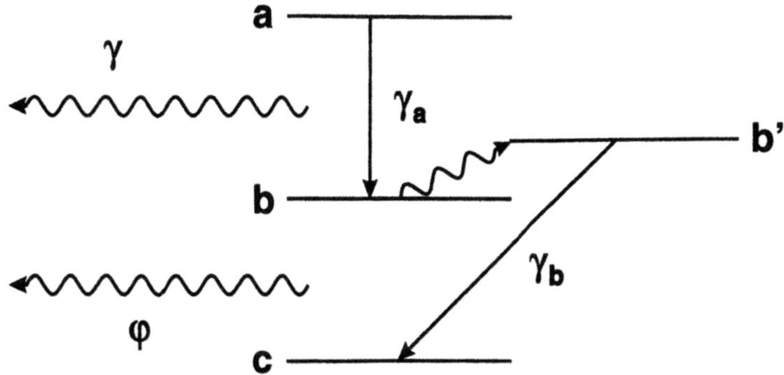

Fig. 5. Figure illustrating decay of atom excited to state at a rate γ_a to non-decaying level b. Upon detection of $a \to b$ photon, population in level b is transfered to b' by means of an external field indicated by wavy line. Level b' decays to c at rate γ_b .

$$|\gamma\rangle = \sum_{\vec{k}} \frac{g_{a,k}e^{-i\vec{k}\cdot\vec{r}}}{(\omega_{ab} - ck) - i\gamma_a}\left|1_{\vec{k}}\right\rangle. \tag{18}$$

Likewise the second photon, associated with the $b' \to c$ transition, is given in the long time limit by

$$|\phi\rangle = \sum_{\vec{q}} \frac{g_{b,q}e^{-i(\vec{q}\cdot\vec{r} - cqt_0)}}{(\omega_{ac} - cq) - i\gamma_b}\left|1_{\vec{q}}\right\rangle, \tag{19}$$

where t_0 is the time of detection of γ photon and the transfer from $b \to b'$.

With Eq's (18) and (19) in hand it is an easy matter to calculate the photon-photon correlation function, or equivalently the $\psi^{(2)}(r_1,t_1;r_2,t_2)$ function as defined by Eq.(11). We find

$$\psi^{(2)}(r_1,t_1;r_2,t_2) = \psi_\gamma(r_1,t_1)\psi_\phi(r_2,t_2) + \psi_\phi(r_1,t_1)\psi_\gamma(r_2,t_2), \tag{20}$$

where

$$\psi_\gamma(r_i,t_i) = \frac{\varepsilon_\gamma}{\Delta r_i}\Theta\left(t_i - \frac{\Delta r_i}{c}\right)e^{-\gamma_a\left(t_i - \frac{\Delta r_i}{c}\right)}e^{-i\omega_{ab}\left(t_i - \frac{\Delta r_i}{c}\right)} \tag{21a}$$

and

$$\psi_\phi(r_i,t_i) = \frac{\varepsilon_\phi}{\Delta r_i}\Theta\left(t_i - t_0 - \frac{\Delta r_i}{c}\right)e^{-\gamma_b\left(t_i - t_0 - \frac{\Delta r_i}{c}\right)}e^{-i\omega_{bc}\left(t_i - t_0 - \frac{\Delta r_i}{c}\right)} \tag{21b}$$

with $i = 1,2$.

It is worthwhile at this point to consider the special case in which $\gamma_a \gg \gamma_b$, i.e., the atom decays promptly to level b via emission of a γ photon. After a time $\frac{\Delta r_i}{c}$, $i = 1,2$, the γ photon is detected by the ith detector and we immediately switch the atom to the b' state at t_0. In the case of rapid decay, $\frac{\Delta r_i}{c} \gg \gamma_a^{-1}$, as in Fig. 6, we have $t_0 \cong 2\Delta r_i/c$; and Eqs. (20), (21a) and (21b) yield

$$\psi^{(2)}(r_1,t_1;r_2,t_2) = \frac{\varepsilon_\gamma \varepsilon_\phi}{\Delta r_1 \Delta r_2}\left[e^{-\left(i\omega_{ab}+\gamma_a\right)\left(t_1-\frac{\Delta r_1}{c}\right)}\Theta\left(t_1 - \frac{\Delta r_1}{c}\right)\right.$$

$$\left. \times \ e^{-\left(i\omega_{bc}+\gamma_b\right)\left(t_2-2\frac{\Delta r_1}{c}-\frac{\Delta r_2}{c}\right)}\Theta\left(t_2 - \frac{2\Delta r_1}{c} - \frac{\Delta r_2}{c}\right)\right] + (1 \leftrightarrow 2). \tag{22}$$

An alternate switching arrangement is depicted in Fig. 7 involving no time delay. In this case the atom would recoil by an amount

$$p = \hbar k = \frac{\hbar v}{c} \tag{23}$$

and so the velocity of the atom would be given by

$$\frac{v}{c} = \frac{\hbar v}{mc^2}. \tag{24}$$

Comparing (22) to (17) we see that the results are quite different. For example the frequency ω_{ab} (not ω_{ac}) appears in the first term in Eq.(22).

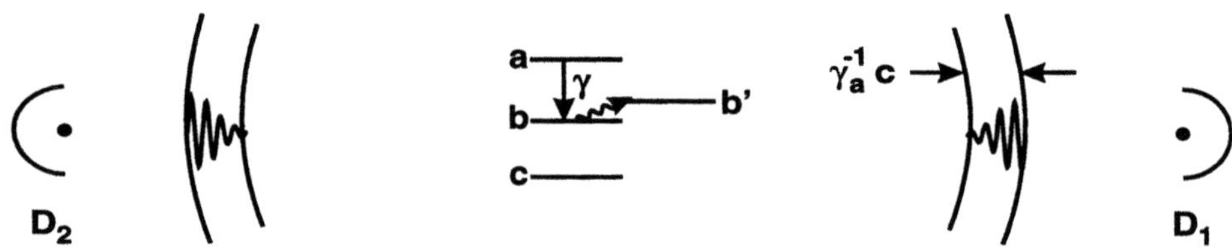

Fig. 6. Figure illustrating γ pulse wave packet is of short duration relative to $\Delta r_i/c$. Signal is sent back to the atom and after round trip time $2\Delta r_i/c$, the atom is put in b'.

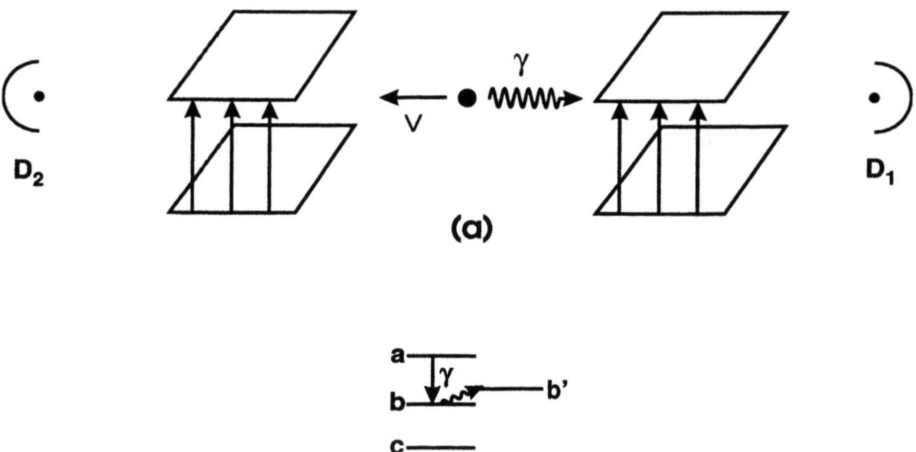

Figure 7. Atom emits photon and recoils into DC field region (7a) where it is switched into b' by interaction with field.

For a 10 eV photon and an atomic rest mass of $\approx 10^9$ eV we see that $v \approx 10^{-8}c = 300$ cm/sec. Hence if we take the plates to be separated by 100 Å it would take a time $\Delta t = (10^{-6} cm)/(300 \text{ cm/sec}) \approx 3 \times 10^{-9}$ sec for the atom to move into the field region and switch to state b'. In this case t_0 as it appears in Eq. (21b) can be made much smaller than $\Delta r_i / c$ and therefore effectively set to zero.

Hence, for this example we may write the interrupted or switched $\psi^{(2)}(r_1,t_1;r_2,t_2)$ as

$$\psi^{(2)}(r_1,t_1;r_2,t_2) = \frac{\varepsilon_r \varepsilon_\phi}{\Delta r_1 \Delta r_2}\left[e^{-(i\omega_{ab}+\gamma_a)\left(t_1-\frac{\Delta r_1}{c}\right)}\Theta\left(t_1-\frac{\Delta r_1}{c}\right) \right.$$

$$\left. \times e^{-(i\omega_{bc}+\gamma_b)\left(t_2-\frac{\Delta r_2}{c}\right)}\Theta\left(t_2-\frac{\Delta r_2}{c}\right) + (1 \leftrightarrow 2) \right]. \tag{25}$$

IV. CONCLUSION

Having established the simple form of $\psi^{(2)}_{SD}$ as given by (23) we conclude by making contact between (23) and the more general expression given by Eq. (17). In particular we note that Eq. (17) is generally valid for all γ_a and γ_b. Considering the case $\gamma_a \gg \gamma_b$, we note that $\psi^{(2)}(r_1,t_1;r_2,t_2)$ is nonzero only for times $t_1 = \frac{\Delta r_1}{c} + o\left(\gamma_a^{-1}\right)$ and write (17) as

$$\psi^{(2)}(r_1,t_1;r_2,t_2) = \frac{-\kappa}{\Delta r_1 \Delta r_2} e^{-[(i\omega_{ac}+\gamma_a)-(i\omega_{bc}+\gamma_b)]\left(t_1-\frac{\Delta r_1}{c}\right)} \Theta\left(t_1-\frac{\Delta r_1}{c}\right)$$

$$\times\ e^{-(i\omega_{bc}+\gamma_b)\left(t_2-\frac{\Delta r_2}{c}\right)} \Theta\left[t_2-\frac{\Delta r_2}{c}+o\left(\gamma_a^{-1}\right)\right] + (1 \leftrightarrow 2)$$

$$= -\frac{-\kappa}{\Delta r_1 \Delta r_2} e^{-(i\omega_{ab}+\gamma_a)\left(t_1-\frac{\Delta r_1}{c}\right)} \Theta\left(t_1-\frac{\Delta r_1}{c}\right)$$

$$\times\ e^{-(i\omega_{bc}+\gamma_b)\left(t_2-\frac{\Delta r_2}{c}\right)} \Theta\left(t_2-\frac{\Delta r_2}{c}\right) + (1 \leftrightarrow 2). \tag{26}$$

Hence in the limit $\gamma_a \gg \gamma_b$, Eq. (17) reduces to (25), as noted in the introduction.

REFERENCES

1. For a nice overview see D. Greenberger, M. Horne and A. Zeilinger, Physics Today 46:22 (1993).
2. An analysis having much in common with the present paper has been carried out by H. Huang and G. S. Agarwal; see also H. Huang and J. H. Eberly, J. Mod. Opt. 40:915 (1993).
3. Marlan O. Scully and K. Druhl, Phys. Rev. A25:2208 (1982).
4. M. Hillery and Marlan O. Scully, in *Quantum Optics, Experimental Gravity, and Measurement Theory*, edited by P. Meystre and Marlan O. Scully (1983).
5. L. Mandel, Phys. Rev. A28:929 (1983).
6. P. Kwiat, A. Steinberg, and R. Chiao, Phys. Rev. A49:61 (1994).
7. R. Ingraham, To Be Published.
8. J. Wheeler in *Problems in the Foundation of Physics* ed. by G.T. di Francia (North Holland 1979).
9. J. Franson, Phys. Rev. Lett. 62:2205 (1989); ibid. 67:2290 (1991).
10. M. Horne, A. Shimoney and A. Zeilinger, Phys. Rev. Lett. 62:2209 (1989).
11. Marlan O. Scully in *Laser Spectroscopy V*, p. 41, edited by A. McKellar and B. Stoicheff (1981).
12. Y.H. Shih and C.O. Alley, Phys. Rev. Lett. 61:2921 (1988).
13. Z.Y. Ou and L. Mandel, Phys. Rev. Lett. 61:50 (1988).
14. G. Rarity and P.R. Tapster, Phys. Rev. Lett. 64:318 (1991).
15. Z.Y. Ou, L.J. Wang, X.Y. Zou, and L. Mandel, Phys. Rev. Lett. 67:318 (1991).
16. U. Eichmann, J.C. Bergquist, J.J. Bollinger, J.M. Gilligan, W.M. Itano, D.J. Wineland, and M.G. Raizen, Phys. Rev. Lett. 70:2359 (1993).
17. Marlan O. Scully, R. Shea and J. McCullen, Phys. Reports 43:486 (1978).
18. We refer here especially to the experiments of Mandel et. al, Chiao et.al, and Franson as referred to in Ref. 1.

EINSTEIN CAUSALITY IN INTER-ATOM MICROCAVITY-CONFINED TRANSVERSE QUANTUM CORRELATIONS

Francesco De Martini and Marco Giangrasso

Dipartimento di Fisica, Universita' "La Sapienza", Roma 00185, Italy

INTRODUCTION

Because of its fundamental dynamical role, the concept of relativistic causality and its theoretical formalization is today at the center of the debate over the foundations of Quantum Mechanics and of Quantum Field Theory (QFT) [1]. Historically, the first impulse to this speculation came from the field theoretical work of Schroedinger and, of course from the subtle implications of the quantum superposition principle expressed in 1935 by the celebrated argument by Einstein, Podolsky and Rosen (EPR) [2]. In more recent years the causality issue emerged within a definite theoretical model, first investigated by Enrico Fermi back in 1932, involving two atoms suitably prepared at time t=0 and then interacting by photon exchange at later times [3]. In the last two decades this model was taken by various theoretical physicists as a paradigmatic example apt to show the evidence (or, on the contrary, the failure of evidence) of the role played by the concept of relativistic causality within the the formal apparatus of quantum mechanics [4-11]. Very recently Gerhard C. Hegerfeldt has re-considered the problem and demonstrated formally that, in spite of the conclusions reached by the work of Fermi through an apparently non orthodox mathematical procedure, a violation of the Einstein causality should in fact be expected according to his model and that this is a formal consequence of some basic operator properties commonly attributed to the Hamiltonian of the dynamical system [12]. The scope of the present work is twofold. First, the problem of relativistic causality within a two-atom coupling problem is considered from a theoretical viewpoint with the close perspective of an actual test of the causality process which is presently in progress in the Quantum Optics Laboratory at the University of Roma. Second, in order to avoid any excessive theoretical abstraction, the given theory refers explicitly to the relevant physical conditions which determine the dynamics of the atoms in our experiment. In this experiment the atoms are trapped in a transparent host at low temperature within the electromagnetic boundaries of an optical Fabry-Perot plane microcavity having a "finesse" $f >$ >0 and a longitudinal dimension $d=m\lambda_0/2$, being m the cavity-order and λ_0 the resonance atomic wave-length. There the atoms are mutually interacting through the field of a common, dominant microcavity mode with transverse coherence extension $l_c=2\lambda_0(mf)^{1/2}$ [13]. Furthermore the trapped atoms are excited by a couple of externally generated, mutually delayed femtosecond laser pulses to provide a time-resolved test of the relativistic causality of the atomic interactions established over a macroscopic distance of order l_c [14].

Advances in Quantum Phenomena, Edited by E.G. Beltrametti
and J.-M. Lévy-Leblond, Plenum Press, New York, 1995

DYNAMICS OF THE INTERATOMIC COUPLING WITHIN A MICROCAVITY

A couple of excited atoms, A and B, are confined between two parallel plane, lossless mirrors with high reflectivity $R_j \equiv |r_j|^2 \approx 1$, $|r_j|^2 + |t_j|^2 = 1$, placed at a mutual "microscopic" distance d, i.e., within a Fabry-Perot resonant microcavity with a small mode-order $m \equiv 2d/\lambda \equiv dk_0/\pi =$ integer odd number ≥ 1. By assuming a two-level model for the atoms by the corresponding spin operators $\hat\sigma_A$, $\hat\sigma_B$ defined at the space-positions r_A, $r_B \equiv r_A + R$ in the cavity, the hamiltonian of the field-atom system may be expressed by:

$$\hat H = (\tfrac{1}{2}\hbar k_{0C})(\hat\sigma_{Az} + \hat\sigma_{Bz}) + \int dk\, c\hbar k \Sigma_j (\hat a^\dagger_{jk}\hat a_{jk} + \hat a'^\dagger_{jk}\hat a'_{jk}) + i\int dk\, [\hbar kc/(16\pi\varepsilon_0)]^{1/2}\Sigma_j D_j^{-1}\exp(-ickt)\,\mu\bullet$$
$$\{\{\hat a_{kj}\,[\varepsilon(k_+j)\,t_j\,\exp(ik_+\cdot r_A) + \varepsilon(k_-j)\,t_j\,r_j\,\exp(ik_-\cdot r_A + i\,kdC)] +$$
$$\hat a'_{kj}\,[\varepsilon(k_-j)\,t_j\exp(ik_-\cdot r_A) + \varepsilon(k_+j)\,t_j\,r_j\,\exp(ik_+\cdot r_A + i\,kdC)]\}(\hat\sigma_{A+} + \xi\hat\sigma_{A-}) +$$
$$\{\hat a_{kj}\,[\varepsilon(k_+j)\,t_j\,\exp(ik_+\cdot r_B) + \varepsilon(k_-j)\,t_j\,r_j\,\exp(ik_-\cdot r_B + i\,kdC)] +$$
$$\hat a'_{kj}[\varepsilon(k_-j)\,t_j\,\exp(ik_-\cdot r_B) + \varepsilon(k_+j)\,t_j\,r_j\,\exp(ik_+\cdot r_B + i\,kdC)]\}(\hat\sigma_{B+} + \xi\hat\sigma_{B-}) + \text{h.c.}\} \qquad (1)$$

where μ is the relevant atomic electric dipole-moment vector, the the symbols j=1,2 and the unit-vector ε represents the field polarization, $C \equiv \cos\theta$, θ, φ are the spherical polar angles of the field's k-vector respect to the symmetry-axis (z) of the microcavity, assumed of infinite transverse extent, k_+, k_- the plane-wavevectors of the field originating at $z=+\infty$ and $-\infty$ respectively, $D_j \equiv [1 - |r_j|^2 \times \exp(i2\pi mC)]$ the cavity Airy factor (Fig.1). The field operators $\hat a_{kj}$, $\hat a'_{kj}$ correspond to the mode-functions $U_{kj}(r)$, $U'_{kj}(r)$ respectively, ξ is a real diagnostic parameter equal to 0 or 1 [15]. According to our experimental conditions expressed by Fig.1, the atoms lie on the plane z=0, the symmetry plane of the cavity orthogonal to the z-axis with the corresponding, mutually parallel, dipoles μ_A, μ_B. We assume first that these ones are paral-lel to the x-axis. The atoms are then locally interacting with the field $\hat c_{kj} = (\hat a_{kj} + \hat a'_{kj})$, with com-mutation property: $[\hat c_{kj}, \hat c^\dagger_{k'j'}] = 2\delta_{jj'}\delta(k-k')$. The interaction term in (1) is now expressed in the simpler form:

$$\hat H_I = -\int dk\,\Sigma_j\,g_j(C) \times \exp[-i(ckt - k\cdot r_A)]\hat c_{kj}\{[\hat\sigma_{A+} + \hat\sigma_{B+}\exp(ik\cdot R)] + \xi[\hat\sigma_{A-} + \hat\sigma_{B-}\exp(ik\cdot R)]\}$$
$$+\text{h.c.}$$

with the coupling parameter expressed by: $g_j(C) \equiv [\hbar kc/(16\pi\varepsilon_0)]^{1/2}G_m(kC)(\mu\bullet\varepsilon_j)$ and: $G_m(kC) \equiv (1 - |r|^2)^{1/2}/[1 + |r| \times \exp(i\pi mC)] \approx 2(f/\pi)^{1/2}$ for $f \gg 1$. We neglect for simplicity the j-dependence of the reflectivity and take, with the new conditions: $|k_+| = |k_-|$. We may now investigate the time behaviour of the z-component of the spin operator of one atom, say A, in the Heisenberg's representation:

$$\hbar\frac{d\hat\sigma_{Az}}{dt} = -i[\hat\sigma_{Az}, \hat H] = i\int dk\Sigma_j 2g_j(C)\exp(ik\cdot r_A)\hat c_{kj}(\hat\sigma_{A+} - \xi\hat\sigma_{A-}) + \text{h.c.}$$

The time evolution of the operators of the r.h.s. of this equation is obtained by the formal solution of an analogous equation involving the field operator $\hat c_{kj}$ in the Markov approximation [15] Evaluate now the average of the spin operator over the state of the field+atom system: $|\{n_{kj}\}, \Psi_{A,B}\rangle$. Assume here, for simplicity the case of the field's vacuum-state, viz., with all field occupation numbers: $n_{kj} = 0$. In other words, consider the process of spontaneous emission of the inter-correlated atoms A or B, placed within the spatial extension of a single transverse microcavity mode and suitably prepared in a state $\Psi_{A,B}$ The solution of the dynamical equation is reached by a previous solution of the analogous equation for the operator $\hat c_{kj}(\hat\sigma_{A+} - \xi\hat\sigma_{A-})$. The final result, reached after some algebra, is found to be:

$$\frac{d\langle\hat\sigma_{Az}\rangle}{dt} = -\Gamma_A\langle\hat\sigma_{Az}\rangle_t - i\int dk\Sigma_j|g_j|^2\langle\hat\sigma_{A+}\hat\sigma_{B-}\rangle_0 \times \{\zeta(k_0 - k)\exp(-ik\cdot R) - \xi^2\zeta(k_0 + k)\exp(ik\cdot R)\}$$
$$+ \text{h.c.} \qquad (2)$$

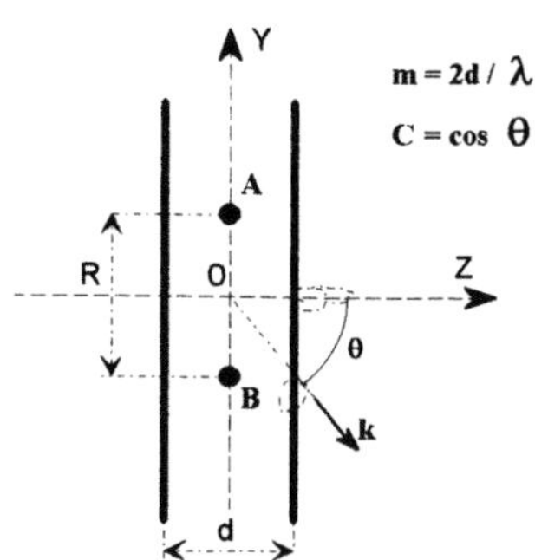

Fig. 1. Two interacting atoms A and B in an optical microcavity

The parameter Γ_A expresses the standard SpE contribution due to each atom's self-interaction: its formal expression may be found in Ref.15. The second term expresses the superradiant inter-atomic correlations. Equation (2) should include an additional (causal) correlation term, structurally similar to the one appearing at the r.h.s. above and oscillating at the high frequency $2ck_0$. It may be taken to give an average zero contribution within the time Γ_A^{-1} and then it can be neglected. The ζ-function in the r.h.s. above is defined as: $\zeta(x) \equiv [1-\exp(itx)]/x$. Very important, we may note here that the adoption of the diagnostic value $\xi=0$ in (2), corresponding to the standard adoption of a Dicke Hamiltonian as the interaction term appearing in (1), *does not* lead to a relativistically causal solution of 2). However we shall see that causality is restored by adopting the *complete* expression of $\hat{H}_I$, i.e., by taking $\xi=1$ in (2) (Ref.10). The conceptual content of the present paper consists precisely of this result together with the single-transverse mode interaction condition. Let's now carry out the integral in (2) related to the differential: $dk \equiv d\Omega \times dk = k^2 dk d\phi dC$, $C \equiv \cos\theta$, for the various possible physical dipole μ-orientations within the micocavity. Assume here, with no relevant loss of generality, that the two equal μ-vectors are oriented parallel to the mirrors and to the x-axis (Fig.1). The integral over ϕ for that dipole orientation can be carried out exactly leading to an expression proportional to a sum of integer-order Bessel functions J_n [16]. The ϕ-integral is: $I''(C) =$

$$\int_{-\pi}^{+\pi} (\sin^2\varphi + C^2\cos^2\varphi)\exp(\pm i\mathbf{k}\cdot\mathbf{R})\,d\varphi = 2\pi\left[J_0(\beta) + \beta(kR)^{-2}J_1(\beta)\right] = 2\pi\left[J_0(\beta) + (k^2R)^{-1}\frac{dJ_0(\beta)}{dR}\right]$$

with: $\beta \equiv kR(1-C^2)^{1/2}$, $|\mathbf{k}| \equiv k$. The integration over the θ-variable is more complex and is carried out exactly by a previous Fourier-series expansion of the Airy-function: $|G_m(kC)|^2 = (1-\eta^2)^{1/2} \times [1+\eta\cos(yC)]^{-1}$, with $\eta \equiv [2|r|/(1+|r|^2)]$ and: $y \equiv (\pi m/k_0)$. That expansion is: $|G_m(kC)|^2 =$

$$\left[1 + 2\sum_{n=1}^{+\infty}(-1)^n\chi^n\cos(nyC)\right]$$

with: $\chi \equiv |[(1-\eta^2)^{1/2}-1]/\eta| = |r|$. The asymptotic values of the parameter are then $\chi=0$ for $|r|=0$, $f=0$, free space and $\chi=[1-(\pi/2f)]$ for $f \gg 0$. The integration of $I''(C)$ over the C-variable concludes the procedure of spatial integration and leads to the following result given in terms of *spherical Bessel* functions j_k [16]. The result is: $I'_m(k) \equiv$

$$\int_{-1}^{+1} dC\,|G_m(kC)|^2 I''(C) = \int_{4\pi} d\Omega\,|G_m(kC)|^2\left[\exp(\pm i\mathbf{k}\cdot\mathbf{R})\right](\sin^2\varphi + C^2\cos^2\varphi) =$$

$$= 4\pi\left\{\left[j_0(kR) - (kR)^{-1}j_1(kR)\right] + 2\sum_{n=1}^{\infty}(-1)^n\chi^n\left[j_0(kR_n) - (kR_n)^{-1}j_1(kR_n)\right]\right\} \quad (3)$$

with [16]: $(R_n)^2 \equiv R^2+(ny)^2$, $j_0(x) \equiv \sin x/x$, $j_1(x) \equiv -dj_0(x)/dx \equiv (\sin x/x^2) - \cos x/x$, $j_2(x) \equiv -j_0(x) + 3j_1(x)/x$.

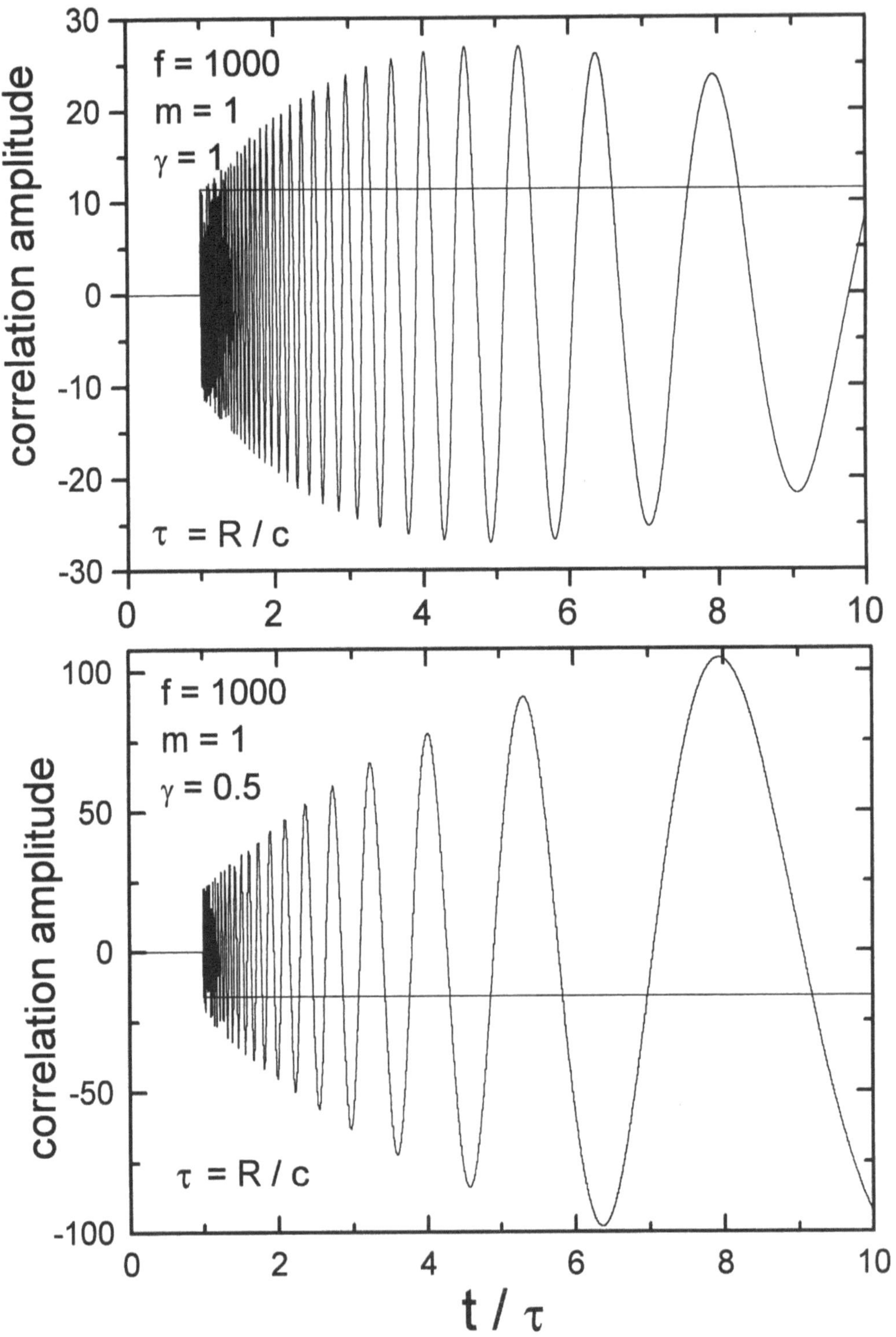

Fig . 2.   Microcavity confined correlation amplitude for two values of the
interatomic distance $\gamma = R / lc$

For the sake of completeness we give also the result appropriate to the physical condition according to which the two, mutually parallel atomic dipoles μ_A and μ_B, are taken to be parallel to the y-axis (Fig.1). The final result of the spatial integration may be expressed again in terms of (higher-order) spherical Bessel functions:

$$I'_m(k) = \int_{4\pi} d\Omega \, |G_m(kC)|^2 \exp(\pm i\mathbf{k}\cdot\mathbf{R})(\cos^2\varphi + C^2\sin^2\varphi) = 8\pi(kR)^{-1}j_1(kR) +$$

$$+ 8\pi \sum_{n=1}^{\infty} (-1)^n \chi^n \left\{ 2(kR_n)^{-1}j_1(kR_n) - \beta_n^2(R_n)^{-2}j_2(kR_n) \right\} \quad ; \beta_n \equiv n(\pi m k_0^{-1}) \quad \textbf{(3')}$$

In case of absence of the cavity, i.e., $|r|=0$, $f=0$, the expressions of (3) and (3') with $\chi=0$ reproduce the well known free-space correlation results [17].

CAUSAL GREEN-FUNCTIONS FOR CONFINED ATOMIC CORRELATIONS

The k-integration of the expressions (3) and (3') multiplied appropriately by the $\zeta(k\pm k_0)$ appearing in (2), for k ranging from $-\infty$ to from $+\infty$ leads finally to the generalized Green functions of the inter-atomic correlations in the microcavity. In the present work we refrain from giving the details of the integration procedure limiting ourselves to present the final results for the relevant experimental physical conditions.

(a) Consider first the condition of mutually parallel atomic dipoles μ_A, μ_B confined by a micro-cavity with finesse f and mode-order m, and parallel to the x-axis of Fig.2. The appro-priate Green function is evaluated correspondingly by taking into account the expressions (2) and (3) and by carrying out the integral:

$$\int_{-\infty}^{+\infty} dk \frac{1 - \exp[ict(k-k_0)]}{(k-k_0)} k^3 \left[j_0(kR) - (kR)^{-1}j_1(kR) \right].$$

In order to treat the formal divergence due to the k^3 term in the integrand, this one is multipli-ed, as usual, by a factor $\exp(-\alpha|k|)$ where $\alpha>0$ is set to zero in the final result of the integration. An appropriate choice of the integration contour and a careful consideration of the poles leads to the result with no further complications. If we consider the infinite sum over the n-terms in (3) the final result of the integration give the generalized Green function related to the stated physical condition. It is expressed by: $D(R,t) \equiv D_{ret}(t - R_n/c) =$

$$4\pi^2 \left\{ \frac{k_0^2}{R} - \frac{ik_0}{R^2} + \frac{1}{R^3} \right\} \exp(ik_0 R) \times \Theta(ct-R) + 8\pi^2 \sum_{n=1}^{+\infty} (-1)^n \chi^n \left\{ \frac{k_0^2}{R_n} - \frac{ik_0}{R_n^2} + \frac{1}{R_n^3} \right\} \exp(ik_0 R_n) \times \Theta(ct-R_n)$$

where the retardation, and then the relativistic causality, is expressed by the Heaviside step function $\theta(z)=1$, for $z>0$ and $\theta(z)=0$, $z<0$. Note in this connection that in the theory no preselection of retarded vs advanced propagation solutions has been made. Note also that $R_n^2 \geq R^2$. The first term in the above expression corresponds to a direct inter-atom correlation over the distance R in free space, i.e., in absence of the cavity, $|r|=0$. The second term is the superposition, with alternate signs, of differently retarded addenda leading to the peculiar quasi-oscillatory response shown in Fig.2. Such behaviour has been found to be strongly dependent on the cavity length, m, on the "finesse" f and on R. The causal retardation effect and this last R-dependence is shown in Fig.2. There the R-dependence is expressed by the parameter $\gamma \equiv R/l_c$ being $l_c \equiv 2\lambda_0(mf)^{1/2}$ the cavity *transverse coherence length* [13]. Fig.2 shows also, superimposed to the (real-part) of the correlation-intensity curve, the corresponding free-space contribution, viz., the correlation-intensity that is expected when the cavity confinement is made to vanish. By this comparison the enhancing cavity effect on the strength of the inter-atom correlation in spontaneous emission may be appreciated.

(b) Consider now that the mutually parallel atomic dipoles μ_A, μ_B confined by a microcavity with finesse f and mode-order m, are taken parallel to the y-axis of Fig.2. For this case we may carry out an analogous contour integral of (3') leading to the following result: $D(R,t)=$

$$8\pi^2\left\{\frac{ik_0}{R^2}-\frac{1}{R^3}\right\}\exp(ik_0R)\times\theta(ct-R)+8\pi^2\sum_{n=1}^{\infty}(-1)^n\chi^n\left\{\beta_n^2\left[\frac{k_0^2}{R_n^3}-i\frac{k_0}{R_n^4}+\frac{1}{R_n^5}\right]+2\left[\frac{ik_0}{R_n^2}-\frac{1}{R_n^3}\right]\right\}\times$$

$$\times\exp(ik_0R_n)\times\theta(ct-R_n)$$

Note for this dynamical condition the absence of long-range interactions, proportional to R^{-1}.

CONCLUSIONS

In the present work we limit ourselves to report the basic solution of the simplest quantum mechanical problem related to the interaction in a microcavity of two atoms with the vacuum field's state. In other words we have presented the theory of the two-atoms superfluorescent spontaneous emission in a cavity. For this rather old problem [3-9] we show, through an explicit and complete formal solution of the Heisenberg's equations, that the Einstein causality is indeed preserved for inter-atom interactions even within the context of the nonrelativistic formulation of quantum mechanics. The key argument leading to this arguably non irrelevant result, consists of a sound, and indeed straigthforward, choice of the Hamiltonian of the problem, a procedure already suggested in work [10]. Our work represents a significant step toward a more complete account of the causality problem involving nonlocal quantum correlations, stimulated emission and high-order nonlinear processes. This more complete work, that will be reported in a forthcoming paper, represents the theoretical support of the experimental test of quantum causality now in progress at the Università di Roma "La Sapienza".

ACKNOWLEDGEMENTS

We acknowledge the support by Istituto Nazionale di Fisica Nucleare and Istituto Nazionale di Fisica della Materia.

REFERENCES

1. M.Cushing in, "Philosophical Foundations of Quantum Field Theory", H.Brown and R.Harré eds., Clarendon , Oxford (1990)
2. M.Redhead. "Incompleteness, Nonlocality and Realism", Clarendon, Oxford (1987)
3. E.Fermi, *Rev.Mod.Phys.* 4, 87 (1932)
4. M.I.Shirokov, *Yad.Fiz.* 4, 1077 (1966) [Sov. J. Nucl. Phys. 4, 774 (1967)]
5. W.Heitler and S. T. Ma, *Proc. R. Ir.Acad.* 52, 123 (1949)
6. J.Hamilton, *Proc. Phys. Soc.* A 62, 12 (1950).
7. M.Fierz, *Helv. Phys. Acta* 23, 731 (1950).
8. P.W.Milonni and P.L.Knight, *Phys. Rev.* A 10, 1096 (1974).
9. M.H.Rubin, Phys. Rev. D 35, 3836 (1987).
10. A.Biswas, G.Compagno, G.Palma, R.Passante and F.Persico, *Phys. Rev.* A 42, 4291 (1990)
11. A.Valentini, *Phys. Lett.* A 153, 321 (1991).
12. G.C.Hegerfeldt, *Phys. Rev. Lett.* 72, 596 (1994).
13. F.De Martini, M.Marrocco and D.Murra, Phys.Rev.Lett. 65,1853 (1990)
14. A.Aiello, F.De Martini, M.Giangrasso, P.Mataloni and D.Murra, in: "Ultrafast Phenomena IX", W.Knox ed., Springer-Verlag, Berlin (1994).
15. F.De Martini, M.Marrocco, P.Mataloni, L.Crescentini and R.Loudon, *Phys.Rev.*A, 43, 2480 (1991)
16. "Handbook of Mathematical Functions", M.Abramowitz , I.Stegun, eds. Dover, New York (1965)
17. D.P.Craig, T.Thirunamachandran, "Molecular Q.E.D.",Academic Press, New York (1984)

THREE COMMENTS ON THE AHARONOV-BOHM EFFECT[*]

Michael Berry

H H Wills Physics Laboratory
Tyndall Avenue
Bristol BS8 1TL, UK

The discovery by Aharonov and Bohm in 1959 (and its partial anticipation ten years earlier by Ehrenberg and Siday) was surprising and important and has been a fertile source of further discoveries. Neverthless some aspects of it have since become the subject of unnecessary mystification, embodied in three commonly held views on which I will comment.

1. THE AB EFFECT CONTRADICTS COMMON SENSE.

The 'common sense' is that fields are physical entities, whereas the potentials the AB effect relies upon (to describe nonlocal influences of fields) are mere mathematical conveniences. This opinion is ahistorical because it ignores the fact that fields were themselves originally conceived as mathematical conveniences - to help describe the distant (that is nonlocal) effects of charges and gravitating masses. These actions at a distance were certainly not common sense, and children who play with magnets or are told that the tides are caused by the moon find these phenomena miraculous today (so do I). One of AB's achievements was to contribute to the process whereby yesterday's mathematical constructions prove their worth by pointing to new phenomena, and so become regarded as physical entities today.

2. THE AB EFFECT IS NONCLASSICAL.

This is doubly misleading. First, although distant fields play no part in Newtonian classical mechanics (which deals only with forces), they contribute in Hamiltonian classical mechanics via the action (which depends on canonical rather than kinetic momentum and so involves the vector potential). Second - and in a deeper sense - the effect is a general one of wave physics, not confined to quantum mechanics. Any wave on a current, that is in a medium moving with velocity $u(r)$ (much smaller than the wave speed), satisfies an equation with the same structure as Schrödinger's, in which $u(r)$ plays the role of a vector potential. It follows that the analogue of a vector potential which describes a field vanishing everywhere outside a line is a flow for which $\nabla \times u(r)=0$ outside the line, and this is the flow of an irrotational vortex. Therefore one way to display the AB wavefunction is to let ripples on the surface of water encounter a bathtub vortex. (The analogue of the dimensionless AB quantum flux parameter is the circulation of the vortex, divided by the product of the

---

ripples' wavelength and group velocity.) This experiment was successfully carried out (Berry et al 1980). Taking a wider perspective, there are four sorts of AB effect, depending on whether the wave or the flux is quantum or classical. In AB's original proposal, and most subsequent experiments, the wave was quantum and the flux classical. In the more recent experiments by Tonomura, involving superconducting rings, both the wave and flux were quantum. V. Steinberg (private communication) is exploring the 'classical-quantum' case, in which waves in liquid helium would encounter rotons (quantized vortices). And of course the bathtub experiment described above is the 'classical-classical' case.

3. THE AB EFFECT DEPENDS ON IDEALISATIONS THAT DO NOT APPLY IN REALITY.

The idealisations were that the electron wave does not penetrate into the magnetic flux, that the solenoid is infinitely long (so that there is no return flux) and that the flux has been switched on forever (so that there was never any electric field induced outside by the changing flux). In any real experiment these ideal conditions are achieved only approximately (even in Tonomura's experiments with superconducting rings there must be some penetration of the flux by the electrons - of order 10^{-23}, or perhaps even $\exp\{-10^{+23}\}$, but strictly nonzero). But when there is the slightest violation of the idealisations it is possible (as several people have shown) to choose a formulation in which the fringe shift arises entirely from the tiny real fields that coexist with the electrons. Therefore there is a sense in which the AB effect - nonlocal influences of distant fields - does not exist in reality. However, there is a heavy price for maintaining this view. As the idealisation-violating fields get smaller, the fringe shifts approach that calculated by AB. Therefore those who maintain that the shifts arise from real fields (as opposed to the gauge-invariant phase factor involving the vector potential) must tolerate finite causes produced by infinitesimal effects. Elsewhere (Berry 1986), I have discussed this point, and its relation to other limiting processes in physics, in a little more detail.

References

Berry M.V., Chambers R.G., Large M.D., Upstill C. and Walmsley J.C., 1980, Wavefront dislocations in the Aharonov-Bohm effect and its water-wave analogue, *Eur. J. Phys.* 1: 154-162.

Berry M.V., 1986, The Aharonov-Bohm effect is real physics, not ideal physics, in *NATO ASI series* 144: 319-320, *Fundamental aspects of quantum theory* (eds: V.Gorini and A. Frigerio; Plenum).

PROTECTIVE MEASUREMENTS

Yakir Aharonov[a,b] and Lev Vaidman[a]

[a] School of Physics and Astronomy
Raymond and Beverly Sackler Faculty of Exact Sciences
Tel-Aviv University
Tel-Aviv, 69978 ISRAEL

[b] Physics Department, University of South Carolina
Columbia, South Carolina 29208, U.S.A.

The content of this lecture has already appeared in several proceedings so here we will present only a short abstract together with the list of references.

At present, the commonly accepted interpretation of the Schrödinger wave is that which leads to the probability density. This interpretation stems from the belief that the Schrödinger wave can only be tested for an ensemble of particles. We have proposed new type of measurements: "protective measurements" which allow direct measurement of the Schrödinger wave density on a single particle. We have shown that one can simultaneously measure the density and the current of the Schrödinger wave in seveal positions. The results of these measurements then allow to reconstruct the Schrödinger wave.

As an example of a simple protective measurement, let us consider a particle in a discrete nondegenerate energy eigenstate $\Psi(x)$. The standard von Neumann procedure for measuring the value of an observable A in this state involves an interaction Hamiltonian, $H = g(t)PA$, coupling the system to a measuring device, or pointer, with coordinate and momentum denoted, respectively, by Q and P. The time-dependent coupling $g(t)$ is normalized to $\int g(t)dt = 1$, and the initial state of the pointer is taken to be a Gaussian centered around zero. In standard impulsive measurements, where $g(t) \neq 0$ for only a very short time interval, the interaction term dominates the rest of the Hamiltonian, and the time evolution $e^{-\frac{i}{\hbar}PA}$ leads to a correlated state: eigenstates of A with eigenvalues a_n are correlated to measuring device states in which the pointer is shifted by these values a_n. By contrast, the protective measurements of interest here utilize the opposite limit of extremely slow measurement. We take $g(t) = 1/T$ for most of the time T and assume that $g(t)$ goes to zero gradually before and after the period T. We choose the initial state of the measuring device such that the canonical conjugate P (of the pointer variable Q) is bounded. We also assume that P is a constant of motion not only of the interaction Hamiltonian, but of the whole Hamiltonian. For $g(t)$ smooth enough we obtain an adiabatic process

Advances in Quantum Phenomena, Edited by E.G. Beltrametti
and J.-M. Lévy-Leblond, Plenum Press, New York, 1995

in which the particle cannot make a transition from one energy eigenstate to another, and, in the limit $T \to \infty$, the interaction Hamiltonian does not change the energy eigenstate. For any given value of P, the energy of the eigenstate shifts by an infinitesimal amount given by first order perturbation theory: $\delta E = \langle H_{int} \rangle = \langle A \rangle P/T$. The corresponding time evolution $e^{-iP\langle A \rangle /\hbar}$ shifts the pointer by the average value $\langle A \rangle$. This result contrasts with the usual (strong) measurement in which the pointer shifts by one of the eigenvalues of A. By measuring the averages of a sufficiently large number of variables A_n, the full Schrödinger wave $\Psi(x)$ can be reconstructed to any desired precision.

The main idea is presented in Ref. 4 and elaborated in Ref. 5. In Ref. 1 an apparent contradiction with causality is resolved. In Refs. 6-9 there are (partially critical) discussions of the proposal, and our reply appears in Ref. 10. More discussions and theoretical analyses of possible realistic experiments are presented in Refs. 11-12. Refs. 13-14 present experimental work which is close to what we propose. Experiments with single trapped atoms, which conceptually are exact realization of protective measurements of the Schrödinger wave, are given in Ref. 15. However, the resolution in these experiments is still one order of magnitude too large, so the observed "cloud" is not the wave of the atom. Our proposal to a a further development of the idea, the protective measurements of a two-state vector, will be presented in Ref. 16.

REFERENCES

1. Y. Aharonov and L. Vaidman, in *Quantum Control and Measurement*, H. Ezawa and Y. Murayama (eds.), Elsevier Publ., Tokyo, 99 (1993).
2. Y. Aharonov and L. Vaidman, in *Fundamental Problems in Quantum Physics*, Oviedo 1993, M. Ferrero and A. van der Merwe (eds.), Denver Publ., to be published.
3. Y. Aharonov and L. Vaidman, in *Nanostructures and Quantum Effects*, H. Sakaki and H. Noge (eds.), Springer-Verlag, Heidelberg, to be published.
4. Y. Aharonov and L. Vaidman, *Phys. Lett.* **A 178**, 38 (1993).
5. Y. Aharonov, J. Anandan, and L. Vaidman, *Phys. Rev.* **A 47**, 4616 (1993).
6. C. Rovelli, *Phys. Rev.* **A**, to be published.
7. W. G. Unruh, *Phys. Rev.* **A**, to be published.
8. D. Freedman, *Science*, **259**, 1542 (1993).
9. J. Samuel and R. Nityananda, e-board: gr-qc/9404051.
10. Y. Aharonov, J. Anandan, and L. Vaidman, The Meaning of Protective Measurements, TAUP-2195-94.
11 J. Anandan, *Found. Phys. Lett.* **6**, 503 (1993).
12 S. Nussinov, "Realistic Experiments for Measuring the Wave Function of a Single Particle", UCLA preprint UCLA/94/TEP/2.
13. T.P. Spiller, T.D. Clark, R.J. Prance, and A. Widom, *Prog. Low Temp. Phys.* XIII (1992) 219.
14. T.P. Spiller, T.D. Clark, R.J. Prance, and A. Widom, *Jap. J. Appl. Phys.*, Series **9**, 53 (1993).
15. G. Birkl, S Kassner, and H. Walter, *Nature* **357** 310 (1992).
16. Y. Aharonov and L. Vaidman, in *Fundamental Problems in Quantum Theory*, D. Greenberger, (ed.) NYAS, to be published.

WEAK MEASUREMENTS

Lev Vaidman

School of Physics and Astronomy
Raymond and Beverly Sackler Faculty of Exact Sciences
Tel-Aviv University
Tel-Aviv, 69978 ISRAEL

1. PRE- AND POST-SELECTED QUANTUM SYSTEMS

In 1964 Aharonov, Bergmann and Lebowitz[1] considered measurements performed on a quantum system between two other measurements, results of which were given. They proposed describing the quantum system between two measurements by using two states: the usual one, evolving towards the future from the time of the first measurement, and a second state evolving backwards in time, from the time of the second measurement. If a system has been prepared at time t_1 in a state $|\Psi_1\rangle$ and is found at time t_2 in a state $|\Psi_2\rangle$, then at time t, $t_1 < t < t_2$, the system is described by

$$\langle\Psi_2|e^{i\int_{t_2}^{t} Hdt} \quad \text{and} \quad e^{-i\int_{t_1}^{t} Hdt}|\Psi_1\rangle.$$

For simplicity, we shall consider the free Hamiltonian to be zero; then the system at time t is described by the two states $\langle\Psi_2|$ and $|\Psi_1\rangle$, see Fig. 1. In order to obtain such a system we prepare an ensemble of systems in the state $|\Psi_1\rangle$, perform measurement of desired variable using separate measuring devices for each system in the ensemble, and perform the post-selection measurement. If the outcome of the post-selection was not the desired result, we discard the system and corresponding measuring device. We look only on measuring devices corresponding to the systems post-selected in the state $\langle\Psi_2|$.

The basic concepts of the two-state approach, weak measurement and weak values, were developed several years ago.[2,3] The weak value of any physical variable A in the time interval between pre-selection of the state $|\Psi_1\rangle$ and post-selection of the state $|\Psi_2\rangle$ is given by

$$A_w \equiv \frac{\langle\Psi_2|A|\Psi_1\rangle}{\langle\Psi_2|\Psi_1\rangle} \quad . \tag{1}$$

Let us present the main idea by way of a simple example. We consider, at time t, a quantum system which was prepared at time t_1 in the state $|B = b\rangle$ and was found at time t_2 in the state $|C = c\rangle$, $t_1 < t < t_2$. The measurements at times t_1 and t_2 are

Advances in Quantum Phenomena, Edited by E.G. Beltrametti
and J.-M. Lévy-Leblond, Plenum Press, New York, 1995

complete measurements of, in general, noncommuting variables B and C. The free Hamiltonian is zero, and therefore, the first quantum state at time t is $|B = b\rangle$. In the two-state approach we characterize the system at time t by backwards-evolving state $\langle C = c|$ as well. Our motivation for including the future state is that we know that if a measurement of C has been performed at time t then the outcome is $C = c$ with probability 1. This intermediate measurement, however, destroys our knowledge that $B = b$, since the coupling of the measuring device to the variable C can change B. The idea of weak measurements is to make the coupling with the measuring device sufficiently weak so B does not change. In fact, we require that both quantum states do not change, neither the usual one $|B = b\rangle$ evolving towards the future nor $\langle C = c|$ evolving backwards.

During the whole time interval between t_1 and t_2, both $B = b$ and $C = c$ are true (in some sense). But then, $B + C = b + c$ must also be true. The latter statement, however, might not have meaning in the standard quantum formalism because the sum of the eigenvalues $b + c$ might not be an eigenvalue of the operator $B + C$. An attempt to measure $B + C$ using a standard measuring procedure will lead to some change of the two quantum states and thus the outcome will not be $b + c$. A weak measurement, however, will yield $b + c$.

When the "strong" value of an observable is known with certainty, i.e., we know the outcome of an ideal (infinitely strong) measurement with probability 1, the weak value equal to the strong value. Let us analyze the example above. The strong value of B is b, its eigenvalue. The strong value of C is c, as we know from *retrodiction*. From the definition (1) immediately follows: $B_w = b$ and $C_w = c$. But weak values, unlike strong values, are defined not just for B and C but for *all* operators. The strong value of the sum $B + C$ when $[B, C] \neq 0$ is not defined, but the weak value of the sum is: $(B + C)_w = b + c$.

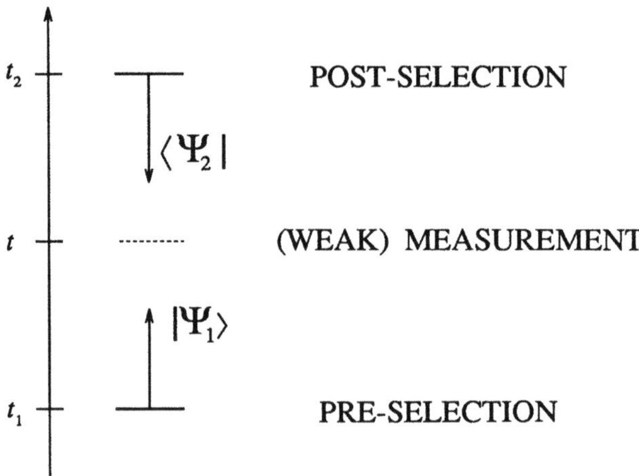

Figure 1. Pre- and post-selected quantum system. The system is considered only if it was found at time t_2 in the state $|\Psi_2\rangle$ after being pre-selected at time t_1 in the state $|\Psi_1\rangle$. The two states yield proper description of the system for analysis of various measurements at time t.

2. QUANTUM MEASURING PROCEDURE

In the standard approach to measurements in quantum theory, we measure observables which correspond to Hermitian operators. The latter have eigenvalues and a (good) measurement must yield one of these eigenvalues. If the state of a quantum system is not an eigenstate of the measured operator, then one can predict only probabilities for different outcomes of the measurement. The state of the system invariably "collapses" to an outcome corresponding to one eigenvalue. A standard measurement of a variable A is modeled in the von Neumann theory of measurement[4] by a Hamiltonian

$$H = g(t)PA \quad , \tag{2}$$

where P is a canonical momentum, conjugate to the pointer variable Q of the measuring device. The function $g(t)$ is nonzero only for a very short time interval corresponding to the measurement, and is normalized so that $\int g(t)dt = 1$. During the time of this impulsive measurement, the Hamiltonian (2) dominates the evolution of the measured system and the measuring device. Since $[A, H] = 0$, the variable A does not change during the measuring interaction. The initial state of the pointer variable is usually modeled by a Gaussian centered at zero:

$$\Phi_{in}(Q) = (\Delta^2 \pi)^{-1/4} e^{-Q^2/2\Delta^2}. \tag{3}$$

Therefore, if the initial state of the system is a superposition $|\Psi_1\rangle = \Sigma \alpha_i |a_i\rangle$, then after the interaction (2) the state of the system and the measuring device is:

$$(\Delta^2 \pi)^{-1/4} \Sigma \alpha_i |a_i\rangle e^{-(Q-a_i)^2/2\Delta^2}. \tag{4}$$

If the separation between various eigenvalues a_i is much larger than the width of the Gaussian Δ, we obtain strict correlation between the values of the variable A and nearly orthogonal states of the measuring device. The measuring procedure continues with an amplification scheme which yields effective (or, according to some physicists, real) collapse to one of the pointer positions and the corresponding eigenstate $|a_i\rangle$. In this model the only possible outcomes of the measurement of the quantum variable A are the eigenvalues a_i. This fact perfectly matches the premise that the only values which can be associated with A are the a_i.

When a quantum system is in a state $|\Psi\rangle$, one can associate, mathematically, to a variable A the the value $\langle \Psi|A|\Psi\rangle$. However, it was commonly believed that $\langle \Psi|A|\Psi\rangle$ is unmeasurable for a single system, and it has physical meaning only for an ensemble of identical systems prepared in the state $|\Psi\rangle$. For the ensemble, $\langle \Psi|A|\Psi\rangle$ is interpreted as statistical average over the results of measuring A on this ensemble. We, however, claim that $\langle \Psi|A|\Psi\rangle$ is more than just a statistical concept. It has physical meaning, since it can be measured directly on a single system (and not just calculated from the statistics of the results a_i).

It is clear why standard measurements cannot yield $\langle \Psi|A|\Psi\rangle$. The expectation value is a property of the quantum state and the state itself is significantly changed during the measuring interaction (2). Thus, in order to obtain $\langle \Psi|A|\Psi\rangle$ on a single system we need a *weak* coupling to the measuring device. And indeed, under certain conditions, including a weakened coupling, von Neumann procedure for a measurement of A yields $\langle \Psi|A|\Psi\rangle$. This procedure is discussed in our other lecture.[5] Here we want to discuss weak measurements on pre- and post-selected quantum systems. The outcomes of such measurements are the weak values defined in Eq. (1).

3. THE WEAK VALUE IS THE OUTCOME OF WEAK MEASUREMENTS

The system at time t in a pre- and post-selected ensemble is defined by two states, the usual one evolving from the time of the preparation and the state evolving backwards in time from the post-selection. We may neglect the free Hamiltonian if the time between the pre-selection and the post-selection is very short. Consider a system which has been pre-selected in a state $|\Psi_1\rangle$ and shortly afterwards post-selected in a state $|\Psi_2\rangle$. The weak value of any physical variable A in the time interval between the pre-selection and the post-selection is given by Eq. (1). Let us show briefly how weak values emerge from a measuring procedure with a sufficiently weak interaction.

We consider a sequence of measurements: a pre-selection of $|\Psi_1\rangle$, a (weak) measurement interaction of the form of Eq. (2), and a post-selection measurement finding the state $|\Psi_2\rangle$. The state of the measuring device after this sequence is given (up to normalization) by

$$\Phi(Q) = \langle\Psi_2|e^{-iPA}|\Psi_1\rangle e^{-Q^2/2\Delta^2} \quad . \tag{5}$$

After simple algebraic manipulation we can rewrite it (in the P-representation) as

$$\tilde{\Phi}(P) = \langle\Psi_2|\Psi_1\rangle \; e^{-iA_w P} \; e^{-\Delta^2 P^2/2}$$
$$+ \langle\Psi_2|\Psi_1\rangle \sum_{n=2}^{\infty} \frac{(iP)^n}{n!} [(A^n)_w - (A_w)^n] e^{-\Delta^2 P^2/2} \, . \tag{6}$$

If Δ is sufficiently large, then we can neglect the second term of (6) when we Fourier transform back to the Q-representation. Large Δ corresponds to weak measurement in the sense that the interaction Hamiltonian (2) is small. Thus, in the limit of weak measurement, the final state of the measuring device (in the Q-representation) is

$$\Phi(Q) = (\Delta^2\pi)^{-1/4} e^{-(Q-A_w)^2/2\Delta^2} \quad . \tag{7}$$

This state represents a measuring device pointing to the weak value, A_w.

Weak measurements on pre- and post-selected ensembles yield, instead of eigenvalues, a value which might lie far outside the range of the eigenvalues. Although we have showed this result for a specific von Neumann model of measurements, the result is completely general: any coupling of a pre- and post-selected system to a variable A, provided the coupling is sufficiently weak, results in effective coupling to A_w. This weak coupling between a single system and the measuring device will not, in most cases, lead to a distinguishable shift of the pointer variable, but collecting the results of measurements on an ensemble of pre- and post-selected systems will yield the weak values of a measured variable to any desired precision.

When the strength of the coupling to the measuring device goes to zero, the outcomes of the measurement invariably yield the weak value. To be more precise, a measurement yields the real part of the weak value. Indeed, the weak value is, in general, a complex number, but its imaginary part will contribute only a phase to the wave function of the measuring device in the position representation of the pointer. Therefore, the imaginary part will not affect the probability distribution of the pointer position which is what we see in a usual measurement. However, the imaginary part of the weak value also has physical meaning. It expresses itself as a change in the conjugate momentum of the pointer variable.[3]

4. AN EXAMPLE: SPIN MEASUREMENT

Let us consider a simple the Stern-Gerlach experiment: measurement of a spin component of a spin-1/2 particle. We shall consider a particle prepared in the initial state spin "up" in the $\hat{x}$ direction and post-selected to be "up" in the $\hat{y}$ direction. At the intermediate time we measure, weakly, the spin component in the $\hat{\xi}$ direction which is bisector of $\hat{x}$ and $\hat{y}$, i.e., $\sigma_\xi = (\sigma_x + \sigma_y)/\sqrt{2}$. Thus $|\Psi_1\rangle = |\uparrow_x\rangle$, $|\Psi_2\rangle = |\uparrow_y\rangle$, and the weak value of σ_ξ in this case is:

$$(\sigma_\xi)_w = \frac{\langle \uparrow_y | \sigma_\xi | \uparrow_x \rangle}{\langle \uparrow_y | \uparrow_x \rangle} = \frac{1}{\sqrt{2}} \frac{\langle \uparrow_y | (\sigma_x + \sigma_y) | \uparrow_x \rangle}{\langle \uparrow_y | \uparrow_x \rangle} = \sqrt{2} \ . \tag{8}$$

This value is, of course, "forbidden" in the standard interpretation where a spin component can obtain the (eigen)values ± 1 only.

An effective Hamiltonian for measuring σ_ξ is

$$H = g(t) P \sigma_\xi \quad . \tag{9}$$

Writing the initial state of the particle in the σ_ξ representation, and assuming the initial state (3) for the measuring device, we obtain that after the measuring interaction the quantum state of the system and the pointer of the measuring device is

$$\cos{(\pi/8)} |\uparrow_\xi\rangle e^{-(Q-1)^2/2\Delta^2} + \sin{(\pi/8)} |\downarrow_\xi\rangle e^{-(Q+1)^2/2\Delta^2} \quad . \tag{10}$$

The probability distribution of the pointer position, if it is observed now without post-selection, is the sum of the distributions for each spin value. It is, up to normalization,

$$prob(Q) = \cos^2{(\pi/8)} e^{-(Q-1)^2/\Delta^2} + \sin^2{(\pi/8)} e^{-(Q+1)^2/\Delta^2} \quad . \tag{11}$$

In the usual strong measurement $\Delta \ll 1$. In this case, as shown on Fig. 2a, probability distribution of the pointer is localized around -1 and $+1$ and it is strongly correlated to the values of the spin, $\sigma_z = \pm 1$.

Weak measurement corresponds to Δ which is much larger than the range of the eigenvalues, i.e., $\Delta \gg 1$. Fig. 2b shows that the pointer distribution has a large uncertainty, but it is peaked between the eigenvalues, more precisely, at the expectation value $\langle \uparrow_x | \sigma_\xi | \uparrow_x \rangle = 1/\sqrt{2}$. An outcome of an individual measurement usually will not be close to this number, but it can be found from an ensemble of such measurements, see Fig. 2c. Note, that we have not yet considered the post-selection.

In order to simplify the analysis of measurements on the pre- and post-selected ensemble, let us assume that we first make the post-selection of the spin of the particle and only then look on the pointer of the device that weakly measures σ_ξ. We must get the same result as if we first look at the outcome of the weak measurement, make the post-selection, and discard all readings of the weak measurement corresponding to the cases in which the result is not $\sigma_y = 1$. The post-selected state of the particle in the σ_ξ representation is $|\uparrow_y\rangle = \cos{(\pi/8)}|\uparrow_\xi\rangle - \sin{(\pi/8)}|\downarrow_\xi\rangle$. The state of the measuring device after the post-selection of the spin state is obtained by projection of (10) onto the post-selected state:

$$\Phi(Q) = \mathcal{N} \left(\cos^2{(\pi/8)} e^{-(Q-1)^2/2\Delta^2} - \sin^2{(\pi/8)} e^{-(Q+1)^2/2\Delta^2} \right) \ , \tag{12}$$

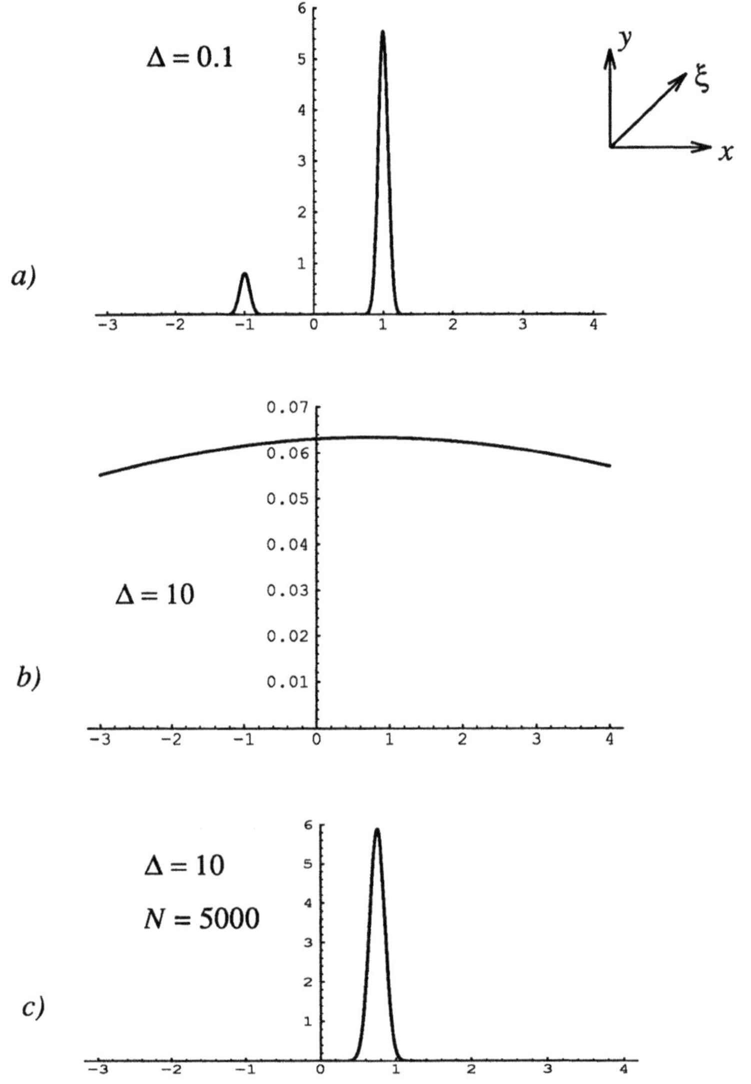

Figure 2. Spin component measurement without post-selection. Probability distribution of the pointer variable for measurement of σ_ξ when the particle is pre-selected in the state $|\uparrow_x\rangle$. (a) Strong measurement, $\Delta = 0.1$. (b) Weak measurement, $\Delta = 10$. (c) Weak measurement on the ensemble of 5000 particles. The original width of the peak, 10, is reduced to $10/\sqrt{5000} \simeq 0.14$. In the strong measurement (a) the pointer is localized around the eigenvalues ± 1, while in the weak measurements (b) and (c) the peak is located in the expectation value $\langle \uparrow_x | \sigma_\xi | \uparrow_x \rangle = 1/\sqrt{2}$.

where $\mathcal{N}$ is a normalization factor. The probability distribution of the pointer variable is given by

$$prob(Q) = \mathcal{N}^2 \left(\cos^2\left(\pi/8\right) e^{-(Q-1)^2/2\Delta^2} - \sin^2\left(\pi/8\right) e^{-(Q+1)^2/2\Delta^2} \right)^2 . \qquad (13)$$

If the measuring interaction is strong, $\Delta \ll 1$, then the distribution is localized around the eigenvalues ± 1 (mostly around 1 since the pre- and post-selected probability to find $\sigma_\xi = 1$ is more than 85%), see Figs. 3a, 3b. But when the strength of the coupling is weakened, i.e., Δ is increased, the distribution gradually changes to a single broad peak around $\sqrt{2}$, the weak value, see Figs. $3c - 3e$.

The width of the peak is large and therefore each individual reading of the pointer usually will be pretty far from $\sqrt{2}$. The physical meaning of the weak value, in this case, can be associated only with an ensemble of pre- and post-selected particles. The accuracy of defining the center of the distribution goes as $1/\sqrt{N}$, so increasing N, the number of particles in the ensemble, we can find the weak value with any desired precision, see Fig. $3f$.

In our example, the weak value of the spin component is $\sqrt{2}$, which is only slightly more than the maximal eigenvalue, 1. By appropriate choice of the pre- and post-selected states we can get pre- and post-selected ensembles with arbitrarily large weak value of a spin component. One of the first proposals[6] was to obtain $(\sigma_\xi)_w = 100$. In this case the post-selected state is nearly orthogonal to the pre-selected state and, therefore, the probability to obtain appropriate post-selection becomes very small. While in the case of $(\sigma_\xi)_w = \sqrt{2}$ the (pre- and) post-selected ensemble was just half of the pre-selected ensemble, in the case of $(\sigma_\xi)_w = 100$ the post-selected ensemble will be smaller than the original ensemble by the factor of $\sim 10^{-4}$.

5. WEAK MEASUREMENTS ON A SINGLE SYSTEM

We have shown that weak measurements can yield very surprising values which are far from the range of the eigenvalues. However, the uncertainty of a single weak measurement (i.e., performed on a single system) in the above example is larger than the deviation from the range of the eigenvalues. Each single measurement separately yields almost no information and the weak value arises only from the statistical average on the ensemble. The weakness and uncertainty of the measurement goes together. Weak measurement corresponds to small value of P in the Hamiltonian (2) and, therefore, the uncertainty in P has to be small. This requires large Δ, the uncertainty of the pointer variable. Of course, we can construct measurement with large uncertainty which is not weak at all, for example, by preparing the measuring device in a mixed state instead of a Gaussian, but no precise measurement with weak coupling is possible. So, usually, a weak measurement on a single system will not yield the weak value with a good precision. However, there are special cases when it is not so. Usual strength measurement on a single pre- and post-selected system can yield "unusual" (very different from the eigenvalues) weak value with a good precision. Good precision means that the uncertainty is much smaller than the deviation from the range of the eigenvalues.

Our example above was not such a case. The weak value $(\sigma_\xi)_w = \sqrt{2}$ is larger than he highest eigenvalue, 1, only by ~ 0.4, while the uncertainty, 1, is not sufficiently large for obtaining the peak of the distribution near the weak value, see Fig. 3c. Let us modify our experiment in such a way that a single experiment will yield meaningful surprising result.

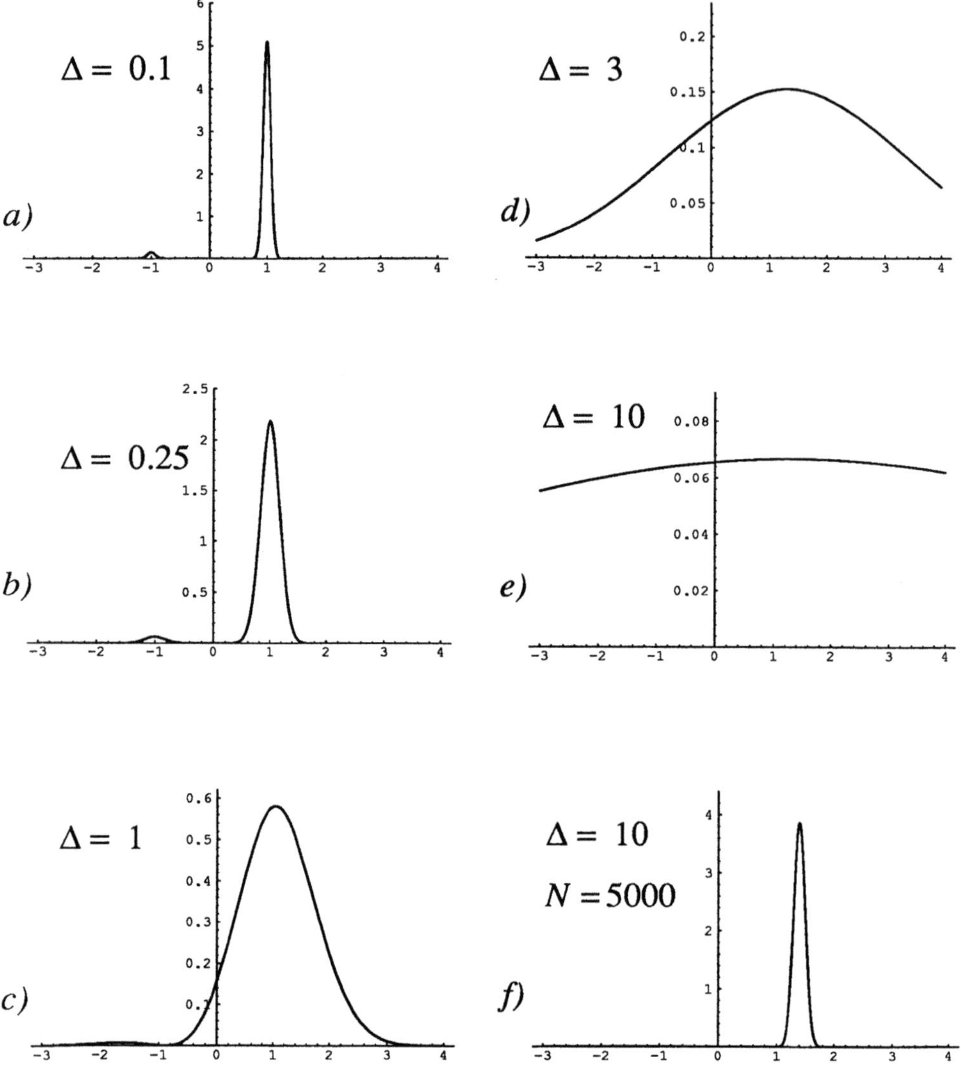

Figure 3. **Measurement on pre- and post-selected ensemble.** Probability distribution of the pointer variable for measurement of σ_ξ when the particle is pre-selected in the state $|\uparrow_x\rangle$ and post-selected in the state $|\uparrow_y\rangle$. The strength of the measurement is parameterized by the width of the distribution Δ. (a) $\Delta = 0.1$; (b) $\Delta = 0.25$; (c) $\Delta = 1$; (d) $\Delta = 3$; (e) $\Delta = 10$. (f) Weak measurement on the ensemble of 5000 particles; the original width of the peak, $\Delta = 10$, is reduced to $10/\sqrt{5000} \simeq 0.14$. In the strong measurements (a)-(b) the pointer is localized around the eigenvalues ± 1, while in the weak measurements (d)-(f) the peak of the distribution is located in the weak value $(\sigma_\xi)_w = \langle\uparrow_y|\sigma_\xi|\uparrow_x\rangle/\langle\uparrow_y|\uparrow_x\rangle = \sqrt{2}$. The outcomes of the weak measurement on the ensemble of 5000 pre- and post-selected particles, (f), are clearly outside the range of the eigenvalues, (-1,1).

We consider a system of N spin-1/2 particles all prepared in the state $|\uparrow_x\rangle$ and post-selected in the state $|\uparrow_y\rangle$, i.e., $|\Psi_1\rangle = \prod_{i=1}^{N} |\uparrow_x\rangle_i$ and $|\Psi_2\rangle = \prod_{i=1}^{N} |\uparrow_y\rangle_i$. The variable which is measured at the intermediate time is $A \equiv (\sum_{i=1}^{N}(\sigma_i)_\xi)/N$. The operator A has $N+1$ eigenvalues equally spaced between -1 and $+1$, but the weak value of A is

$$A_w = \frac{\prod_{k=1}^{N}\langle\uparrow_y|_k \ \sum_{i=1}^{N}((\sigma_i)_x + (\sigma_i)_y) \ \prod_{j=1}^{N}|\uparrow_x\rangle_j}{\sqrt{2}\ N(\langle\uparrow_y|\uparrow_x\rangle)^N} = \sqrt{2} \ . \tag{14}$$

The interaction Hamiltonian is

$$H = \frac{g(t)}{N} P \sum_{i=1}^{N}(\sigma_i)_\xi \ \ . \tag{15}$$

The initial state of the measuring device defines the precision of the measurement. When we take it to be the Gaussian (3), it is characterized by the width Δ. For a meaningful experiment we have to take Δ small. Small Δ corresponds to large uncertain P, but now, the strength of the coupling to each individual spin is reduced by the factor $1/N$. Therefore, for large N, both the forward-evolving state and the backward-evolving state are essentially not changed by the coupling to the measuring device. Thus, this single measurement yields the weak value. In Ref. 7 it is proven that if the measured observable is an average on a large set of systems, $A = (\sum_i^{N} A_i)/N$, then we can always construct a single, good-precision measurement of the weak value. Here let us present just numerical calculations of the probability distribution of the measuring device for N pre- and post-selected spin-1/2 particles. The state of the pointer after the post-selection for this case is

$$\mathcal{N} \sum_{i=1}^{N}(-1)^i\left(\cos^2(\pi/8)\right)^{N-i}\left(\sin^2(\pi/8)\right)^i e^{-(Q-\frac{(2N-i)}{N})^2/2\Delta^2} \ \ . \tag{16}$$

The probability distribution for the pointer variable Q is

$$prob(Q) = \mathcal{N}^2\left(\sum_{i=1}^{N}(-1)^i\left(\cos^2(\pi/8)\right)^{N-i}\left(\sin^2(\pi/8)\right)^i e^{-(Q-\frac{(2N-i)}{N})^2/2\Delta^2}\right)^2 \ \ . \tag{17}$$

The result for $N = 20$ and different values of Δ are presented in Fig. 4. We see that for $\Delta = 0.25$ and larger, the obtained results are very good: the final probability distribution of the pointer is peaked at the weak value, $\left((\sum_{i=1}^{N}(\sigma_i)_\xi)/N\right)_w = \sqrt{2}$. This distribution is very close to that of a measuring device measuring operator O on a system in an eigenstate $|O=\sqrt{2}\rangle$. For N large, the relative uncertainty can be decreased almost by a factor $1/\sqrt{N}$ without changing the fact that the peak of the distribution points to the weak value.

Although our set of particles pre-selected in one state and post-selected in another state is considered as one system, it looks very much as an ensemble. In quantum theory, measurement of the sum does not necessarily yield the same result as the sum of the results of the separate measurements, so conceptually our measurement on the set of particles differs from the measurement on an ensemble of pre- and post-selected particles. However, in our example of weak measurements, the results are the same.

Less ambiguous case is the example considered in the first work on weak measurements.[2] In this work a single system of a large spin N is considered. The system is pre-selected in the state $|\Psi_1\rangle = |S_z=N\rangle$ and post-selected in the state $|\Psi_2\rangle = |S_y=N\rangle$.

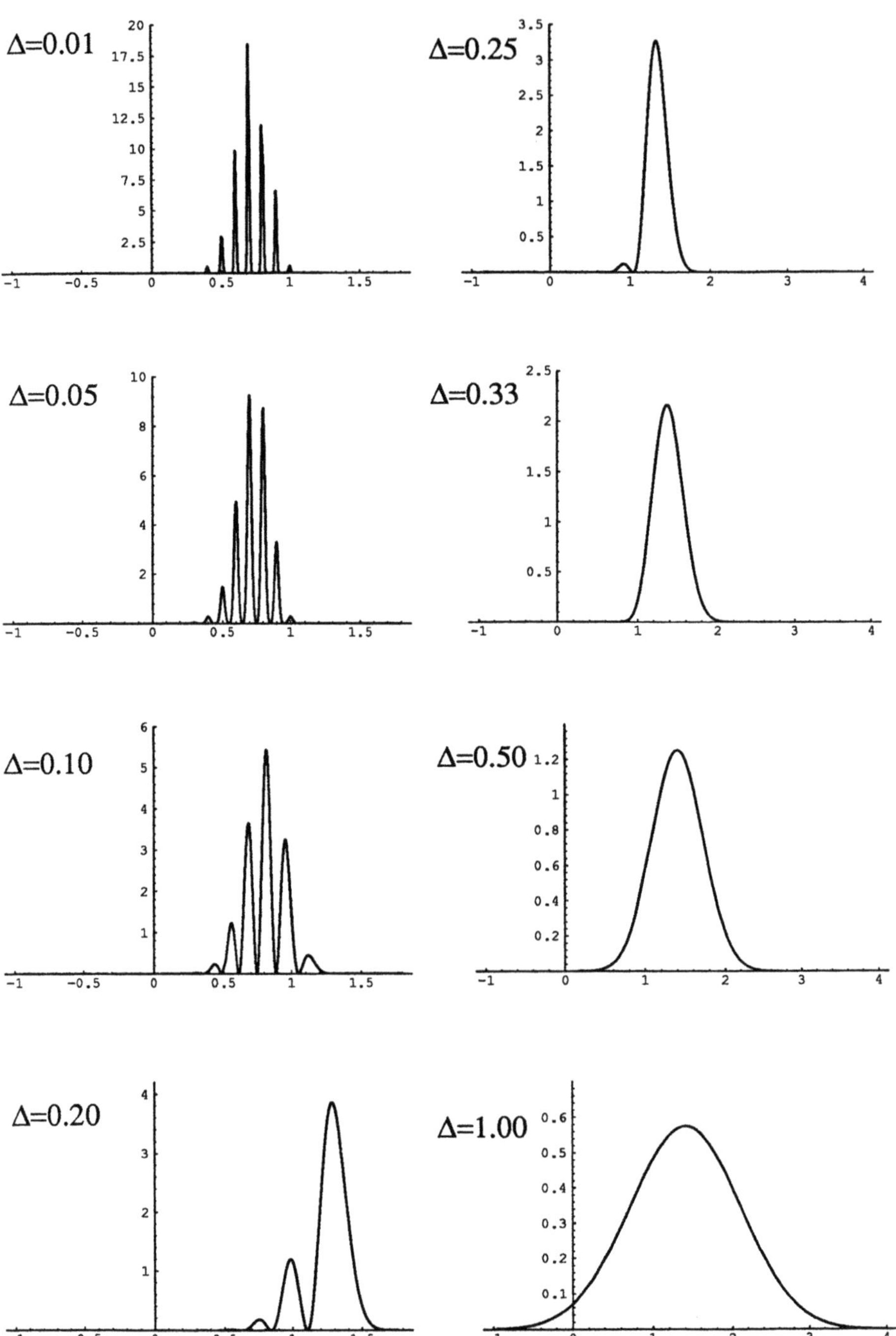

Figure 4. Measurement on a single system. Probability distribution of the pointer variable for measurement of $A = (\sum_{i=1}^{20} (\sigma_i)_\xi)/20$ when the system of 20 spin-1/2 particles is pre-selected in the state $|\Psi_1\rangle = \prod_{i=1}^{20} |\uparrow_x\rangle_i$ and post-selected in the state $|\Psi_2\rangle = \prod_{i=1}^{20} |\uparrow_y\rangle_i$ .The strength of the measurement is parameterized by the width of the distribution Δ. While in the very strong measurements, $\Delta = 0.01 - 0.05$, the peaks of the distribution located at the eigenvalues, starting from $\Delta = 0.25$ there is essentially a single peak at the location of the weak value, $A_w = \sqrt{2}$.

At an intermediate time the spin component S_ξ is weakly measured and again the "forbidden" value $\sqrt{2}N$ is obtained. The uncertainty has to be only slightly larger than $\sqrt{N}$. The probability distribution of the results is centered around $\sqrt{2}N$, and for large N it lies clearly outside the range of the eigenvalues, $(-N, N)$. Unruh[8] made computer calculations of the distribution of the pointer variable for this case and got results which are very similar to what is presented on Fig. 4.

An even more dramatic example is a measurement of the kinetic energy of a tunneling particle.[9] We consider a particle prepared in a bound state of a potential well which has negative potential near the origin and vanishing potential far from the origin; $|\Psi_1\rangle = |E{=}E_0\rangle$. Shortly later, the particle is found far from the well, inside a classically forbidden tunneling region; this state can be characterized by vanishing potential $|\Psi_2\rangle = |U{=}0\rangle$. At an intermediate time a measurement of the kinetic energy is performed. The weak value of the kinetic energy in this case is

$$K_w = \frac{\langle U{=}0|K|E{=}E_0\rangle}{\langle U{=}0|E{=}E_0\rangle} = \frac{\langle U{=}0|E - U|E{=}E_0\rangle}{\langle U{=}0|E{=}E_0\rangle} = E_0 \quad . \tag{18}$$

The energy of the bound state, E_0, is negative, so the weak value of the kinetic energy is negative. In order to obtain this negative value the coupling to the measuring device need not be too weak. In fact, for any finite strength of the measurement we can choose the post-selected state sufficiently far from the well to ensure the negative value. Therefore, for appropriate post-selection, usual measurement of a positive definite operator invariably yields negative result!

How do we get this paradoxical outcome? One can interpret it as a game of errors. Any realistic experiment must have errors. Measurement of kinetic energy must have a spread, so sometimes it might show negative outcomes. Of course, the dial of the measuring device might have a pin preventing negative readings, but we consider the device without such a pin. In our pre- and post-selection measurement a peculiar interference effect of the pointer takes place: destructive interference in the whole "allowed" region and constructive interference of the tails in the "forbidden" negative region. The initial state of the measuring device $\Phi(Q)$, due to the measuring interaction and the post-selection, transforms into a superposition of shifted wave functions. The shifts are by the (possibly small) eigenvalues, but the superposition is approximately equal to the original wave function shifted by (large and/or forbidden) weak value

$$\sum_i c_i \Phi(Q - a_i) \simeq \Phi(Q - A_w) \quad . \tag{19}$$

The example of a single weak measurement on the system of 20 pre- and post-selected spin-1/2 particles which was considered above demonstrates this effect for a Gaussian wave function of the measuring device, but we have proved[7] that "miraculous" interference (19) occurs not just for the Gaussians, but for a large class of functions. The only requirement is that their Fourier transform must be essentially bounded.

It is possible to use this idea for constructing a quantum time machine, a device which can make a cat out of a kitten in a minute.[7,10] The superposition of quantum states shifted by small periods of time can yield a large shift in time; and it even can be a shift to the past.

These surprising, even paradoxical effects are really gedanken experiments. The reason is that, unlike weak measurements on an ensemble, these are extremely rare events. For yielding unusual weak value, a single pre-selected system needs extremely improbable outcome of the post-selection measurement. Let us compare this with a weak measurement on an ensemble. In order to get N particles in a pre- and post-selected ensemble which yield $(\sigma_\xi)_w = 100$, we need $\sim N10^4$ particles in the pre-selected ensemble. But, in order to get a single system of N particles yielding $(S_\xi)_w = 100N$, we need $\sim 10^{4N}$ systems of N pre-selected particles. In fact, the probability to obtain unusual value by error is much larger than the probability to obtain the proper post-selected state. We will see a negative reading of the device measuring kinetic energy much faster than we will find the particle in a deep tunneling region. What makes these rare effects interesting is that there is strong (although only one-way) correlation: every time we find in the post-selection measurement the particle outside the well, we know that the result of the kinetic energy is negative, and not just negative: it is equal to the weak value, $K_w = E_0$, with a good precision.

It is not that weak measurement on a single pre- and post-selected system cannot be measured in a laboratory. These are the experiments with very dramatic results which are not feasible to perform. But an experiment in which the weak value is only slightly outside the range of the eigenvalues, performed on particles which can be identically prepared in millions, is possible.

Although we call it a weak measurement on a single system, in practice the experiment is performed on a large (pre-selected) ensemble. We prepare many systems, couple each system to a separate measuring device, and make the post-selection measurement waiting for the desired result. This is an experiment on a pre-selected ensemble, but it is an experiment on a *single* pre- and post-selected system. Indeed, we discard everything connected to other systems. Only the reading of the measuring device of one system is considered. If we are "lucky" and the first particle gets the right result in the post-selection measurement, then the experiment is completed, and only one system has been involved. This property does not hold for usual measurement on an ensemble: even if by chance the first result yields exactly the measured expectation value, we cannot stop here because we cannot know yet that this is, indeed, the correct value.

6. EXPERIMENTAL REALIZATIONS OF WEAK MEASUREMENTS

Weak measurements have three basic elements: preparation, (weak) coupling to the measuring device, and post-selection. The preparation part is the same as in all usual experiments, so it does not require any special consideration except that we need, for getting large effects, a very good precision in the preparation of quantum state. The second stage too does not present special experimental difficulties: this is a standard measuring procedure with weakened coupling. What limits the feasibility of a weak measurement is the possibility of an effective post-selection. In order to obtain interesting results in weak measurement, the post-selection needs to be very precise, and it has to fulfill a special requirement specified below.

Realistic weak measurements (on an ensemble) involve preparation of a large pre-selection ensemble, coupling to the measuring devices of each element of the ensemble, post-selection measurement which, in all interesting cases, selects only a small fraction of the original ensemble, selection of corresponding measuring devices, and statistical analysis of their outcomes. In order to obtain good precision, this selected ensemble of the measuring devices has to be sufficiently large. Although there are significant technological developments in "marking" particles running in an experiment, clearly the most effective solution is that the particles themselves serve as measuring devices. The information about measured variable is stored, after the weak measuring interaction, in their other degree of freedom. In this case the post-selection of the particles in the required final state automatically yield selection of measuring devices. The requirement for the post-selection measurement is, then, that there is no coupling between the variable in which the result of the weak measurement is stored and the post-selection device.

An example of such a case is the Stern-Gerlach experiment where the shift in the momentum of a particle, translated into a spatial shift, yields the outcome of the spin measurement. Post-selection measurement on a spin in a certain direction can be implemented by another (this time strong) Stern-Gerlach coupling which splits the beam of the particles. The beam corresponding to the desired value of the spin is then analyzed for the result of the weak measurement. The requirement of non-disturbance of the results of the weak measurement by post-selection can be fulfilled by arranging the shifts due to the two Stern-Gerlach devices to be orthogonal to each other. The details are spelled out in Ref. 6.

A weak measurement of a spin component is very difficult to perform in a laboratory. We need very precise pre- and post-selection of spin polarization and Stern-Gerlach experiment is very far from being precise. But analogous experiments can be performed on other two-state systems. The simplest analog of the Stern-Gerlach measurement is an optical polarization experiment. A birefringent prism splits an optical beam according to its polarization modeling the inhomogeneous magnetic field which splits the beam of particles with a spin. And high precision polarization filters serve as excellent devices for pre- and post-selection. We can define a polarization operator

$$Q \equiv |x\rangle\langle x| - |y\rangle\langle y| \ , \tag{20}$$

where $|x\rangle$ and $|y\rangle$ designate photon linear polarization states. The eigenstates of the polarization operator are ± 1 but if the initial state is $|\Psi_1\rangle = \cos\alpha \, |x\rangle + \sin\alpha \, |y\rangle$, and the final state is $|\Psi_2\rangle = \cos\alpha \, |x\rangle - \sin\alpha \, |y\rangle$, then the weak value of the polarization operator is

$$Q_w = \frac{(\cos\alpha\langle x| - \sin\alpha\langle y|) \, (|x\rangle\langle x| - |y\rangle\langle y|) \, (\cos\alpha \, |x\rangle + \sin\alpha \, |x\rangle)}{\cos^2\alpha - \sin^2\alpha} = \frac{1}{\cos(2\alpha)} \tag{21}$$

The initial and final states are chosen by placing an appropriate linear polarization filters. If the polarizations are almost orthogonal, $\alpha \simeq \pi/4$, the weak value of the polarization operator becomes arbitrary large.

An analysis of realistic experiment which can yield large weak value Q_w appears in Ref. 11. Duck, Stevenson, and Sudarshan[12] proposed slightly different optical realization which uses birefringent plate instead of a prism. In this case the measured information is stored directly in the spatial shift of the beam without being generated by the shift in the momentum. Ritchie, Story, and Hulet adopted these scheme and performed the first successful experiment measuring weak value of the polarization operator.[13] Their results are in very good agreement with theoretical predictions. They obtained weak values which are very far from the range of the eigenvalues, $(-1, 1)$, their highest reported result is $Q_w = 100$. The discrepancy between calculated and observed weak value was 1%. The RMS deviation from the mean of 16 trials was 4.7%. The width of the probability distribution was $\Delta = 1000$ and the number of pre- and post-selected photons was $N \sim 10^8$, so the theoretical and experimental uncertainties were of the same order of magnitude. Their other run, for which they showed experimental data on graphs (which fitted very nicely theoretical graphs), has the following characteristics: $Q_w = 31.6$, discrepancy with calculated value 4%, the RMS deviation 16%, $\Delta = 100$, $N \sim 10^5$.

Suter, M. Ernst and R. Ernst reported experimental realization of quantum time-translation machine.[14] We may disagree about their experiment being a model of our proposed time machine,[7,10] but it seems that they indeed performed a weak measurement. The experiment was performed on $^{13}C - \,^1H$ spin pair of chloroform. The heteronuclear J coupling of the two spins S and I is given by

$$H_{SI} = -2\pi J S_z I_z \ . \tag{22}$$

For a particular state of the spin I, the spin S precesses due to this spin-spin interaction. In the experiment, an appropriate pre- and post-selection on the states of spin I were performed, and it was observed that the spin S precession was 4 times faster than the one corresponding to the maximal eigenvalue of the second spin, $I_z = 1$. Rather than interpreting it as a time machine for S we see this experiment as a weak measurement of I_z, the measurement in which the weak value is 4 times larger than the maximal eigenvalue, $(I_z)_w = 4$.

There are numerous experiments on pre- and post-selected systems. Post- selection might lead to very dramatic effects pointed out as early as 1935 by Schrödinger.[15] Not all measurements on pre- and post-selected systems are weak measurements. Since some "weak" measurements are not really weak, and some weak couplings are really strong measurements, it is not easy to find a rigorous definition for weak measurements. A possible criterion is that measurements yielding consistently weak values are weak measurements. Thus, another run of the experiment of Ritchie, Story, and Hulet[13] which shows dramatic effect is not a weak measurement. They considered a post-selection to a state which is orthogonal to the initial state. The probability for this post-selection was not zero due to the intermediate weak coupling. However, since $|\Psi_2\rangle$ is orthogonal to $|\Psi_1\rangle$, the weak value is not defined in this case.

Another system which is a good candidate for weak measurements, due to a well developed technology of preparation and selection of various quantum states, is a Rydberg two-level atom. Between the pre- and post-selection the atom can have weak coupling with a resonant field in a microwave cavity.[16,17]

There are many experiments measuring escape time of tunneling particles. Tunneling is a pre- and post-selection experiment: a particle is pre-selected inside the bounding potential and post-selected outside. Recently, Steinberg[18] suggested that many of these experiments are indeed weak measurements.

We believe that the field of experimental realization of weak measurements is far from being exhausted. The next section explains the potential applications of this procedure.

6. WEAK MEASUREMENT AS AN AMPLIFICATION SCHEME

Let us consider an experiment of a weak measurement of A not as a measurement of the weak value of A but as a measurement of a certain parameter of the measuring device. Indeed, when we consider known initial state $|\Psi_1\rangle$ and known final state $|\Psi_2\rangle$, the weak value (1) is known prior to the measurement, and our experiment yields no new information. But we can perform the weak measuring procedure when the strength of the weak coupling is not known. Then, from the result of the weak measurement we can find the strength of the coupling.

The Hamiltonian of the Stern-Gerlach experiment measuring z component of a spin is

$$H = -g(t)\mu \frac{\partial B_z}{\partial z} z \sigma_z \ . \tag{23}$$

We can prepare the state of the spin, $\sigma_z = 1$, then, assuming that the gradient of the magnetic field is known, our experiment is a von Neumann measurement of the magnetic moment μ. (Or, if μ is known, it is a measurement of the gradient.) Indeed, Eq. (23) has the form of Eq. (2) with P replaced by $-z$ which corresponds to the pointer variable p_z. The shift in the momentum is later transformed into the shift in the position of the particle. If we perform, instead, the pre- and post-selected measurement with the initial and final states of the spin corresponding to, for example, $(\sigma_z)_w = 100$, than our procedure is a measurement of μ which is 100 times more sensitive! The shift in the peak of the pointer position distribution is 100 times larger, while the width of the peak is practically unchanged.

Of course, for increasing the shift of the pointer we have to pay some price. First, we cannot work with narrow peaks. We have to be in a regime of weak measurements, so the initial distribution of z has to be well localized around zero, and this requires wide pointer distribution (p_z). And second, we lose the intensity. For obtaining amplification by factor M we need a post-selection which will reduce the number of systems of the pre-selected ensemble by the factor of $1/M^2$, so in our example the intensity will be reduced by the factor 10^{-4}.

Still, we believe that there are measurements for which this amplification scheme might be helpful. In many experiments intensity is not a problem. If the output of the measurement is a picture on a photographic plate, then the restriction is on the total number of photons which were absorbed by the plate (before it was saturated) while the number of pre-selected photons coming from a light source is practically unlimited. This is the situation in the optical analog of the Stern-Gerlach experiment discussed above. In optical experiment, instead of the magnetic moment μ, we measure the degree of optical activity of the crystal, i.e., the difference between indices of refraction for orthogonal polarization, $n_x - n_y$. It seems that if one wants to measure this difference using a birefringent prism and the given light beam is suitable for weak measurements, then the post-selection certainly increases the precision of the measurement. If, however, our equipment allows us to make the incoming beam well collimated, then the measurement without post-selection has an advantage since it is easier to find the center of a more narrow peak. The analysis of an optimal measurement has not been performed yet. But irrespective of the results of this theoretical analysis, we are convinced that for realistic tasks when the equipment is given, the scheme of weak pre- and post-selected measurement will prove itself useful for improvement of the sensitivity of some measurements.

6. SUMMARY

Weak measurement is a certain measuring procedure which includes post-selection. There are many peculiar effects due to various post-selections, but weak measurements play a special role among them. The outcomes of weak measurements, weak values, are not just peculiar because they are very different from the outcomes of standard measurements: they are part of new simple and rich structure existing in quantum world. The concept of weak values is simple and universal. Weak values are defined for all variables and for all possible histories of quantum systems. They manifest themselves in all couplings which are sufficiently weak.

The two basic elements of our approach were investigated separately. The theory of "unsharp" measurements developed by Bush[19] has the element of weakness of the interaction. Popular today, the "consistent histories" approach,[20] originated by

Griffiths,[21] includes the idea of pre- and post-selection. But it is the combination of the two which created our formalism. It allowed us to see peculiar features of quantum systems. The quantum time machine, the method of increasing sensitivity using post-selection, and other surprising phenomena were inaccessible within the framework of the standard formalism. Neither consistent histories nor unsharp measurements provided tools to see these effects, although they might be helpful for analyzing these phenomena.[22]

The formalism of weak measurement can also be helpful in describing existing peculiar effects. The controversy of superluminal motion of tunneling particles can be resolved by recognizing that the experiments showing superluminal motion are weak measurements.[18] We have showed how, under conditions of weak measurements, the post-selection leads to superluminal motion of light wave packets (Sec. VIII of Ref. 3).

Among applications of the weak value concept is a proposal to study the back reaction of a quantum field on a particle-antiparticle pair created by the field.[23] The weak value of the field is considered between the (initial) vacuum state and the (final) state which includes the particle-antiparticle pair. The aim of this proposal is the analysis of particle creation by a black hole and the problem of what happens in the final stages of black hole evaporation.[24,25] One key to this problem is the back reaction of the pair to the gravitational field that created it, and here the application of weak values signals the possibility of a major breakthrough.

We have shown that the concepts of weak measurements and weak values are useful tools. But one can speculate that it has more meaning than this.[26,27] The formalism of weak values is a candidate for describing reality of quantum systems. It is well known that the usual concepts of reality developed from centuries of thoughts based on classical physics fail to describe our world which includes observed quantum phenomena. It has been shown that there are severe difficulties in defining relativistically invariant elements of reality in quantum theory. Weak values are Lorentz invariant. So, this might be the right course of action to identify weak values as elements of reality.[28] The weak values include the elements of reality as defined by Einstein, Podolsky and Rosen[29] as well as those recently defined by Redhead.[30] In addition to universal applicability, weak values do have desirable features such as the sum rule: if $C = A + B$ then $C_w = A_w + B_w$. However, there is no product rule: $C = AB$ does not lead to $C_w = A_w B_w$. But may be this is how our world really is.[31]

REFERENCES

1. Y.Aharonov, Bergmann, and J.L. Lebowitz, *Phys. Rev.* **B 134**, 1410 (1964).
2. Y. Aharonov, D. Albert, A. Casher, and L. Vaidman, *Phys. Lett.* **A 124**, 199 (1987).
3. Y.Aharonov and L. Vaidman, *Phys. Rev.* **A 41**, 11 (1990); *J. Phys.* **A 24**, 2315 (1991).
4. J. von Neumann, " Mathematical Foundations of Quantum Theory," Princeton University Press, New Jersey (1983).
5. Y. Aharonov and L. Vaidman, this volume.
6. Y. Aharonov, D. Albert, and L. Vaidman, *Phys. Rev. Lett.* **60**, 1351 (1988).
7. Y. Aharonov, J. Anandan, S. Popescu, and L. Vaidman, *Phys. Rev. Lett.* **64**, 2965 (1990).
8. W.G. Unruh, "Time gravity and quantum mechanics," E-board: gr-qc/9312027.

9. Y. Aharonov, S. Popescu, D. Rohrlich, and L. Vaidman, *Phys. Rev.* **A 48**, 4084 (1993); *Jap. Jour. App. Phys.* **9**, 41 (1993).

10. L. Vaidman, *Found. Phys.* **21**, 947 (1991).

11. J.M. Knight and L. Vaidman *Phys. Lett.* **A 143**, 357 (1990).

12. M.Duck, P.M. Stevenson, and E.C.G. Sudarshan, *Phys. Rev.* **D 40**, 2112 (1989).

13. N.W.M. Ritchie, J.G. Story, and R.G. Hulet, *Phys Rev. Lett.* **66**, 1107 (1991); J.G. Story, N.W.M. Ritchie, and R.G. Hulet, *Mod. Phys. Lett.* **B 5**, 1713 (1991).

14. D. Suter, M.Ernst, and R.R. Ernst, *Molec. Phys.* **78**, 93 (1990).

15. E. Schrödinger, *Proc. Camb. Phil. Soc.* **31**, 555 (1935).

16. Y. Aharonov, L. Davidovich, and N. Zagury, *Phys. Rev.* **A 48**,1687 (1993).

17. N. Zagury and A.F.R. de Toledo Piza, *Phys. Rev.* **A**, to be published.

18. A.M. Steinberg, in *Fundamental Problems of Quantum Theory*, D. Greenberger, ed., NYAS (1994).

19. P. Bush, *Phys. Rev.* **D 33**, 2253 (1986).

20. M. Gell-Mann and J.B. Hartle, Phys. Rev. **D 47**, 3345 (1993).

21. R. B. Griffiths *J. Stat. Phys.* **70**, 2201 (1993).

22. P. Bush, *Phys. Lett.* **A 130**, 323 (1988).

23. R. Brout, S. Massar, S. Popescu, R. Parentani, and Ph. Spindel, "Quantum source of the back reaction on a classical field," E-board: hep-th/ 9311019 (1993).

24. F. Englert, S. Massar, R. Parentani, "Source vacuum fluctuations of black hole radiance," E-board: gr-qc/ 9404026 (1994).

25. C.R. Stephens, G. 't Hooft and B.F. Whiting, "Black hole evaporation without information loss," E-board: gr-qc/9310006 (1993).

26. L. Vaidman, "The problem of the interpretation of relativistic quantum theories," *Ph.D. Thesis*, Tel-Aviv University (1987).

27. Y. Aharonov and D. Rohrlich, in *Quantum Coherence*, J. Anandan, ed., World Scientific, 221 (1990).

28. L. Vaidman, *Symposium on the Foundations of Modern Physics*, P.J. Lahti, P. Bush, and P. Mittelstaedt, eds., World Scientific, 406 (1993).

29. A. Einstein, B. Podolsky, and W. Rosen, *Phys. Rev.* **47**, 777 (1935).

30. M. Redhead, *Incompleteness, Nonlocality, and Realism* (Clarendon, Oxford, 1987) p. 72.

31. L. Vaidman, Phys. Rev. Lett. **70**, 3369 (1993).

ERRATUM

Equation (46) on page 108, in the chapter "Interferometry with Particles of Non-Zero Rest Mass: Topological Effects," by Geoffrey I. Opat, did not print in its entirety. The equation should have appeared as

$$N_3 = N_o \left(a_3 + b_3 \cos(\Delta\Phi_n - \Delta\Phi_{A\ell})\cos(\Delta\Phi_{AB} - \Delta\Phi_m)\right) \ . \tag{46}$$

Advances in Quantum Phenomena   0-306-45072-0
Edited by Enrico G. Beltrametti and Jean-Marc Lévy-Leblond Plenum Press, New York, 1996

MIX
Papier aus verantwortungsvollen Quellen
Paper from responsible sources
FSC® C105338

If you have any concerns about our products,
you can contact us on
ProductSafety@springernature.com

In case Publisher is established outside the EU,
the EU authorized representative is:
Springer Nature Customer Service Center GmbH
Europaplatz 3, 69115 Heidelberg, Germany

Printed by Libri Plureos GmbH
in Hamburg, Germany